Springer Series on Environmental Management

More information about this series at http://www.springer.com/series/412

The Springer Series on Environmental Management is concerned with humanity's stewardship of the environment, our use of natural resources, and the ways in which we can mitigate environmental hazards and reduce risks. The main focus of the series is on applied ecology in the widest sense of the term, in theory and in practice, and above all in the marriage of sound principles with pragmatic innovation. It focuses on the definition and monitoring of environmental problems and the search for solutions to them at scales that vary from global to local according to the scope of analysis. No particular academic discipline dominates the series, for environmental problems are interdisciplinary almost by definition. The volumes include a wide variety of specialties, from oceanography to economics, sociology to silviculture, toxicology to policy studies.

The series contributes to the immense effort by ecologists of all persuasions to nurture an environment that is both stable and productive. Increasing rate of resource use, population growth, and armed conflict have tended to magnify and complicate environmental problems that were already difficult to solve a century ago. Attempts to modify nature for the benefit of humankind have often had unintended consequences, especially in the disruption of natural equilibria. Yet, at the same time human ingenuity has developed a new range of sophisticated and powerful techniques for solving environmental problems, such as pollution monitoring, restoration ecology, landscape planning, risk management, and impact assessment. The Springer Series on Environmental Management sheds light on the problems of the modern environment and contributes to the further development of solutions.

Matthew J. Germino • Jeanne C. Chambers
Cynthia S. Brown
Editors

Exotic Brome-Grasses in Arid and Semiarid Ecosystems of the Western US

Causes, Consequences, and Management Implications

Springer

Editors
Matthew J. Germino
US Geological Survey
Forest and Rangeland Ecosystem Science Center
Boise, ID, USA

Jeanne C. Chambers
USDA Forest Service
Rocky Mountain Research Station
Reno, NV, USA

Cynthia S. Brown
Department of Bioagricultural Sciences and Pest Management
Colorado State University
Fort Collins, CO, USA

ISSN 0172-6161
Springer Series on Environmental Management
ISBN 978-3-319-24928-5 ISBN 978-3-319-24930-8 (eBook)
DOI 10.1007/978-3-319-24930-8

Library of Congress Control Number: 2015958775

Springer Cham Heidelberg New York Dordrecht London

Printed on acid-free paper

Springer International Publishing AG Switzerland is part of Springer Science+Business Media (www.springer.com)

Contents

Contributors

Edith B. Allen Department of Botany and Plant Sciences, University of California, Riverside, Riverside, CA, USA

Sheryl Y. Atkinson Department of Bioagricultural Sciences and Pest Management, Colorado State University, Fort Collins, CO, USA

Jeffrey L. Beck Department of Ecosystem Science and Management, University of Wyoming, Laramie, WY, USA

Julie Beckstead Department of Biology, Gonzaga University, Spokane, WA, USA

Jayne Belnap US Geological Survey, Southwest Biological Science Center, Moab, UT, USA

Bethany A. Bradley Department of Environmental Conservation, University of Massachusetts, Amherst, MA, USA

Matthew L. Brooks US Geological Survey, Western Ecological Research Center, Oakhurst, CA, USA

Cynthia S. Brown Department of Bioagricultural Sciences and Pest Management, Colorado State University, Fort Collins, CO, USA

Mark W. Brunson Department of Environment and Society, Utah State University, Logan, UT, USA

Jeanne C. Chambers USDA Forest Service, Rocky Mountain Research Station, Reno, NV, USA

Caroline A. Curtis Graduate Program in Organismic and Evolutionary Biology, University of Massachusetts, Amherst, MA, USA

Christina Czembor Knight Piésold Ltd., Vancouver, BC, Canada

Carla M. D'Antonio Environmental Studies Program, University of California, Santa Barbara, Santa Barbara, CA, USA

Mark Eiswerth Department of Economics, University of Northern Colorado, Greeley, CO, USA

Rebecca Epanchin-Niell Resources for the Future, Washington, DC, USA

Leonardo Frid Apex Resource Management Solutions Ltd., Bowen Island, BC, Canada

Matthew J. Germino US Geological Survey, Forest and Rangeland Ecosystem Science Center, Boise, ID, USA

Stuart P. Hardegree US Department of Agriculture, Agricultural Research Service, Northwest Watershed Research Center, Boise, ID, USA

Rebecca A. Hufft Denver Botanic Gardens, Denver, CO, USA

Halley Kartchner Department of Environment and Society, Utah State University, Logan, UT, USA

Jon E. Keeley US Geological Survey, Western Ecological Research Center, Three Rivers, CA, USA

Richard N. Mack School of Biological Sciences, Washington State University, Pullman, WA, USA

Brian A. Mealor Sheridan Research and Extension Center, University of Wyoming, Sheridan, WY, USA

Susan E. Meyer US Department of Agriculture, Forest Service, Rocky Mountain Research Station, Provo, UT, USA

Thomas A. Monaco US Department of Agriculture, Agricultural Research Service, Forage and Range Research Laboratory, Utah State University, Logan, UT, USA

Jeffrey T. Morisette US Geological Survey, North Central Climate Science Center, Fort Collins, CO, USA

Stephen J. Novak Department of Biological Sciences, Boise State University, Boise, ID, USA

JanaLynn Pearce Department of Plant and Wildlife Sciences, Brigham Young University, Provo, UT, USA

Mike Pellant US Department of Interior, Bureau of Land Management, Boise, ID, USA

Susan Phillips US Geological Survey, Forest and Rangeland Ecosystem Center, Corvallis, OR, USA

Louis Provencher The Nature Conservancy, Reno, NV, USA

David A. Pyke US Geological Survey, Forest and Rangeland Ecosystem Science Center, Corvallis, OR, USA

Benjamin M. Rau US Department of Agriculture, Forest Service, Southern Research Station, Aiken, SC, USA

Kimberly Rollins Department of Economics, University of Nevada, Reno, Reno, NV, USA

Eugene W. Schupp Department of Wildland Resources, Utah State University, Logan, UT, USA

Samuel B. St. Clair Plant and Wildlife Sciences, Brigham Young University, Provo, UT, USA

John M. Stark Department of Biology, and the Ecology Center, Utah State University, Logan, UT, USA

Michael H. Taylor Department of Economics, University of Nevada, Reno, Reno, NV, USA

Tamara J. Zelikova Department of Botany, University of Wyoming, Laramie, WY, USA

Chapter 1
Introduction: Exotic Annual *Bromus* in the Western USA

Matthew J. Germino, Jeanne C. Chambers, and Cynthia S. Brown

Abstract The spread and impacts of exotic species are unambiguous, global threats to many ecosystems. A prominent example is the suite of annual grasses in the *Bromus* genus (*Bromus* hereafter) that originate from Europe and Eurasia but have invaded or are invading large areas of the Western USA. This book brings a diverse, multidisciplinary group of authors together to synthesize current knowledge, research needs, and management implications for *Bromus*. Exotic plant invasions are multifaceted problems, and understanding and managing them requires the biological, ecological, sociological, and economic perspectives that are integrated in this book. Knowing how well information from one geographic or environmental setting can transfer to another is a key need for broadly distributed *Bromus* species especially given ongoing climate change. Thus, the chapters in the book compare and contrast invasibility of different ecoregions and invasiveness of different *Bromus* species. A universal theme is managing for ecosystems that are resilient to disturbance and resistant to invasion which will be essential for adaptation to the human-caused problem of *Bromus* in the Western USA.

Keywords Resistance • Resilience • Exotic annual *Bromus* grasses • Western USA

M.J. Germino (✉)
US Geological Survey, Forest and Rangeland Ecosystem Science Center,
Boise, ID 83706, USA
e-mail: mgermino@usgs.gov

J.C. Chambers
USDA Forest Service, Rocky Mountain Research Station,
Reno, NV 89512, USA
e-mail: jchambers@fs.fed.us

C.S. Brown
Department of Bioagricultural Sciences and Pest Management, Colorado State University,
Fort Collins, CO 80523-1177, USA
e-mail: cynthia.s.brown@colostate.edu

M.J. Germino et al. (eds.), *Exotic Brome-Grasses in Arid and Semiarid Ecosystems of the Western US*, Springer Series on Environmental Management,
DOI 10.1007/978-3-319-24930-8_1

1.1 The Spectrum of Exotic Annual *Bromus*

Exotic annual grass invasions are transforming arid and semiarid landscapes worldwide. Mediterranean annual grasses have become the dominant species following European settlement in areas as widely separated as Chile, Australia (Rossiter et al. 2003; Gulmon 1977), and the Western USA (Mack 1981; D'Antonio and Vitousek 1992). The problem is particularly acute for arid and semiarid uplands of the Western USA, which are being invaded by several exotic annual species in the genus *Bromus* (*Bromus* hereafter) that originated from Europe or Eurasia. Settlement of these semiarid landscapes occurred over a century ago, and since then, urban, agronomic, and energy development have led to open, disturbed habitat, and land use practices such as overgrazing by livestock have depleted native perennial herbaceous species—which are the best competitors with *Bromus* (Mack 1981; Billings 1990). Not only are the *Bromus* species readily invading these disturbed areas, but because they are highly adapted to the relatively wet winters and drier summers that characterize much of the area, they also are expanding into native ecosystems with favorable climatic regimes. In many cases, increases in the abundance and continuity of fine fuels caused by these annual grasses are decreasing fire return intervals and resulting in the progressive conversion of native ecosystems to *Bromus* dominance.

The life history and ecophysiological traits of *Bromus*, such as fall or early spring germination, rapid growth, and high reproduction, make them highly competitive with native species and greatly increase the difficulty of management and restoration. *Bromus* now dominate millions of acres and are present as co- or subdominant species in many low- to mid-elevation ecosystems in the Western USA. Although traditional land uses facilitated initial invasion, expanding land uses, climate change, rising atmospheric carbon dioxide, and nitrogen (N) deposition have accelerated the process.

In contemporary terminology, many of the affected ecosystems have lost or are losing "resistance" to invasive *Bromus* grasses and "resilience" to disturbances, such as altered fire and climatic regimes (Chambers et al. 2014). Conversion of native ecosystems to annual grass dominance can change soil physical and chemical properties and alter ecosystem processes like nutrient cycling and soil water flux and storage (Wilcox et al. 2012). The progressive invasion of *Bromus* species and the global change process that facilitate invasion act at large spatial scales and cross conventional boundaries for land planning and management.

The risks and problems associated with *Bromus* have been known in the USA for at least a century. There has been as much or more research and management attention on *Bromus* species, such as *B. tectorum* L. (cheatgrass or downy brome), as on other exotic invasive plant species in the USA and elsewhere. A Google Scholar search of literature on the *Bromus* genus reveals more than 10,000 primary research publications in the last century. Moreover, many thousands of seeding, fencing, and rangeland "weed" treatments at scales ranging from a few to over 10,000 acres each and collectively covering millions of acres are on record for the decades leading up to 2015 in the Western USA (US Geological Survey's Land Treatment Digital Library). Despite this relatively high number of studies and publications on *Bromus* and the many attempts by managers to control them, their spread and impacts continue. It is noteworthy that much of the ecological research on *Bromus* has focused on a limited number of explanatory variables and been conducted at local or site scales, while most of the management-oriented research has emphasized treatments targeting control of these species.

1.2 Synthesis of Patterns, Challenges, and Opportunities

A critical need exists for research, planning, and management that focuses on ecoregional and larger scales. Scientists, policy specialists, and land managers would like to understand:

- Effects of environmental factors on *Bromus* species distributions
- Ecosystem attributes and processes that influence resistance to invasion
- Traits of *Bromus* species that contribute to invasiveness
- Impacts of *Bromus* invasions on ecosystems
- Effects of pathogens on *Bromus* invasions and their potential for biocontrol
- Effects of land uses on *Bromus* invasions
- Management options for exotic annual *Bromus* and their application
- Socioeconomic drivers and patterns of human response to *Bromus* invasion

This book addresses these points for the Western USA. The book is timely because concern for the *Bromus* problem and efforts to address it at broad spatial scales have increased to unprecedented levels. For example, annual grasses and *B. tectorum* specifically are major aspects of the US Department of Interior (DOI) Secretarial Order #3336 issued in 2015, which mandates one of the largest conservation efforts in US history on behalf of *Centrocercus urophasianus* Bonaparte (greater sage-grouse). New efforts to evaluate *Bromus* and prioritize areas for protection or restoration are now occurring at ecoregional and larger scales, such as in the DOI's Rapid Ecological Assessments (released in the last 2 years) and the Fire and Invasive Assessment Tool (2015; http://www.blm.gov/wo/st/en/prog/more/sagegrouse/documents_and_resources.html).

Herein, the available literature is synthesized to develop a more complete picture of the factors that influence invasibility, invasiveness, impacts, and management of annual *Bromus* invaders. A framework is used that is based on ecosystem resilience to disturbance and resistance to invasion. Resilience is defined as the capacity of an ecosystem to *regain* its fundamental structure, processes, and functioning when altered by stressors like drought and disturbances like overgrazing by livestock and altered fire regimes (Holling 1973; Walker et al. 2004). Resistance is defined as the capacity of an ecosystem to *retain* its fundamental structure, processes, and functioning despite stresses, disturbances, or invasive species (Folke et al. 2004). Resistance to invasion by exotic annual grasses is a subset of resistance and is a function of the abiotic and biotic attributes and ecological processes of an ecosystem that limit the population growth of an invading species (D'Antonio and

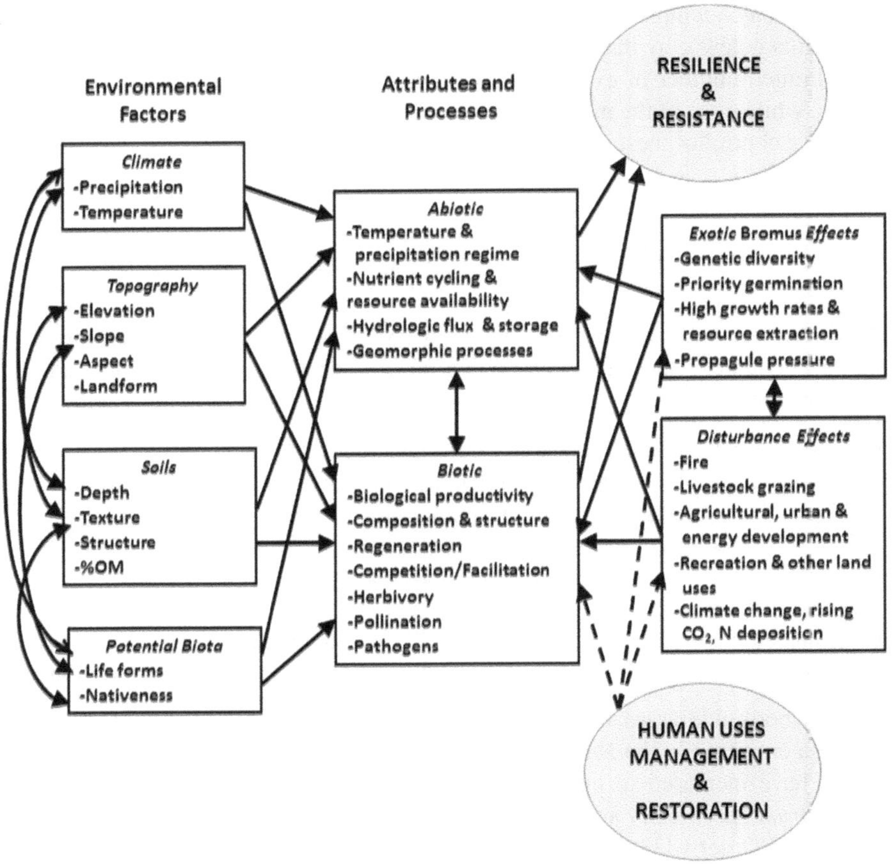

Fig. 1.1 Conceptual model of the environmental factors and ecosystem attributes and processes that affect resilience and resistance. Human activities including land uses and management actions influence invasion by exotic annual *Bromus* and the type and magnitude of disturbance which in turn can affect ecosystem attributes and processes and, ultimately, resilience and resistance (Adapted from Chambers et al. 2014)

Thomsen 2004). The environmental factors and ecosystem attributes that determine resilience of native ecosystems to stress and disturbance and resistance to invasion can be illustrated with a simple conceptual model (Fig. 1.1). The model shows that environmental factors, including climate, topography, and soils, determine the abiotic and biotic attributes of ecosystems. *Bromus* and other invaders, disturbance and stress, and global change factors act on the abiotic and biotic attributes of an ecosystem and influence its relative resilience and resistance over time. Management actions usually target the biotic attributes and processes of ecosystems and can influence the effects of *Bromus* and disturbance. This model contains those ecosystem attributes and processes suggested for evaluating ecosystem sustainability (Chapin et al. 1996). It also includes both the environmental factors and ecosystem attributes used to develop the ecological site descriptions and state-and-transition models that are increasingly used by managers to evaluate options for managing *Bromus* (Briske et al. 2008; Caudle et al. 2013). All chapters in this book relate to key attributes of the model (Fig. 1.1).

There are nearly 150 species in the *Bromus* genus globally, and they vary considerably in their invasiveness and impacts across the vast Western USA. Chapters in this book range from incorporating all *Bromus* species (Chap. 6) to only one species (Chap. 5). Where chapters focus on fewer species, it is generally because the chapters have a relatively thorough treatment of better-studied species like *Bromus rubens* L. (red brome) and especially *B. tectorum* which is a widespread invader in the floristic Great Basin and one of the most studied invasive species globally. Several chapters also incorporate *B. diandrus* Roth (ripgut brome) and *B. hordeaceus* L. (soft brome) of California, the next most invasive and impactful exotic annual *Bromus*. Several chapters evaluate these species across regional and local gradients. The invasion of California grasslands is widespread, and *Bromus* was considered entrenched there nearly a century ago. Moving eastward, extensive invasion of the floristic Great Basin was underway over a century ago but is still occurring. Farther east in areas like the Front Range of Colorado, *Bromus* invasions are considered a relatively new occurrence. This east-to-west gradient of invasion is overlaid onto a gradient of increasing summer precipitation, changes in land use, and changes in plant community composition. *Bromus* invasions also tend to exhibit strong elevation patterns. Chapters that compare and contrast *Bromus* invasions among ecoregions (Chap. 2) or aim to explain evolutionary aspects or invasiveness in the genus (Chaps. 4 and 6) have the greatest taxonomic breadth. By comparing and contrasting *Bromus* species and regions, we hope that the book will produce generalized, transferable concepts and information or illustrate the limits of transferability by highlighting unique aspects of invasiveness or invader impacts.

Chapters in this book parse into four themes, organized as sections. The first section addresses patterns and impacts of invasion, generating a "35,000 foot view" of where, when, and how invasion by the dominant exotic annual *Bromus* species has varied among ecoregions in the Western USA (Chap. 2, Brooks et al.). This analysis reveals that *Bromus* has had very different impacts in different ecoregions. Germino et al. (Chap. 3) then examine ecosystem effects of *Bromus* in further detail, focusing most on *B. tectorum* and providing a basis for why *Bromus* is a concern.

The second section explores some of the organismal traits that contribute to invasiveness of the dominant annual *Bromus* species in the USA. The broad evolutionary, reproductive, and biogeographic traits and patterns affecting the genetic diversification and colonization of the Western USA by *Bromus*, particularly *B. tectorum*, are reviewed next with an emphasis on evidence from allozyme variation (Chap. 4, Novak and Mack). Many recent studies show that a capacity for rapid evolution or ability to withstand a wide range of environmental conditions through phenotypic plasticity has favored many invasive species. Ecological-genetics issues are explored for *B. tectorum* in Chap. 5 with an emphasis on molecular markers and common garden evidence (Hufft and Zelikova). Shifting back to the entire *Bromus* genus, Atkinson and Brown (Chap. 6) compare and contrast organismal and ecological traits conferring invasiveness among all *Bromus* species using a database approach that incorporates many types of information. Theories about "natural enemies" are often invoked to explain invasiveness of species, and Meyer et al. (Chap. 7) evaluate fungal pathogen systems involving *Bromus*, with a focus on *B. tectorum* because it is the subject of current microbial biocontrol research and thus the most is known about its relations to pathogens.

The third section evaluates soil, climate, and plant community controls on *Bromus* to characterize key aspects of invasibility of sites, plant communities, or regions. Belnap et al. (Chap. 8) evaluate how the interaction of soils, nutrients, and climate variability influences invasion patterns, focusing on the greater information available for *B. tectorum* primarily and also *B. rubens*. In Chap. 9, Bradley et al. evaluate large-scale climate controls on *B. tectorum* and *B. rubens* in the Western USA using the species distribution modeling approach. Next, the interactive effects of climate, plant community attributes, and *Bromus* traits on plant community resistance to *Bromus* invasions are evaluated in Chap. 10 by Chambers et al.

The last section explores human and economic dimensions and management options related to both spread and management of *Bromus*. Chapters in this section generate concepts from relatively well-studied *Bromus* species but have some broader applicability to other exotic annual grasses and even to invasive plants in general. Pyke et al. (Chap. 11) evaluate humans as a driver and responder to *Bromus*, focusing on interactions among key land uses such as grazing and other drivers such as fire. A comprehensive review of management treatments directed at *Bromus* and their effectiveness is then provided by Monaco et al. (Chap. 12). The focus is on restoring and maintaining sustainable ecosystems, and a case study is provided on *B. tectorum* in the Great Basin. The vast scale of invasions and size of western landscapes has posed a major challenge to land managers, and modeling frameworks are increasingly important for allowing land managers to evaluate options and future invasion risks, such as with the state-and-transition simulation models that are described next by Provencher et al. (Chap. 13). Management actions are outcomes of the social context, and Brunson and Karcher (Chap. 14) describe how human perceptions, education, and regulatory aspects of *Bromus* and other invasive grasses have changed over the course of the invasion in the USA. Economics are a major driver of efforts to address the causes of *Bromus* invasion, and Chap. 15 by Eiswerth et al. explores how economic models can help inform management.

A few other points need to be noted about the book and the project underlying it. First, the book is an outcome of the US Department of Agriculture's *Bromus* REENet (Research, Extension, and Education Network) which granted funding to bring a diverse range of agency and university specialists from around the USA together to generate the ideas and direction for this book. Second, readers will notice some uniqueness in reporting conventions as well as some overlap in content. Because each chapter will be available as a separate article, we wished to find a balance between interdependence and uniformity among chapters while also enabling each chapter to stand alone. In several cases, chapters cover aspects of the same theme from slightly different perspectives. We believe that bringing these perspectives from different science disciplines and geographic backgrounds together is an essential part of solving the ever-growing problem of *Bromus* invasions and the spread of annual grasses globally.

Acknowledgments Jessica Vander Veen provided invaluable assistance in preparing the manuscript for submission. Sue Phillips provided comments on many chapters. Funding was provided by USDA REENet Grant #03083. Plant taxonomy follows the USDA Natural Resources Conservation Service, PLANTS Database (http://plants.usda.gov) in all but Chap. 6 which uses a broader range of sources. Any use of trade names is for descriptive purposes only and does not imply endorsement by the US Government.

References

Billings WD (1990) *Bromus tectorum*, a biotic cause of ecosystem impoverishment in the Great Basin. The Earth in Transition. Cambridge University Press, New York, NY, pp 301–322

Briske DD, Bestelmeyer BT, Stringham TK et al (2008) Recommendations for development of resilience-based state-and-transition models. Rangel Ecol Manag 61:359–367

Caudle D, DiBenedetto J, Karl M et al (2013) Interagency ecological site handbook for rangelands. http://jornada.nmsu.edu/files/InteragencyEcolSiteHandbook.pdf. Accessed 2 Jun 2015

Chambers JC, Bradley BA, Brown CS et al (2014) Resilience to stress and disturbance, and resistance to *Bromus tectorum* L. invasion in cold desert shrublands of western North America. Ecosystems 17:360–375

Chapin FS III, Torn SE, Tateno M (1996) Principles of ecosystem sustainability. Am Nat 148:1016–1037

D'Antonio CM, Thomsen M (2004) Ecological resistance in theory and practice. Weed Technol 18:1572–1577

D'Antonio CM, Vitousek PM (1992) Biological invasions by exotic annual grasses, the grass/fire cycle, and global change. Ann Rev Ecol Syst 23:63–87

Folke C, Carpenter S, Walker B et al (2004) Regime shifts, resilience, and biodiversity in ecosystem management. Ann Rev Ecol Syst 35:557–581

Gulmon SL (1977) A comparative study of the grassland of California and Chile. Flora 166:261–278

Holling CS (1973) Resilience and stability of ecological systems. Ann Rev Ecol Syst 4:1–2

Mack RN (1981) Invasion of *Bromus tectorum* L. into Western North America: an ecological chronicle. Agro-Ecosystems 7:145–165

Rossiter NA, Setterfield SA, Douglas MM et al (2003) Testing the grass–fire cycle: alien grass invasion in the tropical savannas of northern Australia. Divers Distrib 9:169–176

Walker BH, Holling CS, Carpenter SR et al (2004) Resilience, adaptability and transformability in social-ecological systems. Ecol Soc 9:5

Wilcox BP, Turnbull L, Young MH et al (2012) Invasion of shrublands by exotic grasses: ecohydrological consequences in cold versus warm deserts. Ecohydrology 5:160–173

Part I
Environmental Impacts of *Bromus* Species

Chapter 2
Exotic Annual *Bromus* Invasions: Comparisons Among Species and Ecoregions in the Western United States

Matthew L. Brooks, Cynthia S. Brown, Jeanne C. Chambers, Carla M. D'Antonio, Jon E. Keeley, and Jayne Belnap

Abstract Exotic annual *Bromus* species are widely recognized for their potential to invade, dominate, and alter the structure and function of ecosystems. In this chapter, we summarize the invasion potential, ecosystem threats, and management strategies for different *Bromus* species within each of five ecoregions of the western United States. We characterize invasion potential and threats in terms of ecosystem resistance to *Bromus* invasion and ecosystem resilience to disturbance with an emphasis on the importance of fire regimes. We also explain how soil temperature and moisture regimes can be linked to patterns of resistance and resilience and provide a

M.L. Brooks (✉)
US Geological Survey, Western Ecological Research Center, Oakhurst, CA 93644, USA
e-mail: matt_brooks@usgs.gov

C.S. Brown
Department of Bioagricultural Sciences and Pest Management, Colorado State University, Fort Collins, CO 80523-1177, USA
e-mail: cynthia.s.brown@colostate.edu

J.C. Chambers
USDA Forest Service, Rocky Mountain Research Station, Reno, NV 89512, USA
e-mail: jchambers@fs.fed.us

C.M. D'Antonio
Environmental Studies Program, University of California, Santa Barbara, Santa Barbara, CA 93106, USA
e-mail: dantonio@es.ucsb.edu

J.E. Keeley
US Geological Survey, Western Ecological Research Center, Three Rivers, CA 93271, USA
e-mail: jon_keeley@usgs.gov

J. Belnap
US Geological Survey, Southwest Biological Science Center, Moab, UT 84532, USA
e-mail: jayne_belnap@usgs.gov

M.J. Germino et al. (eds.), *Exotic Brome-Grasses in Arid and Semiarid Ecosystems of the Western US*, Springer Series on Environmental Management,
DOI 10.1007/978-3-319-24930-8_2

conceptual framework that can be used to evaluate the relative potential for invasion and ecological impact of the dominant exotic annual *Bromus* species in the western United States.

Keywords Fire • Resilience • Resistance • Management • Moisture regime • Temperature regime

2.1 Introduction

Numerous *Bromus* species occupy a wide range of ecosystems around the globe (Atkinson and Brown 2015; Clayton et al. 2006), including 58 species within North America (Clayton et al. 2006), approximately half of which are exotic species that are nonnative to the continent (Pavlik 1995). Exotic annual *Bromus* (*Bromus* hereafter) are of significant concern to ecologists and land managers alike (Brooks and Pyke 2001). Much of what has been reported about the environmental and ecological controls on *Bromus* and their interactions with land uses and disturbance regimes have focused on the Cold Deserts ecoregion of the western United States (Fig. 2.1) and on *Bromus tectorum* L. (cheatgrass) in particular. This chapter reaches beyond that specific ecoregion and individual species to examine the invasion potential, ecosystem threats, and management strategies for the dominant *Bromus* species within each of five major ecoregions of the western United States (Table 2.1; Fig. 2.1). Other chapters within this book that are relevant to the topics in this chapter include Chambers et al. (2015), Germino et al. (2015), Monaco et al. (2015), and Pyke et al. (2015).

2.1.1 Resistance and Resilience to Bromus *Invasions*

The invasion potential, ecosystem threats, and management strategies for *Bromus* all vary based on ecosystem resistance to invasion and resilience to disturbance. Resistant ecosystems have the capacity to retain their fundamental structure, processes, and functioning (or remain largely unchanged) despite stressors, disturbances, or invasive species (Folke et al. 2004). Resilient ecosystems have the capacity to regain their fundamental structure, processes, and functioning following stessors like drought and disturbances like wildfire and then reorganize without crossing thresholds to alternative ecosystem states (Holling 1973; Walker et al. 2004). We use resistance and resilience as explanatory concepts throughout this chapter, along with temperature and precipitation gradients which are key factors affecting resistance and resilience (Brooks and Chambers 2011; Chambers et al. 2014a, b; Fig. 1.1).

Resistance to invasion is particularly important in the context of this chapter and is related to abiotic and biotic attributes and ecological processes that limit the population growth of invading species (D'Antonio and Thomsen 2004). Ecosystem resistance to invading species is a function of both the fundamental niche, which is largely related to climate suitability, and the realized niche, which is related

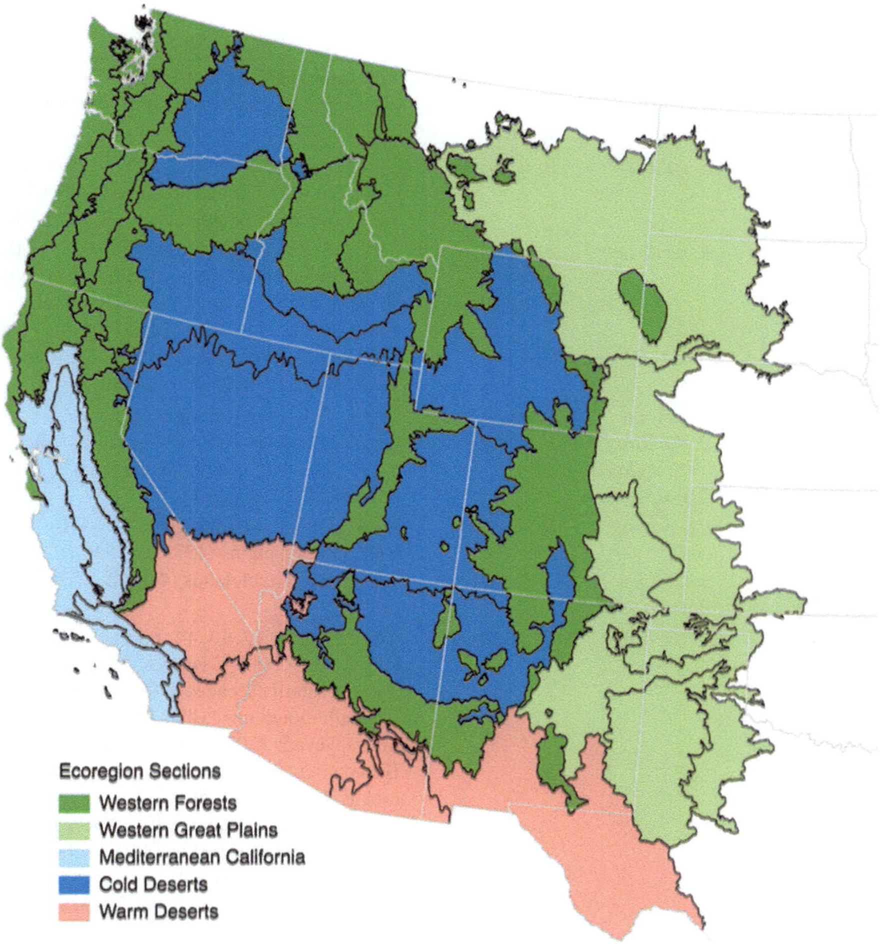

Fig. 2.1 Ecoregions of the western United States that are characterized by distinct temperature and soil moisture regimes and influence resistance to *Bromus*. Each ecoregion is an amalgam of various level I, II, and III Ecoregions of North America (http://www.epa.gov/wed/pages/ecoregions/na_eco.htm, accessed 17 Jun 2014; Wiken et al. 2011). Their specific delineations were based on breaks in soil temperature and moisture (amount and seasonality) regimes associated with ecological resistance to, and ecosystems threats posed by, *Bromus* invasions, and the availability of published information on these topics. The Western Forests ecoregion encompasses Level I ecoregions 6.0 and 7.0 (the Northwestern Forested Mountains and the Marine West Coast Forest). Mediterranean California corresponds to the Level I ecoregion 11.0 which bears the same name. The Cold Deserts and Warm Deserts correspond to Level II ecoregions 10.1 and 10.2, respectively. The Western Great Plains includes three level III ecoregions (Table 2.1) representing parts of two Level II ecoregions, the West-Central Semi-Arid Prairies and South-Central Semi-Arid Prairies (ecoregions 9.3 and 9.4)

to species interactions, both of which need to be understood to evaluate *Bromus* invasions (Chambers et al. 2014a). Temperature and precipitation amounts, and precipitation seasonality, are key attributes that influence ecosystem resistance

Table 2.1 Major ecoregions of the western United States and their most widespread dominant and subdominant exotic annual *Bromus* species

Ecoregion	Dominant *Bromus* species	Subdominant *Bromus* species	Associated EPA III Ecoregions (mean and range of annual precipitation in mm) (Wiken et al. 2011)
Warm Deserts	*B. rubens*	*B. tectorum*	10.2.1, 2, 4: Mojave (167, 50–900), Sonoran (206, 75–560), Chihuahuan (340, 200–635) 12.1.1 Madrean Archipelago (421, 250–950)
Cold Deserts	*B. tectorum*	*B. rubens*	10.1.2–8: Columbia Plateau (334, 150–600), Northern Basin and Range (351, 150–1000), WY Basin (296, 130–500), Central Basin and Range (277, 4–1000), CO Plateaus (298, 130–800), AZ/NM Plateau (293, 125–380), Snake River Plain (316, 110–350)
Mediterranean California	*B. rubens* *B. diandrus* (= *B. rigidus*) *B. hordeaceus*	*B. madritensis* *B. tectorum*	11.1.1–3: CA coastal sage/chaparral/woodlands (548, 200–1400), central valley (443, 125–760), southern California mountains (525, 220–1250)
Western Great Plains	*B. arvensis* (= *B. japonicus*) *B. tectorum*	None	9.3.3: NW great plains (393, 250–510) 9.4.1, 3: high plains (433, 305–530), southwestern tablelands (449, 255–710)
Western Forests	*B. tectorum*	*B. diandrus*	6.2.5, 7, 8, 11, 12, 13, 14: north cascades (1761, 300–6000), Cascades (824, 1150–3600), eastern cascades (649, 500–3500), Klamath mountains (1438, 500–3000), Sierra Nevada (1070, 150–2500), Wasatch/Uinta mountains (602, 150–1400), southern rockies (588, 255–1750) 7.1.7–9: Puget lowland (1400, 300–2500), coast range (2149, 1000–5000), Willamette valley (1228, 900–1600) 13.1.1: AZ/NM mountains (477, 270–2000)

and define ecological niche space (Davis et al. 2000; Miller et al. 2013; Chambers et al. 2014a, b, 2015).

Site occurrence information for *Bromus* gleaned from the GrassPortal website (www.grassportal.org, Osborne et al. 2011) indicates that resistance to invasion should vary among species based on mean annual minimum and maximum temperature and precipitation (Fig. 2.2). *Bromus rubens* L. (red brome) is clearly found at the hottest and driest sites, whereas *Bromus arvensis* L. (field brome) and to some degree *B. tectorum* occur at the coldest and wettest sites in the database. Variability among species is low at the lower end of the precipitation gradient, but very high at the upper end of the gradient (Fig. 2.2c). The preponderance of outlier sites on the high end of the precipitation gradient may represent occurrences at early successional or otherwise disturbed sites embedded within high precipitation vegetation types, especially within the Western Forests ecoregions, that would otherwise have

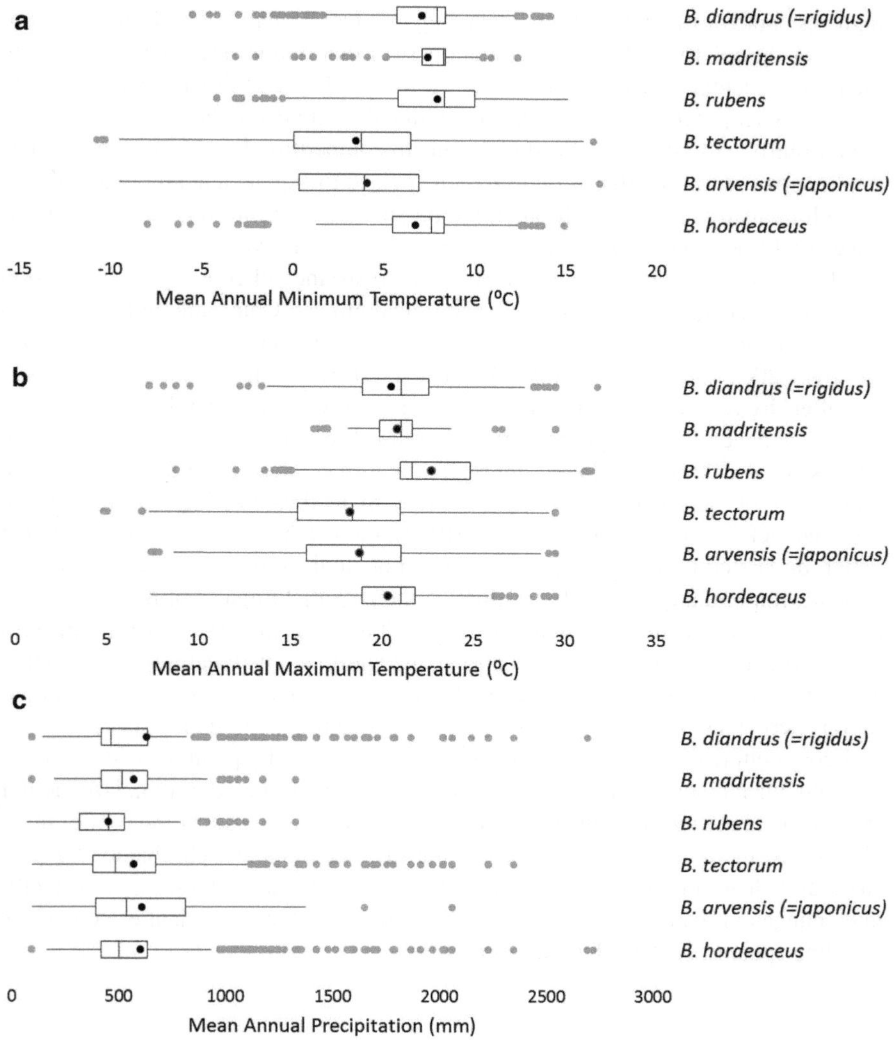

Fig. 2.2 (**a**) Annual minimum temperature, (**b**) annual maximum temperature, and (**c**) annual precipitation for site locations of *Bromus* in the western United States. Data were summarized for states intersecting the ecoregional range map (Fig. 2.1), and downloaded from www.grassportal.org (accessed 16 Feb 2015, Osborne et al. 2011). Species were only included if the database was represented by site locations in states throughout the species range (www.plants.usda.gov). Numbers of sites were as follows: *B. diandrus* (and *B. rigidus*), $n=1241$; *B. madritensis*, $n=226$; *B. rubens*, $n=1568$; *B. tectorum*, $n=2004$; *B. arvensis* (and *B. japonicus*), $n=897$; *B. hordeaceus*, $n=1188$. Climate data are 30-year mean annual values from 1961 to 1991; box plots show quartiles 1, 2, and 3, and upper and lower limits. Mean climate values are shown as *solid dots* and outliers as *gray dots*. Data sources and other database information are listed in Table 2.2

high resistance to *Bromus*. These data suggest that *Bromus* may be firmly established at the relatively lower precipitation sites that define one boundary of both their fundamental and realized niches. However, they additionally suggest there may be higher precipitation sites that are within their fundamental niche, but that are not currently within their realized niche due to competition from other species. The implications of this are that *Bromus* may be poised to expand and potentially dominate climatically suitable high precipitation sites in the event that land use or climate conditions change the competitive balance in their favor.

There is great interest in understanding the resistance of ecosystems to invasion by *Bromus* because of their negative ecological effects. Chief among these effects are altered fire regimes (Whisenant 1990; D'Antonio and Vitousek 1992; Brooks and Pyke 2001; Brooks et al. 2004; Balch et al. 2013; Germino et al. 2015). *Bromus* can alter fire regimes by changing fuel structure in ways that can lengthen the fire season, enhance ignition probabilities, increase fire extent, and alter fire severity (Brooks et al. 2004). The altered fire regime in turn creates conditions that promote dominance of the species that created the new fuels, thus creating a positive grass/fire feedback (D'Antonio and Vitousek 1992), the mechanisms of which are much better understood than perhaps any other potential ecosystem threat posed by *Bromus*. Important questions for evaluating the potential impacts of *Bromus* among ecoregions include the following (Brooks 2008). Has the invasion introduced a novel structure or property into an ecosystem? Has that novel feature in turn altered ecosystem processes or their controls (disturbance regimes)? Do the resultant conditions perpetuate a new, persistent ecosystem state (e.g., alternative state)?

Ecosystem resilience, and thus ecological risks posed by plant invasions, can be affected when disturbance regimes (perturbation type, timing, duration, size, and/or intensity) are altered from historical norms. Land uses are often at the root of these alterations, including inappropriate grazing of grasslands and shrublands by livestock and clear-cut harvests of forests. Plant invasions themselves may also facilitate changes in disturbance regimes in ways that increase ecological effects of the invading species, such as the grass/fire cycle.

2.1.2 Species, Ecoregions, and Chapter Structure

We focus on six widespread annual exotic *Bromus* taxa within the western United States, specifically *B. arvensis*, *Bromus diandrus* Roth (ripgut brome), *Bromus hordeaceus* L. (soft brome), *Bromus madritensis* L. (compact brome), *B. rubens*, and *B. tectorum* (Table 2.1). At times in the past, botanical taxonomic authorities have considered *B. arvensis* and *Bromus japonicus* Thunb. (Japanese brome) to be distinct species, but the taxonomic authority used in this book, namely the USDA, Natural Resources Conservation Service, Plants Database, currently considers them to be one in the same species within *B. arvensis* (http://plants.usgs.gov, accessed 21 Jul 2015). Similarly, *B. diandrus* and *Bromus rigidus* Roth (ripgut brome) have been considered by some as separate species, but are both currently subsumed within

B. diandrus in the Plants Database. Thus, where we reference *B. arvensis* or *B. diandrus* throughout this chapter we are also referencing their Plants Database synonyms.

Bromus madritensis and *B. rubens* are currently considered distinct species in the Plants Database (http://plants.usgs.gov, accessed 21 Jul 2015), although they intergrade so much in their taxonomic characteristics that they are often difficult to distinguish and may be misidentified in some publications. For example, an analysis of 28 field samples from central and southern California and southern Nevada (M. Brooks, unpublished data, 2015) indicated that every sample contained taxonomic characteristics of both species, although those of *B. rubens* were much more frequent (mean 9.1, median 9.0, of 11 characteristics).

We also identify five major ecoregions in the western United States (Fig. 2.1), each of which is characterized by a unique set of dominant and subdominant *Bromus* (Table 2.1). For each ecoregion, we recognize 3–5 major vegetation types which can be placed along hot-dry to cool-moist soil temperature and moisture gradients and which vary in their resistance to *Bromus* and resilience to disturbance. We use soil temperature and moisture regimes as indicators of relative ecosystem resistance and resilience to various *Bromus* species within each ecoregion (Chambers et al. 2014a, b, c).

Each of the five ecoregional sections in this chapter follows a common outline. For each ecoregion, we (1) describe the invasion potential of each widespread *Bromus* species; (2) explain the ecosystem threats posed by *Bromus* invasions; and (3) summarize the efficacy of typical management strategies for *Bromus*. The majority of citations focus on published studies conducted within each ecoregion. The relative resistance to invasion and resilience to fire are illustrated in a series of conceptual models for each ecoregion focusing on the dominant *Bromus* species in the ecoregion. We acknowledge that the amount of supporting science that underpins these figures varies greatly among ecoregions, and we refer to them as hypothetical resistance and resilience models.

2.2 Warm Deserts

Environmental conditions in the Warm Deserts ecoregion are the hottest and driest of anywhere in western United States (Fig. 2.3). Soil temperature regimes are either thermic (mean annual temperature 15–22 °C) or hyperthermic (≥22 °C). Soil moisture regimes are mostly aridic, meaning that the soil is dry for at least half of the growing season and moist for less than 90 consecutive days, which supports mostly desert shrubland vegetation types. Some areas are classified as ustic aridic, meaning they have slightly longer periods of moist soil during the growing season due to higher amounts of summer precipitation which is conducive to the growth of semi-arid desert perennial grasslands. These warm and dry conditions result in limited anthropogenic land uses compared to other ecoregions, although mining, livestock grazing, off-highway vehicle use, and most recently wind and solar energy developments can have broad landscape effects with implications for plant invasion (Lovich and Bainbridge 1999; Brooks and Pyke 2001; Brooks 2009).

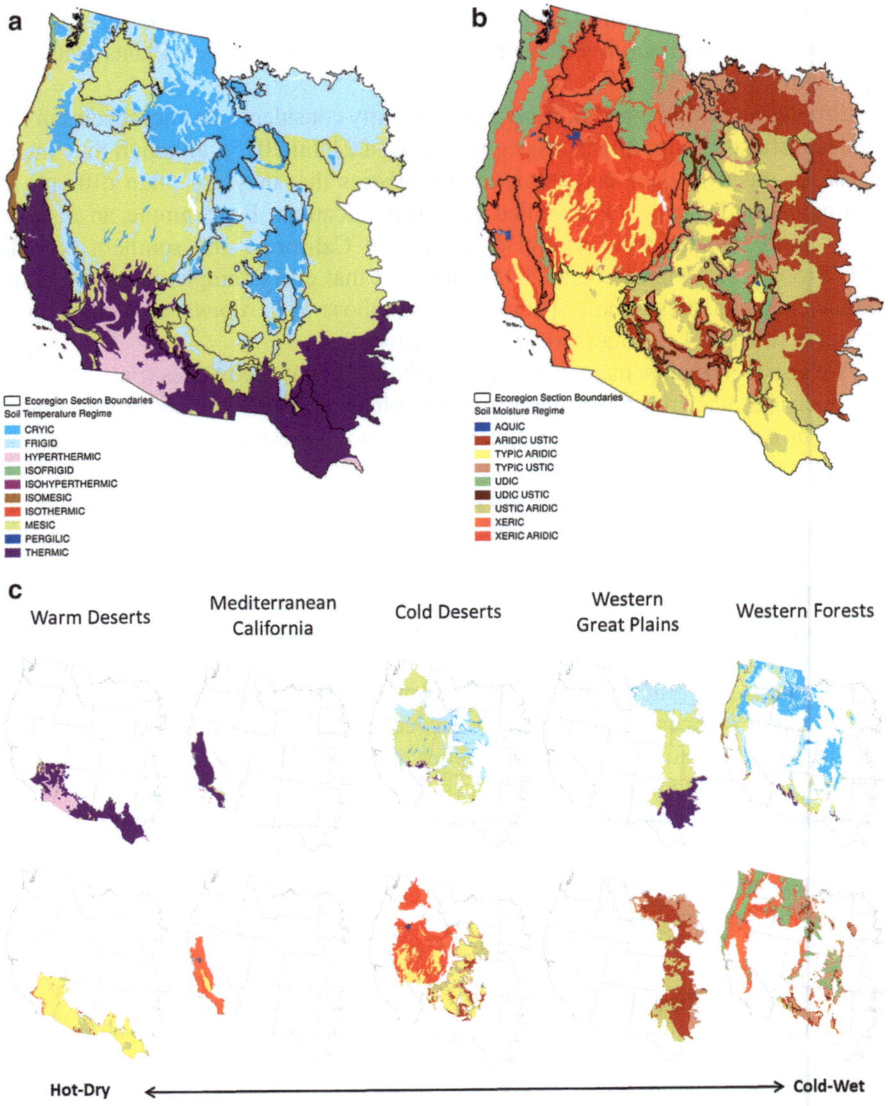

Fig. 2.3 (**a**) Soil temperature regimes, (**b**) soil moisture regimes, and (**c**) comparisons among the six major ecoregions of the western United States

Bromus rubens is the most ubiquitous *Bromus* species in the Warm Deserts ecoregion. It can occur in all but the most hyperthermic and xeric regions, but is most prevalent in middle elevations typified by creosote bush scrub and blackbrush shrubland which have the lowest resistance to its dominance (Fig. 2.4a) (Brooks and Berry 2006; Brooks 2009; Klinger et al. 2011a). *Bromus tectorum* can also be locally abundant, but is much more restricted in its geographic distribution in Warm Deserts than is *B. rubens* (www.calflora.org, accessed 8 Apr 2014). Although *B. tectorum* can

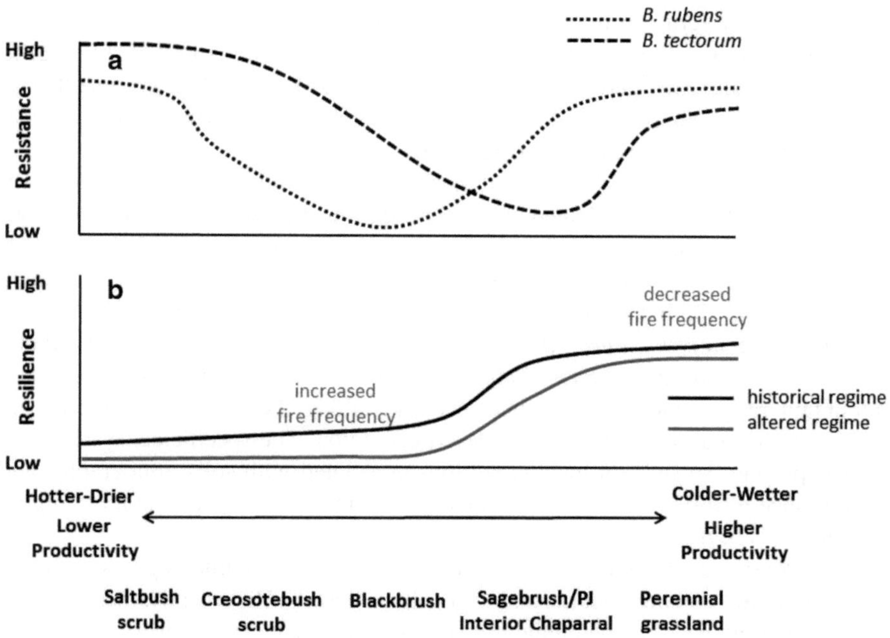

Fig. 2.4 Hypothetical (**a**) resistance to common *Bromus* species and (**b**) resilience to historical and altered fire regimes for the Warm Deserts ecoregion (primary altered regime characteristic labeled in *gray*)

appear in lower elevation areas with locally higher soil moisture such as riparian zones and roadsides (Brooks 2009), it only occurs as a dominant *Bromus* species at higher elevations, such as those within the upper blackbrush ecotone, sagebrush, piñon–juniper, and interior chaparral (Fig. 2.4b) (Klinger et al. 2011a). *Bromus trinii* Desv. (Chilean chess) has also been reported from some locations, generally within the same ecological range as *B. rubens* (www.calflora.org, accessed 8 Apr 2014), but its relative dominance is typically very low (Brooks 2009; exception is Brooks 1999). *Bromus diandrus* can also be found in the Warm Deserts (www.calflora.org, accessed 8 Apr 2014), but almost always in localized areas of higher soil moisture such as along roadsides, in riparian areas, or associated with agricultural or urban developments (Dudley 2009; M. Brooks, pers. obs.).

2.2.1 Invasion Potential of Warm Deserts by Bromus

The lowest elevation ecosystems of the Warm Deserts are characterized by hyperthermic, typic aridic soils (Fig. 2.3c) and sparse cover of creosote bush scrub or saltbush scrub. Creosote bush scrub is dominated by the type species, *Larrea tridentata* (DC.) Coville (creosote bush), and is the most widespread vegetation type. Saltbush scrub is dominated by one or more *Atriplex* species, including *Atriplex polycarpa* (Torr.) S. Watson (cattle saltbush), *Atriplex spinifera* J.F. Macbr. (spinescale saltbush), and *Atriplex canescens* (Pursh) Nutt. (fourwing saltbush). Productivity is strongly limited by low annual precipitation, and resistance to invasion by *Bromus* is generally high (Fig. 2.4a).

Blackbrush occurs on thermic and mesic soils at middle elevations above creosote bush scrub and is dominated by the type species, *Coleogyne ramosissima* Torr. (blackbrush), but may also contain *L. tridentata*, *Atriplex confertifolia* (Torr. & Frém.) S. Watson (shadscale saltbush), *Yucca brevifolia* Engelm. (Joshua tree), and *Juniperus* spp. (juniper). This vegetation type is the least resistant to invasion by *Bromus*, especially *B. rubens* (Fig. 2.4a). At higher elevations, sagebrush, piñon–juniper woodland, and interior chaparral occur on the more mesic soils of desert mountain ranges and at ecoregional ecotones along the western, northern, and north-eastern margins of the Warm Desert ecoregion. The specific ecosystem type that is present depends on local conditions associated with soils, topography, and disturbance history, but in most cases the dominant *Bromus* species at these higher elevations is *B. tectorum*.

Desert grasslands dominate the thermic soils in the monsoonal regions of the far eastern Mojave and the Sonoran and Chihuahuan deserts. Historically, these ecosystems were dominated by perennial grasses and a fire frequency of every 5–15 years (Humphrey 1958; McPherson 1995; Wright and Bailey 1982). Vigorous stands of perennial grasses with their extensive shallow roots very likely provided high ecosystem resistance to invasion by *Bromus* which share the same shallow rooting zone. During the twentieth century, overgrazing and fire suppression have allowed deeper rooting woody species to invade (McPherson and Weltzin 2000), reducing the vigor of perennial grasses and potentially the ecosystem resistance to *Bromus*. Although the exotic annuals *B. rubens* and *B. tectorum* occur in desert grasslands, they have not exhibited the widespread dominance observed for exotic perennial grasses such as *Eragrostis* von Wolf spp. (lovegrasses) and *Pennisetum ciliaris* (L.) Link (buffelgrass) which have replaced native grasses in many areas and maintain a fairly high level of resistance to *Bromus* invasion.

Bromus rubens was present at 44 % of 126 low elevation (149–1222 m) sites in the eastern Mojave Desert near Lake Mead National Recreation Area (Abella et al. 2012). It was twice as prevalent at middle elevation (800–1222 m) sites, which tended to have more precipitation, were more species rich, and were characterized by mixed perennial communities, compared to the lowest elevation sites (<800 m) which were drier, more species poor, and dominated by either *Atriplex* L. spp. or *L. tridentata*. Similar results were reported from the western Mojave Desert in lower to middle elevations ranging from 683 to 1402 m (Brooks and Berry 2006). Similar patterns were also found in post-fire landscapes of the eastern Mojave Desert (Klinger et al. 2011a), where *B. rubens* was the dominant *Bromus* species in lower elevation communities, with peak densities between 800 and 1200 m, whereas *B. tectorum* dominated higher elevation communities, with a peak density at 1800 m. Averaged over both low and high elevation sites in the eastern Mojave Desert, peak densities of

B. rubens were almost twice as high as those of *B. tectorum* (Klinger et al. 2011a). The prevalence of *B. rubens* at lower and middle elevation and *B. tectorum* at high elevations is likely a function of the lower minimum temperature tolerance of the latter (Fig. 2.2a), but may be modified by variation in the realized niches of these two species along this elevation gradient as a result of interspecific competition.

Livestock grazing may decrease resistance to *Bromus* by reducing vigor of perennial plants particularly grasses (Brooks and Pyke 2001; Brooks et al. 2007). Short-term, high-intensity grazing has been used to reduce fine fuel loads and, if applied in early spring or late summer, it can preferentially impact *B. tectorum* populations in Cold Deserts (Strand and Launchbaugh 2013). However, in most cases dominance of *Bromus* increases following cessation of grazing in both Warm Deserts and Cold Deserts, and the net long-term effects of livestock grazing are generally favorable for invasive plants (Brooks and Pyke 2001; Brooks et al. 2007).

Atmospheric nitrogen deposition may also reduce landscape resistance to *Bromus* species in Warm Deserts where deposition levels are high downwind from major urban or agricultural areas (Brooks 2003; Allen et al. 2009; Rao and Allen 2010). These studies explain that naturally low soil nitrogen levels in Warm Desert regions, along with low soil moisture availability, can hinder *Bromus* invasions, and elevated levels can reduce this resistance. Increased biomass of *Bromus* and other annual plants facilitated by nitrogen deposition may increase the potential for fire, which may further promote *Bromus* dominance (Rao et al. 2010, 2015).

2.2.2 *Ecosystem Threats to Warm Deserts from* Bromus

The role of *Bromus* in creating continuous biomass cover in the interspaces between perennial plants and increasing the size and frequency of fires represents its primary ecosystem threat in the Warm Deserts ecoregion (Brooks and Pyke 2001; Brooks and Esque 2002). This threat is most prevalent in areas with low resistance to invasion by *Bromus* and resilience of native vegetation to disturbance, typically at middle elevations dominated by mixed woody scrub and blackbrush communities, and to some degree in creosote bush scrub (Brooks and Minnich 2006; Brooks et al. 2013) (Fig. 2.4). Native perennial cover in blackbrush is already at the threshold of what can carry fire (Fig. 2.5a), and adding *Bromus* biomass can tip the balance to a more fire-prone landscape (Fig. 2.5b) (Brooks and Matchett 2006). *Bromus* dominance following fire can increase substantially after only a single fire in saltbush scrub and creosote bush communities (Fig. 2.5c), although subsequent fires can sequentially reduce cover and diversity of native perennials and promote recurrent fire (Fig. 2.5d) (Brooks 2012). This reduced resilience caused by frequent fires is operative primarily in saltbush, creosote bush scrub, and blackbrush, although it can also affect even perennial grasses in cases of extreme fire frequency (Fig. 2.4b).

This grass/fire cycle has been occurring in the Mojave Desert since at least the early 1900s, when post-fire blackbrush landscapes were recognized to be dominated

Fig. 2.5 (**a**) Unburned blackbrush stand uninvaded by *Bromus*, (**b**) unburned blackbrush stand invaded by *B. rubens*, (**c**) once-burned blackbrush stand dominated by *B. rubens*, and (**d**) multiple-burned blackbrush stand dominated by *B. rubens*

by *B. rubens* and land managers were concerned about the potential for recurrent fire (Fig. 2.6a) (Holmgren 1960; Brooks et al. 2007). Dominance by *B. rubens* in this region can be established after a single fire and may persist at least a half century even without subsequent burning (Fig. 2.6b). By some definitions this condition can be considered a type conversion from native shrubland to exotic annual grassland, even after a single fire. These fires are much more prevalent in the northeastern Mojave Desert (Brooks and Esque 2002), especially in areas of both high winter and high summer rainfall (Tagestad et al. in press). Modeled estimates of precipitation through the year 2100 forecast increased interannual variability including higher peak rainfall years (Tagestad et al. in press) which are conditions associated with increased probability of fire occurrence (Brooks and Matchett 2006). Higher elevation sagebrush, piñon–juniper, and interior chaparral ecosystems with mesic to warm frigid soils have relatively low resistance to invasion by *B. tectorum* (Fig. 2.4a), but they also have high resilience to disturbance and thus are at relatively low risk of conversion to a grass/fire cycle (Fig. 2.4b). Fire occurrences at these high elevation sites are also not associated with interannual rainfall amounts, providing further evidence that they are not significantly affected by altered fire regimes associated with exotic annual grasses (Brooks and Matchett 2006).

Fig. 2.6 (**a**) 7-year post-fire scene dominated by a carpet of *B. rubens* following a stand-replacing fire in a blackbrush shrubland that burned once in 1939 (photo by A. Croft, 12 May 1946). (**b**) A 63-year post-fire scene showing *B. rubens* persisting with early successional shrubs following a half century of no additional fire (photo by D. Oldershaw, 9 May 2002). The view is looking southwest inside the mouth of Horse Spring Basin in the northeastern Mojave Desert, Lincoln County, NV

Bromus rubens has also been shown to effectively compete with native annual (Brooks 2000; DeFalco et al. 2003) and perennial plants in the Mojave Desert (DeFalco et al. 2007). These competitive effects can influence community dynamics and resilience following disturbance (Brooks and Chambers 2011). For example, during the first 3 years following the 2005 Southern Nevada Fire Complex woody plant seedlings were almost nonexistent except where densities of exotic annuals, primarily *B. rubens* and *Erodium cicutarium* (L.) L'Hér. ex Aiton (redstem stork's bill), were exceedingly low (Klinger et al. 2011a). Reduced resilience of native perennials is a key element of the process by which a grass/fire cycle becomes established (Brooks 2008). The reduced productivity and diversity of native annuals and loss of the habitat structure provided by perennial shrubs can negatively affect wildlife, such as the federally Threatened *Gopherus agassizii* Cooper (desert tortoise) (Brooks and Esque 2002).

2.2.3 *Management Strategies for* Bromus *in Warm Deserts*

Bromus are already so wide ranging in the Warm Deserts that early detection and eradication is not a relevant management strategy for these taxa. Even when direct control efforts such as herbicide use are implemented, they are typically economically feasible to implement for only a couple of years and in relatively small areas. In addition, even if dominance of one *Bromus* species is effectively reduced, other exotic species may fill the void if there is no plan for facilitating the growth of other more desirable species to compete with them. In one dramatic case from the riparian zone in Zion Canyon of SW Utah at the regional ecotone between the Warm Deserts, Cold Deserts, and Western Forests, an herbicide treatment effectively controlled *B. diandrus*, but inadvertently led to dominance of *B. tectorum* by the following year (Fig. 2.7). This is an important reminder that any effort to reduce abundance of

Fig. 2.7 Effects of fall Imazapic treatment of *B. diandrus* on subsequent increase in dominance by *B. tectorum* the following spring in a riparian woodland in Zion Canyon, Utah (photo by M. Brooks, Spring 2006)

Bromus in any ecosystem should consider the likely net response of the plant community and include effective contingency plans.

Restoration guidelines in Warm Deserts generally focus on maximizing abundance and diversity of native species, diversity of functional types, and groups of species important for critical aspects of ecosystem function (Brooks et al. 2013). Thus, the most effective management approach for *Bromus* in the Warm Deserts is to maximize ecosystem resilience to disturbance. In doing so, landscapes will be less prone to ecosystem type conversions and can increase resistance to dominance by *Bromus* and other invasive plant species (Brooks and Chambers 2011; Chambers et al. 2014a). This can be done proactively by minimizing the extent and frequency of major disturbances such as wildfire (Brooks et al. 2007). Once these disturbances occur, management options become much more limited. Post-fire seeding is a standard management tool in other ecoregions to promote recovery of native species and minimize dominance of undesirable invasives such as *Bromus*. However, its long-term efficacy has recently been shown to vary depending on relative resistance and resilience in both the Cold Deserts (Beyers 2004; Shinneman and Baker 2009; Pyke et al. 2013; Knutson et al. 2014) and Warm Deserts (Klinger et al. 2011b). Thus, the most effective fire management strategy is to prevent fires from starting, and suppress fires once they start, especially at middle elevations characteristic of the upper creosote bush scrub ecotones and blackbrush shrublands (Brooks et al. 2013) where resistance and resilience are both low (Fig. 2.4).

2.3 Mediterranean California

The central valley and surrounding foothills, coastal mountains, and coastal plains of California have a Mediterranean climate with a relatively cool and wet winter and spring and a hot and dry summer and autumn. Soil temperatures are mostly thermic and similar to much of the warm desert region, but moisture regimes are mostly classified as xeric, rather than drier typic aridic which characterizes the Warm Deserts (Fig. 2.3c).

The grasslands of Mediterranean California can be divided into various ecotypes dominated by different *Bromus* species, often with co-dominance by exotic annual grasses from other genera. The more arid southern and interior grasslands are defined by an abundance of *B. rubens*, although its taxonomic similarity with *B. madritensis* (Sect. 2.1.2) makes it difficult to attribute dominance to one or the other. By contrast, the coast range grassland is defined by dominance of *B. diandrus* and *B. hordeaceus* (Jackson and Bartolome 2002; Keeler-Wolf et al. 2007). Serpentine grasslands can be dominated by *B. hordeaceus*, especially following periods of high rainfall. The coastal prairie is more defined by native and invasive perennial grasses, but *B. hordeaceus* and *B. diandrus* are also common. Disturbed areas within the region's coastal sage scrub and chaparral shrublands tend to be dominated by *B. rubens*, often in association with *B. diandrus* and *B. hordeaceus.* In somewhat moister woodlands and riparian zones, these species are often replaced entirely by *B. diandrus*, and in the higher inland mountain forests with colder winters, they are replaced by *B. tectorum.*

2.3.1 *Invasion Potential of Mediterranean California by* Bromus

Ecosystems within the Mediterranean California ecoregion are diverse and vary in their resistance to *Bromus* invasion. Sage scrub is at the hottest and driest end of the productivity gradient in the Mediterranean California ecoregion and tends to have the lowest resistance to *Bromus* invasion particularly when it is disturbed or stressed by nitrogen addition (Fig. 2.8a). Perennial grasslands and oak woodlands tend to have moderate resistance to *Bromus* invasion (Fig. 2.8a), but only where disturbance regimes have not shifted outside of the range of historic variation due to grazing or increased fire frequency (Fig. 2.8b). Oak woodlands and oak savannas that have been grazed for many decades often have an understory of *B. diandrus* which increases in dominance after grazing ceases (Rice and Nagy 2000; Stahlheber and D'Antonio 2013). Chaparral and mixed conifer forests display the highest resistance to *Bromus* invasion due to their high canopy cover which shades the soil surface impeding annual plant growth.

Bromus species in Mediterranean California thrive best where perennial cover is slow to recover following disturbance, and their dominance is often transitory where

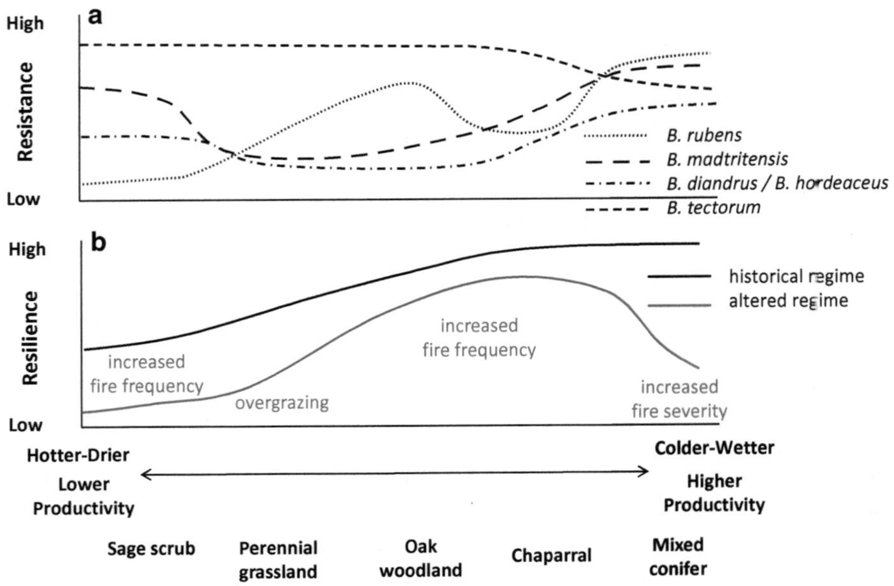

Fig. 2.8 Hypothetical (**a**) resistance to common *Bromus* and (**b**) resilience to historical and altered fire regimes for the Mediterranean California ecoregion (primary altered regime characteristic labeled in *gray*)

perennial cover recovers more quickly. In a study of 250 chaparral sites (Keeley et al. 2008), *B. madritensis* cover was negatively correlated with pre-fire stand age and previous fire severity. In addition, 5 years post-fire cover of exotic species, of which *Bromus* was a major contributor, was best predicted by shrub cover and estimated exotic seed bank at the time of fire. In essence, the extent to which *Bromus* and other exotics can invade disturbed Mediterranean shrublands is a race between exotic propagules getting to the site (since most that were there before are killed by high-intensity fire) and shrub canopy reestablishment. Under historical disturbance regimes, productivity and resilience covaried (Fig. 2.8b). However, various anthropogenic factors have altered disturbance regimes and reduced ecosystem resilience, especially at the lowest and highest ends of the productivity gradient. Because chaparral tends to be more species rich with more post-fire resprouting species compared to sage scrub, it tends to be more resilient to increased fire frequency (Fig. 2.8b).

Exotic annual grasslands, most of which have a strong component of *Bromus*, owe their origins to either (1) disturbance that converts native woody vegetation or mixed native perennial grass/shrub stands or (2) direct invasion of native grasslands (Keeley 1990). These exotic grasslands are markedly unlike native perennial grasslands in that the former are lower in functional diversity than the latter, being dominated entirely by annual grasses and forbs with a different composition of native forbs (Molinari and D'Antonio 2014). Where remnants of native grasslands persist within larger exotic annual grasslands, *Bromus* are usually only co-dominants (e.g., Molinari and D'Antonio 2014).

Fig. 2.9 Chaparral with three different fire histories in southern California. The entire scene burned in the 1970 Laguna Fire. The front and middle third burned in the 2001 Viejas Fire, and the front third burned a third time in the 2003 Cedar Fire. The background illustrates mature chaparral dominated by the native shrub *Adenostoma fasciculatum*, the middle third early seral chaparral dominated by the native shrub *Lotus scoparius*, and the front portion a 1-year post-fire scene dominated by the nonnative annual grass *B. madritensis* (photo by Richard Halsey, Spring 2004)

Fire history plays a critical role in the prevalence of exotic grasses within shrubland ecosystems of the region. Short intervals between fires can thin shrub density, which can both increase grass abundance and reduce native shrub seed banks and the vigor of potential resprouts (Zedler et al. 1983; Haidinger and Keeley 1993; Talluto and Suding 2008; Keeley and Brennan 2012). These thinned stands lead to reduced fire intensity which may further favor survivorship of exotic propagules (Keeley et al. 2008). A classic example of this dynamic is illustrated in Fig. 2.9 where a chaparral stand that burned three times over three decades was dominated by *B. madritensis* compared to areas burned only once or twice. However, in a region-wide remote sensing study only a few shrublands appeared to be undergoing type conversion via fires that occurred two times within 8 years (Meng et al. 2014), so other factors may interact with fire frequency to affect vegetation trajectories.

Atmospheric pollution in the form of nitrogen deposition has been implicated in the invasion of semiarid shrublands by annual grasses including *Bromus* (Westman 1979; Allen et al. 1998). In desert scrub, openings between widely spaced shrubs are readily invaded by invasive grasses and this is enhanced by nitrogen pollution (Brooks 2003). In contrast, within closed canopy shrublands there is no evidence that pollution alone can promote type conversion from shrubs to these grasses, but there is evidence that following a disturbance that opens up the canopy (such as fire or human disturbance), *B. diandrus* and *B. rubens* can quickly establish and suppress native shrub seedlings (Eliason and Allen 1997). Recent work suggests that nitrogen enhances competitive suppression of native shrub seedlings in the presence of abundant *Bromus* at least in part due to interactions with mycorrhizae (E. Allen, pers. comm.). Also in coastal prairies, nitrogen addition such as by cyclical growth of native lupines enhances *B. diandrus* cover (Maron and Connors 1996; Maron and Jeffries 1999; Suttle et al. 2007) at the expense of native grassland species. In serpentine grassland, nitrogen addition enhances *B. hordeaceus* dominance at the expense of native species (Huenneke et al. 1990).

Sites dominated by native perennial grasses today appear to have some resistance to invasion by annual grass species including some *Bromus*. This has been demonstrated experimentally as well as by observation (Seabloom et al. 2003; Corbin and D'Antonio 2004; Lulow 2006). Invasive bromes in these settings include B. *diandrus*, *B. hordeaceus*, and *B. madritensis*. Resistance was disrupted historically by conversion of landscape to crop agriculture and by animal disturbance such as by livestock. The interaction of grazing with native versus exotic grasses in California is controversial and varies by region. However, a meta-analysis showed that exotic grass cover, including largely *Bromus*, decreases with livestock grazing (Stahlheber and D'Antonio 2013) whereas seasonal grazing can enhance native grass cover particularly in the more coastal regions of California. Exotic forb cover tends to increase with grazing while native forb richness (but not cover) increases with grazing. Despite the use of grazing as a tool to enhance native species and reduce exotic annual grasses such as *B. diandrus*, these grasses remain an important component of virtually all of the grazing lands in California (Jackson and Bartolome 2007).

Natural disturbance factors such as gopher excavation have also been shown to promote *Bromus* abundance in grassland settings (Hobbs and Mooney 1995) as do excavations by feral pigs (Cushman et al. 2004). These *Bromus* also invade ruderal environments created by various types of disturbance that remove native vegetation and disturb the soil. Invasion success is complicated and tied to soil nutrients and belowground root competition (Brown and Rice 2010). On severely disturbed sites, *Bromus* and other exotic annuals may persist for 70 years or more in what otherwise would have been sage scrub (Stylinski and Allen 1999).

In ecosystems such as oak woodlands and mixed conifer forests that are characterized mostly by frequent surface fire regimes, decreasing fire frequency can lead to understory fuel accumulations and increased potential for higher severity crown fire and lower post-fire resilience (Fig. 2.8b). The loss of forest canopy cover can change microclimates in ways that significantly reduce resilience and improve conditions for dominance by *Bromus*. In Oak savannas, the understory of trees is well known to promote dominance of *B. diandrus* over other annual species including other *Bromus* (Rice and Nagy 2000), and this is true over a wide latitudinal range of California oak savannas (Stahlheber 2013). Thus, loss of tree cover may change the relative dominance of specific *Bromus* in this vegetation type.

2.3.2 *Ecosystem Threats to Mediterranean California from* Bromus

Bromus invasions have had a multitude of impacts in Mediterranean California. Their success has largely been at the cost of losing native shrublands, primarily chaparral, sage scrub, and oak woodlands, although perennial grasslands have also been affected (Molinari and D'Antonio 2014). When this type conversion occurs it affects biodiversity, hydrology, fire regimes, and the global carbon balance. However, the direction of such effects can be variable. For example, *Bromus* invasion of coastal sage scrub in southern California has been linked to increased soil carbon storage if the shrublands remain unburned. If they burn and convert to grasslands, it is likely that soil carbon storage will decline (Wolkovich et al. 2009a).

Where type conversion has occurred, the dominant functional type has been changed from deep-rooted shrubs to shallow fibrous rooted annuals. This impacts soil water holding capacity and increases erosion and debris flows following heavy winter rains but also has the potential for enhancing stream flow (Meixner and Wohlgemuth 2003). Type conversion also impacts fire regimes in a couple of important ways. Length of the fire season is greatly expanded because fuel moisture in drying herbaceous biomass is lost rapidly and this can provide available fuel during all 12 months of the year. In addition, a substantial proportion of fires in the region ignite in herbaceous fuels (Syphard and Keeley 2015), which provides a wick that carries fires into woody vegetation, promoting repeat fires that further degrades mixed grass and shrub mosaics (Fig. 2.9) (Keeley et al. 2012; Monaco et al. 2015).

This grass/fire cycle where fire regimes are altered by grass invasion is well documented for several ecosystems (D'Antonio and Vitousek 1992). However, it is important to recognize that there are very diverse ways grass invasion impacts fire regimes depending on the natural fire regime in the region. For example, fire-prone perennial grass invasion into fire-sensitive tropical forests results from increased fire intensity, which acts to open up the forest to further invasion. California annual grass invasions by *Bromus* spp., *Avena* L. spp., and others are quite different because these annuals are invading a fire-prone shrubland ecosystem, one with a natural fire regime of high-intensity fires, and such fires are in some respects detrimental to grass invasion as they kill seeds of the invaders. However, these native shrublands require decades to recover and if grasses invade in the early post-fire stages they can produce flashy fuels that readily ignite and carry repeat fires when the native vegetation is still not mature. These short-interval low-intensity fires can result in stand thinning, further grass invasion, and increased probability of a repeat fire (Fig. 2.9). Sage scrub is especially susceptible to alteration due to repeat short-interval fire compared to chaparral, largely because productivity and levels of resilience are lower (Fig. 2.8b). In these stands, fire severity is lower and exotic seed survivorship is higher. As a general rule as stand age decreases, fire severity decreases and exotic seed survivorship and vegetative dominance increase after fire (Keeley et al. 2008).

One of the predominant concerns of reserve managers in California grassland and oak savanna habitats is the apparent decline of native wildflowers (Minnich 2008) and native biodiversity in general. *Bromus*-dominated grasslands are depauperate in diversity of plant functional types and animal communities (Rochester et al. 2010; Molinari and D'Antonio 2014). An ongoing study suggests that it is the accumulated litter of *Bromus* that suppresses native forb germination and not direct competition for soil nutrients between *Bromus* and native forbs (Molinari 2014). Molinari and D'Antonio (2014) demonstrate that there are several native forbs that can coexist with *B. diandrus* and that although native forb richness is depressed where this species becomes dominant, several native forbs preferentially associate with this grass. The influence of *Bromus* invasion on native animals has been little studied. One study found that despite enhancing primary production, *Bromus* invasion into coastal shrublands caused either a decrease in or no effect on arthropod assemblages (Wolkovich et al. 2009b). Key native ant species declined as litter of *Bromus* increased (Wolkovich et al. 2009b).

2.3.3 Management Strategies for Bromus *in Mediterranean California*

Preventing further invasions of *Bromus* species is largely dependent on reducing the incidence of disturbance and maximizing ecosystem resilience (Fig. 2.8). In grasslands, this may involve manipulating livestock grazing patterns or removal of feral animals such as pigs. Where exotic grasses have invaded fire-tolerant or even fire-dependent ecosystems, it is important to appreciate that these shrublands require a particular fire regime that typically involves long fire-free intervals and high-intensity fire. Humans currently cause over 95 % of all fires in the foothills and coastal plain of California and have greatly increased the natural fire frequency (Safford and Van de Water 2014). Keeping unwanted ignitions out of these systems is a major challenge. A more tractable problem is reducing planned ignitions and resource managers in the region are beginning to appreciate that prescription burning, while possibly having some fire-hazard reduction benefit, is adding to the readily ignitable fuel load on these landscapes.

Bromus dominance may decline under passive management for some landscapes whereas other landscapes require active restoration. Sage scrub displaced by grasslands from various disturbances can return within a few decades if disturbance is reduced and native seed sources are in close proximity (Freudenberger et al. 1987; DeSimone and Zedler 1999; Callaway and Davis 1993), although some sites have proved resistant to passive restoration (Stylinski and Allen 1999). The type of disturbance associated with annual grassland appears important as soil disturbance that affects biological soil crusts (if present) can favor exotic annuals such as *Bromus* and inhibit native shrub seedlings (Hernandez and Sandquist 2011).

Native perennial grasslands invaded by exotic grasses show some promise of being restored to a greater native grass and forb composition through active management. In particular, winter/spring grazing has been shown to enhance dominance of the bunch grass *Nassella pulchra* (Hitchc.) Barkworth (purple needlegrass) while reducing dominance of *Bromus* (Stahlheber and D'Antonio 2013). It has also been suggested that spring fires, which kill the exotic seed bank while still on the plant but do not diminish the resprouting capacity of native perennial bunch grasses and forbs, have potential (DiTomaso et al. 2006), although native grasses can still be highly susceptible to mortality from fire (Marty et al. 2005). One experiment showed that high-intensity fires in mixed grasslands can improve the ratio of native versus exotic (*Bromus* and *Avena* spp.) grasses for at least 2 years (Keeley and Brennan 2015). However, a key to the success of this effort was that these native-dominated grasslands possessed substantial fuel loads because it had been many years since the last fire or other major form of disturbance. Large fuel loads contributed to extraordinary flame lengths and presumably high soil temperatures that were sufficient to kill the exotic seed bank but allowed for the survival of native perennial grasses.

Studies from other vegetation types also illustrate the importance of the fuel load in creating fire behavior that can reduce *Bromus* dominance. For example, an investigation of spring burning on *B. tectorum* understory in a ponderosa pine forest failed to displace this exotic invader (Keeley and McGinnis 2007). In that case the low nutrient availability of the soils produced rather sparse grass populations and field measures of temperatures during the fire revealed that they failed to reach high enough levels to kill seed banks. The importance of high fuel loading to make this approach effective has been noted in studies of other *Bromus* (Sweet et al. 2008). Even when prescription burning to eliminate *Bromus* has been effective for the target species, they are replaced by other exotic annuals (Keeley et al. 2011). Other studies have also shown that effects of a single fire on annual grass reduction can be short-lived and repeated burning or burning followed by grazing may be necessary for sustained reductions in annual grasses (Corbin et al. 2004).

2.4 Cold Deserts

Environmental conditions in the Cold Deserts are on average moderately warm and dry in comparison to the range of soil temperature and moisture regimes across the western United States (Fig. 2.3). However, the topographic diversity of this ecoregion results in high temperature and moisture variability associated with local elevation gradients. The lower elevations dominated by salt desert and Wyoming big sagebrush ecosystems are relatively hot and dry whereas the higher elevations typified by mountain sagebrush and mountain brush are cold and moist. The most widespread invasive *Bromus* in the Cold Deserts ecoregion is *B. tectorum* which occurs over a broad range of temperature and precipitation conditions (Fig. 2.2). *Bromus rubens* also occurs in the Cold Deserts but appears less cold tolerant (Fig. 2.2a) and occurs at lower elevations and latitudes than *B. tectorum*. Colder winter conditions, in particular, severe and sudden freeze events, are lethal to *B. rubens* but are not lethal to *B. tectorum* (Bykova and Sage 2012).

2.4.1 *Invasion Potential of Cold Deserts by* Bromus

Salt desert vegetation types typically occur at relatively low elevations and have the warmest (mesic) and driest (aridic) soil temperature and moisture regimes in the Cold Deserts. Vegetation types that characterize the salt desert are typically dominated by members of the Chenopodiaceae such as *Atriplex* L. spp. and Sarcobatus Nees spp. (West 1983a, b). Resistance to *B. tectorum* and presumably *B. rubens*, which has received less study in this vegetation type, is low to moderate (Fig. 2.10a). Under the warmest and driest regimes, establishment of *B. tectorum* is limited by frequent low and sporadic precipitation years, and longer term persistence requires either the necessary environmental conditions to maintain viable populations or high propagule pressure (Meyer et al. 2001). Resilience to disturbance is generally low due to low productivity, infrequent plant establishment, and slow recovery (Fig. 2.10b, and see Haubensak et al. 2009). Recent expansion of *B. tectorum* and *B. rubens* into marginal salt desert environments has been attributed to favorable years for establishment during El Niño–Southern Oscillation (ENSO) events (Meyer et al. 2001; Salo 2005). Rising CO_2 levels (Smith et al. 1987) and high genetic variability (Ramakrishnan et al. 2006; Haubensak et al. 2014) also may help explain increased establishment and spread of *B. tectorum* and likely *B. rubens*. Invasion of *B. tectorum* has increased fine fuels and fuel continuity in these types

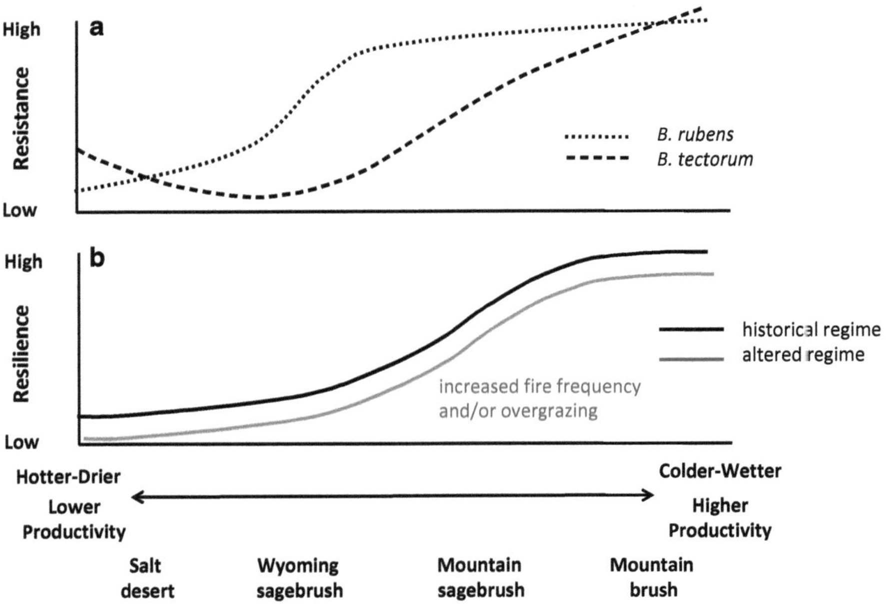

Fig. 2.10 Hypothetical (**a**) resistance to common *Bromus* species and (**b**) resilience to historical and altered fire regimes for the Cold Deserts ecoregion (primary altered regime characteristic labeled in *gray*). Figure modified from Chambers et al. (2014a)

and wildfires are burning for the first time in recorded history (Haubensak et al. 2009; Balch et al. 2013).

Low to mid elevations have warm and dry (mesic/aridic) to warm and moist (mesic/xeric) soil temperature and moisture regimes and are characterized by *Artemisia tridentata* Nutt. ssp. *wyomingensis* Beetle and Young (Wyoming big sagebrush) and to a lesser degree *A. tridentata* Nutt. ssp. *tridentata* (basin big sagebrush) vegetation types (West 1983a, b; Miller et al. 2011). Resistance to *B. tectorum* is low under these regimes, largely due to high climatic suitability for establishment and persistence (Fig. 2.10a) (Chambers et al. 2007, 2014a; Davies et al. 2012). *Bromus rubens* can occur on warmer and drier sites, especially at the lowest elevations (Salo 2005), but its distribution and relative abundance have not been well quantified. Resilience to disturbance and management treatments is low to moderately low (Fig. 2.10b) (Miller et al. 2013; Davies et al. 2012; Chambers et al. 2014a, b). Because the Cold Deserts represent more moderate climates, they also have high levels of anthropogenic disturbance (agricultural, urban and energy development, livestock, wild horse and burro grazing, off-highway vehicle and recreation use, etc.) which has increased the spread and dominance of *B. tectorum* (Knick et al. 2011). Progressive expansion of *B. tectorum* has increased fire frequency and size (Balch et al. 2013), and threshold transitions to annual grass dominance are highest in these vegetation types (Chambers et al. 2014a).

Upper elevations typically have cool and moist (frigid/xeric) to cold and moist (cryic/xeric) regimes and are characterized by *A. tridentata* Nutt ssp. *vaseyana* (Rydb.) Beetle (mountain big sagebrush) and mountain brush (e.g., *Symphoricarpos* Duham. spp. [snowberry.], *Purshia tridentata* (Pursh) DC. [antelope bitterbrush]) vegetation types. Colder soil temperature regimes limit the amount of suitable niche space and increase resistance to *B. tectorum* invasion (Fig. 2.10a) (Chambers et al. 2007; Davies et al. 2012). Although *B. tectorum* can germinate at relatively low temperatures (Roundy et al. 2007), growth and reproduction under cool to cold (frigid to cryic) regimes are limited as indicated by low growth and reproduction (Chambers et al. 2007). At the warmer end of the gradient depletion of perennial herbaceous species due to either inappropriate livestock grazing or expansion of piñon and juniper trees can promote *B. tectorum* establishment following fire, but few of these systems become *B. tectorum* dominated (Chambers et al. 2014b). Favorable conditions for establishment and growth and relatively high levels of productivity result in moderate to high resilience on all but the coldest high elevation sites (Chambers et al. 2014a).

Slope, aspect, and soil characteristics modify soil temperature and water availability and influence occurrence of *B. tectorum* at landscape to plant community scales (Chambers et al. 2007; Condon et al. 2011; Reisner et al. 2013). At higher elevations with cooler soil temperatures, *B. tectorum* is found primarily on south facing slopes (Kulpa et al. 2012) with more solar radiation (Condon et al. 2011; Lovtang and Riegel 2012).

Seasonality of precipitation within the ecoregion also appears to influence establishment and persistence of *B. tectorum*. In the western portion of the ecoregion, most precipitation arrives in winter and spring months and climate suitability to

B. tectorum is high. Life history and physiological traits such as germination in early fall through winter (Mack and Pyke 1983) and high growth and nutrient uptake rates (James et al. 2011) make *B. tectorum* highly competitive with native plants. In the eastern portion of the ecoregion, summer precipitation is higher (ustic soil moisture regimes) and this competitive advantage appears to be greatly diminished (Bradley 2009), although establishment and persistence still can occur following disturbance (Bradford and Lauenroth 2006).

2.4.2 Ecosystem Threats to Cold Deserts from Bromus

The most widely cited effect of *Bromus* invasions in the Cold Deserts ecoregion is the alteration of fire regimes (Whisenant 1990; Brooks and Pyke 2001; Balch et al. 2013). The mechanisms of these changes in Cold Deserts are largely the same as those described for Warm Desert in Sect. 2.2 and the impacts are explained in more detail in Germino et al. (2015). Invasion of *B. tectorum* can also alter the size and dynamics of soil N pools, or the activity of the microbial community, with the type and magnitude of the effect depending on factors such as soil characteristics, plant community composition, and time since invasion (Rimer and Evans 2006; Hooker et al. 2008; Schaeffer et al. 2012). *Bromus* invasions can alter hydrologic conditions resulting in increased soil water recharge (Wilcox et al. 2012) which can decrease resistance to secondary invasion by deep-rooted forbs that mature later in the growing season (Kulmatiski et al. 2006). After wildfires, runoff and erosion can increase, particularly on steep slopes (>15 %) (Wilcox et al. 2012). Wind erosion also can occur after fire, resulting in redistribution or loss of nutrient-enriched soil fines and, under shorter fire return intervals caused by *Bromus*, homogenization of soil properties in Cold Deserts shrublands (Sankey et al. 2009).

Both resilience from disturbance and resistance to *Bromus* in the Cold Deserts are reduced by disturbances that decrease perennial species abundance and increase resource availability. Such disturbances include excessive livestock grazing and altered fire regimes, specifically increased fire frequency. Many of the dominant shrubs (e.g., *Artemisia* spp.) are not fire tolerant in the warmer and drier ecological types that are least resistant to *Bromus*. Plant establishment is often sporadic (Miller et al. 2013) even when these species are seeded (Knutson et al. 2014). More frequent fire coupled with increased competition due to invasion of *Bromus* and following livestock grazing too soon after reseeding can prevent native regeneration (Eiswerth and Shonkwiler 2006). Populations of *B. tectorum* often increase rapidly following fire or other disturbances that remove perennial native species and seed banks can grow to ≥20,000 seeds/m^2 within a few years (Humphrey and Schupp 2001; Meyer et al. 2007; Chambers et al. 2015). Typically small seed banks of most native species (Hassan and West 1986; Allen et al. 2008), coupled with low competitive ability of native seedlings with *B. tectorum* (James et al. 2011; Mazzola et al. 2011), can greatly decrease seedling establishment. Mature individuals of perennial herbaceous species, especially those with similar phenologies, are often strong competitors (Booth et al. 2003), but large individuals or populations of *B. tectorum* can reduce growth and seed production of these species. Decreases in perennial native species, especially grasses and forbs, biological soil crusts, and the distance between perennial herbaceous species (gaps) due to inappropriate livestock grazing or other disturbances are strongly associated with increases in *B. tectorum* across a range of ecological types (Chambers et al. 2007; Dettweiler-Robinson et al. 2013; Reisner et al. 2013; Pyke et al. 2015). Increasing plant community dominance of *B. tectorum* can alter species interactions, influence trophic interactions, and reduce species diversity (Germino et al. 2015). For example, increasing *B. tectorum* cover has a negative influence on nesting of *Centrocercus urophasianus* Bonaparte (greater sage-grouse) which has been considered for listing under the US Endangered Species Act of 1972 (Federal Register, 50 CFR Part 17), and on the abundance of *Spermophilus townsendii* Bachman (Townsend's ground squirrel) which is important prey for raptors and other predators (Yensen et al. 1992). Complete conversion to *B. tectorum* results in loss of shrub structure and shrubland-associated species like *Amphispiza belli* Cassin (sage sparrow) and an overall decrease in native species diversity (Earnst and Holmes 2012).

Ecosystem resilience to disturbances increases with increasing elevation, latitude, and associated primary productivity (Fig. 2.10). Because resistance to invasion follows a similar pattern, especially for *B. tectorum*, the net threat posed by this species differs with seasonality of precipitation, but is generally highest at low to mid productivity, and lowest in areas of higher productivity (Fig. 10.1, Chambers et al. 2015).

2.4.3 *Management Strategies for* Bromus *in Cold Deserts*

In the Cold Deserts, a suite of different vegetation management treatments are used to increase resilience to fire and other disturbances and to enhance resistance to invasive annuals (Chambers et al. 2014a). Primary objectives are to reduce woody fuel loads and thus fire severity and extent, decrease exotic annual species abundance and spread, and increase perennial herbaceous species dominance which both promote recovery after disturbance and compete with exotic annual plants. Treatments include various combinations of prescribed fire, mechanical treatment, and herbicide application to decrease sagebrush, piñon, and juniper abundance, herbicide application and grazing to control *Bromus*, and seeding with perennial herbaceous species and shrubs to restore native plant communities (Monsen et al. 2004; Pyke 2011). However, responses to these treatments often vary due to inherent differences in site resilience and resistance (Miller et al. 2013) and subsequent land use regimes (Eiswerth and Shonkwiler 2006). Knowledge of environmental factors, ecosystem attributes and processes, and disturbance and land use history effects that influence resilience and resistance (Fig. 1.1) can be used to determine appropriate management strategies at both site and landscape scales (Chambers et al. 2014a; Miller et al. 2013).

Fig. 2.11 (**a**) A fire that started along a major highway and burned through a sagebrush ecosystem with *B. tectorum* in the understory and into a Jeffrey pine ecosystem, Washoe County, Nevada. (**b**) Landscape conversion of a sagebrush ecosystem to *B. tectorum* and other exotic annual dominance following repeated fire, Humboldt County, Nevada. Photos by Nolan E. Preece

The degree of uncertainty increases and management options become increasingly limited with decreases in resilience to fire and management treatments and resistance to invasion. In cool and dry to cool and moist regimes with moderate to high resilience and resistance, a variety of management treatments like carefully timed grazing, prescribed fire, mechanical treatments, and herbicide applications can be used to maintain or improve ecological conditions. However, in warm and dry to warm and moist regimes, widespread invasion and increasing dominance of *Bromus* and other invaders have altered vegetation dynamics and often limit options. Largely irreversible thresholds can occur following either fire or management treatments if sufficient perennial herbaceous species for recovery are lacking (Fig. 2.11).

2.5 Western Great Plains

The Western Great Plains comprises shortgrass steppe and northern and southern mixed-grass prairie. The foothills of the Rocky Mountains form the western border (Lauenroth and Milchunas 1992; Coupland 1992) and tallgrass prairie the eastern border, which is a vegetation transition zone (Sims and Risser 2000) at approximately the 100th meridian (Van Dyne 1975). Temperatures in the ecoregion are cooler in the north than the south (frigid, mesic, thermic from north to south; Fig. 2.3c). The mean annual temperature in shortgrass steppe is 8.6 °C (Lauenroth 2008) and ranges from 8.2 °C in northern Colorado to 17.7 °C in western Texas and southeastern New Mexico (Lauenroth and Milchunas 1992). Likewise, mean annual temperatures in the mixed-grass prairie increase from north to south (3.6 °C in Saskatchewan to 12.9 °C in central Kansas) (Coupland 1992). A precipitation gradient runs perpendicular to the temperature gradient; it is drier in the west than the east (ustic aridic, aridic ustic, typic ustic from west to east; Fig. 2.3c). Annual precipitation in the shortgrass steppe of northern Colorado ranges from 300 to 400 mm on the west side to 500–600 mm on the east side (Lauenroth and Milchunas 1992) with mean annual precipitation of 321 mm (Lauenroth et al. 2008). Mean annual precipitation in the mixed-grass prairie ranges from around 300 mm near the shortgrass steppe to around 600 mm in the eastern transition zone to tallgrass prairie (Coupland 1992; Moran et al. 2014). There is a smaller gradient of precipitation from south to north ranging from around 430 mm in southern mixed-grass prairie to 340 mm in the northern mixed-grass prairie (Van Dyne 1975).

B. tectorum and *B. arvensis* are the two most common invasive *Bromus* that occur in the shortgrass steppe, although they are largely restricted to roadsides and disturbed areas such as old fields because intact native vegetation is highly resistant to invasion (Kotanen et al. 1998; Milchunas et al. 1992) (Fig. 2.13a). The shortgrass steppe is dominated by the short-stature, warm-season species *Bouteloua gracilis* Willd. ex Kunth) Lag. ex Griffiths (blue grama) and *Bouteloua dactyloides* (Nutt.) J.T. Columbus (buffalograss). The mid-height, cool-season grass *Pascopyrum smithii* (Rydb.) Á. Löve (western wheatgrass) is commonly associated with *B. gracilis* in the northern portion of the shortgrass steppe (Lauenroth and Milchunas 1992). This shifts to *B. gracilis* in association with *Sporobolus airoides* (Torr.) Torr. (alkali sacaton), *Pleuraphis jamesii* Torr. (James' galleta), and *B. dactyloides* toward the south (Lauenroth and Milchunas 1992).

Bromus tectorum and *B. arvensis* are also the most common invasive *Bromus* species in northern and southern mixed-grass prairie and are joined by *Bromus catharticus* Vahl (rescuegrass) and *Bromus secalinus* L (rye brome), but these species are of secondary dominance. *Bromus gracilis* remains a major component of the mixed-grass prairie, which comprises cool-season and warm-season species from tallgrass prairie and shortgrass steppe (Sims and Risser 2000). *Pascopyrum smithii* and *Hesperostipa* (Elias) Barkworth spp. (needlegrass) are codominant species with *B. gracilis* in the northern mixed-grass prairie (Van Dyne 1975). Other cool-season species such as *Carex* L. spp., *Elymus lanceolatus* (Scribn. & J.G. Sm.) Gould ssp. *lanceolatus* (thickspike wheatgrass), *Koeleria macrantha* (Ledeb.) Schult. (prairie Junegrass), and *Nassella viridula* (Trin.) Barkworth (green needlegrass) become more abundant in particular microsites. In the southern mixed-grass prairie, taller dominants are *P. smithii*, *Andropogon gerardii* Vitman (big bluestem), *Schizachyrium scoparium* (Michx.) Nash (little bluestem), and *Bouteloua curtipendula* (Michx.) Torr. (sideoats grama), with *B. gracilis* and *B. dactyloides* as a lower layer in the canopy (Van Dyne 1975).

2.5.1 Invasion Potential of Western Great Plains by Bromus

The relative resistance of Western Great Plains ecosystems to invasion by exotic annual *Bromus* species can be attributed to adaptation of the native plants to the climate and characteristic disturbances of drought, fire, and grazing. Of the three subregions of the Western Great Plains, northern mixed-grass prairie appears to be least resistant to exotic annual *Bromus* invasion (Fig. 2.12). *Bromus arvensis* (Heitschmidt et al. 1995; Karl et al. 1999) and *B. tectorum* (Heitschmidt et al. 1995) can be dominant species in northern mixed-grass prairie. In Wind Cave National Park in South Dakota, occurrence of exotic annual *Bromus* species was associated with cool-season grasses such as *P. smithii* and *N. viridula* and the shrub *Symphoricarpos occidentalis* Hook. (western snowberry) and negatively associated with warm-season grasses, especially *A. gerardii*, *B. curtipendula*, and *S. scoparium*, and the shrub *Rhus aromatica* Aiton (fragrant sumac) (Ogle and Reiners 2002).

Data from the U.S. National Park Service Northern Great Plains Network indicate that higher cover of *Bromus* can be associated with lower native species richness, but only in the South Prairie region of the network where *Bromus* cover is highest (Fig. 2.14). This is especially true at Scotts Bluff National Monument in northwestern Nebraska, Badlands National Park in southwestern South Dakota, and Fort Laramie National Historic Site in southeastern Wyoming where *Bromus* cover was repeatedly measured in excess of 30 % (Fig. 2.14) as illustrated in the foreground of Fig. 2.13d. This pattern was not exhibited at Agate Fossil Beds National

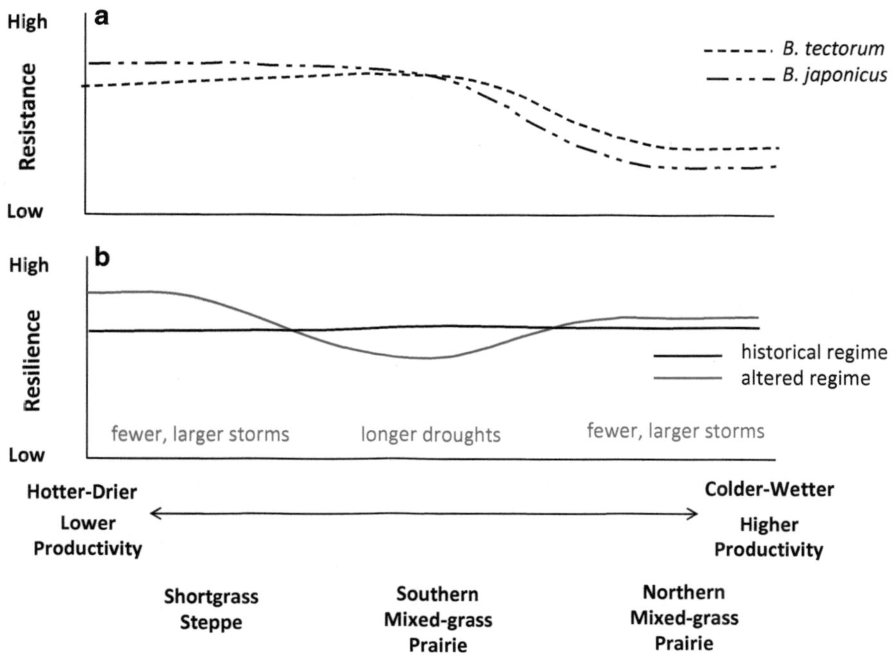

Fig. 2.12 Hypothetical (**a**) resistance to common *Bromus* species and (**b**) resilience to historical and altered fire regimes for the Western Great Plains ecoregion (primary altered regime characteristic labeled in *gray*)

Monument (Fig. 2.13c), or elsewhere in the Northern Great Plains where *Bromus* cover rarely exceeded 10 % *Bromus* cover (Fig. 2.14).

Resistance to invasion by exotic annual *Bromus* species of southern mixed-grass prairie may be as high as in shortgrass steppe (Fig. 2.12). Although *B. catharticus*, *B. arvensis*, and *B. tectorum* occur in Oklahoma (Bidwell et al. 2004), none of the fact sheets for managing invasive species published by the Oklahoma Cooperative Extension Service are about exotic annual grasses or mention them as being problems (http://www.oces.okstate.edu/extension-fact-sheets, accessed 6 Jan 2015). The invasive species of concern are all shrubs and subshrubs (Cummings et al. 2007; Bidwell et al. 2009).

Drought has been a common feature of the Western Great Plains climate (Coupland 1992), where potential evaporation exceeds precipitation (Sims and Risser 2000). Many long-term studies found weather to be a more important influence on vegetation response than grazing or burning treatments (Teague and Dowhower 2003; Teague et al. 2004, 2010; Heitschmidt et al. 2005; White and Loftin 2000; Vermeire et al. 2008, 2014). Modeling efforts by Bradford and Lauenroth (2006) indicate that climate is a stronger influence on invasion of *B. tectorum* than soil characteristics. Bradley (2009) found that two of the strongest predictors of improved habitat suitability for *B. tectorum* in the Great Basin region

Fig. 2.13 (**a**) Typical *B. tectorum* invasion of roadsides in the shortgrass steppe ecosystem, Weld County, Colorado. Photo by CS Brown. (**b**) Conservation Reserve Program land 3 years after seeding with native species. Increased beige vegetation on *left* side of fence is *B. tectorum*. Photo by M Vandever. (**c**) Little or no *Bromus* at Agate Fossil Beds National Monument. Photo by US Geological Survey. (**d**) Aerial cover at this site in Scotts Bluff National Monument is 60 % *B. tectorum* and 10 % *B. arvensis*. Photo by US National Park Service

were decreased annual and summer precipitation. This suggests that the normal pattern of summer precipitation in the Western Great Plains creates conditions unfavorable to *Bromus* and drought could improve their success. However, it does not appear that drought alone (see Van Dyne 1975; Munson and Lauenroth 2009; Moran et al. 2014) or in combination with grazing (Heitschmidt et al. 1999, 2005; Eneboe et al. 2002) has reduced resistance of these grasslands to invasion by *Bromus*.

Grazing of ungulates has always been important in the Great Plains, which was once inhabited by 2–4 million bison (*Bison bison*) and other grazing and browsing ungulates (Van Dyne 1975). As a result, this ecoregion is resilient to grazing (Vermeire et al. 2008). The abundances of *Bromus* and other weedy plants can be reduced (Fig. 2.13b) (Milchunas et al. 1992, 2011; Harmoney 2007; Vermeire et al. 2008) or remain unaffected by grazing, even in combination with drought (Heitschmidt et al. 1999, 2005).

Fire is a natural disturbance in the Western Great Plains and has either no effect on or reduces aboveground productivity (Scheintaub et al. 2009). Herbaceous productivity of shortgrass steppe (Ford and Johnson 2006; Augustine and Milchunas

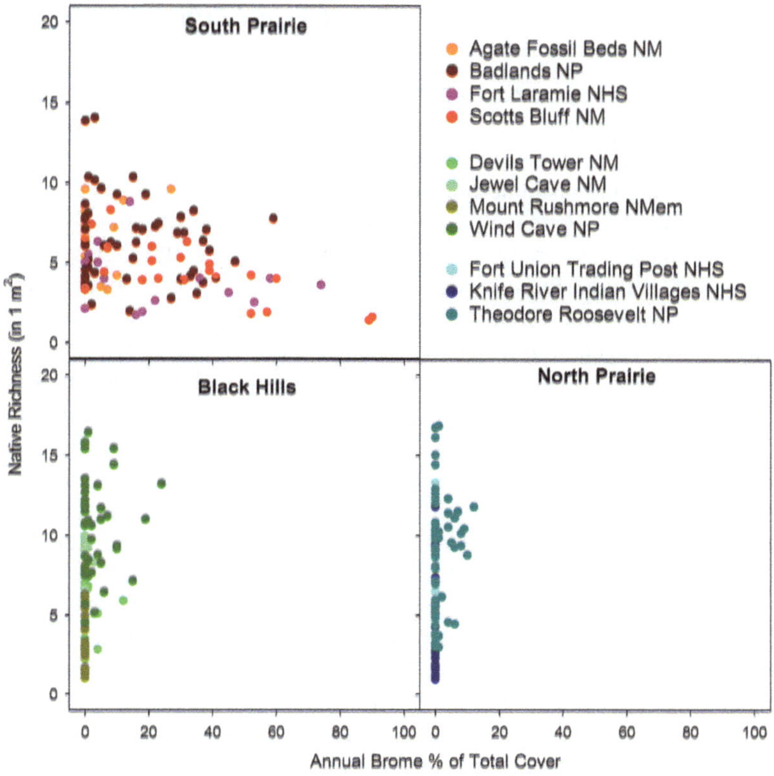

Fig. 2.14 In southern Northern Great Plains Network prairie parks, plots with high cover of *Bromus* have lower native plant species richness than plots with little *Bromus* cover. NM = National Monument; NP = National Park; NHS = National Historic Site; NMem = National Memorial. (Courtesy of the National Park Service, Northern Great Plains Network Inventory and Monitoring Program)

2009; Augustine 2010), northern mixed-grass prairie (Whisenant and Uresk 1990; Vermeire et al. 2011), and southern mixed-grass prairie (Whisenant et al. 1984; Ford and White 2007; Teague et al. 2008, 2010; Vermeire et al. 2014) may be unaffected by fire or can recover productivity and species composition within 6 months to 2 years of burning, depending on precipitation amounts during the recovery period (Teague et al. 2008; Augustine and Milchunas 2009). Grazing is often reported not to alter the effects of fire (Augustine et al. 2010; Vermeire et al. 2014). Not only do these ecosystems have high resilience to fire, but fire either reduces (Whisenant and Uresk 1990; Scheintaub et al. 2009; Teague et al. 2010; Vermeire et al. 2011, 2014) or does not affect (Teague and Dowhower 2003; Augustine and Milchunas 2009; Augustine et al. 2014) abundance of *Bromus*. However, in some cases when *Bromus* were initially reduced by spring burning, they increased in abundance the second year after the burn (Whisenant et al. 1984).

2.5.2 *Ecosystem Threats to Western Great Plains from* Bromus

Although several *Bromus* are considered invasive in the Western Great Plains, they are neither as widespread nor have they had the extensive, ecosystem altering effects documented in some other ecoregions. Even though the Western Great Plains is relatively resistant and resilient to invasion by *Bromus*, invasions can effect change in some circumstance. For example, *Bromus* can reduce the growth of native plants in northern mixed-grass prairie (Haferkamp et al. 1997, 1998; Ogle et al. 2003). Ogle et al. (2003) found *Bromus*-dominated northern mixed-grass prairie had lower above- and belowground biomass, slower decomposition, and sometimes greater litter than native vegetation. Modeling of *B. arvensis* and *B. tectorum* in northern mixed-grass prairie found non-significantly greater carbon storage with high *Bromus* cover, but the differences continued to diverge after 50 years and may become large in the future (Ogle et al. 2004).

Resilience of ecosystems to disturbance and resistance to invasion can be altered when historically normal patterns of disturbance change (Germino et al. 2015). The absence rather than presence of grazing, fire, or both represents disturbances for Western Great Plains grasslands (Milchunas et al. 1992; Haferkamp et al. 1993; Adler and Lauenroth 2000; Brockway et al. 2002; Ford and Johnson 2006; Vermeire et al. 2011) and may facilitate invasion by *Bromus* (Milchunas et al. 1992, 2011). Fire frequencies in the Western Great Plains may increase as a result of accumulation of fuels due to fire suppression (Brockway et al. 2002; Garfin et al. 2014) and increased lightning strikes due to climate warming (Romps et al. 2014).

As explained above, weather is an important determinant of productivity and composition of Western Great Plains grasslands. Thus, changes in climate may influence resistance of Western Great Plains ecosystems to *Bromus* invasion (Fig. 2.12). More frequent high temperatures and extreme heat are predicted for the region (Walsh et al. 2014). *Bromus tectorum* reportedly increases with temperature in northern mixed-grass prairie (Blumenthal et al., unpublished data). Longer periods between storms are predicted for much of Texas and Oklahoma (Walsh et al. 2014), which may decrease its resilience to disturbance and resistance to *Bromus* invasion (Fig. 2.12). Models that assume continued increases in heat trapping greenhouse gases predict that winter and spring precipitation and days with heavy rain will increase in the northern part of the Great Plains (Walsh et al. 2014). Changes such as these are not likely to reduce resilience to disturbance and may increase resilience in more water-limited shortgrass steppe (Fig. 2.12).

The drought in the early 2000s resulted in conditions similar to predicted future climates (Moran et al. 2014). Western Great Plains grasslands had linear responses to current year total precipitation and previous year productivity (Moran et al. 2014). No new species assemblages (e.g., dominance of *Bromus*) emerged in response to the drought (Moran et al. 2014). The system appears to be buffered against changes predicted in the next 50 years (Symstad et al. 2014), but it remains to be seen how far the climate system can be pushed before the biotic components of the ecosystem can no longer respond sufficiently to be sustained.

2.5.3 *Management Strategies for* Bromus *in Western Great Plains*

Bromus invasions can be prevented and minimized throughout the ecoregion by conserving intact grasslands and applying reasonable levels of grazing and burning to maintain their integrity and resistance to invasion. The Western Great Plains ecosystems are currently most threatened by large-scale disturbances such as conversion to cultivated agriculture, urban development, and oil and gas exploration and development, all of which may foster *Bromus* dominance.

Management beyond good stewardship of the ecosystem can focus first on containing and then eliminating *Bromus* where it occurs. Herbicides that control *Bromus* with minimal negative effects on native species can be effectively used alone (Hewlett et al. 1981; Haferkamp et al. 2001) or in combination with burning (Masters et al. 1992; Calo et al. 2012). *Bromus* populations can be managed by reducing seed production through well-timed grazing, mowing, or burning (e.g., Fig. 2.13b) (Vermeire et al. 2008; Milchunas et al. 2011) or application of growth regulator herbicides used for broadleaf weed control (Rinella et al. 2010a, b, 2013).

2.6 Western Forests

The Western Forests range from the Marine West-Coast Forest to the Northwestern Forested Mountains Level I EPA ecoregions and are characterized by a very diverse range of environmental conditions. Collectively, they represent some of the coolest and wettest conditions in the western United States (Fig. 2.3). Soil temperature regimes are generally cryic or frigid in the east and north, but can also be mesic and even thermic to the west and south. These conditions combined with high precipitation are conducive to the growth of forests with relatively high canopy and surface cover of perennial vegetation and high accumulations of litter and duff. Shrublands or grasslands occur as early successional patches following disturbances that remove forest canopy cover, and where productivity levels are relatively low due to shallower soils or warmer and drier soil temperature and moisture regimes (Fig. 2.15).

Bromus species are generally not considered among the highest priority invasive plants in the Western Forests. In fact, one recent review of the impacts of exotic invasive species in US forests did not even mention *Bromus* (Moser et al. 2009). The exception are patches of chaparral and grasslands (Fig. 2.15) within ponderosa pine and Jeffrey pine mixed conifer forests (Pierson and Mack 1990; Keeley and McGinnis 2007). In these situations, soils are more frequently exposed to sunlight and warmer and drier conditions facilitate establishment, growth, and reproduction of *Bromus*. *Bromus tectorum* is the most widespread and abundant *Bromus* in these ecoregions, but *B. diandrus* can also occur within grassland and chaparral forest openings.

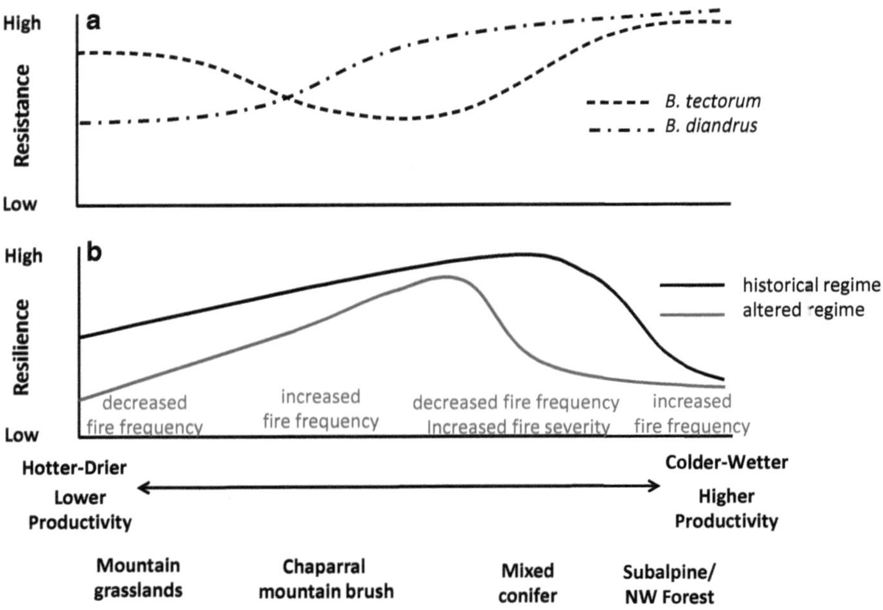

Fig. 2.15 Hypothetical (**a**) resistance to common *Bromus* species and (**b**) resilience to historical and altered fire regimes for the Western Forests ecoregion (primary altered regime characteristic labeled in *gray*)

2.6.1 Invasion Potential of Western Forests by Bromus

Forests with cool to cold temperature regimes have low climatic suitability to *Bromus*. Also, forests with high amounts of canopy cover that accumulate large amounts of litter and duff in the understory and create low light conditions at the soil surface have limited establishment and growth of annual plant species. Drier and warmer mixed conifer, Jeffrey pine, and ponderosa pine forests that have less canopy cover and lower litter and duff accumulation are generally less resistant to *Bromus* than subalpine and northwestern forests due to lower levels of productivity and higher frequency of fire (Fig. 2.15).

Fire events that remove forest canopy, especially high severity fire, are often associated with decreased resistance to *Bromus* in climatically suitable forests (Crawford et al. 2001; Keeley and McGinnis 2007; McGlone et al. 2009; Fornwalt et al. 2010; Kerns et al. 2011). Decreased resistance is partly due to increased light availability at the soil surface, but also may be affected by increased availability of soil nutrients. Biomass of *B. tectorum* beneath *Pinus ponderosa* Lawson and C. Lawson (ponderosa pine) trees can be limited by low N availability. Fire can increase available nutrients due to deposition of ash onto the soil surface, release of available P and N from organic matter, and decomposition of belowground biomass and result in increases in *B. tectorum* growth and reproduction (Covington and Sackett 1992; Monleon et al. 1997; Gundale et al. 2005, 2008). Also, charcoal generated by fire may adsorb phenolic compounds from pine litter inputs which can otherwise impede understory plant growth (Gundale and DeLuca 2006, 2007), and increase nitrification rates, further increasing the amount of plant available N in the soil following fire (DeLuca et al. 2006).

High severity human disturbances such as clear-cut logging can similarly decrease resistance of landscapes. Recruitment, survival, and reproduction of *B. tectorum* can all be higher in clear-cuts across a range of western forest types including those dominated by *P. ponderosa*, *Pseudotsuga menziesii* (Mirb.) Franco (Douglas-fir), *Abies grandis* (Douglas ex D. Don) Lindl. (grand fir), and *Thuja plicata* Donn ex D. Don (western redcedar) (Pierson and Mack 1990). Lower severity understory thinning can also decrease resistance to *B. tectorum* in these forest types, but only where significant understory cover of *B. tectorum* was present before thinning. Even lower severity human disturbances such as cattle grazing occurring at light intensity and short duration can lead to increased cover of *B. tectorum* in ponderosa pine forests under drought conditions (Sorensen and McGlone 2010), possibly due to competitive release of other vegetation such as *Elymus elymoides* (Raf.) Swezey (bottlebrush squirreltail) and *P. smithii* caused by preferential grazing by cattle.

Removal of litter and coarse woody debris can increase exposure of mineral soil and the probability that *Bromus* seeds can make contact with the soil and germinate (Pierson and Mack 1990; Keeley and McGinnis 2007; Gundale et al. 2008). This is partly due to the litter creating a physical barrier preventing seeds from coming into contact with mineral soil, but also may be due to phenolytic compounds in new pine litter that can inhibit germination.

Lower temperature limits may limit the upper elevation sites attainable by *Bromus* species (Fig. 2.2a), but climate change is increasingly recognized as a mechanism by which high elevation habitats such as Western Forests may become increasingly more invasible (Pauchard et al. 2009). There is evidence suggesting that *B. tectorum* already exhibits the ability to adapt to higher elevation sites (Rice and Mack 1991; Leger et al. 2009), and climatic warming may further reduce the resistance of these habitats to *Bromus* invasion (Compagnoni and Adler 2014). Rising CO_2 has also been shown to increase *B. tectorum* productivity through increased water use efficiency in Cold Deserts shrublands (Smith et al. 1987), and the operative mechanism could further promote growth at higher elevations as they warm.

2.6.2 *Ecosystem Threats to Western Forests from* Bromus

It is unclear how much of a threat *Bromus* pose to Western Forests compared to other ecoregions of the western United States. No doubt *Bromus* can have some of the same effects documented for other ecoregions, such as direct competition with other plant species. They may even affect landscape susceptibility to fire by producing continuous fuelbeds at least where the forest canopy does not overshadow and inhibit their growth (e.g., Fig. 2.16). However, the potential for *Bromus* species to alter fire regimes (i.e., fire frequency, severity, seasonality, etc.) in Western Forests is more difficult to discern.

Many of the lower elevation forests that are most susceptible to invasion by *Bromus* (e.g., ponderosa pine) are also evolutionarily suited to fire return intervals as short as once every 2 years (reviewed by Hurteau et al. 2013). So if *Bromus* species, *B. tectorum* in particular, has the potential to increase landscape flammability and promote recurrent fire, is it reasonable to think that they might promote such short intervals that even these fire-adapted forests cannot persist? The general consensus is that they could, especially in areas where anthropogenic activities increase ignition sources and promote productivity of *Bromus*, thus enhancing their competitive suppression of regenerating forest species and their contributions to highly flammable fuelbeds (McGinnis et al. 2010; Hurteau et al. 2013). As explained above in Sect. 2.6.1, as the climate increasingly warms, resistance of Western Forests to invasion by *Bromus* will likely decrease. When this decreased resistance is coupled with increased frequency of large fires (Westerling and Bryant 2008) and length of the fire season (Westerling et al. 2006), the potential for *Bromus* to alter fire regimes in Western Forests may become more likely.

Fig. 2.16 *Bromus tectorum* dominating a landscape 6 years after a high severity fire in a previous conifer forest stand (Star Fire, Tahoe National Forest, 2001). Note the conifer recruitment in the *upper right* side of this image despite the high cover of *B. tectorum* (photo by Jon Keeley, 22 June 2007)

2.6.3 *Management Strategies for* Bromus *in Western Forests*

Prevention and control of *Bromus* in the Western Forests have not be evaluated to the degree they have in other ecoregions, most notably in the Cold Deserts and Mediterranean California. Strategies developed in those regions should be generally applicable to the Western Forests, although there are some unique characteristics of Western Forests that lend themselves to approaches only possible in this ecoregion. For example, litter and duff can rapidly accumulate on the forest floor, which can suppress *B. tectorum* germination (Keeley and McGinnis 2007). However, some of the principles that have been applied to restore historical fire regimes where fire suppression and fuel accumulation have altered them may reduce litter and forest canopy cover, which could inadvertently improve conditions for *Bromus* species. Some of these include reducing surface and ladder fuels, decreasing crown density, and maintaining widely spaced large trees. Forest managers are challenged to find an effective balance between the restoration of historical fire regimes and the suppression of invasive plants such as *Bromus*, thus maximizing resilience to fire and resistance to invasion. Management of *Bromus* invasion should be linked with maintaining historical ecosystem structure and function, which may entail suppressing or facilitating fire, or reducing fire severity (e.g., by pre-fire fuels manipulations), as dictated by the historical fire regimes and current condition of local areas.

2.7 Management Implications

This chapter explained that invasion potential, ecosystem threats, and management strategies for *Bromus* vary both within and among ecoregions and vegetation types, and among *Bromus* species. In particular, it should not be assumed that just because a particular *Bromus* is documented to pose a significant threat in one ecoregion or vegetation type, that other *Bromus* will pose similar threats to other ecoregions or vegetation types. Variation among ecoregions and vegetation types in ecosystem resistance to *Bromus* invasion, and resilience to disturbance, can be used to explain the conditions associated with high probability of ecosystem impacts such as an altered fire regime caused by a grass/fire cycle.

A strategic, landscape-scale management approach that includes protection, prevention, and restoration (e.g., Brooks and Chambers 2011; Chambers et al. 2014a) can be used to better focus management activities across the environmental/productivity gradients in the western United States. Protection focuses on maintaining or increasing the resilience and resistance of ecosystems with high conservation value by eliminating or minimizing current and future stressors. Protection is a viable strategy for intact communities within ecosystem types that have inherently low resistance and resilience such as blackbrush ecosystems in the Warm Deserts and salt desert and warmer Wyoming big sagebrush ecosystems in the Cold Deserts. Protection may also be a viable strategy for critical habitat for threatened and endangered species that are sensitive to *Bromus* invasions and their associated impacts.

Prevention involves increasing both resilience and resistance of ecosystems that have not crossed thresholds, but that exhibit declining conditions and are at risk of high severity fire and invasion (Miller et al. 2013). The focus is on ecosystems with higher resilience, like cooler and moister *A. tridentata* ecosystems in the Cold Deserts (Fig. 2.10) and mixed conifer ecosystems of the Western Forests (Fig. 2.15). Management objectives are to improve or maintain ecosystem attributes and processes by reducing woody species dominance and promoting a functionally diverse community of shrubs, perennial grasses, and forbs.

Restoration involves increasing resilience and resistance of disturbed, degraded, or invaded areas by reestablishing functionally diverse plant communities. Integrated management strategies that include using herbicides to control/suppress *Bromus* and seeding perennial herbaceous and shrub species adapted to local conditions may be used to restore priority areas dominated by *Bromus* and other annual invaders (Monaco and Sheley 2012). Restoration activities that target relatively moister sites are likely to be more successful, and even then repeated intervention may be required. As the climate warms, it may be necessary to consider transformative restoration in areas with novel climates in which species are established that are climatically suitable but that have not occurred on a given landscape historically (Bradley et al. 2010).

2.8 Research Needs

This chapter indicates that patterns of ecological resistance to *Bromus* invasions have been well documented in the western United States, especially related to presence/absence of *B. tectorum* and *B. rubens* in the Cold Deserts, Mediterranean California, and Warm Desert Ecoregions. What is less understood are the patterns of relative abundance (e.g., cover or biomass) of these and other *Bromus*, and the ecological mechanisms associated with these patterns. Also, the range of conditions associated with the grass/fire cycle have only been studied in depth within these same three ecoregions, and even within them there is limited capacity to predict when and where they will establish. A framework exists for objectively evaluating the grass/fire cycle potential (Brooks 2008) which can be systematically applied to each major vegetation types within each ecoregion to better understand when and where *Bromus* invasions pose the threat of altering fire regimes. This information is needed to more reliably predict potential future conditions of spread, dominance, and ecological impact by *Bromus* species within each ecoregion of the western United States.

The relative resistance to invasion by the dominant *Bromus* species and resilience to fire of each of five major ecoregions in the western United States were presented in Figs. 2.4, 2.7, 2.9, 2.11, and 2.14. In the introduction to this chapter, we emphasized that the degree to which scientific studies supported these models varies widely among ecoregions, and we suggested that the conservative approach would be to consider them hypothetical models. We feel confident with the scientific support for the Cold Desert and Warm Desert models, but encourage others to develop studies to test the predictions of all five ecoregional models. The refinement and validation of these models would improve their reliability and utility in evaluating the relative potential for invasion and ecological impact of *Bromus* in the western United States.

Appendix

See Table 2.2.

Table 2.2 GrassPortal (www.grassportal.org, accessed 16 Feb 2015, Osborne et al. 2011) site localities used to construct Fig. 2.2

	B. arvensis (= *japonicus*)	*B. hordeaceus*	*B. diandrus* (= *rigidus*)	*B. madritensis*	*B. rubens*	*B. tectorum*
Arizona	82	36	108	4	377	183
California	104	979	934	220	1109	664
Colorado	59	3	1	0	1	121
Idaho	31	16	2	0	10	48
Kansas	220	7	0	0	0	266
Montana	45	14	0	0	0	77
Nebraska	50	0	0	0	0	98
Nevada	8	7	9	0	4	25
New Mexico	45	1	9	0	11	61
North Dakota	0	0	0	0	0	34
Oklahoma	36	0	0	0	0	22
Oregon	40	55	71	2	35	141
South Dakota	7	1	0	0	0	36
Texas	89	8	6	0	4	30
Utah	31	10	9	0	13	45
Washington	27	12	92	0	4	113
Wyoming	23	39	0	0	0	40
Sum	897	1188	1241	226	1568	2004

Data Provider	**Dataset**
USDA PLANTS	USDA PLANTS Database
Consortium of California Herbaria	Consortium of California Herbaria
University of Arizona Herbarium	UA Herbarium
Arizona State University, International Institute for Species Exploration	Arizona State University Vascular Plant Herbarium
University of Connecticut	CONN GBIF data
Berkeley Natural History Museums	University and Jepson Herbaria DiGIR provider
New Mexico Biodiversity Collections Consortium	New Mexico Biodiversity Collections Consortium database

(continued)

Table 2.2 (continued)

Data Provider	Dataset
Missouri Botanical Garden	Missouri Botanical Garden
Utah Valley State College (UVSC)	Utah Valley State College Herbarium
University of Kansas Biodiversity Research Center	Botany Vascular Plant Collection
Oregon State University	Vascular Plant Collection
University of Washington Burke Museum	Vascular Plant Collection – University of Washington Herbarium (WTU)
Bernice Pauahi Bishop Museum	Bishop Museum Natural Sciences Data
University of Alabama Biodiversity and Systematics	Herbarium (UNA)
Colorado State University Herbarium (CSU)	Colorado State University Herbarium
National Museum of Natural History	NMNH Botany Collections
Canadian Museum of Nature	Canadian Museum of Nature Herbarium
University of Colorado Museum of Natural History	Specimen Database of Colorado Vascular Plants
The New York Botanical Garden	Herbarium of The New York Botanical Garden

The 17 states represent the ecoregions of the western United States (Fig. 2.1). The seven *Bromus* species are represented by state site records spanning their current distributional range as indicated by the USDA Plants database (www.plants.usda.gov, accessed 18 Feb 2015). The 19 data providers represent the original sources of information for the database used in Fig. 2.2

References

Abella SR, Embrey TM, Schmid SM et al (2012) Biophysical correlates with the distribution of the invasive annual red brome (*Bromus rubens*) on a Mojave Desert landscape. Invasive Plant Sci Manag 5:47–56

Adler PB, Lauenroth WK (2000) Livestock exclusion increases the spatial heterogeneity of vegetation in Colorado shortgrass steppe. Appl Veg Sci 3:213–222

Allen E, Chambers JC, Nowak RS (2008) Immediate and longer-term effects of a spring prescribed-burn on the soil seed bank in an encroaching semi-arid woodland. West N Am Nat 68:265–277

Allen EB, Padgett PE, Bytnerowicz A et al (1998) Nitrogen deposition effects on coastal sage vegetation of southern California. In: Proceedings of the international symposium on air pollution and climate change effects on forest ecosystems. Gen Tech Rep PSW-GTR-166. USDA, Forest Service, Pacific Southwest Research Station, Albany, CA, pp 131–140

Allen EB, Rao LE, Steers RJ et al (2009) Impacts of atmospheric nitrogen deposition on vegetation and soils in Joshua Tree National Park. The Mojave Desert: Ecosystem processes and sustainability. University of Nevada Press, Las Vegas, NV

Atkinson SY, Brown CS (2015) Attributes that confer invasiveness and impacts across the large genus *Bromus* – lessons from the *Bromus* REEnet database. In: Germino MJ, Chambers JC, Brown CS (eds) Exotic brome-grasses in arid and semiarid ecosystems of the western USA: causes, consequences, and management implications. Springer, New York, NY (Chapter 6)

Augustine DJ (2010) Spatial versus temporal variation in precipitation in a semiarid ecosystem. Landscape Ecol 25:913–925

Augustine DJ, Brewer P, Blumenthal DM et al (2014) Prescribed fire, soil inorganic nitrogen dynamics, and plant responses in a semiarid grassland. J Arid Environ 104:59–66

Augustine DJ, Derner JD, Milchunas DG (2010) Prescribed fire, grazing, and herbaceous plant production in shortgrass steppe. Rangel Ecol Manag 63:317–323

Augustine DJ, Milchunas DG (2009) Vegetation responses to prescribed burning of grazed shortgrass steppe. Rangel Ecol Manag 62:89–97

Balch JK, Bradley BA, D'Antonio CM et al (2013) Introduced annual grass increases regional fire activity across the arid western USA (1980–2009). Glob Change Biol 19:173–183

Beyers JL (2004) Postfire seeding for erosion control: Effectiveness and impacts on native plant communities. Conserv Biol 18:947–956

Bidwell TG, Engle DM, Moseley ME et al (2009) Invasion of Oklahoma rangelands and forests by eastern redcedar and ashe juniper, vol E947. Oklahoma Cooperative Extension Service, Stillwater, OK

Bidwell TG, Masters RE, Tyrl RJ (2004) A checklist of prairie, shrubland, and forest understory plants of Oklahoma: characteristics and value to deer, auail, turkey, and cattle. Fact sheet NREM-2872. Oklahoma Cooperative Extension Service, Stillwater, OK

Booth MS, Caldwell MM, Stark JM (2003) Overlapping resource use in three Great Basin species: implications for community invasibility and vegetation dynamics. J Ecol 91:36–48

Bradford JB, Lauenroth WK (2006) Controls over invasion of *Bromus tectorum*: the importance of climate, soil, disturbance and seed availability. J Veg Sci 17:693–704

Bradley BA (2009) Regional analysis of the impacts of climate change on cheatgrass invasion shows potential risk and opportunity. Glob Change Biol 15:196–208

Bradley BA, Blumenthal DM, Wilcove DS et al (2010) Predicting plant invasions in an era of global change. Trends Ecol Evol 25:310–318

Brockway DG, Gatewood RG, Paris RB (2002) Restoring fire as an ecological process in shortgrass prairie ecosystems: initial effects of prescribed burning during the dormant and growing seasons. J Environ Manag 65:135–152

Brooks ML (1999) Habitat invasibility and dominance by alien annual plants in the western Mojave Desert. Biol Inv 1:325–337

Brooks ML (2000) Competition between alien annual grasses and native annual plants in the Mojave Desert. Am Midl Nat 144:92–108

Brooks ML (2003) Effects of increased soil nitrogen on the dominance of alien annual plants in the Mojave Desert. J Appl Ecol 40:344–353

Brooks ML (2008) Plant invasions and fire regimes. In: Zouhar K, Kapler-Smith J, Sutherland S et al (eds) Wildland fire in ecosystems: fire and nonnative invasive plants. Gen Tech Rep RMRS-GTR-42-volume 6. USDA, Forest Service, Rocky Mountain Research Station, Ogden, UT, pp 33–46

Brooks ML (2009) Spatial and temporal distribution of non-native plants in upland areas of the Mojave Desert. In: Webb RH, Fenstermaker LF, Heaton JS et al (eds) The Mojave Desert: ecosystem processes and sustainability. University of Nevada Press, Reno, NV, pp 101–124

Brooks ML (2012) Effects of high fire frequency in creosote bush scrub vegetation of the Mojave Desert. Int J Wildl Fire 21:61–68

Brooks ML, Berry KH (2006) Dominance and environmental correlates of alien annual plants in the Mojave Desert. J Arid Environ 67:100–124

Brooks M, Chambers J (2011) Invasive plants that alter fire regimes in the deserts of North America. Rangel Ecol Manag 64:431–438

Brooks ML, Chambers JC, McKinley RA (2013) Fire history, effects, and management in southern Nevada. In: Chambers JC, Brooks ML, Pendleton BK et al (eds) The southern Nevada agency partnership science and research synthesis: science to support land management in southern Nevada. Gen Tech Rep RMRS-GTR-303. USDA, Forest Service, Rocky Mountain Research Station, Fort Collins, CO, pp 75–96

Brooks ML, D'Antonio CM, Richardson DM et al (2004) Effects of invasive alien plants on fire regimes. Bioscience 54:677–688

Brooks ML, Esque TC (2002) Alien annual plants and wildfire in desert tortoise habitat: status, ecological effects, and management. Chelonian Conserv Biol 4:330–340

Brooks ML, Esque TC, Duck T (2007) Creosotebush, blackbrush, and interior chaparral shrublands. In: Hood S, Miller M (eds) Fire ecology and management of the major ecosystems of Southern Utah. Gen Tech Rep RMRS-GTR-202. USDA, Forest Service, Rocky Mountain Research Station. Fort Collins, CO, pp 97–110

Brooks ML, Matchett JR (2006) Spatial and temporal patterns of wildfires in the Mojave Desert. 1980–2004. J Arid Environ 67:148–164

Brooks ML, Minnich RA (2006) Southeastern Deserts Bioregion. In: Sugihara NG, van Wagtendonk JW, Shaffer KE et al (eds) Fire in California's Ecosystems. US Press, Berkeley, pp 391–414

Brooks ML, Pyke D (2001) Invasive plants and fire in the deserts of North America. In: Galley K, Wilson T (eds) Proceedings of the invasive species workshop: the role of fire in the control and spread of invasive species. Fire conference 2000: the first national congress on fire ecology, prevention and management. Miscellaneous publications No. 11, Tall Timbers Research Station, Tallahassee, FL, pp 1–14

Brown CS, Rice KJ (2010) Effects of belowground resource use complementarity on invasion of constructed grassland plant communities. Biol Invasions 12:1319–1334

Bykova O, Sage RF (2012) Winter cold tolerance and the geographic range separation of *Bromus tectorum* and *Bromus rubens*, two severe invasive species in North America. Glob Change Biol 18(12):3654–3663

Callaway RM, Davis FW (1993) Vegetation dynamics, fire, and the physical environment in coastal central California. Ecology 74:1567–1578

Calo A, Brause S, Jones S (2012) Integrated treatment with a prescribed burn and post-emergent herbicide demonstrates initial success in managing cheatgrass in a northern Colorado natural area. Nat Area J 32:300–304

Chambers JC, Germino MJ, Belnap J et al (2015) Plant community resistance to invasion by *Bromus* species – the roles of community attributes, *Bromus* interactions with plant communities, and *Bromus* traits. In: Germino MJ, Chambers JC, Brown CS (eds) Exotic brome-grasses in arid and semiarid ecosystems of the Western USA: causes, consequences, and management implications. Springer, New York, NY (Chapter 10)

Chambers JC, Bradley BA, Brown CS et al (2014a) Resilience to stress and disturbance, and resistance to *Bromus tectorum* L. invasion in cold desert shrublands of western North America. Ecosystems 17:360–375

Chambers JC, Miller RF, Board DI et al (2014b) Resilience and resistance of sagebrush ecosystems: implications for state and transition models and management treatments. Rangel Ecol Manag 67:440–454

Chambers JC, Pyke DA, Maestas JD et al (2014c) Using resistance and resilience concepts to reduce impacts of invasive annual grasses and altered fire regimes on the sagebrush ecosystem and greater sage-grouse: a strategic multi-scale approach. Gen Tech Rep RMRS-GTR-326. USDA, Forest Service, Rocky Mountain Research Station, Fort Collins, CO, p 73

Chambers JC, Roundy BA, Blank RR et al (2007) What makes Great Basin sagebrush ecosystems invasible by *Bromus tectorum*? Ecol Monogr 77:117–145

Clayton W, Vorontsova MS, Harman KT, Williamson H (2006) GrassBase – The Online World Grass Flora. http://www.kew.org/data/grasses-db.html. Accessed 20 Mar 2014

Compagnoni A, Adler PB (2014) Warming, soil moisture, and loss of snow increase *Bromus tectorum*'s population growth. Elem Sci Anth 2:000020

Condon L, Weisberg PJ, Chambers JC (2011) Abiotic and biotic influences on *Bromus tectorum* invasion and *Artemisia tridentata* recovery after fire. Int J Wildl Fire 20:597–604

Corbin J, D'Antonio CM (2004) Competition between native perennial and exotic annual grasses: Implications for an historical invasion. Ecology 85:1273–1283

Corbin JD, D'Antonio CM, Bainbridge SJ (2004) Tipping the balance in restoration of native plants. In: Gordon MS, Bartol SM (eds.) Experimental Approaches to conservation biology. UC Press, Berkeley, pp 154–179
Coupland R (1992) Mixed prairie. In: Coupland R (ed) Ecosystems of the World. Elsevier, Netherlands, pp 151–182
Covington WW, Sackett SS (1992) Soil mineral nitrogen changes following prescribed burning in ponderosa pine. Forest Ecol Manag 54:175–191
Crawford JA, Wahren CHA, Kyle S et al (2001) Responses of exotic plant species to fires in Pinus ponderosa forests in northern Arizona. J Veg Sci 12:261–268
Cummings DC, Bidwell TG, Medlin CR et al (2007) Ecology and management of *Sericea lespedeza*. Division of Agricultural Sciences and Natural Resources, Oklahoma State University
Cushman JH, Tierney TA, Hinds JM (2004) Variable effects of feral pig disturbances on native and exotic plants in a California grassland. Ecol Appl 14:1746–1756
D'Antonio CM, Thomsen M (2004) Ecological resistance in theory and practice. Weed Technol 18:1572–1577
D'Antonio CM, Vitousek PM (1992) Biol invasions by exotic grasses, the grass fire cycle, and global change. Ann Rev Ecol Syst 23:63–87
Davies GM, Bakker JD, Dettweiler-Robinson E et al (2012) Trajectories of change in sagebrush steppe vegetation communities in relation to multiple wildfires. Ecol Appl 22:1562–1577
Davis MA, Grime JP, Thompson K (2000) Fluctuating resources in plant communities: a general theory of invasibility. J Ecol 88:528–534
DeFalco LA, Bryla DR, Smith-Longozo V et al (2003) Are Mojave Desert annual species equal? Resource acquisition and allocation for the invasive grass *Bromus madritensis* subsp. *rubens* (Poaceae) and two native species. Am J Bot 90:1045–1053
DeFalco LA, Fernandez GCJ, Nowak RS (2007) Variation in the establishment of a non-native annual grass influences competitive interactions with Mojave Desert perennials. Biol Invasions 9:293–307
DeLuca TH, MacKenziea MD, Gundalea MJ et al (2006) Wildfire-produced charcoal directly influences nitrogen cycling in forest ecosystems. Soil Sci Soc Am J 70:448–453
DeSimone SA, Zedler PH (1999) Shrub seedling recruitment in unburned Californian coastal sage scrub and adjacent grassland. Ecology 80:2018–2032
Dettweiler-Robinson E, Bakker JD, Grace JB (2013) Controls of biological soil crust cover and composition shift with succession in sagebrush shrub-steppe. J Arid Environ 94:96–104
DiTomaso JM, Brooks ML, Allen EB et al (2006) Control of invasive weeds with prescribed burning. Weed Technol 20:535–548
Dudley TL (2009) Invasive plants in Mojave Desert riparian areas. In: Webb RH, Fenstermaker LF, Heaton JS et al (eds) The Mojave Desert: ecosystem processes and sustainability. University of Nevada Press, Reno, NV, pp 125–155
Earnst SL, Holmes AL (2012) Bird-habitat relationships in interior Columbia Basin shrubsteppe. Condor 114:15–29
Eiswerth ME, Shonkwiler JS (2006) Examining post-wildfire reseeding on arid rangeland: a multivariate tobit modelling approach. Ecol Model 192:286–298
Eliason SA, Allen E (1997) Exotic grass competition in suppressing native shrubland re-establishment. Restor Ecol 5:245–255
Eneboe EJ, Sowell BF, Heitschmidt RK et al (2002) Drought and grazing: IV. Blue grama and western wheatgrass. J Range Manag 55:197–203
Folke C, Carpenter S, Walker B et al (2004) Regime shifts, resilience, and biodiversity in ecosystem management. Ann Rev Ecol Syst 35:557–581
Ford P, White CS (2007) Effects of dormant-season fire at three different fire frequencies in shortgrass steppe of the southern Great Plains. In: Masters R, Galley KEM (eds) 23rd Tall timbers fire ecology conference: fire in grassland and shrubland ecosystems. Tall Timbers Research Station, Tallahassee, FL, p 71

Ford PL, Johnson GV (2006) Effects of dormant vs. growing-season fire in shortgrass steppe: biological soil crust and perennial grass responses. J Arid Environ 67:1–14

Fornwalt PJ, Kaufmann MR, Stohlgren TJ (2010) Impacts of mixed severity wildfire on exotic plants in a Colorado ponderosa pine–Douglas-fir forest. Biol Invasions 12:2683–2695

Freudenberger DO, Fish BE, Keeley JE (1987) Distribution and stability of grasslands in the Los Angeles Basin. Bull South Calif Acad Sci 86:13–26

Garfin G, Franco G, Blanco H et al (2014) Sounthwest. In: Melillo J, Richmond TC (eds) Climate change impacts in the United States: the third national climate assessment. US global change research program, pp 461–486

Germino MJ, Belnap J, Stark JM et al (2015) Ecosystem impacts of exotic annual invaders in the genus *Bromus*. In: Germino MJ, Chambers JC, Brown CS (eds) Exotic brome-grasses in arid and semiarid ecosystems of the Western USA: causes, consequences, and management implications. Springer, New York, NY (Chapter 3)

Gundale MJ, DeLuca TH (2006) Temperature and source material influence ecological attributes of ponderosa pine and Douglas-fir charcoal. Forest Ecol Manag 231:86–93

Gundale MJ, DeLuca TH (2007) Charcoal effects on soil solution chemistry and growth of *Koeleria macrantha* in the ponderosa pine/Douglas-fir ecosystem. Biol Fertil Soils 43:303–311

Gundale MJ, DeLuca TH, Fiedler CE et al (2005) Restoration management in a Montana ponderosa pine forest: effects on soil physical, chemical, and biological properties. Forest Ecol Manag 213:25–38

Gundale MJ, Sutherland S, DeLuca TH (2008) Fire, native species, and soil resource interactions influence the spatio-temporal invasion pattern of *Bromus tectorum*. Ecography 31:201–210

Haferkamp M, Volesky J, Borman M et al (1993) Effects of mechanical treatments and climatic factors on the productivity of northern great-plains rangeland. J Range Manag 46:346–350

Haferkamp MR, Grings EE, Heitschmidt RK et al (2001) Suppression of annual bromes impacts rangeland: animal responses. J Range Manag 54:663–668

Haferkamp MR, Heitschmidt RK, Karl MG (1997) Influence of Japanese brome on western wheatgrass yield. J Range Manag 50:44–50

Haferkamp MR, Heitschmidt RK, Karl MG (1998) Clipping and Japanese brome reduce western wheatgrass standing crop. J Range Manag 51:692–698

Haidinger TL, Keeley JE (1993) Role of high fire frequency in destruction of mixed chaparral. Madroño 40:141–147

Harmoney KR (2007) Grazing and burning Japanese brome (*Bromus japonicus*) on mixed grass rangelands. Rangel Ecol Manag 60:479–486

Hassan M, West N (1986) Dynamics of soil seed pools in burned and unburned sagebrush semi-deserts. Ecology 67:269–272

Haubensak KA, D'Antonio CM, Saundra Embry S et al (2014) A comparison of *Bromus tectorum* growth and mycorrhizal colonization in salt desert vs. sagebrush habitats. Rangel Ecol Manag 67:275–284

Haubensak K, D'Antonio C, Wixon D (2009) Effects of fire and environmental variables on plant structure and composition in grazed salt desert shrublands of the Great Basin (USA). J Arid Environ 73:643–650

Heitschmidt R, Grings E, Haferkamp M et al (1995) Herbage dynamics on 2 Northern Great Plains range sites. J Range Manag 48:211–217

Heitschmidt RK, Haferkamp MR, Karl MG et al (1999) Drought and grazing: I. Effects on quantity of forage produced. J Range Manag 52:440–446

Heitschmidt RK, Klement KD, Haferkamp MR (2005) Interactive effects of drought and grazing on Northern Great Plains rangelands. Rangel Ecol Manag 58:11–19

Hernandez RR, Sandquist DR (2011) Disturbance of biological soil crust increases emergence of exotic vascular plants in California sage scrub. Plant Ecol 212:1709–1721

Hewlett D, Johnson J, Butterfield R et al (1981) Japanese brome response to atrazine in combination with nitrogen fertilizer in the mixed prairie. J Range Manag 34:22–25

Hobbs RJ, Mooney HA (1995) Spatial and temporal variability in California annual grassland: results from a long term study. J Veg Sci 6:43–56

Holling CS (1973) Resilience and stability of ecological systems. Ann Rev Ecol Syst 4:1–2

Holmgren R (1960) Inspection tour of old blackbrush burns in BLM District N-5, southern Nevada. USDA, Forest Service, Intermountain Forest and Range Experiment Station, Reno Research Center, Reno, NV

Hooker TD, Stark JM, Norton U et al (2008) Distribution of ecosystem C and N within contrasting vegetation types in a semiarid rangeland in the Great Basin, USA. Biogeochemistry 90:291–308

Huenneke LF, Hamburg SP, Koide R et al (1990) Effects of soil resources on plant invasion and community structure in Californian serpentine grassland. Ecology 71:478–491

Humphrey LD, Schupp EW (2001) Seed banks of *Bromus tectorum*-dominated communities in the Great Basin. West N Am Nat 61:85–92

Humphrey RR (1958) The desert grassland. Bot Rev 24:193–253

Hurteau MD, Bradford JB, Fulé PZ et al (2013) Climate change, fire management, and ecological services in the southwestern US. Forest Ecol Manag 327:280–289

Jackson RD, Bartolome JW (2002) A state-transition approach to understanding non-equilibrium plant community dynamics in California grasslands. Plant Ecol 162:49–65

Jackson RD, Bartolome JW (2007) Grazing ecology of California grasslands. In: Stromberg M, Corbin J, D'Antonio CM (eds) California grasslands: ecology and management. UC Press, Berkeley, CA, pp 197–206

James JJ, Drenovsky RE, Monaco TA et al (2011) Managing soil nitrogen to restore annual grass-infested plant communities: effective strategy or incomplete framework? Ecol Appl 21:490–502

Karl MG, Heitschmidt RK, Haferkamp MR (1999) Vegetation biomass dynamics and patterns of sexual reproduction in a northern mixed-grass prairie. Am Midl Nat 141:227–237

Keeler-Wolf T, Evens JM, Solomeshsch AI et al (2007) Community classification and nomenclature. In: Stromberg M, Corbin J, D'Antonio CM (eds) California grasslands: ecology and management. UC Press, Berkeley, CA, pp 21–36

Keeley JE (1990) The California valley grassland. In: Schoenherr AA (ed) Endangered plant communities of southern California. Southern California Botanists, Fullerton, CA, pp 2–23

Keeley JE, Bond WJ, Bradstock RA et al (2012) Fire in Mediterranean ecosystems: ecology, evolution and management. Cambridge University Press, New York, NY, p 528

Keeley JE, Brennan TJ (2012) Fire-driven alien invasion in a fire-adapted ecosystem. Oecologia 169:1043–1052

Keeley JE, Brennan T (2015) Research on the effects of wildland fire and fire management on federally listed species and their habitats on San Clemente Island, California. Unpublished report submitted to the US Navy

Keeley JE, Brennan T, Pfaff AH (2008) Fire severity and ecosystem responses following crown fires in California shrublands. Ecol Appl 18:1530–1546

Keeley JE, Franklin J, D'Antonio C (2011) Fire and invasive plants on California landscapes. In: McKenzie D, Miller C, Falk DA (eds) The landscape ecology of fire. Springer, Netherlands, pp 193–221

Keeley JE, McGinnis TW (2007) Impact of prescribed fire and other factors on cheatgrass persistence in a Sierra Nevada ponderosa pine forest. Int J Wildl Fire 16:96–106

Kerns BK, Buonopane M, Thies WG et al (2011) Reintroducing fire into a ponderosa pine forest with and without cattle grazing. Ecosphere 2:art59

Klinger R, Brooks ML, Frakes N et al (2011a) Establishment of aerial seeding treatments in Blackbrush and Pinyon-Juniper sites following the 2005 Southern Nevada Complex. In: Derasaray L, Frakes N, Gentilcore D et al (eds) Southern Nevada complex emergency stabilization and rehabilitation final report. USDI, Bureau of Land Management, Ely, NV, pp 92–117 (Chapter 5)

Klinger R, Brooks ML, Frakes N et al (2011b) Vegetation trends following the 2005 Southern Nevada Complex Fire. In: Derasaray L, Frakes N, Gentilcore D et al (eds) Southern Nevada complex emergency stabilization and rehabilitation final report. USDI, Bureau of Land Management, Ely, NV, pp 118–194 (Chapter 6)

Knick ST, Connelly JW (2011) Greater sage-grouse: ecology and conservation of a landscape species and its habitats, vol 38, Studies in Avian Biology. UC Press, Berkeley, CA

Knick ST, Hanser SE, Miller RF et al (2011) Ecological pathways of land use in sagebrush. In: Knick ST, Connelly JW (eds) Greater sage-grouse: ecology and conservation of a landscape species and its habitats, vol 38, Studies in Avian Biology. UC Press, Berkeley, CA, pp 203–251 (Chapter 12)

Knutson KC, Pyke DA, Wirth TA et al (2014) Long-term effects of seeding after wildfire on vegetation in Great Basin shrubland ecosystems. J Appl Ecol 51:1414–1424

Kotanen PM, Bergelson J, Hazlett DL (1998) Habitats of native and exotic plants in Colorado shortgrass steppe: a comparative approach. Can J Bot 76:664–672

Kulmatiski A, Beard KH, Stark JM (2006) Exotic plant communities shift water-use timing in a shrub-steppe ecosystem. Plant Soil 288:271–284

Kulpa SM, Leger EA, Espeland EK et al (2012) Postfire seeding and plant community recovery in the Great Basin. Rangel Ecol Manag 65(2):171–181

Lauenroth WK (2008) Vegetation of the Shortgrass Steppe. In: Lauenroth WK, Burke IC (eds) Ecology of the shortgrass steppe: a longterm perspective. Oxford University Press, New York, NY, p 7083

Lauenroth W, Milchunas DG (1992) Short-grass steppe. In: Coupland R (ed) Ecosystems of the world. Elsevier, Netherlands, pp 183–226

Lauenroth W, Burke IC, Morgan JA (2008) The shortgrass steppe. In: Lauenroth W, Burke IC (eds) Ecology of the shortgrass steppe: a long-term perspective. Oxford University Press, New York, NY, pp 3–13

Leger EA, Espeland EK, Merrill KR et al (2009) Genetic variation and local adaptation at a cheatgrass (*Bromus tectorum*) invasion edge in western Nevada. Mol Ecol 18:4366–4379

Lovich JE, Bainbridge D (1999) Anthropogenic degradation of the southern California desert ecosystem and prospects for natural recovery and restoration. Environ Manag 24:309–326

Lovtang SCP, Riegel GM (2012) Predicting the occurrence of downy brome (*Bromus tectorum*) in central Oregon. Invasive Plant Sci Manag 5:83–91

Lulow ME (2006) Invasion by non-native annual grasses: the importance of species biomass, composition and time among California native grasses of the Central Valley. Restor Ecol 14:616–626

Mack RN, Pyke DA (1983) The demography of *Bromus tectorum*: variation in time and space. J Ecol 71:69–93

Maron JL, Connors P (1996) A native nitrogen-fixing shrub facilitates weed invasion. Oecologia 105:302–312

Maron JL, Jeffries RL (1999) Bush lupine mortality, altered resource availability and alternative vegetation states. Ecology 80:443–454

Marty JT, Collinge SK, Rice KJ (2005) Responses of a remnant native bunchgrass population to grazing, burning and climatic variation. Plant Ecol 181:101–112

Masters RA, Vogel KP, Mitchell RB (1992) Response of central plains tallgrass to fire, fertilizer, and atrazine. J Range Manag 45(3):291–295

Mazzola MB, Chambers JC, Blank RR et al (2011) Effects of resource availability and propagule supply on native species recruitment in sagebrush ecosystems invaded by *Bromus tectorum*. Biol Invasions 13:513–526

McGinnis TW, Keeley JE, Stephens SL et al (2010) Fuel buildup and potential fire behavior after stand-replacing fires, logging fire-killed trees and herbicide shrub removal in Sierra Nevada forests. Forest Ecol Manag 260:22–35

McGlone CM, Springer JD, Covington WW (2009) Cheatgrass encroachment on a ponderosa pine Ecol Restor project in northern Arizona. Ecol Restor 27:37–46

McPherson GR (1995) The role of fire in the desert grasslands. In: McClaran MP, Van Devender TR, Thomas R (eds) The desert grassland. University of Arizona Press, Tucson, AZ, pp 130–151

McPherson GR, Weltzin JF (2000) The role and importance of disturbance and climate change in US/Mexico borderlands: a state-of-the-knowledge review. Gen Tech Rep RMRS-GTR-50. USDA, Forest Service, Rocky Mountain Research Station, Fort Collins, CO, p 24

Meixner T, Wohlgemuth PM (2003) Climate variability, fire, vegetation recovery, and watershed hydrology. First interagency conference on research in the watersheds, Benson, AZ, 27–30 October 2003, pp 651–656

Meng R, Dennison PE, D'Antonio CM et al (2014) Remote sensing analysis of vegetation recovery following short-interval fires in southern California shrublands. PLoS One 9, e110637

Meyer SE, Garvin SC, Beckstead J (2001) Factors mediating cheatgrass invasion of intact salt desert shrubland. In: McArthur D, Fairbanks DJ (eds) Shrubland ecosystem genetics and biodiversity. Gen Tech Rep RMRS-P-21. USDA Forest Service, Rocky Mountain Research Station, Fort Collins, CO, pp 224–232

Meyer SE, Quinney D, Nelson DL et al (2007) Impact of the pathogen *Pyrenophora semeniperda* on *Bromus tectorum* seedbank dynamics in North American cold deserts. Weed Res 47:54–62

Milchunas DG, Lauenroth WK, Chapman PL (1992) Plant competition, abiotic, and long-term and short-term effects of large herbivores on demography of opportunistic species in a semiarid grassland. Oecologia 92:520–531

Milchunas DG, Vandever MW, Ball LO et al (2011) Allelopathic cover crop prior to seeding is more important than subsequent grazing/mowing in grassland establishment. Rangel Ecol Manag 64:291–300

Miller R, Chambers JC, Pyke DA, et al (2013) A review of fire effects on vegetation and soils in the Great Basin Region: response and ecological site characteristics. Gen Tech Rep RMRS-GTR-308. USDA, Forest Service, Rocky Mountain Research Station, Fort Collins, CO

Miller RF, Knick ST, Pyke DA et al (2011) Characteristics of sagebrush habitats and limitations to long-term conservation. Stud Avian Biol 38:145–184

Minnich R (2008) California's fading wildflowers: lost legacy and biological invasions. UC Press, Berkeley, CA

Molinari N (2014) Invasion, impact and persistence of an exotic annual grass. PhD dissertation, University of California, Santa Barbara, p 181

Molinari N, D'Antonio CM (2014) Structural, compositional and trait differences between native- and non-native-dominated grassland patches. Funct Ecol 28:745–754

Monaco TA, Hardegree SP, Pellant M et al (2015) Assessing restoration and management needs for ecosystems invaded by exotic annual *Bromus* species. In: Germino MJ, Chambers JC, Brown CS (eds) Exotic brome-grasses in arid and semiarid ecosystems of the Western USA: causes, consequences, and management implications. Springer, New York, NY (Chapter 12)

Monaco TA, Sheley R (2012) Invasive plant ecology and management: linking processes to practice. CABI Invasive Series. CABI

Monleon VJ, Cromack K, Landsberg JD (1997) Short- and long-term effects of prescribed underburning on nitrogen availability in ponderosa pine stands in central Oregon. Can J Forest Res 27:369–378

Monsen SB, Stevens R, Shaw NL (2004) Restoring western ranges and wildlands. Gen Tech Rep RMRS-GTR-1136-vol 1, 2, 3. USDA, Forest Service, Rocky Mountain Research Station, Fort Collins, CO, p 294

Moran MS, Ponce-Campos GE, Huete A et al (2014) Functional response of US grasslands to the early 21st-century drought. Ecology 95:2121–2133

Moser WK, Barnard EL, Billings RF, Crocker SJ et al (2009) Impacts of nonnative invasive species on US forests and recommendations for policy and management. J Forestry 107:320–327

Munson SM, Lauenroth WK (2009) Plant population and community responses to removal of dominant species in the shortgrass steppe. J Veg Sci 20:224–232

Ogle SM, Ojima D, Reiners WA (2004) Modeling the impact of exotic annual brome grasses on soil organic carbon storage in a northern mixed-grass prairie. Biol Invasions 6:365–377

Ogle SM, Reiners WA (2002) A phytosociological study of exotic annual brome grasses in a mixed grass prairie/ponderosa pine forest ecotone. Am Midl Nat 147:25–31

Ogle SM, Reiners WA, Gerow KG (2003) Impacts of exotic annual brome grasses (*Bromus* spp.) on ecosystem properties of northern mixed grass prairie. Am Midl Nat 149:46–58

Osborne CP, Visser V, Chapman S et al (2011) GrassPortal: an online ecological and evolutionary data facility. www.grassportal.org. Accessed 16 Feb 2015

Pauchard A, Kueffer C, Dietz H et al (2009) Ain't no mountain high enough: plant invasions reaching new elevations. Front Ecol Environ 7:479–486

Pavlik LE (1995) Bromus L. of North America. Royal British Columbia Museum, Victoria, BC, p 160

Pierson EA, Mack RN (1990) The population of biology of *Bromus tectorum* in forests: effect of disturbance, grazing, and litter on seedling establishment and reproduction. Oecologia 84:526–533

Pyke D (2011) Restoring and rehabilitating sagebrush habitats. In: Knick S, Connelly JW (eds) Greater sage-grouse: ecology and conservation of a landscape species and its habitats, vol 38. University of California Press, Berkeley, CA, pp 531–548

Pyke DA, Chambers JC, Beck JL et al (2015) Land uses, fire, and invasion – exotic annual *Bromus* and human dimensions. In: Germino MJ, Chambers JC, Brown CS (eds) Exotic brome-grasses in arid and semiarid ecosystems of the Western USA: causes, consequences, and management implications. Springer, New York, NY (Chapter 11)

Pyke DA, Wirth TA, Beyers JL (2013) Does seeding after wildfires in rangelands reduce erosion or invasive species? Restor Ecol 21:415–421

Ramakrishnan AP, Meyer SE, Fairbanks DJ et al (2006) Ecological significance of microsatellite variation in western North American populations of *Bromus tectorum*. Plant Species Biol 21:61–73

Rao LE, Allen EB (2010) Combined effects of precipitation and nitrogen deposition on native and invasive winter annual production in California deserts. Oecologia 162(4):1035–1046

Rao LE, Allen EB, Meixner T (2010) Risk-based determination of critical nitrogen deposition loads for fire spread in southern California deserts. Ecol Appl 20:1320–1335

Rao LE, Matchett JR, Brooks ML et al (2015) Relationships between annual plant productivity, nitrogen deposition and fire size in low-elevation California desert scrub. Int J Wildl Fire 24:48–58

Reisner MD, Grace JB, Pyke DA et al (2013) Conditions favouring *Bromus tectorum* dominance of endangered sagebrush steppe ecosystems. J Appl Ecol 50:1039–1049

Rice KJ, Mack RN (1991) Ecological genetics of *Bromus tectorum*. III. The demography of reciprocally sown populations. Oecologia 88:91–101

Rice KJ, Nagy ES (2000) Oak canopy effects on the distribution patterns of two annual grasses: the role of competition and soil nutrients. Am J Bot 87:1699–1706

Rimer RL, Evans RD (2006) Invasion of downy brome (*Bromus tectorum* L.) causes rapid changes in the nitrogen cycle. Am Midl Nat 156:252–258

Rinella MJ, Haferkamp MR, Masters RA et al (2010a) Growth regulator herbicides prevent invasive annual grass seed production. Invasive Plant Sci Manag 3:12–16

Rinella MJ, Masters RA, Bellows SE (2010b) Growth regulator herbicides prevent invasive annual grass seed production under field conditions. Rangel Ecol Manag 63:487–490

Rinella MJ, Masters RA, Bellows SE (2013) Effects of growth regulator herbicide on Downy brome (*Bromus tectorum*) seed production. Invasive Plant Sci Manag 6:60–64

Rochester CJ, Brehme CS, Clark DR et al (2010) Reptile and amphibian responses to large-scale wildfires in southern California. J Herpetol 44:333–351

Romps DM, Seeley JT, Vollaro D et al (2014) Projected increase in lightning strikes in the United States due to global warming. Science 346:851–854

Roundy BA, Hardegree SP, Chambers JC et al (2007) Prediction of cheatgrass field germination potential using wet thermal accumulation. Rangel Ecol Manag 60:613–623

Safford HD, Van de Water KM (2014) Using fire return interval departure (FRID) analysis to map spatial and temporal changes in fire frequency on national forest lands in California. Gen Tech Rep PSW-RP-266. USDA, Forest Service, Pacific Southwest Research Station, p 59

Salo LF (2005) Red brome (*Bromus rubens* subsp *madritensis*) in North America: possible modes for early introductions, subsequent spread. Biol Invasions 7:165–180

Sankey JB, Germino MJ, Glenn NF (2009) Aeolian sediment transport following wildfire in sagebrush steppe. J Arid Environ 73:912–919

Schaeffer SM, Ziegler SE, Belnap J et al (2012) Effects of *Bromus tectorum* invasion on microbial carbon and nitrogen cycling in two adjacent undisturbed arid grassland communities. Biogeochemistry 111:427–441

Scheintaub MR, Derner JD, Kelly EF et al (2009) Response of the shortgrass steppe plant community to fire. J Arid Environ 73:1136–1143

Seabloom EW, Harpole WS, Riechman OJ et al (2003) Invasion, competitive dominance and resource use by exotic and native California grassland species. Proc Natl Acad Sci U S A 100:13384–13389

Shinneman DJ, Baker WL (2009) Environmental and climatic variables as potential drivers of post-fire cover of cheatgrass (*Bromus tectorum*) in seeded and unseeded semiarid ecosystems. Int J Wildl Fire 18:191–202

Sims P, Risser PG (2000) Grasslands. In: Barbour M, Billings WD (eds) North American terrestrial vegetation, 2nd edn. Cambridge University Press, Cambridge, UK, pp 323–356

Smith SD, Strain BR, Sharkey TD (1987) Effects of CO_2 enrichment on four Great Basin grasses. Funct Ecol 1:139–143

Sorensen CD, McGlone CM (2010) Ponderosa pine understory response to short-term grazing exclusion (Arizona). Ecol Restor 28:124–126

Stahlheber K (2013) The influence of savanna oaks on California grassland composition. PhD dissertation, University of California, Santa Barbara, p 275

Stahlheber K, D'Antonio CM (2013) Using livestock to manage plant composition: a meta-analysis of grazing in California Mediterranean grassland. Biol Conserv 157:300–308

Strand EK, Launchbaugh KL (2013) Livestock grazing effects on fuel loads for wildland fire in sagebrush dominated ecosystems. Great Basin fire science delivery report, p 21

Stylinski CD, Allen EB (1999) Lack of native species recovery following severe exotic disturbance in southern Californian shrublands. J Appl Ecol 36:544–554

Suttle B, Thomsen MA, Power ME (2007) Species interactions reverse grassland responses to changing climate. Science 315:640–642

Sweet SB, Kyser GB, DiTomaso JM (2008) Susceptibility of exotic annual grass seeds to fire. Invasive Plant Sci Manag 1:158–167

Symstad A, Long AJ, Stamm JF et al (2014) Two approaches for incorporating climate change into natural resource management planning at Wind Cave National Park. Gen Tech Rep NPS/WICA/NRTR-2014.918. National Park Service, Fort Collins, CO

Syphard AD, Keeley JE (2015) Location, timing and extent of wildfire vary by cause of ignition. Int J Wildl Fire 24:37–47

Tagestad J, Brooks M, Cullinan V et al (in press) Precipitation regime classification for the Mojave Desert: implications for fire occurrence. J Arid Environ

Talluto MV, Suding KN (2008) Historical change in coastal sage scrub in southern California, USA in relation to fire frequency and air pollution. Landscape Ecol 23:803–815

Teague WR, Dowhower SL (2003) Patch dynamics under rotational and continuous grazing management in large, heterogeneous paddocks. J Arid Environ 53:211–229

Teague WR, Dowhower SL, Baker SA et al (2010) Soil and herbaceous plant responses to summer patch burns under continuous and rotational grazing. Agr Ecosyst Environ 137:113–123

Teague WR, Dowhower SL, Waggoner JA (2004) Drought and grazing patch dynamics under different grazing management. J Arid Environ 58:97–117

Teague WR, Duke SE, Waggoner JA et al (2008) Rangeland vegetation and soil response to summer patch fires under continuous grazing. Arid Land Res Manag 22:228–241

USDA NRCS (2015) The PLANTS Database. http://plants.usda.gov. Accessed 26 Feb 2015

Van Dyne G (1975) An overview of the ecology of the Great Plains grasslands with special reference to climate and its impact. Grassland biome – ecosystem analysis studies. Gen Tech Rep 290. Natural resources ecology laboratory, Colorado State University, Fort Collins, CO

Vermeire LT, Crowder JL, Wester DB (2011) Plant community and soil environment response to summer fire in the Northern Great Plains. Rangel Ecol Manag 64:37–46
Vermeire LT, Crowder JL, Wester DB (2014) Semiarid rangeland is resilient to summer fire and postfire grazing utilization. Rangel Ecol Manag 67:52–60
Vermeire LT, Heitschniidt RK, Haferkamp MR (2008) Vegetation response to seven grazing treatments in the Northern Great Plains. Agric Ecosyst Environ 125:111–119
Walker BH, Holling CS, Carpenter SR et al (2004) Resilience, adaptability and transformability in social-ecological systems. Ecol Soc 9:5
Walsh J, Wuebbles D, Hayhoe K et al (2014) Our changing climate. In: Melillo JM, Richmond TC, Yohe GW (eds) Climate change impacts in the United States: the third national climate assessment. US Global Change Research Program, pp 19–67
West N (1983a) Great Basin-Colorado Plateau sagebrush semi-desert. Temperate Deserts and Semi-deserts. Elsevier, Amsterdam
West N (1983b) Intermountain salt-desert shrubland. Temperate Deserts and Semi-deserts. Elsevier, Amsterdam, pp 375–378
Westerling AL, Bryant BP (2008) Climate change and wildfire in California. Clim Change 87:S231–S249
Westerling AL, Hidalgo HG, Cayan DR et al (2006) Warming and earlier spring increase western U.S. forest wildfire activity. Science 313:940–943
Westman WE (1979) Oxidant effects on Californian coastal sage scrub. Science 205:1001–1003
Whisenant SG (1990) Changing fire frequencies on Idaho's Snake River Plains: ecological and management implications. In: McArthur ED, Romney EM, Smith SD et al (eds) Symposium on cheatgrass invasion, shrub die-off and other aspects of shrub biology and management, 5–7 April 1989, Las Vegas, NV. Gen Tech Rep INT-276. USDA, Forest Service, Intermountain Research Station, Ogden, UT, pp 4–10
Whisenant S, Ueckert D, Scifres C (1984) Effects of fire on Texas wintergrass communities. J Range Manag 37:387–391
Whisenant SG, Uresk DW (1990) Spring burning Japanese brome in a western wheatgrass community. J Range Manag 43:205–208
White CS, Loftin SR (2000) Response of two semiarid grasslands to cool-season prescribed fire. J Range Manag 53:52–61
Wilcox BP, Turnbull L, Young MH et al (2012) Invasion of shrublands by exotic grasses: ecohydrological consequences in cold versus warm deserts. Ecohydrology 5:160–173
Wiken E, Nava FJ, Griffith G (2011) North American Terrestrial Ecoregions—Level III. Commission for Environmental Cooperation, Montreal, Canada
Wolkovich EM, Bolger DT, Holway DA et al (2009a) Invasive grass litter facilitates native shrubs through abiotic effects. J Veg Sci 20:1121–1132
Wolkovich EM, Bolger DT, Holway DA (2009b) Complex responses to invasive grass litter by ground arthropods in a Mediterranean scrub ecosystem. Oecologia 262:697–708
Wright HE, Bailey AW (1982) Fire ecology: United States and Canada. Wiley, New York, NY
Yensen E, Quinney DL, Johnson K et al (1992) Fire, vegetation changes, and population fluctuations of Townsend ground-squirrels. Am Midl Nat 128:299–312
Zedler PH, Gautier CR, McMaster GS (1983) Vegetation change in response to extreme events: the effect of a short interval between fires in California chaparral and coastal scrub. Ecology 64:809–818

Chapter 3
Ecosystem Impacts of Exotic Annual Invaders in the Genus *Bromus*

Matthew J. Germino, Jayne Belnap, John M. Stark, Edith B. Allen, and Benjamin M. Rau

Abstract An understanding of the impacts of exotic plant species on ecosystems is necessary to justify and guide efforts to limit their spread, restore natives, and plan for conservation. Invasive annual grasses such as *Bromus tectorum*, *B. rubens*, *B. hordeaceus*, and *B. diandrus* (hereafter collectively referred to as *Bromus*) transform the structure and function of ecosystems they dominate. Experiments that prove cause-and-effect impacts of *Bromus* are rare, yet inferences can be gleaned from the combination of *Bromus*-ecosystem associations, ecosystem condition before/after invasion, and an understanding of underlying mechanisms. *Bromus* typically establishes in bare soil patches and can eventually replace perennials such as woody species or bunchgrasses, creating a homogeneous annual cover. Plant productivity and cover are less stable across seasons and years when *Bromus* dominates, due to a greater response to annual climate variability. *Bromus*' "flash" of growth followed by senescence early in the growing season, combined with shallow rooting and annual habit, may lead to incomplete use of deep soil water, reduced C

M.J. Germino (✉)
US Geological Survey, Forest and Rangeland Ecosystem Science Center, Boise, ID 83706, USA
e-mail: mgermino@usgs.gov

J. Belnap
US Geological Survey, Southwest Biological Science Center, Moab, UT 84532, USA
e-mail: jayne_belnap@usgs.gov

J.M. Stark
Department of Biology, and the Ecology Center, Utah State University, Logan, UT 84322-5305, USA
e-mail: jstark@biology.usu.edu

E.B. Allen
Department of Botany and Plant Sciences, University of California, Riverside, Riverside, CA, USA
e-mail: edith.allen@ucr.edu

B.M. Rau
US Department of Agriculture, Forest Service, Southern Research Station, Aiken, SC 29803, USA
e-mail: benjaminmrau@fs.fed.us

M.J. Germino et al. (eds.), *Exotic Brome-Grasses in Arid and Semiarid Ecosystems of the Western US*, Springer Series on Environmental Management,
DOI 10.1007/978-3-319-24930-8_3

sequestration, and accelerated nutrient cycling. Litter produced by *Bromus* alters nearly all aspects of ecosystems and notably increases wildfire occurrence. Where *Bromus* has become dominant, it can decrease soil stability by rendering soils bare for months following fire or episodic, pathogen-induced stand failure. *Bromus-invaded* communities have lower species diversity, and associated species tend to be generalists adapted to unstable and variable habitats. Changes in litter, fire, and soil properties appear to feedback to reinforce *Bromus'* dominance in a pattern that portends desertification.

Keywords *Bromus* • Annual exotic grasses • Ecosystems • Desertification • Feedbacks

3.1 Introduction

Exotic annual grasses in the genus *Bromus* that are invading semiarid and arid landscapes in the Western USA (hereafter "*Bromus*") have significant impacts on ecosystem structure and function where they dominate plant community cover (Fig. 1.1 in Germnio et al. 2015). This chapter aims to give an overview of these impacts, with some treatment of *Bromus rubens* L. (red brome), *B. diandrus* Roth (ripgut brome), and *B. hordeaceus* L. (soft brome) but a primary focus on *B. tectorum* L. (cheatgrass, downy brome). *B. tectorum* has become the most widespread exotic annual grass in the Western USA and has also attained the greatest local dominance of communities, often occurring in vast and nearly monocultural patches (Brooks et al. 2015). Nearly 10 % of the many publications on invasive plant impacts are on *Bromus* (specifically *B. tectorum*; Hulme et al. 2013), and an exhaustive review of the studies on ecosystem impacts of *Bromus* invaders is beyond the scope of this chapter. Several previous reviews have suggested ways that *Bromus* can transform ecosystem structure and function (Hulbert 1955; D'Antonio and Vitousek 1992; Allen et al. 2011), particularly *B. tectorum* (Stewart and Hull 1949; Klemmedson and Smith 1964; Mack 1981; Billings 1990). We begin with a short overview of how the plant community is altered by *Bromus* invasion. Then, we describe how disturbance interacts with *Bromus*, focusing on fire because it is a key disturbance for most landscapes in western North America and is a key aspect of *Bromus* invader's impacts (Brooks et al. 2004). We evaluate *Bromus'* impact on fire frequency and extent and the influence of *Bromus* on ecosystem characteristics such as site and soil stability and flow or cycling of energy, water, carbon (C), and nutrients. Last, we discuss impacts on invertebrate and vertebrate communities, including wildlife and domestic livestock.

We focus on comparisons between native- and *Bromus*-dominated communities where there is supporting literature, although few studies can truly separate *Bromus* invader effects from those of native species (Stark and Norton 2015). Many of the native plant communities impacted by *Bromus* have native perennial grasses that increase after disturbance in the absence of *Bromus* invasions, followed by increases in abundance of woody species over time. Thus, we assess *Bromus* impacts by comparing *Bromus* effects relative to both native perennial grasses and later-successional plants, as well as to plant communities created by land treatments (e.g., seedings) designed to counter threats of *Bromus* invasion. As our review demonstrates, no single study has distinguished the influence of *Bromus* from effects of associated land uses and disturbances or compared *Bromus* impacts among various native community states.

3.2 Impacts on Plant Communities

Where *Bromus* occur with native perennials, they often have inverse relationships to density or cover of natives or may have a patchy distribution among perennial plant, bare interspace, or biological soil crust microsites, as shown for sagebrush steppe in the Northern Basin and Range and Wyoming Basin (Anderson and Inouye 2001; Gasch et al. 2013; Reisner et al. 2013), Mojave and Sonoran Deserts (Brooks 2000; Salo et al. 2005; DeFalco et al. 2001, 2007), Colorado Plateau (Belnap and Phillips 2001), and Mediterranean California Grasslands (Corbin et al. 2007). *Bromus* typically attains dominance following disturbances such as fire and/or grazing, in those areas having suitable climate (Chambers et al. 2014, 2015). *Bromus* invasions can transform communities that have a relatively high proportion of perennials or mosses and lichens that cover plant interspaces into exotic annual and ephemeral native communities (Brooks et al. 2015). Such invasions can also change the most basic spatial and temporal structure of the plant community depending on factors such as seasonality of precipitation and community productivity and composition (Fig. 10.1 in Chambers et al. 2015). For example, *Bromus* impacts are greater in the less productive mix of shrubs and forbs in the Mojave Basin and Range compared to the relatively productive shortgrass steppe, as we describe below. Few native species are able to establish in dense stands of *Bromus* (e.g., Humphrey and Schupp 2004), and loss of native species richness and diversity often leads to either *Bromus* monocultures or exotic annual/biennial communities (Fig. 3.1). However, *Bromus* invasion into relatively undisturbed and protected communities or regions does occur, typically at subdominant levels (*Bromus* comprising <20 % relative cover). Examples of undisturbed but invaded areas include perennial grasslands of the Colorado Plateau (Belnap and Phillips 2001), islands of sagebrush steppe in the Northern Basin and Range isolated by rugged/impassable lava where native species richness declined ("kipukas", Bangert and Huntly 2010), a ~1000-km^2 grazing exclosure at the Idaho National Lab (Anderson and Inouye 2001; Bagchi et al. 2013), undisturbed *Coleogyne ramosissima* Torr. (blackbrush) shrubland (Brooks and Matchett 2003), and coastal sage scrub subject to anthropogenic nitrogen (N) deposition in southern California (Cox et al. 2014).

Fig. 3.1 Photos of a Wyoming big sagebrush site that had no *B. tectorum* before fire (*top left*) and the resulting "clean burn" typical of such sites (*top right*), a site dominated by *B. tectorum* before fire (*middle left*), and the resulting "dirty burn" and dense *Bromus* litter (*middle right*). The *bottom panel* shows a close-up view of the charred but incompletely combusted *Bromus* litter occluding the soil. Postfire photos were taken approximately 1 month after summer wildfire. Photo credits: *top row*, M. Germino; lower three, A. Halford

3.2.1 Changes to Above- and Belowground Community Structure

Invasion of *Bromus* in the absence of fire often occurs through establishment in bare soil interspaces between perennials, creating a homogeneous herbaceous layer, such as for *B. tectorum* in grazed sagebrush steppe (Reisner et al. 2013). However, greater establishment under shrubs or trees than interspaces is also evident in some communities, such as for *B. rubens* in the Mojave Basin and Range (Brooks 1999;

Abella et al. 2011), *B. tectorum* in ponderosa pine forest of the Northern Rockies, or sagebrush steppe in the ecotone between the Sierra Nevada and Central Basin and Range (Gundale et al. 2008 and Griffith 2010, respectively). Following fire, many of the dominant shrub species are killed, further promoting the shift toward homogenous cover of *Bromus* (Fig. 3.1). Belowground, the native diversity of deep/tap rooting and shallow/fibrous-rooting plant species and the associated soil-resource partitioning that is fundamental to sagebrush steppe (Ryel et al. 2008) is lost when *Bromus* dominate sites, often leaving a dense, continuous, and shallow dispersion of fine, annual roots (reviewed in Wilcox et al. 2012). Specifically, many of the native communities impacted by *Bromus* have a mix of woody species that have tap roots extending 1–3 or more meters deep into the soil profile, while neighboring herbs and particularly grasses typically have their highest densities of roots in shallower soils (i.e., 10–50-cm depths). Following conversion to *Bromus* grasslands, these sites are left with only a dense proliferation of roots in the shallow soil horizons (typically above 30–40-cm depth although their roots can extend to 1 m or deeper; Wilcox et al. 2012).

Bromus invasions strongly affect the phenology of the plant community because of *Bromus'* capacity for fall germination, rapid growth, and senescence by spring or early summer. *Bromus* cover also exhibits relatively high interannual variability compared to perennial vegetation, primarily due to its response to precipitation in fall and early spring (e.g., for *B. tectorum*, West and Yorks 2006). Communities that are dominated by *Bromus* are often also comprised of other annual or biennial exotics that have highly variable presence from year to year (Piemeisel 1951; Prevey et al. 2010). Few native plants impart spatiotemporal effects as strong as *Bromus*. One example is *Poa secunda* J. Presl (Sandberg bluegrass), a small bunchgrass that can create vast expanses of homogenous cover with early-season green up and senescence, often in sagebrush steppe that has been repeatedly disturbed.

Significant instability in community cover results when *B. tectorum* exhibits large-scale, periodic stand failure in response to fungal pathogens on sites it otherwise dominates (Stewart and Hull 1949; Klemmedson and Smith 1964, Meyer chapter on Pathogens, see maps in Fig. 3.2). The absence of vegetation cover prior to recolonization can result in extensive wind erosion, depletion of topsoil organic matter, and other key attributes of these ecosystems (Fig. 3.3, Sect. 3.4). Die-offs of *B. tectorum* lasting at least 2 years apparently occurred on approximately one million hectares per year from 2000 to 2010 in the Northern Basin and Range, and 10 % of *Bromus*-invaded areas had a >80 % chance of stand failure (Wylie 2012).

3.2.2 *Effects of Litter Accumulation on Community Structure*

Dense layers of fine-textured, straw-like litter (including standing dead tissue) produced from senesced stems, leaves, and reproductive tissue of *Bromus* are an important impact at either patch (interspace) or landscape scales (Fig. 3.1). The litter layers can either be mats laying on soil and/or standing dead tissue. Factors such as

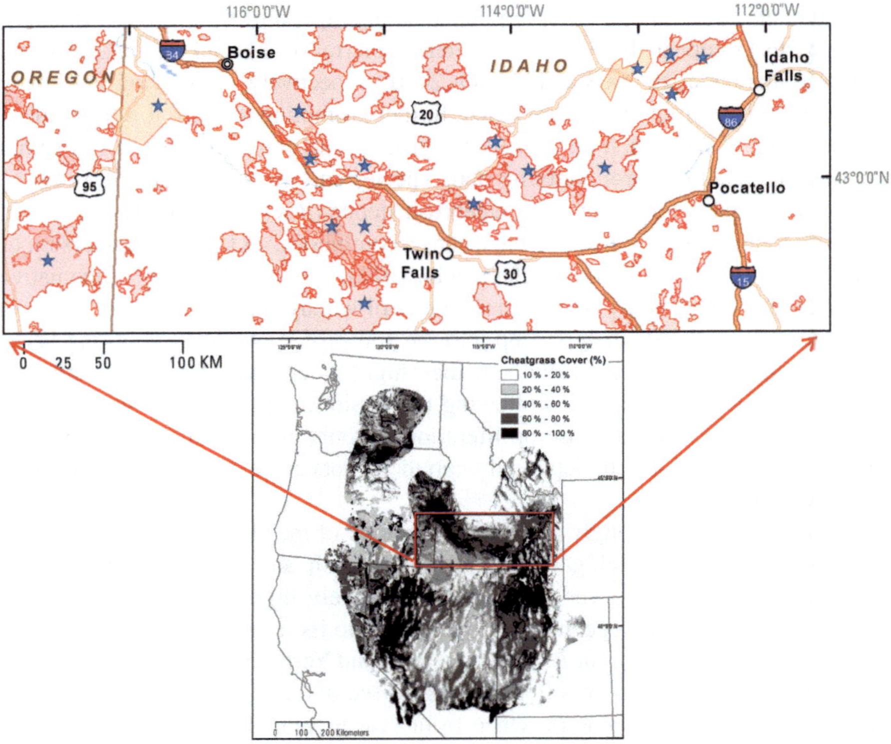

Fig. 3.2 Distribution and abundance of *Bromus tectorum* as detected in satellite imagery from Wylie (2012) and expanded map of wildfire boundaries (2002–2015) of the *red-highlighted region* in the *B. tectorum* map, obtained from the Monitoring Trends in Burn Severity (www.mtbs.gov) and also www.geomac.gov. Stars show fires that were followed by wind erosion based on publications, direct observations, and/or geomorphic indicators

erosion, grazing, climate and soils, and decomposition rates likely contribute to variability in density and form of the litter, but most have not been formally studied. Dense *Bromus* litter acts like mulch and can insulate soil from solar and terrestrial radiation and wind, intercept precipitation, prevent seeds from contacting soil, alter biogeochemistry, and increase combustibility and continuity of wildfire fuels (described further in Sects. 3.3 and 3.7 below). Furthermore, alteration of litter characteristics could promote "legacy effects" of *Bromus* on soil or plant community properties, which in turn may favor *Bromus* and discourages natives. However, evidence is equivocal for the importance of *Bromus'* soil-legacy effects in light of other ways *Bromus* impacts native plants (see Sects. 3.4–3.8; Rowe and Brown 2008; Belnap and Phillips 2001; Evans et al. 2001; Gill and Burke 1999; Sperry et al. 2006; Bansal et al. 2014). Litter produced by nonnatives is typically considered to selectively benefit the nonnatives and often inhibits establishment of native plants due to excessive shading or preventing seed from reaching the soil. However, larger established native shrubs can benefit from litter mulch beneath them due to

Fig. 3.3 Examples of extent and impact of *B. tectorum* die-off in the Northern Basin and Range. The map in the *top right* shows estimated die-off patterns during 2010 based on MODIS imagery [*red* shows areas in which growth potential of *B. tectorum* based on its mapped presence and climate/weather is not materialized, from Wylie (2012)]. *Upper left* photo shows a die-off in 2014 (dark ground is died off, blond grasses survived and senesced normally, photo courtesy of Mindy Wheeler). *Middle right* photo shows large deposition of soil eroded from a 2010 die-off (photo, Megan Hynes). Photo on bottom shows high bare soil exposure and a dust plume originating from a site that experienced stand failure of *B. tectorum*, a site classified as "Loamy with 5–8 inches of precipitation" in August 2012 (photo, Tamzen Strigham). Other BLM staff reported complete loss of visibility on similarly afflicted sites due to dense dust clouds in the same summer (M. Pellant, pers. comm.)

increased soil resources, particularly water (e.g., for *Bromus* species in central California shrublands; Wolkovich et al. 2009). Other studies show biomass production of *B. hordeaceus* to be negatively related to its own litter (Amatangelo et al. 2008). Litter removal increased plant-available N and abundances of *B. tectorum* and other exotic annuals in a cool and wet summer but led to decreases in a warm and dry year (Bansal et al. 2014; Jones et al. 2015a). High variability in conclusions among studies addressing litter effects of *Bromus* may relate to the timing of experimental litter removal treatments, density/thickness of experimental litter layers, nonlinear effects of litter amount (e.g., positive effects of small amounts, negative effects of large amounts), and weather events that affect soil microclimate, plant establishment and productivity, and decomposition and nutrient cycling.

3.3 Impacts on Landscape Disturbance, Specifically Fire

Increased incidence of fire due to greater and more spatially continuous fuel (litter) production by *Bromus* has been recognized since the early 1900s. Fires have been abundant recently in regions that have large areas of *Bromus*, such as the Northern Basin and Range and Snake River Plain or the ecotone between the Central and Mojave Basin and Ranges (Fig. 3.2, Brooks and Matchett 2006).

Within particular ecoregions, fire-return intervals on *B. tectorum*-dominated areas are considered to have been markedly reduced (e.g., Stewart and Hull 1949; Whisenant 1990), although little evidence exists to substantiate the commonly cited 3–5-year fire-return interval. Over the whole 650,000 km^2 of Great Basin dominated by *B. tectorum*, annual probabilities of burning were 1–2 %, compared to <1 % for areas mapped as sagebrush habitat, using USGS fire records from 1980 to 2007 (Balch et al. 2013, estimates vary by data source). Fire-return intervals summarized by decade were 50–82 years for *B. tectorum* sites, compared to 97–313 years for sagebrush sites (Balch et al. 2013). *Bromus tectorum* sites were ~250 % more likely to burn than sagebrush sites, compared to 4 % and 25 % more likely to burn than pinyon-juniper and desert shrub sites, respectively (Balch et al. 2013). Balch et al. (2013) also found that 65 % of fires from 2000 to 2009 started on *B. tectorum-dominated* sites, and a substantial fraction of these spread onto sites that were not dominated by *B. tectorum*. Climate- and weather-fire relationships are strengthened where *B. tectorum* dominates (Knapp 1996). For example, 22–27 % of the variation in fire frequency and fire size in the Great Basin was positively related to precipitation of the previous calendar year on *B. tectorum* sites, compared to only 12–13 % on sagebrush sites or no correlation found over all vegetation types combined (Balch et al. 2013). Probability of increased fire size increased steeply above a threshold of 125 g/m^2 of fine fuel in the Mojave Desert, driven by prior year precipitation and N deposition. In this case, fine fuel included *B. rubens* at higher elevations and the exotic annual grasses *Schismus* spp. P. Beauv (Mediterranean grass) at lower elevations (Rao et al. 2015).

3.3.1 Mechanisms Underlying Altered Fire Regime

Increases in fire occurrence and rate of spread with dominance of *Bromus* are due to the increase in fine fuel abundance and continuity caused by *Bromus* replacing more widely spaced, native perennial fuels (described in Sect. 3.2 and Fig. 3.1, Brooks and Zouhar 2008). More specifically, species such as *B. tectorum* increase surface-to-volume ratio, horizontal continuity (size and abundance of grass fuel patches per unit area), and the packing ratio (amount per unit volume) of fire fuels compared to perennial vegetation, thereby increasing likelihood of ignition and spread (Brooks et al. 2004; Davies and Nafus 2013). Link et al. (2006) furthermore demonstrated that these attributes of *Bromus* increase combustibility, specifically the probability of ignition and fire spread. Rapid curing is also important; litter in *B. tectorum* plots is completely depleted of moisture (reportedly 0 % water content) at least a month prior to several bunchgrasses drying to their minimum seasonal water content (about 20 % of dry mass, Davies and Nafus 2013).

Although *Bromus* increases fire occurrence, it is a common misconception that *Bromus* increases fire intensity in terms of temperatures reached and duration of heating; loss of woody fuel or large herbaceous perennials actually reduced fire intensity (Brooks et al. 2004). Fires on *Bromus*-dominated sites are characterized by low temperatures and often do not completely combust the litter layer (Jones et al. 2015b), leaving what rangeland managers sometimes refer to as a "dirty burn" that contrasts with the completely combusted, charred ground where fire has occurred in uninvaded sagebrush steppe ("clean" burn, Fig. 3.1). "Dirty burn" conditions are perceived to complicate postfire recovery and seeding success, mainly due to insufficient heating during fire to kill *Bromus* seeds and reduced seed-soil contact of seeded native species due to residual litter.

3.4 Impacts on Soil Stability

Soil stability, or resistance to erosion by water or wind, is an important concern for semiarid landscapes because relatively sparse vegetation cover increases exposure of soil and erosion by removing the thin layer of topsoil and the organic matter, nutrients, and seed banks that are concentrated in it (Hasselquist et al. 2011). These losses can push ecosystems toward primary succession conditions. Temporary losses of biological soil crust and plant cover through fire or stand failure (Fig. 3.3) are key ways that *Bromus* increases erosion risks. However, rapid establishment of *Bromus* in disturbed sites with low abundance of native perennial herbaceous species may confer some site stabilization by the second or third year after disturbances like wildfire (Stewart and Hull 1949; Klemmedson and Smith 1964; Miller et al. 2012).

Fire can increase water erosion >100-fold on steep slopes to the detriment of entire watersheds, and two growing seasons are usually required before stabilization is observed (reviewed in Wilcox et al. 2012). Wind erosion in the year after fire can transport several cm or more of topsoil from large burn areas in sagebrush steppe (Sankey et al. 2010), and the resulting dust clouds can be so dense that they exceed measurement capacity of air-quality instrumentation and air-quality standards over vast air sheds (>65 mg/m^3; Wagenbrenner et al. 2013). Postfire dust impacts human health and safety, radiation balance, precipitation, and contaminant transport. The herbicide Oust© applied to inhibit the emergence of *B. tectorum* on burned areas was blown long distances (10–100's of km) with the soil onto downwind crop fields

(Paez 2011). The resulting litigation has discouraged the use of herbicides on many US Bureau of Land Management lands and affected control of *B. tectorum* and other exotics in the critical postfire window across millions of hectares of sagebrush steppe in the Snake River Plain.

3.4.1 Direct Effects on Erosion

Aside from fire or stand failure, there are a number of ways that *Bromus* could decrease or increase water erosion depending on the type and amount of vegetation cover, slope, and soil type. Craddock and Pearse (1938) conducted many simulated rain experiments on steep slopes, and although they did not control for the amount of plant cover, their data suggest that erosion under *B. tectorum* was greater than under *Pseudoroegneria spicata* Pursh. A. Löve (bluebunch wheatgrass) but less than where native or exotic forbs dominated. Wilcox et al. (2012) used simulation models to separate the confounding factors of amount of vegetation cover, slope, and burn severity in determining *B. tectorum* grassland effects on water erosion (HYDRUS to partition rain into infiltration, storage, evapotranspiration, or runoff; MAHLERAN to simulate sediment transport). Their models predicted water erosion to double on steep slopes (20–40 %) dominated by *B. tectorum* (<50 % of community cover) compared to native sagebrush steppe, although high abundances of *B. tectorum* (>66 % of community cover) reduced erosion compared to native sagebrush steppe (46 % community cover) on intermediate slopes (10–20 % pitch). The *Bromus*-induced increase in erosion hinged on the assumption that *Bromus* grasslands altered particle-size distributions and decreased hydraulic conductivity of soil (infiltration)—an assumption based on potentially confounded findings from Boxell and Drohan (2009; discussed in Sect. 3.5 below). Regardless, the water-erosion rates where *Bromus* or natives are present and differences in erosion between them are much smaller than the water-erosion rates after wildfire.

Few data sources are available to assess direct effects of *Bromus* on wind erosion. Wind erosion was greater on a *Bromus*-invaded site compared to native sagebrush steppe in only 1 of 5 years in the Snake River Plain (Fig. 3.4, M. Germino). In that study, total and relative cover of *B. tectorum* sites ranged from 40 to 100 % of ground area across sampling periods, and plant heights were typically about 20 cm. The native site had *A. tridentata* ssp. *wyomingensis* Beetle & Young (Wyoming big sagebrush) with a *P. secunda* and *Elymus. elymoides* Raf. Swezey (squirreltail) understory; its cover in June 2014 was 18 % bare soil, 5 % herbs, 45 % litter/wood, and 30 % shrub, and plant heights were up to 90 cm. The absolute levels of soil movement on both sites were small compared to the very large annual flux observed on a severe postfire wind erosion site using the same measurement devices (442 kg/m/day equates to ~100 gallons of soil flowing through an area the size of a door, daily for a year; Germino, unpublished for the site observed in Wagenbrenner et al. 2013). Like for water erosion, these data suggest that indirect effects of *Bromus* on wind erosion following fire vastly exceed any direct effects of the *Bromus* on erosion.

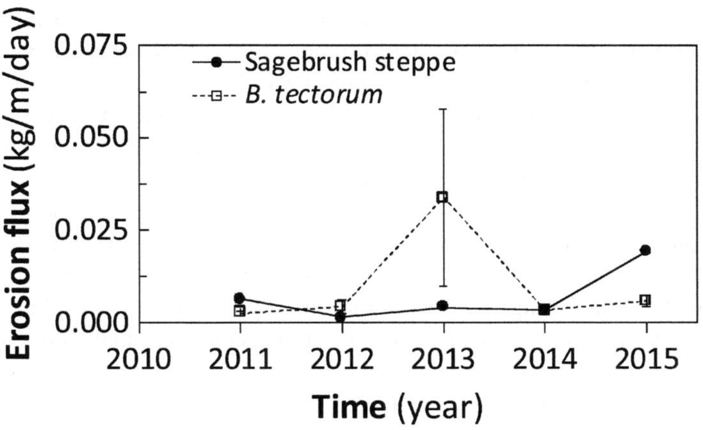

Fig. 3.4 Annual horizontal flux of soil on sites dominated by *B. tectorum* (*open symbols*) or intact sagebrush (*solid symbols*, *Artemisia tridentata* ssp. *wyomingensis*, *Poa secunda*, and *Elymus elymoides*) near Twin Falls, Idaho. Data were collected on four towers of Big Springs Number Eight collectors (5 collectors/tower) in June of each year, and fluxes were determined from the integral of the relationship of mass of soil captured to height aboveground. Measurement of saltation activity in 2013 on the *B. tectorum* sites indicates that the erosion occurred when average wind speeds (5-min periods) were 6–10 m/s. From M.J. Germino, unpublished data

3.5 Impacts on Soil Hydrology and Water Balance

The domain of *Bromus* in the Western USA is semiarid and, by definition, limited by precipitation. Thus, the effects of *Bromus* on how precipitation is partitioned into infiltration, runoff, evapotranspiration, or deep soil-water storage could heavily impact ecosystems. Craddock and Pearse (1938) found greater runoff and thus less infiltration, on sites with more *B. tectorum* than bunchgrasses. The hydrology simulations of Wilcox et al. (2012, see Sect. 3.4.1 above) suggest runoff is about twice as large from *B. tectorum* grasslands compared to sagebrush steppe on slopes greater than 20 %. In addition to reduced plant cover, reductions in soil hydraulic conductivity and infiltration (determined by Boxell and Drohan 2009, Sect. 3.4) were pivotal model parameters predicting greater runoff of *B. tectorum* grasslands compared to native conditions in Wilcox et al.'s (2012) simulations. *Bromus* could decrease infiltration if they reduce organic matter inputs (Gill and Burke 1999; Evans et al. 2001; Norton et al. 2004) because organic matter inhibits formation of hard physical soil crusts and enhances aggregate formation, infiltration, and water retention of desert soils (Singer and Shainberg 2004). Loss of shrub cover and replacement by *Bromus* spp. in southern California led to reduced percolation of water through the 150-cm shrub rooting depth and thus shallow soil-water availability (Wood et al. 2006). Physical soil crusts harden as they dry, and their permeability resumes only after prolonged wetting (Hoover and Germino 2012); thus, drying of surface soils by *Bromus* likely increases the duration that soils are hard and impermeable each year. In contrast, infiltration was greater on burned sites where dense invasion of *B. tectorum* and *B. arvensis* (L. field brome; >80 % total cover of *Bromus*) was observed compared to burned sites where native *A. t. tridentata* recovered, hypothetically due to greater porosity created by the high density of near-surface root channels in *Bromus* roots (Gasch et al. 2013; in the Wyoming Basin; authors referred to *B. japonicus* Thunb., Japanese Brome).

Several considerations for the discrepancies or uncertainties in these studies point to additional research needs, especially accounting for initial soil properties prior to invasion. For example, Boxell and Drohan (2009) examined four areas in the Northern Basin and Range that burned up to 20 years prior to sampling and were subsequently invaded by *B. tectorum* and four adjacent areas that were undisturbed sagebrush steppe. The burned and invaded areas had slightly coarser soil textures, greater stability of soil aggregates, and smoother soil surfaces but had harder and less permeable soil surfaces compared to nearby unburned sagebrush communities (Boxell and Drohan 2009). Fire heating, combustion of soil organic matter, and erosion of topsoil all could have occurred prior to *Bromus* invasion and contributed to the altered soil properties that were attributed to *B. tectorum*—similar to other studies that did not have before/after fire and invasion measurements or other experimental control of confounding factors. Furthermore, Boxell and Drohan (2009) compared soils under shrubs from unburned and uninvaded sites to all microsite types (undershrub plus interspaces) in burned and invaded sites. *Bromus tectorum* can establish on hard, often-impermeable interspace physical crusts where cracks in the crust or litter facilitate its establishment (Evans and Young 1970, 1972), even though it typically exhibits greater growth where shrubs create islands of fertility (like most herbs; Chambers et al. 2007; Hoover and Germino 2012). Thus, the association of *B. tectorum* and altered soil permeability in both Boxell and Drohan's (2009) and Gasch et al.'s (2013) studies could have resulted from fire effects or *Bromus'* microsite selection of initial soil properties, in addition to *Bromus'* influence on the soils. To our knowledge, there is no information available on direct effects of invasive *Bromus* on physical properties of soil surfaces and infiltration/runoff in California grasslands, Mojave Basin and Range, Colorado Plateau, or High Plains.

3.5.1 Spatial and Temporal Patterns of Soil-Water Use

Bromus typically has fibrous and relatively shallow rooting systems (about 5-cm to 35-cm depths), although in some ecosystems (e.g., Palouse prairie, tallgrass prairie, and Great Basin sagebrush), studies have shown that some roots may penetrate to 1–2 m (Hulbert 1955; Hironaka 1961; Peek et al. 2005). This high concentration of *Bromus* roots in surface soil layers combines with their rapid growth and winter-annual phenology to strongly alter community soil-water use (reviewed in Klemmedson and Smith 1964; Reever Morghan et al. 2007; Wilcox et al. 2012). Detailed field data on soil-water profiles in soils under *Bromus* compared to analog native grasses or later seral communities are surprisingly scarce given how commonly the early depletion of soil moisture by *Bromus* is cited.

Bromus tectorum appears less capable of extracting water from dry soils compared to natives; instead this species uses freely available water in shallow soils and then senesces as this resource is depleted. Water status at predawn (i.e., maximum daily hydration) or even midday remained above −1 MPa in *B. tectorum*, which is considerably wetter than the permanent wilting point for many crops −1.5 MPa and for *P. secunda* growing under the same conditions (nearly −3 MPa, Link et al. 1990). Transpiration rates were considerably greater on a per leaf-area basis in *B. tectorum* compared to *P. secunda* (Link et al. 1990), which, combined with a high capacity for leaf-area production, enables rapid soil-water depletion by *B. tectorum*.

Bromus may deplete shallow soil water early in growing seasons, but they do not efficiently deplete shallow water across the entire year nor do they utilize deep soil water efficiently. Norton et al. (2012; Fig. 3.7) found consistently wetter surface soils beneath *B. tectorum* in the Great Basin compared to sagebrush vegetation throughout much of the growing season, which could be due to reduced rainfall interception by foliage (greater throughfall) in the *Bromus* stands. At midsummer, deep soil-water depletion by the *Bromus*-invaded plant community is greatly diminished compared to early or late-seral bunchgrasses and the evergreen sagebrush (sagebrush depletes soil water to −5 MPa or perhaps drier; Cline et al. 1977; Peek et al. 2005; Ryel et al. 2008). The lack of deep soil-water extraction by *Bromus* could lead to increases in deep soil-water accumulation of 60–70 mm/year in coarse soils, as shown in the Central and Northern Basin and Ranges and Central California Valley (reviewed in Wilcox et al. 2012 and Reever Morghan et al. 2007, respectively). Soil-water wetting fronts under *Bromus* grasslands were predicted to become deeper at rates up to 2 m/year in coarse soils or 0.2 m/year in loam soils, in a winter-wet/summer-dry climate typical of the Great Basin (Wilcox et al. 2012). Ultimately, contact of the wetting front could link vadose and groundwater systems in those rare situations where weather, climate, geology, and topography result in shallow water tables,such as in some ancient lacustrine basins.

3.6 Impacts on Ecosystem Fluxes and Energy Partitioning

Plants have a primary role in regulating the flow or storage of C, water, and energy through ecosystems, and field observations suggest that *Bromus* likely impacts net ecosystem exchange of CO_2 (NCE) and water (evapotranspiration, ET) and energy balance (Prater et al. 2006; Prater and DeLucia 2006; Germino et al., unpublished data). Energy balance of plant and soil surfaces refers to the partitioning or dissipation of net radiation (solar and thermal) into latent heat (LE; vapor flux dominated by evapotranspiration, ET), sensible heat (conductive and primarily convective heat exchange), and storage of heat in soil. Semiarid ecosystems generally have high net radiation fluxes as a result of cloudless skies and low ET and thus CO_2 exchange on an annual basis (given water scarcity and the water-for-C tradeoff in leaf gas exchange). The effect of conversion of diverse perennial ecosystems to exotic annual grasslands on these fluxes is of interest because vast areas are likely to have strongly altered radiation due to greater albedo, reduced NCE, and changes to nearly all aspects of energy balance.

Three studies provide insights into the changes in energy balance on *B. tectorum-invaded* sites and intact sagebrush steppe in the Central Basin and Range and Snake River Plain (*A. tridentata* ssp. *wyomingensis*). These studies ranged in spatial and temporal scale as follows: (1) plot-level chamber measurements of NCE and ET over 3 years on a variety of burned and unburned sagebrush sites (Prater et al. 2006), (2) short-term landscape flux-tower measurements that provide indirect, snapshot estimates of ET and associated energy balance on a burned and an unburned site (7 sampling days over a 3-month period, using gradient method, Prater and DeLucia 2006), and (3) long-term landscape flux-tower measurements of NCE, ET, and energy balance on a *B. tectorum*-dominated site (mean *Bromus* cover ranged 40–100 %) and an *A. tridentata* ssp. *wyomingensis* site with *P. secunda* and *E. elymoides* understory (5 years, using replicate eddy covariance towers dispersed over 1–2 km in each site type; Germino et al., unpublished data). Although the Prater and DeLucia studies have been cited as isolating flux impacts of *B. tectorum* (e.g., Bradley et al. 2006; Wilcox et al. 2012), *B. tectorum* tended to comprise only about 20 % of relative cover in both the burned and sagebrush site. Abundance of *Agropyron cristatum* L. Gaertn. (crested wheatgrass) actually varied more consistently and appreciably between Prater and DeLucia's (2006) sites. Their study also relied on the Bowen-ratio approach, in which a determination of the ratio of sensible to latent heat is combined with measurement of often miniscule (and difficult to detect) gradients in temperature and vapor in air aboveground to indirectly estimate evapotranspiration (ET).

Prater and DeLucia's (2006) data suggested overall lower ET occurred in the burned site dominated by exotic grasses, in spite of "flashes" of greater ET during wet-soil periods in late spring. Lower ET in their study appeared to result from lower ability of *Bromus* to extract water from dry soils combined with a hypothetically greater albedo (reflectance) and thermal radiation emitted from warmer soils. Consistent with these hypotheses, reflectance of solar radiation could be up to 50 % greater for the *B. tectorum* compared to sagebrush community at spring and midsummer (mean ± SD daily values were 0.21 ± 0.05 for *Bromus* vs. 0.16 ± 0.03 for the sagebrush community over 11 representative clear-sky days, Fig. 3.5, bottom panel from Germino et al. unpublished data). Ecosystem surface temperatures were 2 °C warmer on the *B. tectorum* compared to sagebrush community across all days and nights (Fig. 3.5, middle panel), which could increase emission of longwave radiation relative to incoming radiation. Consequently, the *Bromus* community had about 10 MJ/m^2 less net radiation (not shown). Over 4 years and in the particular case year shown in Fig. 3.6, evapotranspiration (latent heat flux) was never greater for the *Bromus* community, at least on a monthly basis. The dissipation of greater energy and warmer surfaces contributed almost 50 MJ/m^2 more sensible heat loss to the air (i.e., warming air, not shown) for the *Bromus* compared to sagebrush community during the particular period shown in Fig. 3.5 (mostly senesced *Bromus*). Changes in these surface energy balance parameters have the potential to feedback on climate of the ecosystem, and they relate to hydrological differences between the sites.

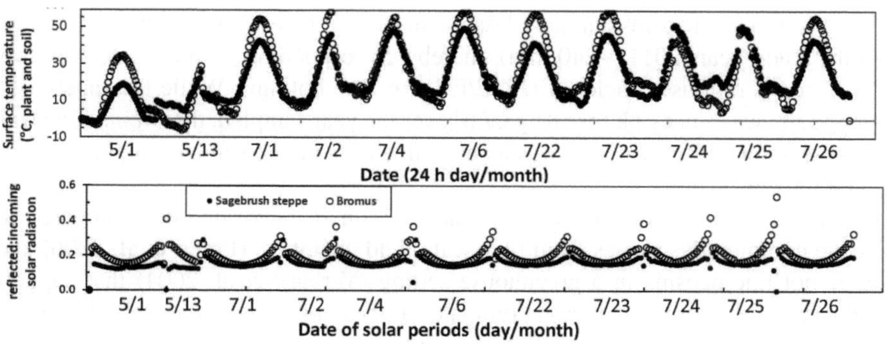

Fig. 3.5 Representative parameters affecting net radiation balance on a *Bromus tectorum* landscape over 11 days during the green (May) and senesced (July) months. Parameters were measured by duplicate sensors separated by about 1 km and positioned at 4-m height and are reported on 30-min intervals. Ecosystem surface temperature of about 10–100-m^2 patches was measured with infrared thermometers (Apogee, narrow and wide focus sensors), and the ratio of solar radiation reflected off the plant/soil surface to incoming (direct and diffuse) solar radiation (i.e., albedo, measured with two Hukseflux NR01 4-way sensors per site). From MJ Germino, R Allen, W Zhao, and C Robison, unpublished data

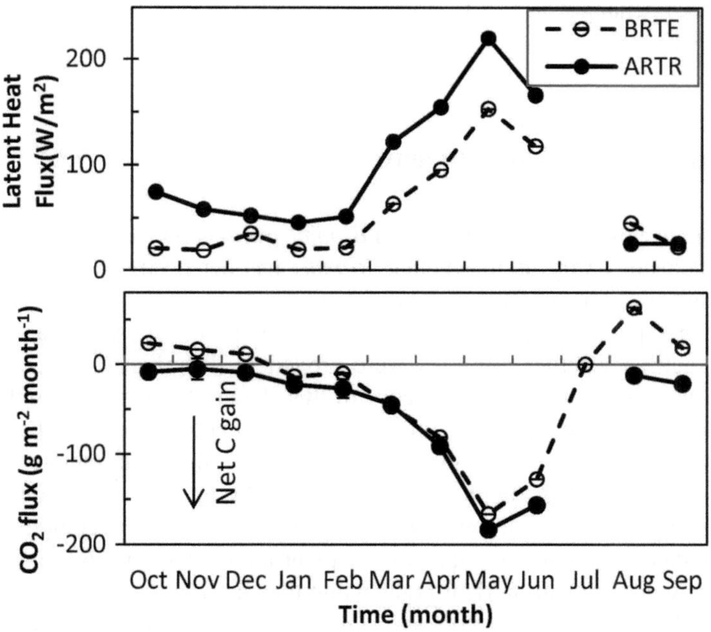

Fig. 3.6 Monthly ecosystem fluxes of latent heat and CO_2 on *Bromus tectorum (BRTE)*-dominated or sagebrush-steppe communities (ARTR). Data are for two RM Young 8100 and Campbell CSAT 3D sonic anemometers and LiCOR 7500 open-path gas analyzers per site. SE are smaller than symbols. MJ Germino, R Allen, W Zhao, J Greth, C Robison, unpublished data

The *Bromus* site had substantially less ET in spite of receiving similar precipitation in the study year (2011; ~240 mm), thereby corresponding with considerably less NCE and C rain-use efficiency (NCE/ET, Fig. 3.6, bottom). While the sagebrush-steppe site was a net sink for 581 g C/m^2 over the year sampled, the *Bromus* site was a net sink for 313 g C/m^2 and was a net source of C in late summer months (calculated for Fig. 3.6).

Further evidence of reduced NCE and ET on *B. tectorum* sites compared to native communities comes from plot-scale field chambers (Prater et al. 2006) and large-pot microcosms in a greenhouse setting (Verburg et al. 2004) that suggest reduced NCE where *B. tectorum* dominates. For example, in Prater et al. (2006), the plot with the greatest cover of *B. tectorum* (up to 70 % cover of *B. tectorum*) had lower NCE and ET than all other plots with less *B. tectorum*. In Verburg et al. (2004), indoor microcosms containing Kansas soils (presumably C-rich loess) and planted with *B. tectorum* had very high respiration rates and negative annual NCE (300 g C m^{-2}). These smaller-scale findings combine with the landscape-level data to suggest C sequestration at the ecoregional level would also be reduced by *Bromus* invasion. The potential switch of ecosystems from sinks to sources of C to the atmosphere is an important impact of *Bromus*. An estimated 8 Tg of the aboveground standing crop of C in perennial vegetation (including wood) has already been released where *B. tectorum* has dominated the Western USA, and another 50 Tg C of losses are likely with its continued invasion into all suitable habitats (Bradley et al. 2006). Changes in soil C storage have also occurred with *Bromus* invasion (Sect. 3.7.1).

3.7 Impacts on Soil Nutrients

An abundance of correlative studies on *Bromus* impacts on soil nutrients exist, and their conclusions vary as to whether *Bromus* has net positive, negative, or null effects on soil nutrients. This variability is likely due to differences in site context (vegetation type, soils, climate), the spatial and temporal patterns of sampling, and factors such as whether top soils were eroded in disturbances associated with *Bromus* invasion. It is beyond the scope of this chapter to identify why the studies might differ, as the sampling regime of each would need to be described, but it is useful to evaluate representative cases. *Bromus* might accelerate elemental cycling—provided their biomass is not removed by grazers or erosion—because they annually return all nutrients to the soil as they senesce and decompose and they might change the quality and quantity of plant inputs into soil. *Bromus* may also temporarily deplete soil nutrient pools while they are in their rapid growth phase. The disturbances and soil factors associated with *Bromus* may also influence soil nutrients (e.g., fire affects net N mineralization, soil texture affects, and soil nutrient content).

3.7.1 Effects on Soil C in Topsoil Versus Subsoil

In ecosystems where NCE and spatial and temporal patterns of litter and root distribution differ between *Bromus* and native communities, corresponding differences in soil organic matter C may be expected. Inorganic and particularly organic C (SIC and SOC, respectively) are important for nutrient cycling. SOC is the primary source of plant-available N and can be an important source of phosphorus (P; Belnap et al. 2015), and it also enhances water infiltration and retention in semiarid soils. In addition to altering litter inputs and organic matter, *Bromus* may influence SOC by altering soil aggregates, which increase C storage by shielding SOC.

Several studies suggest that soil organic matter is reduced and tends to be shallower and cycle more rapidly where *B. tectorum* dominates (Norton et al. 2004; Gill and Burke 1999; Evans et al. 2001). However, other studies, which examined *B. tectorum*-dominated sites that had not recently burned, have shown no difference or greater SOC under *B. tectorum* (Hooker et al. 2008; Norton et al. 2012; Stark and Norton 2015). The immediate effects of fires on *Bromus*-dominated sites are reductions in vegetation and litter C by combustion. Soil C may be lost through erosion of topsoil, while vegetation recovers in those areas that lack a significant litter layer. Aside from these fire or erosion effects, longer-term decreases in litter C contents on *Bromus* sites may occur only with repeated fire, as neither vegetation nor soil C contents decrease over time (Jones et al. 2015a). Fires in *Bromus*-dominated ecosystems are generally characterized by soil temperatures that are too low to volatilize either soil N or C (Jones et al. 2015a), and longer-term variation in C after fires likely relates more to differences in decomposition and nutrient cycling (Jones et al. 2015a).

Deep soil (i.e., deeper than 20–30 cm) had less SOC where *B. tectorum* had invaded compared to uninvaded sagebrush steppe in the Central and Northern Basin and Ranges, as would be expected from loss of deep rooting in the invaded community (Norton et al. 2004; Rau et al. 2011). Reductions in SOC in deep soils with *Bromus* invasion may be a function of the interaction between root inputs, soil aggregates, and decomposition. Austreng (2012) found less SOC in *B. tectorum-dominated* compared to *A. cristatum*-dominated stands, 27 years after fire (SOC was 23 and 33 Mg/ha, respectively, and 44 Mg/ha in undisturbed sagebrush steppe). Notably, soil aggregates >250 μm, an important form of C storage, were absent on Austreng's burned/grassland sites. In contrast, Hooker et al. (2008) found that at a Great Basin rangeland site, 10 years after fire converted sagebrush to *B. tectorum* and adjacent sites were seeded to *A. cristatum*, SOC contents to a 1-m soil depth were 72, 69, and 62 Mg C/ha in *B. tectorum*-, *A. cristatum*-, and *A. tridentata* ssp. *wyomingensis*-dominated stands, respectively. While root C below 20 cm was substantially lower in *B. tectorum* soils, there were no detectable differences in deep SOC 10 years after vegetation conversion. Over longer time periods, replacement of native perennial shrubs and herbaceous vegetation with exotic annual grasses could release 6–9 Mg/ha. of root and SOC over the whole soil profile, more than double the aboveground losses (Rau et al. 2011; Bradley et al. 2006).

Decreases in SOC in deep soils may be an important driver of changes in the overall soil profile, as several studies suggest that SOC can be greater in near-surface soils under *Bromus*. In the Northwestern Great Plains, subtle increases in SOC in shallow soils (top 20 cm) became significant only after 50 years of *Bromus* cover, according to computer simulation models [40 % *B. tectorum* plus *B. arvensis* (authors referred to *B. japonicus*) cover, CENTURY model, Ogle et al. 2004]. Rapid increases in SOC following *B. tectorum* invasion of *Krascheninnikovia lanata* A. Meeuse & Smit (winterfat) stands were detected to 1-m soil depth by Blank (2008) in the Central Basin and Range. Although SOC differences may have been at least partly present before invasion in Blank's sites (2008; see Sect. 3.7.2), they are corroborated by greater SOC in shallow soils of invaded sagebrush steppe of the Northern Great Basin (Norton et al. 2004; Hooker et al. 2008). Similarly, surface SOC (top 10 cm) in southern California chaparral was greater where *B. rubens* and *B. diandrus* invaded and dominated after fire (Dickens and Allen 2014a). By contrast, mean SOC was no different (but was less variable) where *Bromus* had invaded coastal sage scrub in California (Dickens et al. 2013). The differences in SOC-*Bromus* relationships between these ecosystems likely relate to a tendency for chaparral to have coarse and deep roots compared to a high density of fine shallow roots in coastal sage scrub.

3.7.2 Effects on Soil Nitrogen Cycling

Soil N is generally considered the most limiting of macronutrients in semiarid soils. The effect of *Bromus* invasion has been evaluated in many studies, but findings have been mixed. *Bromus* is associated with soils higher in N in a number of correlative field studies, like many invasive plant species (e.g., Liao et al. 2008). Several studies were able to evaluate soil N before and after invasion of undisturbed grasslands of the Colorado Plateau. *Bromus tectorum* did not affect N pools 2–4 years following invasion where *Achnatherum hymenoides* (Roem. & Schult.) Barkworth (Indian ricegrass) and *Stipa comata* (Trin. & Rupr.) Barkworth (needle and thread) dominated. However, *B. tectorum* initially increased total soil N from by 0.04 to 0.12 g/kg at 0–10-cm depth where *Pleuraphis jamesii* Torr. (James' galleta) dominated, which is equivalent to an increase of 13 and 40 kg N/ha/year (Evans et al. 2001; Rimer and Evans 2006; Miller et al. 2006; Schaeffer et al. 2012). In addition, the *P. jamesii* sites invaded by *Bromus* had greater N before the invasion, which was magnified by *Bromus* (Kleiner and Harper 1977). While the higher of the two rates exceeds ecosystem N inputs in this region by four- to tenfold, redistribution of N into the surface soil following extraction of N from deeper in the soil profile might explain the increase in the surface soil in this study (Sperry et al. 2006). Eight years after the invasion, extractable N (NO_3^- plus NH_4^+) was still no greater where *B. tectorum* was present compared to *Stipa*-dominated patches (only 1 of 21 sampling events), whereas it was greater for 5 of 21 sampling events compared to *P. jamesii-dominated* patches (Schaeffer et al. 2012). In southern California where *Bromus* spp. had invaded coastal sage scrub, there were no differences in organic N, but extractable mineral N was higher during the dry season and lower during the growing season (Dickens and Allen 2014b). This result was attributed to high rates of N uptake by rapidly growing *Bromus* during the rainy season and rapid N mineralization upon senescence. In the Central Basin and Range, Blank (2008) found N was 1380 kg/ha greater at a *K. lanata* site that had been invaded by *B. tectorum* 3 years prior to sampling compared to an uninvaded *K. lanata* stand, but the annual inputs of 460 kg N/ha/year required to create the enrichment does not seem possible via *Bromus* alone. Thus, similar to the Utah site, *Bromus* may have selectively invaded N-rich soils.

A few carefully controlled field experiments provide less ambiguous evidence for the direct enhancement of soil N by *Bromus*. Soil net N mineralization, net nitrification, inorganic N concentrations, and denitrifier population size were positively correlated with variations in abundance of *B. rubens*, *B. hordeaceus*, and another exotic annual grass, *Hordeum murinum* L. (mouse barley), among other perennial grasses, in an experiment in which the species were planted 5–7 years prior to evaluating species effects (Parker and Schimel 2010). In sagebrush steppe of northwest Colorado, net N mineralization and net nitrification rates were 50 % and 28 % faster in surface soils of plots seeded 24 years earlier with *B. tectorum* compared to *A. tridentata* ssp. *wyomingensis* and native perennial grasses, respectively (Stark and Norton 2015). Soil nitrate, organic C and N, and respiration were also greater while C:N of organic substrates consumed by microbes was lower in *B. tectorum* compared to native plots (C:N ratios of 7.7:10.4 compared to 9.8:15.6, respectively; Stark and Norton 2015)—all of which are consistent with accelerated N cycling. After just 8 weeks following planting, greenhouse mesocosms of *B. tectorum* contained a third more soil C and N and twice as much N leakage from plant roots into soil than mesocosms containing *A. cristatum* (Morris et al. accepted).

Bromus can affect soil N in multiple ways that may be site specific. Annual return of tissue N early in the growing season will affect N cycling, although this effect likely differs across ecoregions or sites (e.g., on the Colorado Plateau, soil fauna that bury litter is scarce and thus litter input enhancement by *Bromus* may be less). *Bromus* can release lower C:N detritus or exudates, as has been observed for *B. tectorum* in sagebrush steppe of the Central and Northern Basin and Ranges and Wyoming Basin, C3/C4-shortgrass steppe, and Colorado Plateau grasslands (Bolton et al. 1990; Evans et al. 2001; Booth et al. 2003; Saetre and Stark 2005; Hooker and Stark 2008; Adair and Burke 2010). Shifts in phenology and soil-water availability appear to be important modifiers of the influence of *Bromus* on N cycling: increases in NO_3^- are pronounced as seasonal (summer) drying and senescence of *Bromus* occurs (Fig. 3.7; Svejcar and Sheley 2001; Booth et al. 2003; Norton et al. 2004, 2008; Hooker et al. 2008; Dickens and Allen 2014b, but see Schaeffer et al. 2012). *Bromus* may create warmer soil conditions (Sect. 3.7.3) that could increase microbial enzyme kinetics given sufficient available soil water. Also, invasion can alter N inputs through loss of biological soil crusts (Evans and Belnap 1999; Belnap 2003). Moreover, greater total N concentrations are usually associated with faster rates of N cycling induced by recent SOC inputs and associated with greater labile soil N pools such as NH_4^+, NO_3^-, microbial N, and readily mineralizable organic N (Booth et al. 2005; Morris et al. accepted).

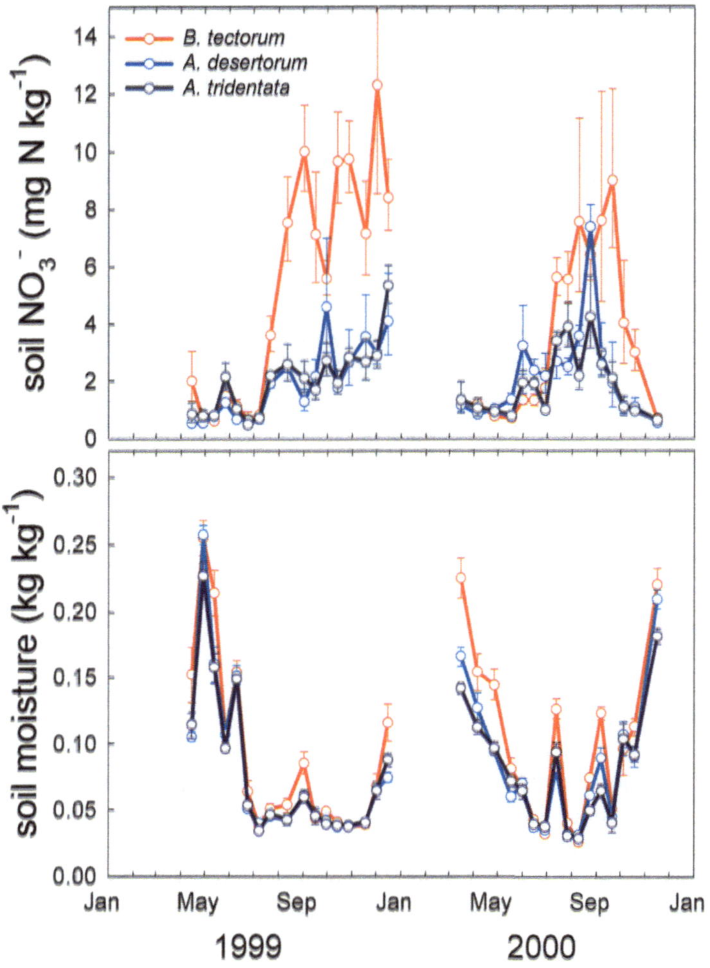

Fig. 3.7 Seasonal changes in soil nitrate concentrations in the surface soils (0–10-cm layer) beneath *Bromus tectorum*, *Agropyron desertorum*, and *Artemisia tridentata* ssp. *wyomingensis* vegetation in Rush Valley, UT. Values represent means and SE ($n = 4$) for each vegetation type (compiled from Stark et al. 2006; Norton et al. 2012; Morris et al. accepted)

Bromus may affect N fixation, redistribution of N in soil, leaching losses of N, and denitrification and could thereby influence longer-term trends of soil N where it has become dominant. High rates of N mineralization and nitrification relative to rates of N immobilization in mesic ecosystems typically lead to increased N loss through leaching or trace N gas emissions (Stark and Hart 1997). However, this pattern may not occur in semiarid and arid ecosystems because high nitrification could conserve N by depleting NH_4^+, which is otherwise readily lost as NH_3 in high pH soils. Also, while nitrification is normally a key step toward denitrifying losses of NO, N_2O, or N_2 gases, denitrification is suppressed by the low C and water availability in semiarid soils (Smart et al. 1999). Moreover, precipitation is scarce enough in the area impacted by *Bromus* to minimize leaching of NO_3^- from soils but is sufficient to redistribute it to depths where denitrification is less likely (Sperry et al. 2006; Hooker et al. 2008). *Bromus* decreases ecosystem fixation of N_2 by displacing N-fixing plants or biological soil crusts (Evans et al. 2001). Promotion of fire by *Bromus* does not appear to result in mineral N losses through volatilization but rather may result in short-term increases in mineral N availability through deposition of N-rich ash (Jones et al. 2015a, b). Weather, specifically precipitation and minimum winter temperature, had stronger effects on soil N availability and *B. tectorum* success than litter removal or seeding a competitor over four consecutive repeated burns (Jones et al. 2015a, b). Thus, studies evaluating effects of *Bromus* on mineral N should monitor weather variables and soil water availability and examine differences among years.

3.7.3 Phosphorus and Other Nutrients

In high pH soils, such as found in deserts, phosphorus (P) and calcium (Ca) combine to form moderately to highly insoluble calcium phosphate minerals that are unavailable for plant uptake (Lajtha and Schlesinger 1988). These Ca-bound forms of refractory P are the dominant mineral P pool in deserts. However, there are multiple lines of evidence showing that the presence of *B. tectorum* can result in the conversion of recalcitrant P (bio-unavailable) to bio-available P. Extractable P was much higher in *Bromus*-invaded plots than adjacent uninvaded plots (Hansen 1999) despite no significant differences in soil P before invasion (Kleiner and Harper 1977, see also Blank 2008). Furthermore, extractable P was positively correlated with *B. tectorum* cover (0 %, 10 %, and >40 % *Bromus* cover associated with 14.6, 19.5, and 28.2 μg P/g soil, respectively; Hansen 1999). In a controlled greenhouse setting, *B. tectorum* decreased recalcitrant P and doubled soil extractable P in several sandy soils (Gopalani 2004). Phosphorus availability is highly correlated with precipitation in the months preceding measurement, and *B. tectorum*'s enhancement of P availability follows wet periods (Belnap 2011). Furthermore, elevated foliar P of *B. tectorum* following experimental watering on loam soils in the Colorado Plateau and correlations of *B. tectorum* growth and pH buffering capacity suggest a role for root exudates in liberating P (Miller et al. 2006). Greater soil moisture following experimental watering would provide *B. tectorum* more resources to increase exudates and stimulate microbial activity. Root exudates could include H^+ ions that acidify the soils, changing solubilities of ions that precipitate P (Frossard et al. 1991), phosphatase enzymes that release organic P, organic chelates that bind metal ions and mobilize P, and/or organic substrates that stimulate microbial activity. Microbes produce phosphatases (Blank 2008; Dighton 1983) and oxalates (Jurinak et al. 1986; Knight 1991; Knight et al. 1992), which can dissolve Ca–P bonds. Alkaline phosphatase was 38 % greater in soils under *B. tectorum* compared to interspace soils occupied by *K. lanata* (Blank et al. 2013).

The mechanisms by which *B. tectorum* could increase the availability of other nutrients are the same as for soil P; root exudates can directly dissolve chemical bonds that make them bio-unavailable, or the release of C can stimulate microbes to indirectly increase their availability. Manganese (Mn) increased while potassium (K), calcium (Ca), and magnesium (Mg) concentrations were unchanged after 7 years of dominance by *B. tectorum* in sandy loam soils (Belnap et al. 2005). In contrast, K, Ca, Fe, Mn, and copper availability doubled in soils on a *B. tectorum-invaded* site (Blank 2008); however, this may be due to preexisting soil differences. Elevated Cl, Na, and Zn also occurred during the growing season following invasion of the *K. lanata* community by *B. tectorum* (Blank et al. 2013).

3.8 Impacts on Soil Microbiological and Microfaunal Communities

Changes in soil communities following *Bromus* invasion seem likely given the biogeochemical changes described above and may be part of a self-perpetuating feedback that *Bromus* imposes on its environment. Alteration of key soil food web components, such as the loss of biological soil crusts, which contribute to soil fertility and stability, and arbuscular mycorrhizal fungi (AMF) that form mutualisms essential for many native plants in the impacted habitats (Allen 1988, reviewed in Knapp 1996), are examples. AMF in the genus *Glomulus* are particularly important for natives such as *A. tridentata* because their glomalin glycoproteins stimulate formation of beneficial soil aggregates (reviewed in Weber et al. 2015).

Several studies have revealed appreciable differences in microbial communities in soil under *Bromus*-invaded compared to non-invaded communities. The ungrazed and unburned grasslands of Belnap and Phillips (2001) invaded by *B. tectorum* had lower species richness and abundances of fungi and invertebrates and greater abundances of active bacteria than uninvaded patches. Their sampling included soil bacteria, fungi, detritivores, algivores, and mites and microarthropods of various trophic levels. Notable shifts with *B. tectorum* invasion were increases in generalist, non-mycorrhizal, saprophytic fungi compared to mycorrhizal fungi (Belnap and Phillips 2001). Furthermore, losses of key taxa such as the AMF found in the *Glomus* genus were revealed in the *B. tectorum* soils using restriction fragment length polymorphism (RFLP, Hawkes et al. 2006). Although *B. tectorum* supports AMF, it is a relatively poor host compared to species such as sagebrush (Busby et al. 2012), perhaps due to lapses in photosynthetic C when *Bromus* is senescent. *Bromus hordeaceus* has also supported very few RFLP types compared to native herbs (Hawkes et al. 2005). Wyoming Basin soils invaded by *Bromus* had very low abundances of all microbial groups according to phospholipid fatty acid analysis, which detects only live microbes (Gasch et al. 2013). Fewer taxonomic orders and a greater presence of pathogenic, opportunistic, and saprotrophic taxa and an absence of AMF *Glomus* fungi occurred in soils dominated by *B. tectorum* following sagebrush removal from a plot 14 years following experimental sagebrush removal (Weber et al. 2015). In studies of coastal sage scrub, *Artemisia californica* Less. (coastal sagebrush) benefitted from a greater diversity of AMF fungi, while *B. rubens* was associated primarily only with the fine endophyte *Glomus tenue* Greenhall (I. R. Hall) (Egerton-Warburton and Allen 2000; Sigüenza et al. 2006). Although *Bromus*-induced changes in soil biota would likely have substantial effects on plants and overall ecosystem functions, such effects have seldom been experimentally identified such as through reciprocal soil transplant studies. Altered soil food webs where *Bromus*-invaded *P. jamesii* grasslands on the Colorado Plateau did not translate to appreciable shifts in decomposition, mineralization, or demographics of *P. jamesii* (Belnap et al. 2005). Conversely, Sigüenza et al. (2006) showed that *G. tenue* was a more effective mutualist for *B. rubens* than the diverse AMF fungi were for *A. californica*.

3.9 Impacts on Higher Trophic Levels, Aboveground

The habitat transformation of diverse plant communities into *Bromus*-invaded grasslands over vast areas leads to marked changes in invertebrate and vertebrate animal populations. Animals dependent on native plant species such as sagebrush or native forbs are lost from sites as a result of *Bromus* invasion or become scarcer, and the resulting animal communities are dominated by generalist species. *Bromus* is cited as a key stressor for several species that have become imperiled. For example, *B. rubens* adversely impacts *Gopherus agassizii* Cooper (Mojave desert tortoise; Brooks and Esque 2002) and *Euphydryas editha quino* Behr (Quino checkerspot butterfly (USFWS 2003), and *B. tectorum* adversely affects *Centrocercus urophasianus* Bonaparte (greater sage grouse), *Brachylagus idahoensis* Merriam (pygmy rabbits), and other sagebrush obligates.

Although *Bromus* have some forage value, they provide relatively unstable habitat for most animals, in terms of food, shelter, and increased fire. Generally, forage quality of *Bromus* declines rapidly following its short green-up period during each year, which combines with its high year-to-year variability to detract from its reliability as a food source. *Bromus tectorum*, *B. rubens*, and *B. diandrus* impose major irritants to animals who contact or ingest floral parts, due to their stiff, barbed awns on florets, and sharp or barbed seeds (e.g., Medica and Eckert 2007). Ailments directly attributed to reproductive parts of *Bromus* include mouth sores, abscesses, corneal abrasions, skin and genital infections, and respiratory distress including coughing and gagging (reviewed in Zouhar 2003).

3.9.1 Impacts on Invertebrates and Small Vertebrates

Changes in diversity and abundance of insects, small mammals, and reptiles form one of the most distinctive and tractable impacts of *Bromus* on ecosystems. Insects and small mammals, particularly ants or ground squirrels, can be important ecosystem engineers due to their herbivory, granivory, and soil disturbances that accompany colony development. Seed-harvesting genera, such as the mound-forming *Pogonomyrmex occidentalis* L (harvester ant), became more abundant in *B. tectorum-dominated* plots compared to native sagebrush steppe and would likely impose seed predation problems for restoration seedings (Ostoja et al. 2009).

Loss of native forbs and the pollinator community associated with them is an important but understudied impact of *Bromus*, considering that loss of pollinator services (e.g., from bees) could greatly impact efforts to restore native plants. Population stability of *Lepidium papilliferum* L.F. Hend. A. Nelson & J.F. Macbr. (Idaho pepperweed) was related to diversity and pollination activity of insect pollinators, in a region of southwest Idaho that is heavily impacted by *Bromus* (Robertson and Klemash 2003).

A number of studies demonstrate negative relationships in overall abundance and species richness and diversity of small mammals with *Bromus* invasion (e.g., Great Basin studies: Larrison and Johnson 1973; Gano and Rickard 1982; Ostoja and Schupp 2009; Hall 2012; see Fig. 3.8 from Freeman et al. 2014). Omnivorous species that can tolerate a wide range of physical conditions and are generalists, such as *Peromyscus maniculatus* Wagner (deer mice), tend to dominate the animal communities of *B. tectorum*-invaded landscapes (it was one of the few species in Bromus-dominated plots in Fig. 3.8).

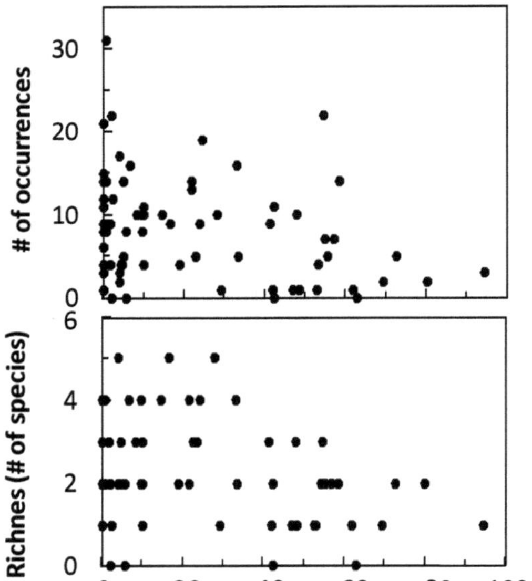

Fig. 3.8 Relationship of *Bromus tectorum* percentage cover to total abundance and species richness of all small mammals detected in 10,437 trap nights on 70 sites throughout NE Utah over 2 years. Animal communities included the following (in decreasing abundance): *Peromyscus maniculatus* (deer mouse), *Dipodomys ordii* (Ord's kangaroo rat), *Dipodomys microps* (chisel-toothed kangaroo rat), *Ammospermophilus leucurus* (white-tailed antelope squirrel), *Chaetodipus formosus* (long-tailed pocket mouse), *Perognathus parvus* (Great Basin pocket mouse), *Onychomys leucogaster* (northern grasshopper mouse), *Neotoma lepida* (desert wood rat), *Perognathus longimembris* (little pocket mouse), *Microtus montanus* (montane vole), *Tamias minimus* (least chipmunk), and *Microdipodops megacephalus* (dark kangaroo mouse). From data in Freeman et al. (2014)

Ground squirrel relationships to *Bromus* are particularly noteworthy because of their close link as important prey (Van Horne et al. 1997). For example, ground squirrels are important food for hawks, *Taxidea taxus* Long (American badger), and snakes and are critical for nesting *Falco mexicanus* Schlegel (prairie falcons, Yensen et al. 1992). Their burrowing strongly affects infiltration, biogeochemistry, and plant community dynamics (e.g., Blank et al. 2013). Burrow densities (number/ha) of *Spermophilus townsendii* Bachman (Townsend's ground squirrels) and abundances and body condition of *Spermophilus mollis* Kennicott (Piute ground squirrels) were inversely related to *B. tectorum* abundances in sites otherwise dominated by *P. secunda* in southern Idaho (Yensen et al. 1992; Steenhof et al. 2006). Abundances of *S. townsendii* were not reduced in *Bromus*-invaded sites, but they were highly variable over time in a "boom-bust" fashion, ranging from 25 to 350 burrows/ha from 1 year to the next compared to consistent <100 burrows/ha per year in intact vegetation (Yensen et al. 1992).

3.9.1.1 Impacts on Nutrition, Shelter, and Locomotion

Although Kelrick and MacMahon (1985) suggested that *Bromus* is not a preferred food, some small mammals consume it (Zouhar 2003), and *B. tectorum* even comprised 100 % of stomach contents of species like *Perognathus parvus* Peale (Great Basin pocket mouse, Richardson et al. 2013). Sylvilagus spp. (cottontail rabbits) prefer *B. tectorum* over other forage options in winter (Turkowski 1975), but ground squirrel preference for *B. tectorum* was very low compared to *P. secunda* and or other shrubs and forbs (Van Horne et al. 1998). *Bromus rubens* is an important food source for *Dipodomys microps* Merriam (chisel-toothed kangaroo rat, Rowland and Turner 1964) and can provide some nutritional benefit when green to *G. agassizii* (Esque et al. 2014), although they generally prefer a diverse diet that is usually not available in *Bromus*-dominated areas (Esque 1994; Jennings 2002). *Bromus rubens* alone is unable to provide the energy and nutrients required by *G. agassizii* and is an irritant to mouth tissues and thus is not preferred over other forages (McArthur et al. 1994; Nagy et al. 1998; Medica and Eckert 2007).

Cover and navigability for small animals is transformed by the loss of relatively tall perennials, such as sagebrush or taller bunchgrasses with high crown-to-basal area ratios, and the bare or crusted soil that normally separates them and facilitates animal movement (Bachen 2014). Spacing between stems or litter in dense *Bromus* canopies becomes smaller than animal sizes, and experiments show movements made for foraging or escaping predation become slower and noisier, and thus risks of predation are increased for a number of rodents such as mice and reptiles such as lizards (Esque et al. 2002; Newbold and Carpenter 2005; Rieder et al. 2010; Bachen 2014). Mice avoid moving through noisy substrates, and acoustic cues are important for hunting by *Vulpes* spp. (fox), *Canis latrans* Say (coyotes), and birds of prey, particularly nocturnal hunters such as owls (reviewed in Bachen 2014). However, tall and dense *Bromus* canopies might also obscure small animals from predation, offsetting negative effects of *Bromus* on small mammals, although few studies have separated these mechanisms.

3.9.2 Changes in Bird and Large-Mammal Communities

Although *Bromus* provides import forage and cover for introduced game birds such as the nonnative *Alectoris chukar* Gray (chukar partridge; reviewed in Zouhar 2003), the loss of perennials such as sagebrush is a notable detriment to many bird species such as greater sage-grouse. Sage-grouse benefits from the structural protection sagebrush provides throughout all life stages (Connelly et al. 2011). The grouse feed exclusively on sagebrush leaves in winter (Patterson 1952; Wallestad and Eng 1975), and chicks and adults depend on forbs often displaced by *Bromus* for a major portion of their diet in spring and early summer (Johnson and Boyce 1990; Schroeder et al. 1999). Direct negative effects of *Bromus* have also been noted on birds including sharp florets of *B. rubens* scratching eyes of hawks and causing infections, loss of vision, and thus starvation (McCrary and Bloom 1984). Birds associated with grasslands tended to use bunchgrass more than *B. tectorum* sites: *Eremophila alpestris* L. (horned lark) and *Ammodramus savannarum* Gmelin. (grasshopper sparrow) visits and densities were greater on *B. tectorum* sites than native perennial grasslands (Earnst and Holmes 2012). Fire effects on birds include loss of nesting and temporary removal of food and protection provided by vegetation.

Bromus can contribute to the diet of native and domestic ungulates but generally are considered low-quality forage. *Odocoileus hemionus* Rafinesque (mule deer) diets can have a large portion of *B. tectorum* in winter if the grasses are accessible, and it is also an important forage before native herbs emerge in spring (reviewed in Zouhar 2003). In Arizona, *B. rubens* comprised <1 % of seasonal diet of desert mule deer, but this grass species is grazed by *Ovis canadensis nelsoni* Merriam (desert bighorn sheep; reviewed in Zouhar 2003). Predators such as coyotes and fox were less abundant on *Bromus*-invaded sites than in sagebrush steppe in Utah (Arjo et al. 2007). Like wildlife, livestock will utilize *Bromus*, especially in winter and particularly spring when livestock preferences and weight gain suggest that it has good nutritional value (for *B. tectorum*: Murray and Klemmedson 1968; Murray 1971). While some livestock operations are able to make extensive use of *Bromus-dominated* pastures in spite of mechanical irritants caused by stiff awns, the forage lost in frequent wildfire and additional year(s) of grazing deferment that often follow wildfire detracts considerably from Bromus' appeal for livestock operations (Pyke et al. 2015).

3.10 Management Implications

Assessments of ecosystem impacts of exotic invaders are important for ensuring that resources are allocated toward the most problematic species and that restoration addresses processes that we have the capacity to modify (Hulme et al. 2013). *Bromus* has biological and physical impacts on the environment that justify efforts to control them where they can dominate the plant community. Impacts of *Bromus* on fire regimes and plant community conditions, hydrologic functioning, and soil stability are difficult to mitigate. These impacts all reduce ecosystem productivity and portend a pattern of desertification. However, some of the perceived impacts of *Bromus* are not clearly and scientifically demonstrated or have been demonstrated only under a narrow set of responses (e.g., as described above for infiltration and erosion).

Allowing *Bromus* to reestablish may seem more desirable in some cases than a bare landscape that can lead to wind or water erosion following its eradication (e.g., with post-fire herbicide). Where *Bromus* are dense, restoration of desired plants will likely only be possible following its eradication and clearing of accumulated litter in order to ameliorate the microclimate and the hydrological and biogeochemical conditions required by native or naturalized restoration species (see Monaco et al. 2015). The homogenization and loss of microsite structure with *Bromus* invasion into otherwise patchy perennial communities can reduce the availability of safe sites for establishment of restoration plantings or seedings (Davies et al. 2009), and techniques to replicate or restore this microsite variability may also enhance restoration. Management practices that conserve or reintroduce appropriate root symbionts, such as mycorrhizae, may overcome putative feedbacks that otherwise inhibit native plant recolonization of *Bromus*-invaded sites.

3.11 Research Needs

Studies that identify how relative cover of *Bromus* compared to that of other exotic invaders affect ecosystem structure and function are scarce in the literature but are needed to identify tolerances for *Bromus* abundance in land use planning. For several ecosystem impacts, such as reduced infiltration, controlled studies that manipulate abundances of *Bromus* and evaluate ecosystem outcomes under a range of soil and climate contexts are still needed. Similarly, comparisons of impacts caused by *Bromus* with the impacts of native or naturalized species that otherwise would dominate the ecosystems in the years as well as decades following disturbance (e.g., *P. secunda* and *Artemisia* spp.) would be most meaningful for both assessing impacts and guiding restoration.

Comparing impacts of *Bromus* to other exotics is important considering the emergence and continuing spread of other annual exotic grasses in the Western USA, such as *Taeniatherum caput-medusae* (L.) Nevski (medusahead) and *Ventenata dubia* (Leers) Coss. (North Africa grass) that could occupy sites currently dominated by *Bromus*. Furthermore, comparison of plant communities resulting from management actions targeting *B. tectorum* (e.g., communities dominated by introduced species such as *A. cristatum* used in restoration) with unmanaged *B. tectorum* communities is increasingly important given the past and future extent of these exotic grasses seedings and the difficulty in restoring them to native communities (Davies et al. 2011).

Overall, research is needed that enables *Bromus*-ecosystem relationships to be both generalized and applied to specific sites. Variability among studies of *Bromus* impacts was apparent for key topics such as *Bromus* effects on infiltration and nutrients. The variability is likely due to local site differences and to temporal factors such as weather or time since disturbance or invasion. Transferability of information on *Bromus* impacts can be done more reliably if studies can (1) incorporate environmental gradients that characterize these largely arid and semiarid ecosystems at both local and larger scales and (2) determine mechanisms by which *Bromus* impact their environment across these gradients. Because of the aridity associated with *Bromus*, research needs to include longer time frames to identify how weather modulates *Bromus* impacts and its implications for management. For example, understanding how drought modulates the impact of *Bromus* on soil fertility will help restoration assessment and planning. Broadscale studies that use similar methodology and hold factors constant across the large spatial extents affected by *Bromus* will typically require collaborative approaches and are clearly needed to provide information useful to both ecologists and managers.

Acknowledgments Patti Haggerty assisted with map production for figures. Any use of trade names is for descriptive purposes only and does not imply endorsement by the US Government.

References

Abella SR, Craig DJ, Chiquoine LP et al (2011) Relationships of native desert plants with red brome (*Bromus rubens*): toward identifying invasion-reducing species. Invasive Plant Sci Manag 4:115–124

Adair EC, Burke IC (2010) Plant phenology and life span influence soil pool dynamics: *Bromus tectorum* invasion of perennial C3–C4 grass communities. Plant Soil 335:255–269

Allen E (1988) Some trajectories of succession in Wyoming sagebrush grassland: implications for restoration. In: Allen E (ed) The reconstruction of disturbed arid lands. West Press, Boulder, CO, pp 89–110

Allen EB, Steers RJ, Dickens SJ (2011) Impacts of fire and invasive species on desert soil ecology. Rangeland Ecol Manag 64:450–462

Amatangelo KL, Dukes JS, Field CB (2008) Responses of a California annual grassland to litter manipulation. J Veg Sci 19:605–612

Anderson JE, Inouye RS (2001) Landscape-scale changes in plant species abundance and biodiversity of a sagebrush steppe over 45 years. Ecol Monogr 71:531–556

Arjo WM, Gese EM, Bennett TJ et al (2007) Changes in kit fox-coyote-prey relationships in the Great Basin Desert, Utah. West N Am Naturalist 67:389–401

Austreng AC (2012) The carbon budget impact of sagebrush degradation. MS thesis, Boise State University, Boise, ID

Bachen D (2014) Cheatgrass invasion of sagebrush steppe: impacts of vegetation structure on small mammals. MS thesis, Montana State University, Bozeman, MO

Bagchi S, Briske DD, Bestelmeyer BT et al (2013) Assessing resilience and state-transition models with historical records of cheatgrass *Bromus tectorum* invasion in North American sagebrush-steppe. J Appl Ecol 50:1131–1141

Balch JK, Bradley BA, D'Antonio CM et al (2013) Introduced annual grass increases regional fire activity across the arid western USA (1980–2009). Glob Change Biol 19:173–183

Bangert R, Huntly N (2010) The distribution of native and exotic plants in a naturally fragmented sagebrush-steppe landscape. Biol Invasions 12:1627–1640

Bansal S, Sheley RL, Blank B et al (2014) Plant litter effects on soil nutrient availability and vegetation dynamics: changes that occur when annual grasses invade shrub-steppe communities. Plant Ecol 215:367–378

Belnap J (2003) Factors influencing nitrogen fixation and nitrogen release in biological soil crusts. In: Belnap J, Lange OL (eds) Biological Soil Crusts: Structure, Function, and Management. Springer-Verlag, Berlin, Germany, pp 241–261

Belnap J (2011) Biological phosphorus cycling in dryland regions. In: Bünemann EK, Oberson A, Frossard E (eds) Phosphorus in action: biological processes in soil phosphorus cycling, vol 26. Springer Science and Business Media, New York, NY, pp 371–406

Belnap J, Phillips SL (2001) Soil biota in an ungrazed grassland: response to annual grass (*Bromus tectorum*) invasion. Ecol Appl 11:1261–1275

Belnap J, Phillips SL, Sherrod SK et al (2005) Soil biota can change after exotic plant invasion: does this affect ecosystem processes? Ecology 86:3007–3017

Belnap J, Stark JM, Rau BJ (2015) Soil moisture and biogeochemical factors influence the distribution of annual *Bromus* species. In: Germino MJ, Chambers JC, Brown CS (eds) Exotic brome-grasses in arid and semiarid ecosystems of the Western USA: causes, consequences, and management implications. Springer, New York, NY (Chapter 8)

Billings W (1990) *Bromus tectorum*, a biotic cause of ecosystem impoverishment in the Great Basin. In: Woodwell G (ed) The earth in transition: patterns and processes of biotic impoverishment. Cambridge University Press, Cambridge, pp 301–310

Blank RR (2008) Biogeochemistry of plant Invasion: a case study with downy brome (*Bromus tectorum*). Invasive Plant Sci Manag 1:226–239

Blank RR, Morgan T, Clements CD et al (2013) *Bromus tectorum* L. invasion: changes in soil properties and rates of bioturbation. Soil Sci 178:281–290

Bolton H, Smith JL, Wildung RE (1990) Nitrogen mineralization potentials of shrub-steppe soils with different disturbance histories. Soil Sci Soc Am J 54:887–891

Booth MS, Stark JM, Caldwell MM (2003) Inorganic N turnover and availability in annual- and perennial-dominated soils in a northern Utah shrub-steppe ecosystem. Biogeochemistry 66:311–330

Booth MS, Stark JM, Rastetter E (2005) Controls on nitrogen cycling in terrestrial ecosystems: a synthetic analysis of literature data. Ecol Monogr 75:139–157

Boxell J, Drohan PJ (2009) Surface soil physical and hydrological characteristics in *Bromus tectorum* L. (cheatgrass) versus *Artemisia tridentata* Nutt. (big sagebrush) habitat. Geoderma 149:305–311

Bradley BA, Houghtonw RA, Mustard JF et al (2006) Invasive grass reduces aboveground carbon stocks in shrublands of the Western US. Glob Change Biol 12:1815–1822

Brooks ML (1999) Habitat invasibility and dominance by alien annual plants in the western Mojave Desert. Biol Invasions 1:325–337

Brooks ML (2000) Competition between alien annual grasses and native annual plants in the Mojave Desert. Am Midl Nat 144(1):92–108

Brooks ML, Esque TC (2002) Alien plants and fire in desert tortoise (*Gopherus agassizii*) habitat of the Mojave and Colorado deserts. Chelonian Conserv Biol 4:330–340

Brooks ML, Matchett JR (2003) Plant community patterns in unburned and burned blackbrush (*Coleogyne ramosissima* Torr.) shrublands in the Mojave Desert. West N Am Nat 63:283–298

Brooks ML, Matchett JR (2006) Spatial and temporal patterns of wildfires in the Mojave Desert, 1980–2004. J Arid Environ 67:148–164

Brooks ML, Zouhar K (2008) Plant invasions and fire regimes. In: Zouhar K, Kapler Smith J et al. (eds) Wildland fire in ecosystems: effects of fire on flora. Gen Tech Rep RMRS-GTR-42. USDA, Forest Service, Rocky Mountain Research Station, Fort Collins, CO, pp 33–45

Brooks ML, D'Antonio CM, Richardson DM et al (2004) Effects of invasive alien plants on fire regimes. Bioscience 54:677–688

Brooks ML, Belnap J, Brown CS et al (2015) Exotic annual *Bromus* invasions – comparisons among species and ecoregions in the Western United States. In: Germino MJ, Chambers JC,

Brown CS (eds) Exotic brome-grasses in arid and semiarid ecosystems of the Western USA: causes, consequences and management implications. Springer, New York, NY (Chapter 2)

Busby RR, Passchke MW, Stromberger ME et al (2012) Seasonal variation in arbuscular mycorrhizal fungi root colonization of cheatgrass (*Bromus tectorum*), an invasive winter annual. J Ecosyst Ecogr S8-001

Chambers JC, Roundy BA, Blank RR et al (2007) What makes Great Basin sagebrush ecosystems invasible by *Bromus tectorum*? Ecol Monogr 77:117–145

Chambers JC, Bradley BA, Brown CS et al (2014) Resilience to stress and disturbance, and resistance to *Bromus tectorum* L. invasion in cold desert shrublands of western North America. Ecosystems 17:360–375

Chambers JC, Germino MJ, Belnap J et al (2015) Plant community resistance to invasion by *Bromus* species – the roles of community attributes, *Bromus* interactions with plant communities, and *Bromus* traits. In: Germino MJ, Chambers JC, Brown CS (eds) Exotic brome-grasses in arid and semiarid ecosystems of the Western USA: causes, consequences, and management implications. Springer, New York, NY (Chapter 10)

Cline J, Uresk D, Rickard W (1977) Comparison of soil water used by a sagebrush-bunchgrass and a cheatgrass community. J Range Manag 30:199–201

Connelly J, Rinkes E, Braun C (2011) Characteristics of greater sage-grouse habitats: a landscape species at micro and macro scales. Stud Avian Biol 38:69–83

Corbin J, Dyer A, Seabloom E (2007) California grasslands: ecology and management. In: Stromberg MR, Corbin JD, D'Antonio CM (eds) California grasslands: ecology and management. UC Press, Berkeley, pp 156–168

Cox RD, Preston KL, Johnson RF et al (2014) Influence of landscape-scale variables on vegetation conversion to exotic annual grassland in southern California, USA. Glob Ecol Conserv 2:190–203

Craddock GW, Pearse CK (1938) Surface run-off and erosion on granitic mountain soils of Idaho as influenced by range cover, soil disturbance, slope, and precipitation intensity. USDA 482:1–14

D'Antonio CM, Vitousek PM (1992) Biological invasions by exotic grasses, the grass fire cycle, and global change. Ann Rev Ecol Syst 23:63–87

Davies KW, Nafus AM (2013) Exotic annual grass invasion alters fuel amounts, continuity and moisture content. Int J Wildland Fire 22:353–358

Davies KW, Bates JD, James JJ (2009) Microsite and herbaceous vegetation heterogeneity after burning *Artemisia tridentata* steppe. Oecologia 159:597–606

Davies KW, Boyd CS, Beck JL et al (2011) Saving the sagebrush sea: an ecosystem conservation plan for big sagebrush plant communities. Biol Conserv 144:2573–2584

DeFalco LA, Detling JK, Tracy CR et al (2001) Physiological variation among native and exotic winter annual plants associated with microbiotic crusts in the Mojave Desert. Plant Soil 234:1–14

DeFalco LA, Fernandez GCJ, Nowak RS (2007) Variation in the establishment of a non-native annual grass influences competitive interactions with Mojave Desert perennials. Biol Invasions 9:293–307

Dickens SJM, Allen EB (2014a) Soil nitrogen cycling is resilient to invasive, exotic annuals following restoration of coastal sage scrub. J Arid Environ 110:12–18

Dickens SJM, Allen EB (2014b) Exotic plant invasion alters chaparral ecosystem resistance and resilience pre- and post-wildfire. Biol Invasions 16:1119–1130

Dickens SJM, Allen EB, Santiago LS et al (2013) Exotic annuals alter the variation in coastal sage scrub soil chemical and biological characteristics. Soil Biol Biochem 58:70–81

Dighton J (1983) Phosphatase production by mycorrhizal fungi. Plant Soil 71:455–462

Earnst SL, Holmes AL (2012) Bird-habitat relationships in interior Columbia Basin shrubsteppe. Condor 114:15–29

Egerton-Warburton LM, Allen EB (2000) Shifts in arbuscular mycorrhizal communities along an anthropogenic nitrogen deposition gradient. Ecol Appl 10:484–496

Esque TC (1994) Diet and diet selection of the desert tortoise (*Gopherus agassizii*) in the northeast Mojave Desert. MS thesis, Colorado State University, Fort Collins, CO

Esque TC, Búrquez AM, Schwalbe CR et al (2002) Fire ecology of the Sonoran Desert tortoise. In: Van Devender TR (ed) The Sonoran Desert tortoise: natural history, biology, and conservation. Arizona-Sonora Desert Museum and University of Arizona Press, Tucson, AZ, pp 312–333

Esque TC, Drake KK, Nussear KE (2014) Water and food acquisition and their consequences on life history and metabolism of North American tortoises. In: Rostal D, Mushinsky H, McCoy ED (eds) Ecology of North American tortoises. Johns Hopkins Press, Baltimore, MD, pp 85–96

Evans RD, Belnap J (1999) Long-term consequences of disturbance on nitrogen dynamics in an arid ecosystem. Ecology 80:150–160

Evans RA, Young JA (1970) Plant litter and establishment of alien annual weed species in rangeland communities. Weed Sci 18:697–703

Evans RA, Young JA (1972) Microsite requirements for establishment of annual rangeland weeds. Weed Sci 20:350–356

Evans RD, Rimer R, Sperry L et al (2001) Exotic plant invasion alters nitrogen dynamics in an arid grassland. Ecol Appl 11:1301–1310

Freeman ED, Sharp TR, Larsen RT et al (2014) Negative effects of an exotic grass invasion on small-mammal communities. Plos One 9:e108843

Frossard E, Brossard M, Hedley MJ et al (1991) Reactions controlling the cycling of P in soils. In: Tiessen H (ed) Phosphorus in the global environment: transfers, cycles, and management. Wiley, England, p 325

Gano K, Rickard W (1982) Small mammals of a bitterbrush-cheatgrass community. Northwest Sci 56:1–7

Gasch CK, Enloe SF, Stahl PD et al (2013) An aboveground-belowground assessment of ecosystem properties associated with exotic annual brome invasion. Biol Fertil Soils 49:919–928

Germnio MJ, Chambers JC, Brown CS (2015) Introduction: exotic annual *Bromus* in the Western US. In: Germino MJ, Chambers JC, Brown CS (eds) Exotic brome-grasses in arid and semiarid ecosystems of the Western USA: causes, consequences, and management implications. Springer, New York, NY (Chapter 1)

Gill RA, Burke IC (1999) Ecosystem consequences of plant life form changes at three sites in the semiarid United States. Oecologia 121:551–563

Gopalani S (2004) Changes in soil phosphorus caused by *Bromus tectorum* in four contrasting Mojave Desert soils and in comparison with *Hilaria jamesii* in a Canyonlands soil. Unpub MS thesis, University of Denver, Denver, CO

Griffith AB (2010) Positive effects of native shrubs on *Bromus tectorum* demography. Ecology 91:141–154

Gundale MJ, Sutherland S, DeLuca TH (2008) Fire, native species, and soil resource interactions influence the spatio-temporal invasion pattern of *Bromus tectorum*. Ecography 31:201–210

Hall LK (2012) Effect of cheatgrass on abundance of the North American deermouse (*Peromyscus maniculatus*). Southwest Nat 57:166–169

Hansen KK (1999) Cheatgrass (*Bromus tectorum* L.) invasion in relation to phosphorus sources and availability in Canyonlands National Park, Utah. PhD dissertation, University of Denver, Denver, CO

Hasselquist NJ, Germino MJ, Sankey JB et al (2011) Aeolian nutrient fluxes following wildfire in sagebrush steppe: implications for soil carbon storage. Biogeosciences 8:3649–3659

Hawkes CV, Wren IF, Herman DJ et al (2005) Plant invasion alters nitrogen cycling by modifying the soil nitrifying community. Ecol Lett 8:976–985

Hawkes CV, Belnap J, D'Antonio C et al (2006) Arbuscular mycorrhizal assemblages in native plant roots change in the presence of invasive exotic grasses. Plant Soil 281:369–380

Hironaka M (1961) The relative rate of root development of cheatgrass and medusahead. J Range Manag 14:263–267

Hooker TD, Stark JM (2008) Soil C and N cycling in three semiarid vegetation types: response to an in situ pulse of plant detritus. Soil Biol Biochem 40:2678–2685

Hooker TD, Stark JM, Norton U et al (2008) Distribution of ecosystem C and N within contrasting vegetation types in a semiarid rangeland in the Great Basin, USA. Biogeochemistry 90:291–308
Hoover AN, Germino MJ (2012) A common-garden study of resource-island effects on a native and an exotic, annual grass after fire. Rangel Ecol Manag 65:160–170
Hulbert LC (1955) Ecological studies of *Bromus tectorum* and other annual bromegrasses. Ecol Monogr 25:181–213
Hulme PE, Pysek P, Jarosik V et al (2013) Bias and error in understanding plant invasion impacts. Trends Ecol Evol 28:212–218
Humphrey LD, Schupp EW (2004) Competition as a barrier to establishment of a native perennial grass (*Elymus elymoides*) in alien annual grass (*Bromus tectorum*) communities. J Arid Environ 58:405–422
Jennings WB (2002) Diet selection by the desert tortoise in relation to the flowering phenology of ephemeral plants. Chelonian Conserv Biol 4:353–358
Johnson GD, Boyce MS (1990) Feeding trials with insects in the diet of sage grouse chicks. J Wildlife Manag 54:89–91
Jones R, Chambers JC, Johnson DW et al (2015a) Effect of repeated burning on plant and soil carbon and nitrogen in cheatgrass (*Bromus tectorum*) dominated ecosystems. Plant Soil 386:47–64
Jones R, Chambers JC, Board DI et al (2015b) Understanding the role of resource limitation in restoration of sagebrush ecosystems dominated by cheatgrass – a mechanistic approach. Ecosphere 6:Art107
Jurinak JJ, Dudley LM, Allen MF et al (1986) The role of calcium-oxalate in the availability of phosphorus in soils of semiarid regions – a thermodynamic study. Soil Sci 142:255–261
Kelrick MI, MacMahon JA (1985) Nutritional and physical attributes of seeds of some common sagebrush-steppe plants – some implications for ecological theory and management. J Range Manag 38:65–69
Kleiner EF, Harper K (1977) Occurrence of four major perennial grasses in relation to edaphic factors in a pristine community. J Range Manag 30:286–289
Klemmedson JO, Smith JG (1964) Cheatgrass (*Bromus tectorum* L.). Bot Rev 30:226–262
Knapp PA (1996) Cheatgrass (*Bromus tectorum* L) dominance in the Great Basin Desert – History, persistence, and influences to human activities. Glob Environ Change 6:37–52
Knight WG (1991) Chemistry of arid region soils. In: Skujins J (ed) Semiarid lands and deserts soil resource and reclamation. Marcel Dekker, New York, NY, pp 111–142
Knight W, Dudley L, Jurinak J (1992) Oxalate effects on solution phosphorus in a calcareous soil. Arid Land Res Manag 6:11–20
Lajtha K, Schlesinger WH (1988) The biogeochemistry of phosphorus cycling and phosphorus availability along a desert soil chronosequence. Ecology 69:24–39
Larrison EJ, Johnson DR (1973) Density changes and habitat affinities of rodents of shadscale and sagebrush associations. Great Basin Nat 33:255–264
Liao C, Peng R, Luo Y et al (2008) Altered ecosystem carbon and nitrogen cycles by plant invasion: a meta-analysis. New Phytol 177:706–714
Link SO, Gee GW, Downs JL (1990) The effect of water stress on phenological and ecophysiological characteristics of cheatgrass and Sandberg's bluegrass. J Range Manag 43:506–513
Link SO, Keeler CW, Hill RW et al (2006) *Bromus tectorum* cover mapping and fire risk. Int J Wildland Fire 15:113–119
Mack RN (1981) Invasion of *Bromus tectorum* L. into Western North America: an ecological chronicle. Agro-Ecosystems 7:145–165
McArthur ED, Sanderson SC, Webb BL (1994) Nutritive quality and mineral content of potential desert tortoise food plants. Research Paper INT-473. USDA, Forest Service, Intermountain Research Station, Ogden, UT, p 26
McCrary MD, Bloom PH (1984) Lethal effects of introduced grasses on red-shouldered hawks. J Wildlife Manag 48:1005–1008
Medica PA, Eckert SE (2007) *Gopherus agassizii*, desert tortoise. Food/mechanical Injury. Natural history notes. Herpetol Rev 38:446–448

Miller ME, Belnap J, Beatty SW et al (2006) Performance of *Bromus tectorum* L. in relation to soil properties, water additions, and chemical amendments in calcareous soils of southeastern Utah, USA. Plant Soil 288:1–18

Miller ME, Bowker MA, Reynolds RL et al (2012) Post-fire land treatments and wind erosion – Lessons from the Milford Flat Fire, UT, USA. Aeolian Res 7:29–44

Monaco TA, Hardegree SP, Pellant M et al (2015) Assessing restoration and management needs for ecosystems invaded by exotic annual *Bromus* species. In: Germino MJ, Chambers JC, Brown CS (eds) Exotic brome-grasses in arid and semiarid ecosystems of the Western USA: causes, consequences, and management implications. Springer, New York, NY (Chapter 12)

Morris KA, Stark JM, Bugbee B et al (accepted) Root exudates and senescence of cheatgrass and crested wheatgrass contribute to nitrogen cycling in the rhizosphere. Oecologia

Murray RB (1971) Grazing capacity, sheep gains: cheatgrass, bunchgrass ranges in southern Idaho. J Range Manag 24:407–410

Murray R, Klemmedson J (1968) Cheatgrass range in southern Idaho: seasonal cattle gains and grazing capacities. J Range Manag 21:308–313

Nagy KA, Henen BT, Vyas DB (1998) Nutritional quality of native and introduced food plants of wild desert tortoises. J Herpetol 32:260–267

Newbold TAS, Carpenter GC (2005) Desert horned lizard (Phrynosoma platyrhinos) locomotor performance: the influence of cheatgrass (*Bromus tectorum*). Southwest Nat 50:17–23

Norton JB, Monaco TA, Norton JM et al (2004) Soil morphology and organic matter dynamics under cheatgrass and sagebrush-steppe plant communities. J Arid Environ 57:445–466

Norton U, Mosier A, Morgan J et al (2008) Moisture pulses, trace gas emissions and soil C and N in cheatgrass and native grass-dominated sagebrush-steppe in Wyoming, USA. Soil Biol Biochem 40:1421–1431

Norton U, Saetre P, Hooker TD et al (2012) Vegetation and moisture controls on soil carbon mineralization in semiarid environments. Soil Sci Soc Am J 76:1038–1047

Ogle SM, Ojima D, Reiners WA (2004) Modeling the impact of exotic annual brome grasses on soil organic carbon storage in a northern mixed-grass prairie. Biol Invasions 6:365–377

Ostoja SM, Schupp EW (2009) Conversion of sagebrush shrublands to exotic annual grasslands negatively impacts small mammal communities. Divers Distrib 15:863–870

Ostoja SM, Schupp EW, Sivy K (2009) Ant assemblages in intact big sagebrush and converted cheatgrass-dominated habitats in Tooele County, Utah. West North Ame Nat 69:223–234

Paez RA (2011) Opinion on Adams et al. vs. United States of America and DuPont, United States Court of Appeals for the Ninth Circuit, No. 10-35611, D.C. No. v. 4:03-cv-00049-BLW OPINION, Filed 8 September

Parker SS, Schimel JP (2010) Invasive grasses increase nitrogen availability in California grassland soils. Invasive Plant Sci Manag 3:40–47

Patterson R (1952) The sage grouse in Wyoming. Sage Books, Denver, CO

Peek MS, Leffler AJ, Ivans CY et al (2005) Fine root distribution and persistence under field conditions of three co-occurring Great Basin species of different life form. New Phytol 165:171–180

Piemeisel RL (1951) Causes affecting change and rate of change in a vegetation of annuals in Idaho. Ecology 32:53–72

Prater MR, DeLucia EH (2006) Non-native grasses alter evapotranspiration and energy balance in Great Basin sagebrush communities. Agr Forest Meteorol 139:154–163

Prater MR, Obrist D, Arnone JA et al (2006) Net carbon exchange and evapotranspiration in post-fire and intact sagebrush communities in the Great Basin. Oecologia 146:595–607

Prevey JS, Germino MJ, Huntly NJ et al (2010) Exotic plants increase and native plants decrease with loss of foundation species in sagebrush steppe. Plant Ecol 207:39–51

Pyke DA, Beck JL, Brooks ML et al (2015) Land uses, fire, and invasion – exotic annual *Bromus* and human dimensions. In: Germino MJ, Chambers JC, Brown CS (eds) Exotic brome-grasses in arid and semiarid ecosystems of the Western USA: causes, consequences, and management implications. Springer, New York, NY (Chapter 11)

Rao LE, Matchett JR, Brooks ML et al (2015) Relationships between annual plant productivity, nitrogen deposition and fire size in low-elevation California desert scrub. Int J Wildland Fire 24:48–58

Rau BM, Johnson DW, Blank RR et al (2011) Transition from Sagebrush steppe to annual Grass (*Bromus tectorum*): influence on belowground carbon and nitrogen. Rangel Ecol Manag 64:139–147

Reever Morghan KJ, Corbin JD, Gerlach J (2007) Water relations. In: Stromberg MR, Corbin JD, D'Antonio CM (eds) California grasslands: ecology and management. UC Press, Berkeley, CA (Chapter 7)

Reisner MD, Grace JB, Pyke DA et al (2013) Conditions favouring *Bromus tectorum* dominance of endangered sagebrush steppe ecosystems. J Appl Ecol 50:1039–1049

Richardson KA, West SD, Gitzen RA (2013) Cheatgrass (*Bromus tectorum*) dominates cheek pouch contents of the great basin pocket mouse (*Perognathus parvus*). West N Am Nat 73:158–167

Rieder JP, Newbold TAS, Ostoja SM (2010) Structural changes in vegetation coincident with annual grass invasion negatively impacts sprint velocity of small vertebrates. Biol Invasions 12:2429–2439

Rimer RL, Evans RD (2006) Invasion of downy brome (*Bromus tectorum* L.) causes rapid changes in the nitrogen cycle. Am Midl Nat 156:252–258

Robertson IC, Klemash D (2003) Insect mediated pollination in slickspot peppergrass, *Lepidium papilliferum* L. (Brassicaceae), and its implications for population viability. West N Am Nat 63:333–342

Rowe HI, Brown CS (2008) Native plant growth and seedling establishment in soils influenced by *Bromus tectorum*. Rangel Ecol Manag 61:630–639

Rowland RH, Turner FB (1964) Correlation of the local distributions of *Dipodomys microps* and *D. merriami* and of the annual grass *Bromus rubens*. Southwest Nat 9:56–61

Ryel RJ, Ivans CY, Peek MS et al (2008) Functional differences in soil water pools: a new perspective on plant water use in water-limited ecosystems. In: Progress in botany, vol 69. Springer, pp 397–422

Saetre P, Stark JM (2005) Microbial dynamics and carbon and nitrogen cycling following re-wetting of soils beneath two semi-arid plant species. Oecologia 142:247–260

Salo LF, McPherson GR, Williams DG (2005) Sonoran Desert winter annuals affected by density of red brome and soil nitrogen. Am Midl Nat 153:95–109

Sankey JB, Glenn NF, Germino MJ et al (2010) Relationships of aeolian erosion and deposition with LiDAR-derived landscape surface roughness following wildfire. Geomorphology 119:135–145

Schaeffer SM, Ziegler SE, Belnap J et al (2012) Effects of *Bromus tectorum* invasion on microbial carbon and nitrogen cycling in two adjacent undisturbed arid grassland communities. Biogeochemistry 111:427–441

Schroeder MA, Young JR, Braun CE (1999) Sage grouse (*Centrocercus urophasianus*). In: Poole A, Gill F (eds) The Birds of North America, vol 425. Birds of North America, Philadelphia, PA, p 28

Sigueenza C, Corkidi L, Allen EB (2006) Feedbacks of soil inoculum of mycorrhizal fungi altered by N deposition on the growth of a native shrub and an invasive annual grass. Plant Soil 286:153–165

Singer MJ, Shainberg I (2004) Mineral soil surface crusts and wind and water erosion. Earth Surf Process 29:1065–1075

Smart DR, Stark JM, Diego V (1999) Resource limitations to nitric oxide emissions from a sagebrush-steppe ecosystem. Biogeochemistry 47:63–86

Sperry LJ, Belnap J, Evans RD (2006) *Bromus tectorum* invasion alters nitrogen dynamics in an undisturbed arid grassland ecosystem. Ecology 87:603–615

Stark JM, Hart SC (1997) High rates of nitrification and nitrate turnover in undisturbed coniferous forests. Nature 385:61–64

Stark JM, Norton JM (2015) The invasive annual cheatgrass increases nitrogen availability in 24-year-old replicated field plots. Oecologia 177:799–809

Stark JM, Hooker T, Saetre P et al (2006) Changes in soil nitrogen and carbon cycling in big sagebrush ecosystems following exotic grass introduction and invasion. Ecol Soc Am Abstracts Merida, Mexico

Steenhof K, Yensen E, Kochert MN et al (2006) Populations and habitat relationships of Piute ground squirrels in southwestern Idaho. West N Am Nat 66:482–491

Stewart G, Hull A (1949) Cheatgrass (*Bromus tectorum* L.) – an ecologic intruder in southern Idaho. Ecology 30:58–74

Svejcar T, Sheley R (2001) Nitrogen dynamics in perennial- and annual-dominated arid rangeland. J Arid Environ 47:33–46

Turkowski FJ (1975) Dietary adaptability of the desert cottontail. J Wildlife Manag 39:748–756

Van Horne B, Schooley RL, Sharpe PB (1998) Influence of habitat, sex, age, and drought on the diet of Townsend's ground squirrels. J Mammal 79:521–537

VanHorne B, Olson GS, Schooley RL et al (1997) Effects of drought and prolonged winter on Townsend's ground squirrel demography in shrubsteppe habitats. Ecol Monogr 67:295–315

Verburg PSJ, Arnone JA, Obrist D et al (2004) Net ecosystem carbon exchange in two experimental grassland ecosystems. Glob Change Biol 10:498–508

Wagenbrenner NS, Germino MJ, Lamb BK et al (2013) Wind erosion from a sagebrush steppe burned by wildfire: measurements of PM10 and total horizontal sediment flux. Aeolian Res 10:25–36

Wallestad R, Eng RL (1975) Foods of adult sage grouse in central Montana. J Wildlife Manag 39:628–630

Weber CF, King GM, Aho K (2015) Relative abundance of and composition within fungal orders differ between cheatgrass (*Bromus tectorum*) and sagebrush (*Artemisia tridentata*)-associated soils. Plos One 10:e0117026

West NE, Yorks TP (2006) Long-term interactions of climate, productivity, species richness, and growth form in relictual sagebrush steppe plant communities. West N Am Nat 66:502–526

Whisenant SG (1990) Changing fire frequencies on Idaho's Snake River Plains: ecological and management implications. In: McArthur ED, Romney EM, Smith SD et al (eds) Symposium on cheatgrass invasion, shrub die-off, and other aspects of shrub biology and management, 5–7 April 1989. Gen Tech Rep INT-276. USDA, Forest Service, Intermountain Research Station, Ogden, UT, pp 4–10

Wilcox BP, Turnbull L, Young MH et al (2012) Invasion of shrublands by exotic grasses: ecohydrological consequences in cold versus warm deserts. Ecohydrology 5:160–173

US Fish and Wildlife Service (2003) Recovery plan for the quino checkerspot butterfly (*Euphydryas editha quino*). www.fws.gov/Carlsbad/SpeciesStatusList/RP/20030811_RP_QCB.pdf

Wolkovich EM, Bolger DT, Cottingham KL (2009) Invasive grass litter facilitates native shrubs through abiotic effects. J Veg Sci 20:1121–1132

Wood YA, Meixner T, Shouse PJ et al (2006) Altered ecohydrologic response drives native shrub loss under conditions of elevated nitrogen deposition. J Environ Qual 35:76–92

Wylie B (2012) Mapping cheatgrass dieoff in the Northern Great Basin using ecosystem performance modeling. Final report to Northwest Climate Science Center. https://www.sciencebase.gov/catalog/item/50477198e4b067bd38f7f4fb

Yensen E, Quinney DL, Johnson K et al (1992) Fire, vegetation changes, and population fluctuations of Townsend's ground squirrels. Am Midl Nat 128:299–312

Zouhar K (2003) *Bromus tectorum*. In: Fire Effects Information System [Online]. USDA, Forest Service, Rocky Mountain Research Station, Fire Sciences Laboratory. http://www.fs.fed.us/database/feis/

Part II
Invasiveness of *Bromus* Species (Emphasis on Biological Attributes of *Bromus*)

Chapter 4
Mating System, Introduction and Genetic Diversity of *Bromus tectorum* in North America, the Most Notorious Product of Evolution Within *Bromus* Section *Genea*

Stephen J. Novak and Richard N. Mack

Abstract The genus *Bromus* (Poaceae) consists of approximately 150–160 species that occur in temperate regions worldwide. Members of section *Genea* in *Bromus* are native to Eurasia, and most are invasive in North America (NA). Most environmentally and economically important among these species is *Bromus tectorum*, which along with *Bromus rubens* and *B. diandrus* are some of the most destructive invasive grasses in the Western United States (USA). Members of section *Genea* form a polyploid series. Molecular data suggest a reticulate speciation pattern (a netlike pattern of diversification) for the polyploid members of *Genea*. Seed set, floral characteristics, sex allocation patterns, and genetic markers reveal that *B. tectorum* is predominately self-pollinating. Historical records and genetic data indicate that *B. tectorum* entered repeatedly into NA and that its current distribution can be attributed to multiple phenomena, including direct introduction from its native range and multidirectional range expansion. Across invasive populations of *B. tectorum*, overall genetic diversity (e.g., allelic richness and number of polymorphic loci) is lower compared to the diversity among native populations. Conversely, invasive populations of *B. tectorum* exhibit, on average, higher within-population genetic diversity than found in native populations, suggesting that multiple introductions have partially offset founder effects within invasive populations. Additionally, invasive populations appear to be genetic admixtures composed of two or more independently derived native genotypes. These admixtures have produced invasive populations with generally less genetic differentiation than native populations. The genetic consequences of multiple introductions have implications for the management of invasive populations of *B. tectorum*, especially in Western NA.

S.J. Novak (✉)
Department of Biological Sciences, Boise State University, Boise, ID 83725-1515, USA
e-mail: snovak@boisestate.edu

R.N. Mack
School of Biological Sciences, Washington State University, Pullman, WA 99164, USA
e-mail: rmack@wsu.edu

M.J. Germino et al. (eds.), *Exotic Brome-Grasses in Arid and Semiarid Ecosystems of the Western US*, Springer Series on Environmental Management,
DOI 10.1007/978-3-319-24930-8_4

Keywords Classical biological control • Enzyme electrophoresis (allozymes) • Founder effects • Genetic admixture • Herbarium specimens • Multiple introductions • Multidirectional range expansion • Reticulate speciation pattern • Self-pollinating mating system • Source populations

4.1 Introduction

Some members of the Poaceae rank among the worst plant invaders, especially in steppe, shrub-steppe, and savannas (Holm et al. 1977; Mack 1989); furthermore, some invasive grasses have become prominent where the forest canopy has been removed, including low-elevation coniferous forest (Sumners and Archibold 2007; D'Antonio et al. 2011). The array of grasses that have become invasive (or at least extensively naturalized) defies straightforward taxonomic classification, but multiple species in *Avena*, *Brachiaria*, *Eragrostis*, *Nassella*, *Pennisetum*, and *Poa* have become major pests in their new ranges (Holm et al. 1977; Parsons and Cuthbertson 1992; Bromilow 1995). Whether outcompeting native species or causing devastating alterations of the fire cycle within invaded communities (D'Antonio and Vitousek 1992), these grass immigrants collectively cause enormous environmental and economic damage worldwide (Pimentel et al. 2005). Among any list of destructive grasses, few genera match the ecological breadth and resulting spread seen for many members of the genus *Bromus* L. (Germino et al. 2015; Brooks et al. 2015).

The introduction and range expansion of species in the genus *Bromus* (brome-grasses) across Western North America (NA) provide abundant examples of the negative consequences of plant invaders. For example, *Bromus rubens* L. (red brome) is the dominant invasive brome-grass in the Southwestern United States (USA) (Beatley 1966; Brooks 1999a, b; Salo 2005), and *Bromus diandrus* Roth. (ripgut brome) and *Bromus hordeaceus* L. (soft brome or chess) are invasive in California, especially the Central Valley (Dyer and Rice 1999; Gerlach and Rice 2003; Keeley et al. 2006; Brooks et al. 2015). In addition, the invasion of *B. tectorum* L. in sagebrush-steppe habitats of the Intermountain West (the Columbia Basin, especially the Snake River Plains, Basin and Range Province, and the Colorado Plateau) has contributed to the loss of native species diversity and community structure, modified ecosystem processes, such as nutrient cycling and productivity patterns, and altered disturbance regimes, especially the frequency, extent, and intensity of rangeland fires (Stewart and Hull 1949; Hulbert 1955; D'Antonio and Vitousek 1992; Knaap 1996; Sperry et al. 2006; Chambers et al. 2007; Mack 2011; Germino et al. 2015).

Three of the four species listed above, *B. diandrus*, *B. rubens*, and *B. tectorum*, as well as several other brome-grasses invasive in NA, occur in section *Genea* of the genus *Bromus* (Sales 1993, 1994). Consequently, we begin this chapter by describing evolutionary relationships among species in section *Genea*, which includes a polyploid series with diploid, tetraploid, hexaploid, and octoploid species. *Bromus tectorum* (cheatgrass, downy brome) will however be our main focus, given its enormous impacts as an invasive species in temperate grasslands in NA. We (1) describe the mating system of *B. tectorum*, based on seed set, floral characteristics, sex allocation patterns, and genetic markers, (2) present historical and genetic data to reconstruct the introduction and pattern of range expansion of *B. tectorum* in NA, and (3) assess the genetic consequences of these events by summarizing studies describing the level and structure of genetic diversity within and among native and invasive populations of *B. tectorum*. The results discussed here indicate that invasive populations of *B. tectorum* throughout much of NA have, on average, higher levels of genetic diversity than native populations, due to multiple introductions. Furthermore, the genetic consequences of multiple introductions have implications for the management of invasive populations of *B. tectorum*, especially in Western NA.

4.2 Evolutionary Relationships among Species in Section *Genea*

A context for assessing life history traits, ecology, and the potential for invasiveness of *B. tectorum*, and other species within section *Genea*, is provided by considering the species' evolutionary history, phylogenetic relationships, and patterns of speciation (see Prinzing et al. 2002 for an interesting discussion of these issues). Such an assessment is particularly useful for the study of diploid species within this section and their roles in the origins of the polyploid taxa of this section.

Within the Poaceae, *Bromus* is classified in the tribe *Bromeae*, which is part of subfamily Pooideae (Catalan et al. 1997; Kellog 2001; Bouchenak-Khelladi et al. 2008; Schneider et al. 2009; Atkinson and Brown 2015). The genus *Bromus* consists of approximately 150–160 annual and perennial species; these species collectively have a worldwide distribution (Smith 1970; Acedo and Liamas 2001; Fortune et al. 2008). Smith (1970) assigned the species of *Bromus* to six sections, *Bromus*, *Ceratochloa* (P. Beauv.) Griseb., *Genea* Dumort., *Neobromus* Shear., *Nevskiella* V. Krecz and Vvevd., and *Pnigma* Dumort. Species within section *Genea* are hypothesized to have originated in the Middle East and the Mediterranean Region (Sales 1993, 1994), where they likely evolved concomitantly with the emergence of agriculture and human disturbance (Balfourier et al. 1998). In addition, these species exhibit much morphological variation, phenotypic plasticity, and polyploidy (Sales 1993, 1994). Seven species that form a polyploid series are recognized in section *Genea*: the diploids *B. fasciculatus* Presl., *B. sterilis* L., and *B. tectorum*, the tetraploids *B. madritensis* L. and *B. rubens*, the hexaploid *B. rigidus* Roth., and the octoploid *B. diandrus* Roth. (Smith 1970; Sales 1993, 1994; Acedo and Liamas 2001; Fortune et al. 2008). These species are native to Eurasia; all are invasive in the USA, with the exception of *B. fasciculatus*.

Fortune et al. (2008) elegantly assessed evolutionary relationships and patterns of speciation among these seven species. DNA sequence data from three gene regions were employed: the low copy nuclear gene *Waxy*, nuclear ribosomal DNA (rDNA) regions ITS1 and ITS2, and the chloroplast spacer regions *trnT-trnL* and *trnL-trnF*. Results from nuclear ITS regions and chloroplast DNA indicate that section *Genea* is monophyletic. The phylogenetic tree based on nucleotide sequence data from *Waxy* however indicates that section *Genea* is paraphyletic because this clade also includes species from section *Bromus* (Fortune et al. 2008). Two of the diploid species within section *Genea*, *B. sterilis* and *B. tectorum*, are sister species (i.e., they are more closely related to each other than any other species in this section); *B. fasciculatus* is more distantly related. Based on chloroplast DNA data, *B. fasciculatus* appears to have served as the maternal parent for the hybridization that produced the tetraploids *B. madritensis* and *B. rubens*, whereas the paternal parents for the two tetraploids appear to be *B. sterilis* and *B. tectorum*, respectively (Fig. 4.1). The hexaploid *B. rigidus* may have originated from a hybridization event involving *B. rubens* and an unknown diploid species, which served as the maternal parent (Fig. 4.1). The octoploid *B. diandrus* appears to have originated from a crossing event involving *B. rigidus* and *B. sterilis* as the maternal and paternal parents, respectively (Fig. 4.1).

The above summarized results point to the reticulate (i.e., a netlike pattern of diversification) origins of the polyploid species in section *Genea* (Fortune et al. 2008). In addition, these data indicate that all seven species in this section have unique evolutionary histories. Thus, all four polyploid species in section *Genea* (*B. madritensis*, *B. rubens*, *B. rigidus*, and *B. diandrus*) are appropriately recognized as distinct, separate species. With the exception of *B. fasciculatus*, which is listed as endangered on Mediterranean islands (Fortune et al. 2008), the remaining six species in section *Genea* are considered weeds across much of their native range

Species	Ploidy	Chlorotype	ITS	*Waxy* 1	*Waxy* 2	*Waxy* 3	*Waxy* 4
B. madritensis	4*x*	F	F	S	F	--	--
B. rubens	4*x*	F	T	T	F	--	--
B. rigidus	6*x*	X?	X?	T	F	X	--
B. diandrus	8*x*	X?	X?	T	F	X	S

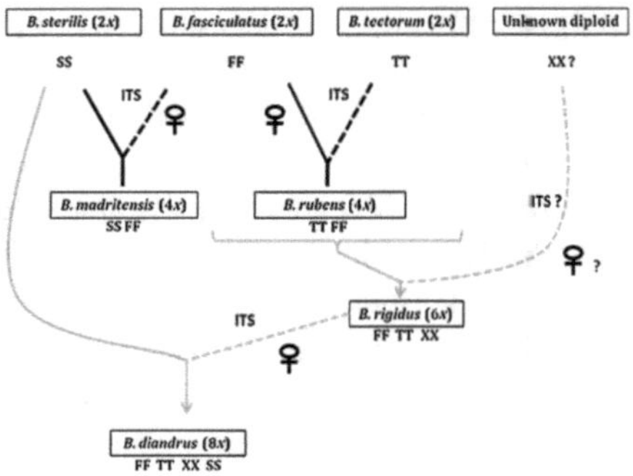

Fig. 4.1 Depiction of hypothesized origins of the polyploid species in section *Genea* of the genus *Bromus*, reproduced from Fortune et al. (2008). Molecular markers used in this analysis are listed in the table and include chlorotype (chloroplast spacer regions *trnT-trnL* and *trnL-trnF*), ITS (nuclear rDNA regions ITS1 and ITS2), and four *Waxy* (a low copy nuclear gene) nucleotide sequences. The table provides the identity of the genomes (F, S, T, and X) detected in the polyploid taxa based on different hybridization events. F=*B. fasciculatus*, S=*B. sterilis*, T=*B. tectorum*, and X=an unknown diploid species. The maternal parent in each cross is indicated (female symbol). See the text for a summary of the results shown in this figure

(Smith 1980). Traits that contributed to their weediness in the native range likely served as preadaptations (*sensu* Futuyma 2005), contributing to their establishment and subsequent invasions, especially in Western NA.

4.3 Mating Systems

In his classic paper "The evolution of weeds," Baker (1974) described the characteristics of the "ideal weed," and whereas no weed species exhibits all these characteristics, *B. tectorum* displays many. We concentrate here on the fifth characteristic in Baker's list, which describes a weed's potential mating system, "Self-compatible but not completely autogamous or apomictic." The mating system (how much outcrossing versus self-pollination occurs) has a large influence on the biology and genetics of plant populations. The outcome of reproduction is directly associated with the mating system; plants with high rates of self-pollination (hereafter referred to as selfing) typically have greater seed set compared with plants with an outcrossing mating system (Harmon-Threatt et al. 2009; Eckert et al. 2010). In addition, plants with a selfing mating system have lower levels of gene flow, through the movement of pollen, within and among populations. Thus, plant mating systems govern the spatial and temporal fluctuation of genotypes within and among populations: populations with high rates of selfing typically contain few individuals with heterozygous genotypes and many individuals with homozygous genotypes, compared to populations with high rates of outcrossing (Stebbins 1957; Allard et al. 1968; Barrett et al. 2008). Furthermore, species with high rates of selfing have lower levels of genetic diversity within populations and higher levels of genetic differentiation among populations, compared to species with high rates of outcrossing (Brown and Burdon 1987; Slatkin and Barton 1989; Ellstrand and Elam 1993; Sork et al. 1999).

4.3.1 *Mating System of* Bromus tectorum*: Seed Set, Floral Characteristics, and Sex Allocation Pattern*

McKone (1985, 1987, 1989) compared the mating systems of five *Bromus* species: *B. inermis* Leyss., *B. latiglumis* (Shear) Hitchc., B. *ciliatus* L., *B. kalmii* A. Gray, and *B. tectorum*. These grasses encompass a range of characteristics; they include species that are native and introduced in NA, perennials and annuals, rhizomatous and non-rhizomatous species, and species with diploid and polyploid cytotypes (see McKone 1985). In order to determine self-fertility, McKone (1985) measured the seed set of bagged inflorescences of all five species. He also documented floral characteristics, including the manner in which the florets of each species opened and the extent to which anthers were exserted. Both the bagged and open-pollinated inflorescences of *B. tectorum* had the highest seed set (McKone 1985). These results are consistent with previous reports for highly selfing plant species (Stebbins 1957) and indicate that *B. tectorum* is self-compatible and highly selfing. In addition, McKone (1985) made observations on floral characteristics of all five *Bromus* species and found that the florets of *B. tectorum* "…rarely (if ever) opened…" (p. 1336), and he "…never found anthers completely exserted from the florets" (p. 1335). These results indicate that *B. tectorum* produces cleistogamous flowers (i.e., they rarely open), which is associated with the production of progeny almost exclusively through selfing (see Campbell et al. 1983).

Sex allocation theory predicts equal paternal (pollen production) and maternal investment (seed production) for hermaphroditic plants (Charlesworth and Charlesworth 1981; Brunet 1992; Campbell 2000). McKone (1987) determined the relative paternal and maternal investment among the same five *Bromus* species listed above by measuring the ratio of pollen to seed production. He also determined the species' absolute paternal and maternal investment by measuring the caloric energy and mineral content (nitrogen, phosphorous, potassium, magnesium, and calcium) of their reproductive structures. Of the five species, *B. tectorum* has the shortest anther length (0.65 mm), smallest pollen diameter (32.43 μm), least volume of pollen per floret ($cm^3 \times 10^{-6}$), intermediate seed weight (1.99 mg), largest value for percent seed set per floret (79.0 %), and smallest ln (ratio of pollen volume to seed weight per floret) (2.28), compared with values for the obligate outcrosser *B. inermis* (4.02 mm, 41.61 μm, 424.1 $cm^3 \times 10^{-6}$, 2.32 mg, 29.9 %, and 6.31, respectively) (McKone 1987). Estimates of sex allocation based on caloric energy and mineral content show that the paternal investment of *B. tectorum* was < 2 %, compared to approximately 50 % for *B. inermis*. Additionally, the cleistogamous flowers of *B. tectorum* produce less pollen relative to number of seeds produced, especially compared to *B. inermis* (McKone 1989). *Bromus tectorum* invests little in pollen production, results that are consistent with the ratio of paternal to maternal investment expected for a highly selfing plant species (Brunet 1992; Campbell 2000).

4.3.2 Mating System of **Bromus tectorum***: Genetic Markers and Outcrossing Rates*

A species' mating system can also be estimated using molecular markers that are expressed through codominance, such as allozymes (enzyme electrophoresis) or microsatellite DNA (Avise 2004; Schlotterer 2004). Using codominant genetic markers, individuals with homozygous and heterozygous genotypes can be clearly identified based on their banding or chromatogram phenotypes (i.e., patterns). Two methods are commonly used to estimate mating system parameters using molecular data: progeny array analysis and Wright's Coefficient of Inbreeding (Wright 1931; Clegg 1980; Brown 1990; Dudash and Fenster 2001). With progeny array analyses, the genotypes of 10–20 seeds from a maternal plant are determined; mating system parameters (outcrossing rate *t* and selfing rate *s*) are estimated from these data,

Table 4.1 Summary of genetic diversity for *Bromus tectorum* populations in the introduced range in North America and the native range in Europe and Southwest Asia

Region	Number of pops.	N	A	$\%P$	H_{exp}	H_{obs}	H_T	G_{ST}
Eastern USA[a]	38	1248	1.01	1.05	0.002	0.0000	0.075	0.560
Central USA[b]	60	1905	1.05	4.73	0.014	0.00003	0.187	0.582
Mid-continent USA[c]	54	1624	1.04	4.07	0.009	0.0002	0.084	0.290
Intermountain West USA[d]	26	961	1.05	4.31	0.011	0.0000	0.119	0.241
Nevada and California[d]	14	546	1.05	5.14	0.014	0.0000	0.192	0.500
California and American Southwest[e]	60	1974	1.03	3.27	0.007	0.000017	0.057	0.277
Eastern Canada[f]	16	495	1.05	5.00	0.013	0.0003	0.265	0.447
Western Canada[f]	44	1397	1.04	3.91	0.012	0.00005	0.131	0.316
Europe[g]	39	1246	1.01	1.64	0.006	0.0001	0.069	0.656
SW Asia[g]	12	484	1.03	3.67	0.005	0.002	0.201	0.735

N is number of individuals analyzed in each region, A is number of alleles per locus, $\%P$ is percent polymorphic loci, H_{exp} is expected mean heterozygosity, H_{obs} is mean observed heterozygosity, H_T is mean total gene diversity, and G_{ST} is proportion of total gene diversity partitioned among populations

[a]Bartlett et al. (2002); [b]Huttanus et al. (2011); [c]Schachner et al. (2008); [d]Novak et al. (1991); [e]Pawlak et al. (2015); [f]Valliant et al. (2007); [g]Novak and Mack (1993) as cited in Huttanus et al. (2011)

typically using a mixed mating system model that assumes both outcrossing and selfing are occurring (Ritland 2002). The Coefficient of Inbreeding, F, can be calculated from population genetic data based on the ratio of mean observed heterozygosity (H_{obs}) to expected mean heterozygosity (H_{exp}), using the equation $F = 1 - H_{obs}/H_{exp}$ (Wright 1931; Dudash and Fenster 2001). The outcrossing rate is calculated as $t_f = (1 - F)/(1 + F)$, and the selfing rate is calculated as $t_s = 1 - t_f$.

Along with our students, we have assessed the genetic diversity within and among populations of *B. tectorum*, using enzyme electrophoresis (Novak et al. 1991; Novak and Mack 1993; Bartlett et al. 2002; Valliant et al. 2007; Schachner et al. 2008; Huttanus et al. 2011; Pawlak et al. 2015). To date, we have analyzed 10,150 individuals from 312 naturalized or invasive populations in NA and 1730 individuals from 51 native populations in Eurasia (Table 4.1), with the genetic diversity of each individual assayed at 25 putative enzyme loci. Heterozygous individuals have been detected (reported as H_{obs} values) within populations from the Central USA, the mid-continent USA, California and the American Southwest, Eastern Canada, Western Canada, Europe, and Southwest Asia (Table 4.1).

Using data from these population genetic surveys, we have estimated mating system parameters (outcrossing rate, t_f, and selfing rate, t_s) for populations of *B. tectorum*, employing Wright's Coefficient of Inbreeding (Table 4.2). These results indicate that for six of the seven regions analyzed, the average outcrossing rate per population ranged from 0.11 % to 1.17 %, and the average selfing rate per population ranged from 98.83 % to 99.89 % (Table 4.2). Estimates of the mating system parameters for

Table 4.2 Percent outcrossing and selfing among populations of *Bromus tectorum* calculated using the Coefficient of Inbreeding (F) method (Wright 1931; Dudash and Fenster 2001). Rates of outcrossing are low for populations from most regions, with the exception of Southwestern Asia

Region	Number of pops.	H_{exp}	H_{obs}	F	t_f, outcrossing rate (%)	t_s, selfing rate (%)
Central USA[a]	60	0.014	0.00003	0.998	0.11	99.89
Mid-continent USA[b]	54	0.009	0.0002	0.978	1.12	98.88
California and American Southwest[c]	60	0.007	0.000017	0.998	0.12	99.88
Eastern Canada[d]	16	0.013	0.0003	0.977	1.17	98.83
Western Canada[d]	44	0.012	0.00005	0.996	0.21	99.79
Europe[e]	39	0.006	0.0001	0.983	0.84	99.16
SW Asia[e]	12	0.005	0.002	0.600	25.00	75.00

H_{exp} is expected mean heterozygosity, and H_{obs} is mean observed heterozygosity.
[a]Huttanus et al. (2011); [b]Schachner et al. (2008); [c]Pawlak et al. (2015); [d]Valliant et al. (2007); [e]Novak and Mack (1993) as cited in Huttanus et al. (2011)

populations from Southwest Asia are in striking contrast; populations from this region had an outcrossing rate of 25 % and a selfing rate of 75 %. This outcome probably arose because low-frequency alleles were only detected in individuals with heterozygous genotypes in several populations in Southwest Asia (Novak and Mack 1993); this result increased the value of H_{obs} relative to that of H_{exp}.

The estimated outcrossing rates (t) published by Meyer et al. (2013) are remarkably similar to the rates reported for six of the seven regions listed in Table 4.2. Meyer et al. (2013) estimated outcrossing rates for four "wild" populations of *B. tectorum* from the Western USA, which ranged from 0.0027 to 0.0133 (i.e., 0.27 % to 1.33 %, with selfing rates that ranged from 98.67 % to 99.73 %). Similar results (t=0.0082, or 0.82 %) were obtained with analysis of individuals grown in a common garden (Meyer et al. 2013). The values of H_{obs} reported in Table 4.2 are generally consistent with observed heterozygosity values previously reported for *B. tectorum* by Ramakrishnan et al. (2006), Kao et al. (2008), Leger et al. (2009), and Scott et al. (2010). In contrast, the high levels of observed heterozygosity reported by Ashley and Longland (2007, 2009) in two of four populations of *B. tectorum* from Northern Nevada are inconsistent with the values reported by other investigators. For instance, using seven microsatellite DNA loci, Ashley and Longland (2009) report an observed heterozygosity value of 0.06 for the Peavine Mountain population. This value is approximately 35- to 54-fold larger than values reported by Meyer et al. (2013) for the Lower Peavine (0.0017) and Upper Peavine (0.0011) populations, even though Meyer et al. (2013) used a more polymorphic marker system, 91–93 single nucleotide polymorphisms (SNPs). This wide discrepancy among these results for populations of *B. tectorum* from a similar locality (Peavine Mountain) remains unresolved.

The preponderance of data indicates that *B. tectorum* reproduction is predominantly through selfing (as reported by McKone 1985); outcrossing rarely occurs within populations from NA. High selfing rates also appear to occur in European populations of *B. tectorum* (Table 4.2); consequently, this trait may form a preadaptation that has contributed to the establishment and invasion of *B. tectorum* in NA. Rare outcrossing events in populations that are comprised of genetic admixtures may generate novel, recombinant genotypes (Novak 2007). Such events may indicate post-immigration evolution (Maron et al. 2004; Lavergne and Molofsky 2007; Schierenbeck and Ellstrand 2009; Xu et al. 2010). Reproductive plasticity in *B. tectorum* (see Meyer et al. 2013) could in turn enhance the species' invasion in NA (Huttanus et al. 2011). Whether or not post-immigration evolution is occurring within populations of invasive species is not solely an intriguing theoretical question; it also holds important implications for the management of invasions (Allendorf and Lundquist 2003; Sakai et al. 2001; Muller-Scharer et al. 2004; Le Roux and Wieczorek 2009).

4.4 Biological Invasions: Propagule Pressure, Preadaptation, and Post-immigration Evolution

Invasive species can have profoundly negative ecological, conservation, economic, and health consequences (D'Antonio and Vitousek 1992; Mack et al. 2000; Sala et al. 2000; Pimentel et al. 2005; Mack and Smith 2011) and have consequently been a focus of much deserved research. Identifying which species will become invasive and which communities are susceptible to invasion however remains elusive (Novak 2007), and our ability to predict which introduced species will become established (and potentially invasive) will likely be facilitated by gaining a better understanding of the invasion process (Estoup and Guillemaud 2010). Consequently, propagule pressure, defined as the number of individuals transported to and introduced into a new range (Lockwood et al. 2005), has emerged as one predictor of naturalization and the likelihood of invasion (Kolar and Lodge 2001; Lockwood et al. 2005; Colautti and MacIsaac 2004; von Holle and Simberloff 2005; Colautti et al. 2006; Simberloff 2009; Ricciardi et al. 2011). Simberloff (2009) described three components of propagule pressure: (1) propagule size, the number of founder individuals in a propagule; (2) propagule number, the rate at which propagules arrive per unit time; and (3) both propagule size and propagule number. Humans' pervasive role as dispersal agents certainly increases the probability of high propagule pressure. With high propagule pressure, alien species are buffered from negative stochastic events during all phases of an invasion (Mack 2000), and large founder populations or multiple introduction events or both could characterize the release of an alien species in a new range.

Propagule pressure also has genetic consequences (Simberloff 2009; Huttanus et al. 2011): with high propagule pressure (e.g., multiple introductions), the overall genetic and phenotypic diversity of a species would likely be increased in its new range, and the potential for severe founder effects would likely decrease (Novak and Mack 2005; Wares et al. 2005; Durka et al. 2005; Lavergne and Molofsky 2007; Dlugosch and Parker 2008). In addition, high propagule pressure may lead to the formation of introduced populations that are admixtures, i.e., introduced populations containing the genetic information of multiple native populations (Kolbe et al. 2007). Propagule pressure can also influence the likelihood of an invasion through the introduction of preadapted genotypes and by increasing the potential for post-immigration evolution.

Preadaptation during invasion involves the chance sampling of genotypes (and phenotypes) in one environment and their subsequent release into a new environment where they contribute to fitness (Futuyma 2005). Increased propagule pressure increases the likelihood of preadapted genotypes arriving in a new range, thus increasing the potential for invasion. The introduction of preadapted genotypes is likely to be the main mechanism leading to the invasion of some plant species, especially species that exhibit uniparental reproduction such as apomixis, vegetative reproduction, and selfing. The introduction of preadapted genotypes is implicated in the invasions of *Capsella bursa-pastoris* (Neuffer and Hurka 1999) in North America, the colonization of *Hypericum canariense* (Dlugosch and Parker 2007) in the Canary Islands, and the invasion of *Chondrilla juncea* (Gaskin et al. 2013) across three continents.

Post-immigration evolution is now widely recognized as an important contributor to invasion. Post-immigration evolution results from multiple introductions; the descendants conjointly produce genetic admixtures, outcrossing or hybridization among individuals with different genotypes, genetic reshuffling or recombination, and generation of novel genotypes, leading to subsequent natural selection (Lee 2002; Cox 2004; Facon et al. 2006; Novak 2007; Suarez and Tsutsui 2008; Prentis et al. 2008; Schierenbeck and Ellstrand 2009). Post-immigration evolution has been demonstrated repeatedly (e.g., Brown and Marshall 1981; Sexton et al. 2002; Leger and Rice 2003; Maron et al. 2004; Blair and Wolfe 2004; Brown and Eckert 2004; Lavergne and Molofsky 2007; Cano et al. 2008; Xu et al. 2010). Sufficient genetic diversity (especially additive genetic variance) within introduced populations is a prerequisite for post-immigration adaptive evolution. High propagule pressure increases the likelihood for high levels of genetic diversity and can set the stage for an increase in invasiveness through post-immigration evolution (Novak and Mack 2005; Wares et al. 2005). The multiple facets of evolution in an invasion suggest that it may be difficult to predict whether an invasion will occur because invasions unfold over widely varying periods of time (Novak 2007).

4.5 Introduction of *B. tectorum* in North America: Historical Information and Genetic Data

As pointed out above, assessing the propagule pressure of a biological invasion requires the number of founder individuals (founder population size), the rate at which propagules arrive (the number of discrete founder events), or both (Lockwood et al. 2005; Simberloff 2009). Such detailed demographic information is usually unavailable, especially for accidentally introduced plants, such as *B. tectorum*. As a result, attempts to determine patterns of introduction and range expansion for an invasive species (its propagule pressure) are almost always constructed retrospectively (Mack 2005). These investigations require multiple sources of data, including historical records and genetic analyses (Estoup and Guillemaud 2010; Estoup et al. 2010; Lombaert et al. 2011; Novak 2011; Gaskin et al. 2013).

We have attempted to reconstruct the introduction history and the concatenation in the range expansion for *B. tectorum* into and across NA by combining reliable historical sources (herbarium specimens and published accounts) with the results of molecular genetic analyses that employ enzyme electrophoresis (allozymes) (Novak et al. 1991, 1993; Novak and Mack 1993, 2001; Bartlett et al. 2002; Valliant et al. 2007; Schachner et al. 2008; Huttanus et al. 2011; Pawlak et al. 2015). Other studies assessing the genetic diversity of *B. tectorum* in NA using molecular markers either have analyzed only 2–10 populations from a limited geographic area (1–2 states) (Ramakrishnan et al. 2004; Ashley and Longland 2007, 2009; Kao et al. 2008; Leger et al. 2009; Scott et al. 2010) or have analyzed larger numbers of populations from across a portion of Western NA (Ramakrishnan et al. 2006; Merrill et al. 2012). Our reconstruction has been assembled through the analysis of 10,150 individuals from 312 populations distributed across NA (the USA and Canada). The main results of these studies are summarized here.

The earliest record of *B. tectorum* in NA of which we are aware is from Lancaster County, Pennsylvania, where the grass was collected no later than 1790 (Muhlenberg 1793). Unfortunately, Muhlenberg's identification cannot be verified because his herbarium specimen has been lost, although he probably employed a European voucher specimen with which to identify the grass (Bartlett et al. 2002). The species was not reported again in the Eastern USA until a specimen was collected approx. 80 km east of Lancaster in West Chester, Pennsylvania, in 1859. This gap in detection could have arisen because populations of the plant were small and overlooked or the populations Muhlenberg saw went extinct and the species' detection in the nineteenth century stems from reintroductions. In any event, the frequency of its collection in the Eastern USA increased markedly after 1860 (see Table 7, Bartlett et al. 2002). Although the geographic spread of these records (1860–1880) suggests that multiple introductions occurred, these data alone do not demonstrate this scenario.

We determined the distribution of multilocus genotypes (MLGs) and genetic diversity among 38 populations of *B. tectorum* from the Eastern USA to assess the grass' pattern of introduction and regional spread (Bartlett et al. 2002). All 25 populations sampled at localities extending from Maine to Delaware are fixed for what we have termed the Most Common Genotype (MCG) (Fig. 4.2). The MCG occurs most frequently in populations from the native and introduced ranges of *B. tectorum* and refers to the most common combination of alleles at the 25 scored loci (Novak et al. 1991; Novak and Mack 1993; Bartlett et al. 2002; Valliant et al. 2007; Schachner et al. 2008; Huttanus et al. 2011; Pawlak et al. 2015). The *Pgm-1a & Pgm-2a* MLG occurs in four Eastern US populations included in this analysis, and we also detected the *Mdh-2b & Mdh-3b* MLG in two populations (Fig. 4.2). Consequently, Eastern US populations of *B. tectorum* appear to have arisen from a minimum of two to three separate introductions. Additional introductions may have taken place but remain undetected because of the preponderance of the MCG in

East Coast populations. The distribution of *B. tectorum* across NA suggests that the grass could have spread westward with European settlers. Alternatively, the species' occurrence throughout NA may stem from separate introductions. These two hypotheses were tested through genetic analysis of populations of *B. tectorum* across the USA and Canada.

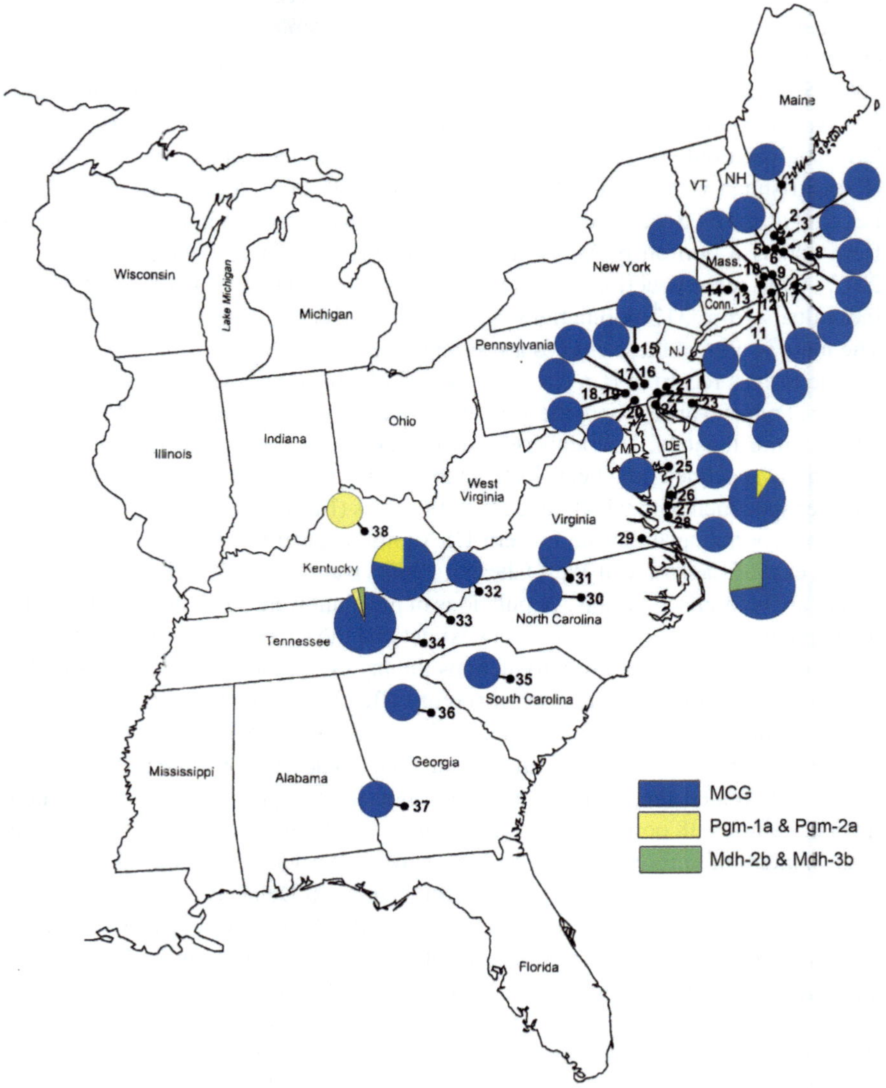

Fig. 4.2 Multilocus genotypes for 38 populations of *Bromus tectorum* in the Eastern USA. Pie diagrams illustrate the proportions of each multilocus genotype in each population. (Sizes of the pie diagrams and any offset segments here and in Figs. 4.3–4.8 vary only to enhance legibility.) Lines accompanying pie diagrams indicate locations of the populations. Numbers are population designations from Bartlett et al. (2002)

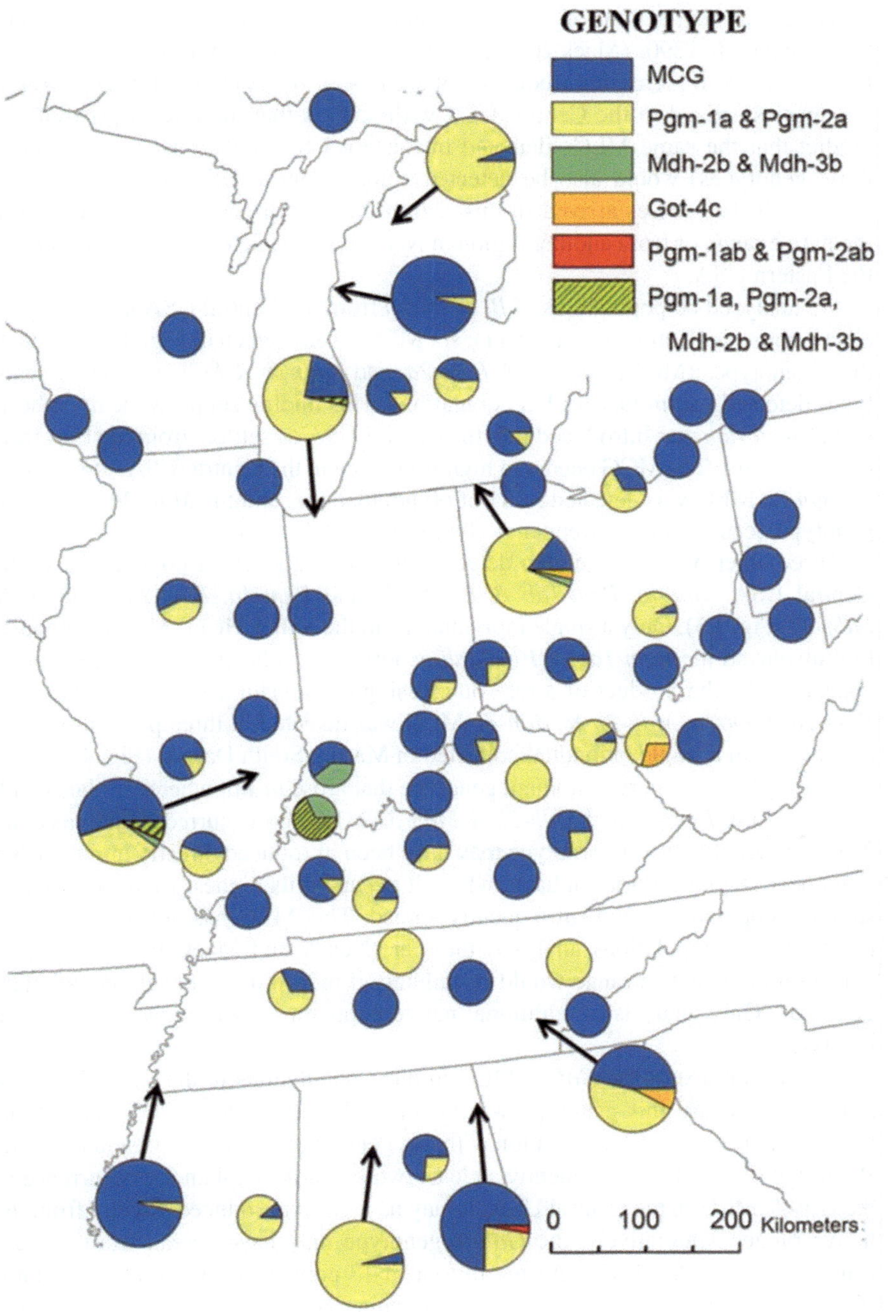

Fig. 4.3 Multilocus genotypes for 60 populations of *Bromus tectorum* in the Central USA. Pie diagrams illustrate the proportions of each multilocus genotype in each population. (See caption for Fig. 4.2 for further explanation of the pie diagrams.) Data are from Huttanus et al. (2011)

The earliest collections of *B. tectorum* in the Central USA were made in the late 1880s and early 1890s (Mack 2011), 20–30 years after collections of the plant were being made with increasing frequency in the Eastern USA (Bartlett et al. 2002). If the species arrived in the Central USA with settlers that migrated westward, we predict that the same MLGs detected in Eastern US populations (or a subset of these genotypes) would also be detected among Central US populations. Alternatively, if the species arrived in this region through direct introduction(s) from the native range or from another region in NA, we would detect MLGs not found in the Eastern USA.

We analyzed 60 populations of *B. tectorum* from the Central USA using enzyme electrophoresis (Huttanus et al. 2011); six MLGs were detected (Fig. 4.3). Three of these genotypes (MCG, *Pgm-1a & Pgm-2a*, and *Mdh-2b & Mdh-3b*) match genotypes detected within Eastern US populations. This finding supports the hypothesis that *B. tectorum* was introduced into the Central USA as settlers from the East traveled westward. The MCG occurs in high frequency in the Central USA, the *Pgm-1a & Pgm-2a* MLG was detected in 34 of 60 populations, and the *Mdh-2b & Mdh-3b* genotype occurred at low frequency (four of 60 populations).

Three other MLGs were also detected at low frequency in populations in the Central USA: *Got-4c*, *Pgm-1ab & Pgm-2ab*, and *Pgm-1a*, *Pgm-2a*, *Mdh-2b & Mdh-3b* (Fig. 4.3). Only a single individual from the Valley Head, Alabama population displayed the *Pgm-1ab & Pgm-2ab* genotype; as a heterozygous genotype, it appears to be the product of a rare outcrossing event (Huttanus et al. 2011). The *Pgm-1a*, *Pgm-2a*, *Mdh-2b & Mdh-3b* MLG was detected in three populations and also occurs in a single individual collected in Martin, South Dakota (Fig. 4.4). This MLG may be a novel, recombinant genotype that arose *in situ* wherever plants with the *Pgm-1a & Pgm-2a* and *Mdh-2b & Mdh-3b* MLGs co-occurred (Huttanus et al. 2011). Alternatively, this genotype may have been introduced directly into the USA (e.g., near Daylight/Earle, Indiana, where it occurs at its highest frequency among sampled populations) and subsequently spread. This MLG has not been reported among native populations analyzed thus far (Novak and Mack 1993, 2001); its detection in the native range would be unlikely if indeed it is a novel, recombinant genotype. Only analysis of additional native populations will allow us to resolve this issue.

The occurrence of the *Got-4c* MLG in three populations in the Central USA is intriguing because this genotype was undetected among Eastern US populations (Fig. 4.2); the closest US population with this genotype is far to the west in Laramie, Wyoming (Fig. 4.4). Consequently, at least two scenarios explain the occurrence of the *Got-4c* MLG in the Central USA. It may have been introduced directly from the native range. Alternatively, the *Got-4c* genotype may have spread from Eastern Canada, where it has been detected in four of 13 populations in Southern Ontario (Fig. 4.5). Great Lakes commerce and shipping between the Central USA and Southeastern Canada predates 1800, and *B. tectorum* may have been dispersed by this traffic (Huttanus et al. 2011). This same traffic may provide an alternative explanation for the occurrence of the *Pgm-1a & Pgm-2a* MLG in the Central USA; this genotype was detected in nine of 13 populations from Southern Ontario

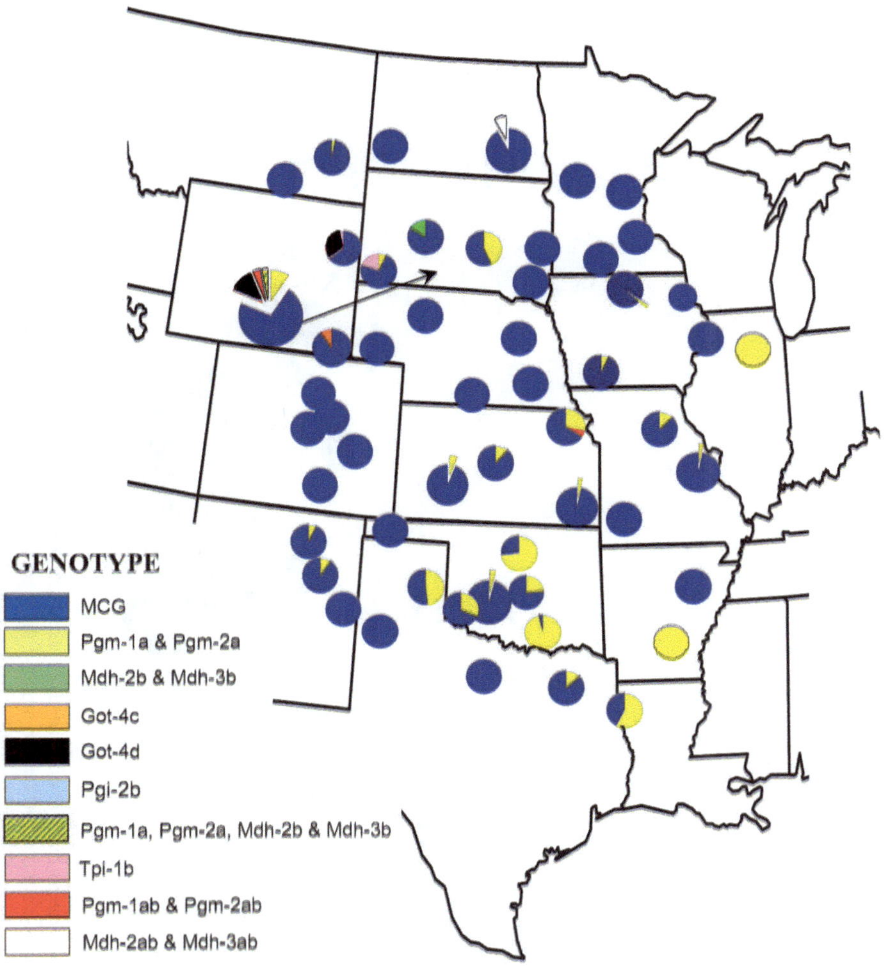

Fig. 4.4 Multilocus genotypes for 60 populations of *Bromus tectorum* in the mid-continent USA. Pie diagrams illustrate the proportions of each multilocus genotype in each population. (See caption for Fig. 4.2 for further explanation of the pie diagrams.) Data are from Schachner et al. (2008)

(Fig. 4.5). Thirty-two of the 60 populations (53.3 %) from the Central USA are genetically polymorphic (i.e., genetic admixtures).

The earliest herbarium specimens of *B. tectorum* from the northern portion of the mid-continent USA were collected in the 1890s, whereas the earliest specimens from the southern portion of this vast region were collected in the 1920s. These sizeable differences in the apparent timing of the entry of *B. tectorum* in the mid-continent USA may reflect different patterns of introduction, patterns that can also be detected through genetic analysis. Among the 54 mid-continent US populations of *B. tectorum* analyzed by Schachner et al. (2008), nine MLGs were detected: the MCG, *Pgm-1a & Pgm-2a*, *Mdh-2b & Mdh-3b*, *Got-4c*, *Got-4d*, *Tpi-1b*, *Pgm-1a*,

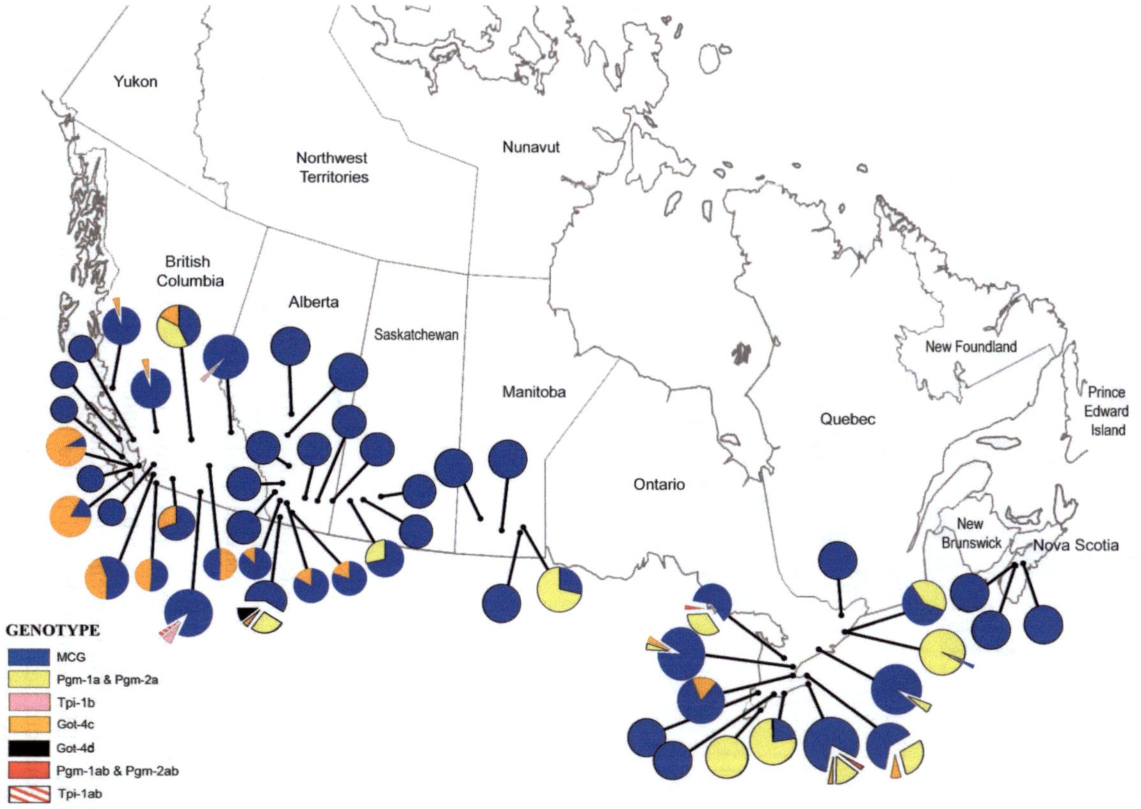

Fig. 4.5 Multilocus genotypes for 51 populations of *Bromus tectorum* in Canada. This figure does not include nine population of *B. tectorum* from central British Columbia (see Fig. 4.8). These nine populations were omitted only to increase the readability of the figure. Pie diagrams illustrate the proportions of each multilocus genotype in each population. (See caption for Fig. 4.2 for further explanation of the pie diagrams.) Data are from Valliant et al. (2007)

Pgm-2a, *Mdh-2b & Mdh-3b*, *Pgm-1ab & Pgm-2ab*, and *Mdh-2ab & Mdh-3ab* (Fig. 4.4). Approximately one-half of the populations in the mid-continent USA (26 of 54) are genetic admixtures, i.e., they consist of two or more genotypes previously detected among native Eurasian populations (Novak and Mack 1993). The MCG occurs at high frequency (52 of 54 populations) in the mid-continent USA. The *Pgm-1a & Pgm-2a* MLG occurs in 24 of 54 populations from the mid-continent USA: 17 of these populations occur in the eight states in the region's southern portion. Only three MLGs (MCG, *Pgm-1a & Pgm-2a*, and *Pgm-1ab & Pgm-2ab*) were detected in the eight states in this southern portion, whereas all nine MLGs found within the mid-continent USA were detected in populations from the eight states in the region's northern portion (Fig. 4.4). These patterns of genotype distributions indicate that events leading to the establishment of *B. tectorum* in the northern and southern portions of the mid-continent USA arose via different scenarios.

The MCG is clearly prominent within the 26 widespread populations we sampled in the northern portion of the mid-continent USA. The *Pgm-1a & Pgm-2a* MLG is a distant second in prominence with detection in seven populations (Fig. 4.4). Other MLGs however have more restricted distributions. For instance, the *Mdh-2b & Mdh-3b* MLG, the *Got-4d*, and the *Tpi-1b* genotypes were found in two populations each. The *Got-4c* genotype was detected in a single population. All populations with these MLGs are located in either South Dakota or Eastern Wyoming. The seemingly scattered occurrence of populations with these MLGs indicates additional sampling is warranted to improve estimates of the regional genetic diversity of populations. In addition, field experimentation with *B. tectorum* populations could be conducted to determine whether this genetic diversity has contributed to greater invasiveness.

Five heterozygous individuals were encountered within three populations in the mid-continent USA (Fig. 4.4). The *Pgm-1ab & Pgm-2ab* heterozygous MLG was detected in two individuals from the Oskaloosa, Kansas, population and one individual from the Martin, South Dakota, population. Two individuals from the Colfax, North Dakota, population had the *Mdh-2ab & Mdh-3ab* genotype. These heterozygous individuals were the product of outcrossing events. The *Pgm-1a*, *Pgm-2a*, *Mdh-2b & Mdh-3b* genotype was observed in a single individual from the Martin, South Dakota population, and as discussed above, several mechanisms potentially explain its occurrence.

Several non-mutually exclusive scenarios potentially explain the occurrence of different MLGs among *B. tectorum* populations in the mid-continent USA, especially those in the region's northern portion. European settlers migrating from the east may have dispersed some of these genotypes (e.g., the MCG, *Pgm-1a & Pgm-2a*, and *Mdh-2b & Mdh-3b*), some genotypes (e.g., *Got-4c*, *Got-4d*, and *Tpi-1b*) may have arrived from the Western USA, where they also occur (Fig. 4.6), and some may have been introduced directly from the native range. Ascertaining the role, if any, of these scenarios in the regional occurrence of genotypes will require a molecular marker with greater resolving power (e.g., microsatellite DNA or SNPs) than allozymes (Avise 2004; Schlotterer 2004; Estoup et al. 2010; Lombaert et al. 2011).

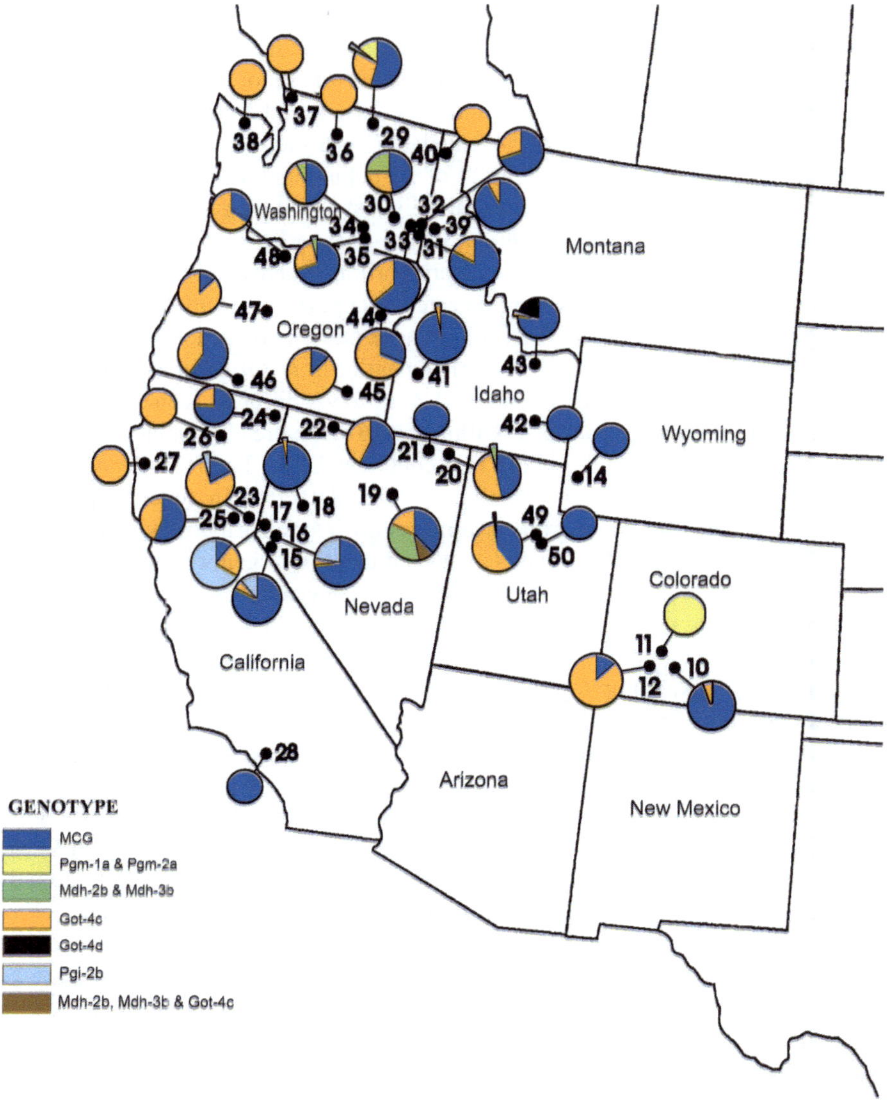

Fig. 4.6 Multilocus genotypes for 40 populations of *Bromus tectorum* in the Western USA. Pie diagrams illustrate the proportions of each multilocus genotype in each population. (See caption for Fig. 4.2 for further explanation of the pie diagrams.) Numbers are the population designations from Novak et al. (1991)

The collection history of *B. tectorum* in the Intermountain West of NA is well documented (Mack 1981). The grass was first collected at Spences Bridge, British Columbia, in 1889; Ritzville, Washington, in 1893; Provo, Utah, in 1894; Pullman, Washington, in 1897, Eastern Colorado in 1897, Boise, Idaho, in 1899, and the Klamath Valley of Oregon in 1902. Similar to the introduction of *B. tectorum* in the

Eastern USA, the chronology in the West suggests multiple introductions. Analysis of 40 populations from this region (which includes 26 populations from the Intermountain West and 14 populations from Nevada and California, see Table 4.1), using enzyme electrophoresis largely confirms this hypothesis (Novak et al. 1991).

Seven MLGs were identified among Western US populations: the MCG, *Pgm-1a & Pgm-2a*, *Mdh-2b & Mdh-3b*, *Got-4c*, *Got-4d*, *Pgi-2b*, and *Got-4c*, *Mdh-2b & Mdh-3b* (Fig. 4.6). Unlike the prominence of genotypes elsewhere in the USA, the MCG and the *Got-4c* genotype are equally common in the West. The *Got-4c* genotype occurs in 34 of 40 populations and is the only genotype (i.e., it is "fixed") in six of these populations. The MCG occurs in 33 of 40 populations and is also fixed in six populations. Other MLGs occur at lower frequency: *Mdh-2b & Mdh-3b* occurs in six populations (four from Eastern Washington and two from Northern Nevada); *Pgm-1a & Pgm-2a* was detected in two populations (Gunnison, Colorado, and Tonasket, Washington), and *Got-4d* was detected in the populations from Provo, Utah, and Dubois, Idaho (Fig. 4.6). The *Got-4c*, *Mdh-2b & Mdh-3b* MLG was found only in the population at Emigrant Pass, Nevada, and appears to be a novel, recombinant genotype. The *Pgi-2b* MLG was detected in three populations near Reno, Nevada, and among several individuals in the population at Truckee, California (Fig. 4.6). In addition, this genotype has been also found at low frequency in seven populations from California and the American Southwest (the Southwestern USA) (Fig. 4.7). No heterozygous individuals were detected among the individuals from the 40 mostly Intermountain West populations we analyzed (Table 4.1). Among these 40 populations, most (28) were genetically polymorphic (i.e., genetic admixtures).

Several scenarios potentially explain the occurrence of different MLGs among *B. tectorum* populations from the Western USA. The *Got-4d* genotype may have been introduced through long distance dispersal into Provo, Utah, an early collection site for *B. tectorum* in the West (Mack 1981). Other genotypes may have arrived from elsewhere in NA. For instance, the occurrence of plants with the *Pgm-1a & Pgm-2a* genotype at Gunnison, Colorado, may be the result of dispersal from mid-continent US populations, and the occurrence of this genotype in a population at Tonasket, Washington, may reflect spread from populations in nearby British Columbia. The *Pgm-1a & Pgm-2a* MLG has a high frequency of occurrence within populations in the southern portion of the mid-continent USA (Fig. 4.4) and in central British Columbia (Fig. 4.8).

The earliest collection of *B. tectorum* in California of which we are aware was made in Siskiyou County in 1899 (Pawlak et al. 2015). Over the next 20 years, other collections were made in northern counties in California (Siskiyou, Shasta, and Humboldt). The conspicuous absence of any collection records well into the twentieth century in the San Francisco Bay area, a locale of intense plant collecting, strongly suggests that ships arriving in San Francisco Bay played no role in its introduction (Pawlak et al. 2015). The earliest collections in Southern California occurred in 1902 in Santa Cruz, Santa Barbara, and Ventura counties. Earliest collections in both northern and southern parts of Nevada were made in the early 1900s (Pawlak et al. 2015). The earliest collection in Arizona was in Coconino County in 1901.The collection record of *B. tectorum* in the rest of the American Southwest is

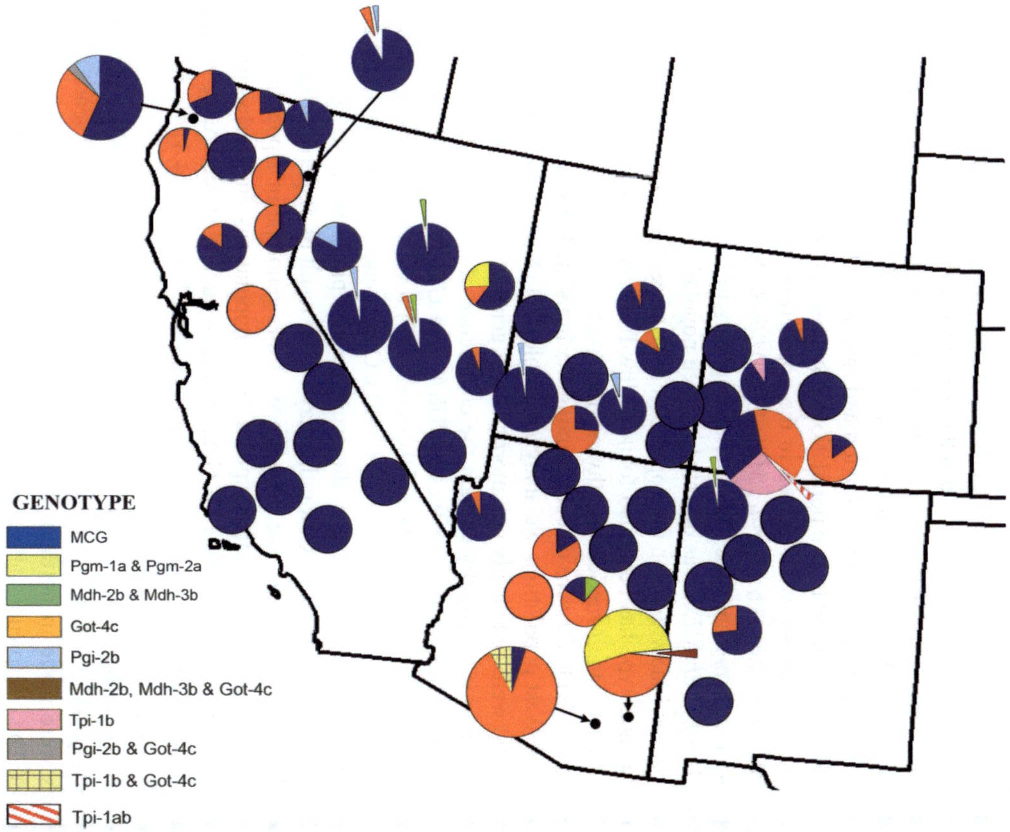

Fig. 4.7 Multilocus genotypes for 60 populations of *Bromus tectorum* in California and the American Southwest. Pie diagrams illustrate the proportions of each multilocus genotype in each population. (See caption for Fig. 4.2 for further explanation of the pie diagrams.) Data are from Pawlak et al. (2015)

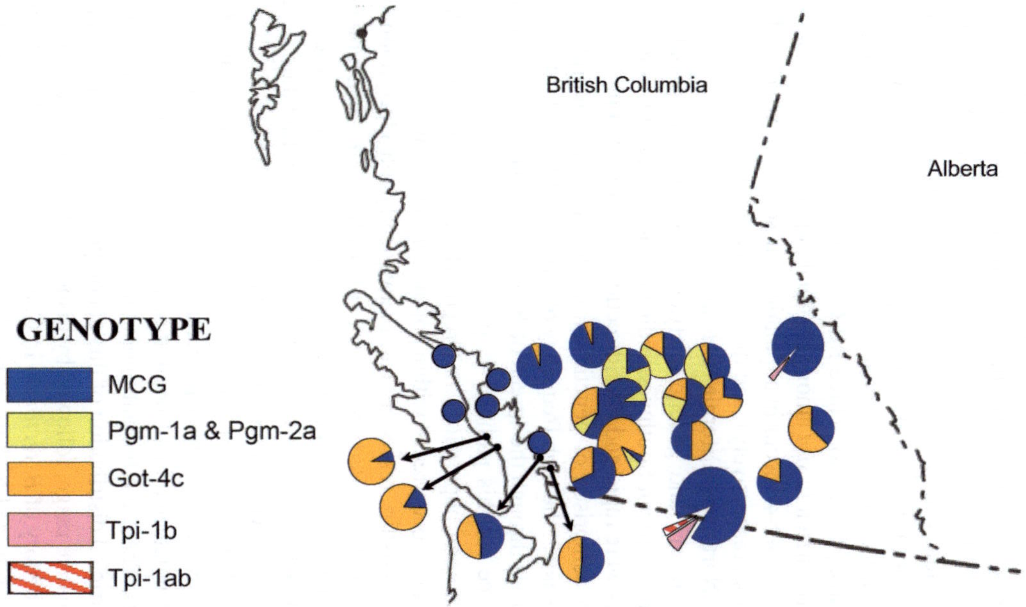

Fig. 4.8 Multilocus genotypes for 25 populations of *Bromus tectorum* in British Columbia. This figure includes nine population of *B. tectorum* from central British Columbia previously reported in Novak et al. (1991). Pie diagrams illustrate the proportions of each multilocus genotype in each population. (See caption for Fig. 4.2 for further explanation of the pie diagrams.) Data are from Valliant et al. (2007)

exceptionally sparse, especially for Colorado and New Mexico. As indicated above, the earliest collection in Utah (1894) occurred in Provo.

Ten MLGs were detected among the 60 populations of *B. tectorum* from California and the American Southwest analyzed by Pawlak et al. (2015). Two MLGs are conspicuous in California and the American Southwest; the MCG occurs in 57 of 60 populations and is predominant in many, whereas the *Got-4c* genotype has a high frequency of occurrence within populations in Northern California and Southern Arizona (Fig. 4.6). The *Pgi-2b* genotype occurs at relatively low frequency in seven populations, three from Northern California, two from Nevada, and two from Utah. The *Pgm-1a & Pgm-2a* and *Mdh-2b & Mdh-3b* MLGs sporadically occur across the region, within three and four populations, respectively. Two populations in southeastern Colorado (Olathe and Hermosa) contain the *Tpi-1b* MLG. The only heterozygous individual we detected in this region (*Tpi-1ab*) was collected at Hermosa, Colorado. Three MLGs believed to be novel, recombinant genotypes may have arisen *in situ* among regional populations: *Mdh-2b*, *Mdh-3b* & Got-4c and *Tpi-1b & Got-4c* occur in populations in Southern Arizona (Turkey Flat and Mount Lemmon, respectively), and the *Pgi-2b & Got-4c* MLG was detected in a population from Northern California (Orleans).

The collection record and genetic analysis of *B. tectorum* populations from California and the American Southwest indicate that the introduction and regional spread of the grass were rapid and multidirectional and resulted in 31 of 60 (52 %) populations being comprised of genetic admixtures (i.e., polymorphic at one or more loci). For instance, the high frequency of *Got-4c* MLG in Northern California (Fig. 4.7) likely arose through the spread of *B. tectorum* southward from the Pacific Northwest; evidence for this pathway is provided by the high frequency of the *Got-4c* genotype among populations in Oregon (Fig. 4.6). The *Got-4c* MLG also has a high frequency of occurrence within populations in the mountains of Southern Arizona (Fig. 4.7). Postfire reseeding can inadvertently introduce nonnative plant species (Keeley et al. 2006). Perhaps *B. tectorum* with the *Got-4c* genotype arrived in Southern Arizona by this mode: forests on Mount Lemmon were extensively burned in 2003, and *B. tectorum* was first detected on the mountain 2 years later (L.A. Brigham, personal communication) (Pawlak et al. 2015).

In contrast, the MCG and other genotypes appear to have spread into California and the American Southwest from populations farther east. Finally, the *Pgi-2b* MLG has only been detected in 11 populations from California, Nevada, and Utah (Figs. 4.6 and 4.7). These are the only populations of *B. tectorum* in NA with this MLG, and the restricted distribution of this genotype likely stems from direct introduction from the native range, followed by dispersal (Pawlak et al. 2015).

The collection history of *B. tectorum* in Canada was described by Valliant et al. (2007). The oldest reported herbarium specimen for the grass in Canada was collected in Kingston, Ontario, in 1886 (see Fig. 4.3, Valliant et al. 2007). Other early collection sites for *B. tectorum* in Southern Ontario are ports along the Great Lakes. In Western Canada, *B. tectorum* was first collected in 1889 from an irrigated field near Spences Bridge, British Columbia, a site more than 3000 km west of Southern Ontario. A second pre-1900 collection of the grass in Western Canada was made on

Vancouver Island. Additional collecting of *B. tectorum* in Southern Ontario in the early twentieth century appears to have lagged, while the grass was collected with increasing frequency in the arid interior of British Columbia (the Okanagan Valley) between 1901 and 1925 (Valliant et al. 2007).

As described earlier, this distinct bicoastal pattern of collection is consistent with multiple introductions for the grass. This hypothesis was tested with an enzyme electrophoresis analysis of 60 *B. tectorum* populations from across Canada (Valliant et al. 2007). In Eastern Canada, the MCG and the *Pgm-1a & Pgm-2a* MLG occur commonly, and two populations in Southern Ontario (Port Colborne and Alliston) contain individuals with the heterozygous genotype *Pgm-1ab & Pgm-2ab* (Fig. 4.5). The *Got-4c* genotype occurs in four populations in Southern Ontario. Although the frequency of this genotype in Eastern Canada is low (Fig. 4.5), it is higher than its frequency in populations in the eastern third of the USA (see Figs. 4.2 and 4.3). As observed in Eastern Canada, the MCG and the *Pgm-1a & Pgm-2a* genotype are commonly found within populations in Western Canada (Fig. 4.5). The *Pgm-1a & Pgm-2a* genotype was detected in populations from Manitoba, Saskatchewan, Alberta, and British Columbia. But its highest frequency (seven populations) occurs in central British Columbia (Fig. 4.8). Several genotypes were found in populations in Western Canada: the *Tpi-1b* MLG in two populations in British Columbia (Golden and Osoyoos) and the *Tpi-1ab* heterozygous genotype in a population at Osoyoos (Fig. 4.8). In Western Canada, the *Got-4d* genotype has been detected only in a population at Waterton, Alberta. In the Western USA, this same genotype had been previously reported from populations in Provo, Utah, and Dubois, Idaho (Fig. 4.6) (Novak et al. 1991, 1993). The area around Waterton, Alberta, was settled by people from Utah at the end of the nineteenth century, and their route to Alberta took them near Dubois, Idaho (Valliant et al. 2007). Whether *B. tectorum* with this genotype may have been spread from Provo, Utah, via Dubois, Idaho, to Waterton, Alberta, by these settlers remains an intriguing hypothesis.

4.6 Source Populations

Although the patterns described above are consistent with multiple introductions, definitive test of such introductions requires detecting the same genotype within native populations. We have conducted an enzyme electrophoresis analysis of 1730 individuals from 51 native population of *B. tectorum* in Eurasia (Novak and Mack 1993), and these data allow us to assess the species' pattern of introduction into NA and identify source populations (or regions) for such introductions. We are not aware of any other published reports that have assessed and compared the genetic diversity (and the distribution of genotypes) of native and invasive populations of *B. tectorum* using the same molecular marker system.

For *B. tectorum*, many of the multilocus genotypes observed in NA are widely distributed across the continent. In contrast, the same genotypes often have restricted geographic distribution among native, Eurasian populations (Novak and Mack 1993).

The exception to this general pattern is the MCG, which occurs across much of the species' native range. If native genotypes have restricted geographic distributions (i.e., geographically structured), accurate identification of the geographic origins of an invasion is facilitated (Novak and Mack 2001; Novak 2011). The *Pgm-1a & Pgm-2a* MLG has been detected in the native range at Bratislava, Slovakia, and Vac, Hungary (Novak and Mack 1993, 2001). The *Got-4c* MLG occurs in some populations in Eastern NA and has its highest prominence in Western NA, whereas in the native range this genotype is known only from Libochovice, Czech Republic, and Bayreuth, Germany. In the native range, we have detected the *Got-4d* genotype in a single population in Vienna, Austria. In Western NA, the *Pgi-2b* genotype has been detected in 11 populations from California, Nevada, and Utah; in the native range, this genotype occurs in populations in France, Spain, and Morocco. In contrast, no published accounts report the *Mdh-2b & Mdh-3b* and *Tpi-1b* genotypes among native Eurasian populations. Based on these data, the geographic origins of the invasion of *B. tectorum* in NA appear to be among populations from Central Europe and the western Mediterranean region (Novak and Mack 1993, 2001). Identification of the sources for other genotypes in NA will require analysis of additional native populations and perhaps a different genetic marker.

4.7 Genetic Diversity within and among Populations of *B. tectorum*

If the above listed demographic data are unobtainable, propagule pressure can be estimated with molecular data (Ficetola et al. 2008; Ross and Shoemaker 2008; Goncalves da Silva et al. 2010; Huttanus et al. 2011). Huttanus et al. (2011) lists four patterns of genetic diversity that may reflect high propagule pressure: (1) the number of multilocus genotypes of a species across its new range would likely be large; (2) genetic admixtures, if propagule pressure results in the establishment of two or more independently derived native genotypes within invasive populations; (3) similar levels of genetic diversity within native and introduced populations and little evidence for founder effects; and (4) if genetic admixtures are common, introduced populations will have similar, or even less, genetic structure than native populations.

The detection of seven homozygous, nonrecombinant multilocus genotypes among NA populations of *B. tectorum* indicates that multiple introductions have occurred, which in turn likely reflect moderate propagule pressure. In addition, 156 of the 312 (50 %) NA populations we have analyzed are genetically diverse, the products of either direct introduction events or dispersal of genotypes among populations. These populations are genetic admixtures.

The genetic diversity of a species, or group of populations, can be hierarchically partitioned across (among) and within populations, and the amount of diversity at these two levels can be compared to assess the genetic consequences of introduction and range expansion (Novak and Mack 2005). At the 25 scored loci, NA populations of *B. tectorum* have lower allelic richness (33 vs. 43) and fewer polymorphic loci (7 vs. 13) than native populations (Novak and Mack 1993, 2005), indicating that NA populations arose through the loss of genetic diversity by founder effects. Despite reduced genetic diversity across invasive populations, the amount of genetic diversity detected, on average, within populations from all regions in NA except one (Eastern USA) is higher than the diversity within populations in either of the major regions we denote in the grass's native range (Table 4.1). Thus, at the within-population level, populations of *B. tectorum* from NA possess more genetic diversity than native populations.

Values of G_{ST} for the populations of *B. tectorum* from all regions of NA (G_{ST}=0.241–0.582) are lower than the values for either European populations (G_{ST}=0.656) or populations from Southwest Asia (G_{ST}=0.735) (Table 4.1). These data indicate that the genetic structure among populations in NA is less than the genetic structure of populations from both regions in the native range, a likely outcome of multiple introductions. The propagule pressure associated with the introduction of *B. tectorum* into NA from genetically diverse source populations and subsequent dispersal events appears to have largely offset the reduction in genetic diversity associated with transcontinental migration.

4.8 Management Implications

The evolutionary history of species within section *Genea*, including *B. tectorum* (Sales 1993, 1994; Balfourier et al. 1998), serves as a guide to management of brome-grasses and *B. tectorum* in particular. First, disturbances (e.g., livestock grazing, off-road vehicle use) should be minimized in habitats within which *Bromus* species do not yet occur or are not yet abundant. Although the link between disturbance and invasion is well known (Hobbs and Huenneke 1992; Hierro et al. 2006), the information presented here places the relationship between the occurrence of these species and disturbance in an evolutionary context.

Results of the genetic analysis of *B. tectorum* across NA reveal the importance of multiple introductions of MLGs from different native source populations to the subsequent spread of these genotypes among NA populations. These events have resulted in genetic admixtures: one-half (156 of 312) of NA populations are comprised of genotypes from different native populations. This diversity complicates the management of *B. tectorum* in NA because no one control prescription is likely to be effective for all genotypes, especially in all habitats. Consequently, we advocate measures that reduce propagule pressure (i.e., reduce or eliminate additional introduction events). These efforts should certainly include prevention through border inspections as well as early detection and rapid response, and eradication while the infested area remains small (Mack et al. 2000; Wittenberg and Cock 2005). In addition, gene flow should be deliberately limited (i.e., reduce propagule dispersal) among populations of *B. tectorum*; gene flow can produce populations with increased genetic diversity and bring together genotypes that have never co-occurred in the same population.

We also believe that the management of *B. tectorum* in NA can be enhanced by assessing the performance (survival, growth and reproduction) of plants with different MLGs across an array of environmental conditions. Based on this information, optimal habitat conditions for specific MLGs may be identified. If a stringent regime for genetic monitoring of *B. tectorum* populations were implemented (see below), land managers could specifically target certain MLGs for control because of the ability of plants with these genotypes to dominate a site. Such information on the performance of different MLGs could also be used to predict changes in the distribution and abundance of *B. tectorum* driven by climate change (see Hufft and Zelikova 2015).

The landscape level management of *B. tectorum* in NA by herbicides, tillage, and grazing has been largely ineffective (Mack 2011); consequently, a classical biological control program, which involves foreign exploration for natural enemies, such as insects or fungi, should be considered. Although use of classical biological control for invasive grasses has been viewed as problematic (McFadyen 1998), this method is gaining more support for particularly noxious grasses. For instance, biological control agents have been sought for *Spartina alterniflora* Loisel. in the USA (Grevstad et al. 2003; Fisher et al. 2005) and Chilean needle grass (*Nassella neesiana* (Trin. and Rupr.) Barkworth in Australia and New Zealand (Anderson et al. 2010, 2011).

The information summarized in this chapter (source populations, reconstructing the introduction and range expansion of *B. tectorum* in its new range, and an assessment of the genetic diversity of invasive populations) is requisite for development of a classical biological control program (Roderick and Navajas 2003; Novak and Sforza 2008; Gaskin et al. 2011). For instance, the overall genetic diversity of *B. tectorum* in NA is low compared with native populations of the grass; consequently, as suggested by Muller-Scharer et al. (2004), only a few highly specialized control agents may be necessary. In such cases, control agents would probably experience fast population buildup and spread as they attack the target species. With multiple introductions, genotypes from several native source populations have been detected across NA, and under these circumstances, agents from different portions of the native range would be required for control (Burdon and Marshall 1981). Furthermore, several control agents may be needed within the same locality (Muller-Scharer et al. 2004), given the high frequency of genetic admixtures among invasive populations of *B. tectorum*. Finally, the geographic origins for the invasion of *B. tectorum* into NA appear to be in Central Europe and the western Mediterranean region. The search for biological control agents that are both specific and effective should be conducted within these two regions (Goolsby et al. 2006; Gaskin et al. 2011).

4.9 Research Needs

The composite picture that emerges from the genetic analysis of *B. tectorum* summarized here is that multiple introductions into NA occurred, and range expansion of the grass across the continent was rapid and multidirectional. Future research should analyze both native and invasive populations of *B. tectorum* using genetic markers with higher resolving power than enzyme electrophoresis (e.g., microsatellite DNA, SNPs) to more precisely determine the number of introductions that have occurred in NA and the location of source populations within the native range. The genetic markers listed above are considered to be neutral to selection; thus care should be exercised when attempting to infer the adaptive significance of these data (Avise 2004). These markers can be instead used more reliably to infer demographic processes (e.g., dispersal and colonization events and the effective population size), detect population admixture, estimate mating system parameters, and assess parental (paternity) and kinship relationships (Schlotterer 2004).

Alternatively, the amount of genetic diversity within populations, especially the level of allelic richness within populations, is associated with a species' evolutionary potential (Fisher 1958; Van Kleunen et al. 2000; Vergeer et al. 2003). To more accurately estimate evolutionary potential of *B. tectorum*, efforts need to be made to determine more precisely the amount of single-locus genetic diversity (measured using genetic markers), quantitative (ecologically important) trait variation (e.g., ecophysiological traits), and phenotypic plasticity within and among native and invasive populations of the grass within the same experimental design. For example, this approach was utilized by Lavergne and Molofsky (2007) to show that invasive populations of *Phalaris arundinacea* L. possess higher single-locus allelic diversity, quantitative trait variability, and phenotypic plasticity, compared with native populations; these features increased the evolutionary potential that contributed to this species' invasion in NA.

Outcrossing events, although rare, have nonetheless led to the production of heterozygous individuals within populations of *B. tectorum* in several regions; legacies of such outcrossing events are novel, recombinant genotypes within populations in the Central USA, mid-continent USA, and California and the American Southwest. These findings suggest that post-immigration evolution is taking place, even within populations of this highly selfing species. Thus, future efforts should be taken to evaluate the extent to which these events influence the invasion of *B. tectorum* by monitoring populations using a genetic approach (see Novak and Rausch 2009). This genetic approach might first focus on marginal habitats or recently invaded sites, as described in Kao et al. (2008).

Future research could assess how the occurrence of heterozygous individuals and recombinant genotypes influence quantitative trait variation and phenotypic plasticity within invasive populations of *B. tectorum*. In the future, this effort could incorporate genomics and proteomics approaches that include DNA sequence analysis, quantitative trait loci (QTL) mapping, and microarray analysis of transcriptional regulation (Basu et al. 2004; Hudson 2008; Stinchcombe and Hoekstra 2008; Nadeau and Jiggins 2010; Ekblom and Galindo 2011). In addition, assessing the role of epigenetic variation and differentiation in the invasion of *B. tectorum* may prove especially important because epigenetic mechanisms can result in phenotypic trait variability, even in the absence of genetic variation (Bossdorf et al. 2008; Richards et al. 2010, 2012). Finally, the manner in which single-locus genetic diversity, quantitative trait variation, phenotype plasticity, and epigenetics, singly and in combination, influences the invasion of *B. tectorum* in the USA under global climate change will need to be assessed (Reusch and Wood 2007; Salamin et al. 2010; Nicotra et al. 2010) if this most noxious member of *Bromus* section *Genea* is to be curbed.

Acknowledgments It would not have possible to conduct the research summarized in this chapter without the assistance of countless people who over the years collected population samples of *B. tectorum* around the world. We thank them all. In addition, we thank our graduate students at Washington State University and Boise State University who have played a major role in many of these studies: Elizabeth Bartlett, Morgan Valliant, Lauren Schachner, Temsha Huttanus, and Angela Pawlak. We gratefully thank Rich Scott for his work on the figures (maps) presented in this chapter, and we especially thank him for his interest in this research over the years. SJN completed portions of this chapter while on sabbatical leave at and visiting the European Biological Control Laboratory, USDA-ARS, Montferrier-sur-Lez, France. He is extremely grateful to the personnel of this laboratory, especially Rene Francois Henri Sforza, Daniel Strickman, and Kim Hoelmer, for the opportunity to work at the lab.

References

Acedo C, Liamas F (2001) Variation of morphological characters of lemma and palea in the genus *Bromus* (Poaceae). Ann Bot Fenn 38:1–14

Allard RW, Jain SK, Workman PL (1968) The genetics of inbreeding populations. Adv Genet 14:55–131

Allendorf FW, Lundquist LL (2003) Introduction: population biology, evolution, and control of invasive species. Conserv Biol 17:24–30

Anderson FE, Barton J, McLaren D (2010) Studies to assess the suitability of *Uromyces pencanus* as a biological control agents for *Nassella neesiana* (Poaceae) in Australia and New Zealand. Australas Plant Pathol 39:69–78

Anderson FE, Diaz ML, Barton J et al (2011) Exploring the life cycles of three South American rusts that have potential as biological control agent of the stipoid grass *Nassella neesiana* in Australia. Fungal Biol 115:370–380

Ashley MC, Longland WS (2007) Microsatellite evidence of facultative outcrossing in cheatgrass (*Bromus tectorum*): implications for the evolution of invasiveness. Plant Species Biol 22:197–204

Ashley MC, Longland WS (2009) Assessing cheatgrass (*Bromus tectorum*) genetic diversity and population structure using RAPD and microsatellite molecular markers. Western N Am Nat 69:63–74

Atkinson SY, Brown CS (2015) Attributes that confer invasiveness and impacts across the large genus Bromus – lessons from the Bromus REEnet database. In: Germino MJ, Chambers JC, Brown CS (eds) Exotic brome-grasses in arid and semiarid ecosystems of the Western US: causes, consequences, and management implications. Springer, New York, NY (Chapter 6)

Avise JC (2004) Molecular markers, natural history, and evolution. Sinauer, Sunderland, MA

Baker HG (1974) The evolution of weeds. Annu Rev Ecol Syst 5:1–24

Balfourier F, Charmet G, Ravel C (1998) Genetic differentiation within and between natural populations of perennial and annual ryegrass (*Lolium perenne* and *L. rigidum*). Heredity 81:100–110

Barrett SCH, Colautti RI, Eckert CG (2008) Plant reproductive systems and evolution during biological invasion. Mol Ecol 17:373–383

Bartlett E, Novak SJ, Mack RN (2002) Genetic variation in *Bromus tectorum* (Poaceae): differentiation in the eastern United States. Am J Bot 89:602–612

Basu C, Halfhill MD, Mueller TC et al (2004) Weed genomics: new tools to understand weed biology. Trends Plant Sci 9:391–398

Beatley JC (1966) Ecological status of introduced brome grasses (*Bromus* spp.) in desert vegetation of southern Nevada. Ecology 47:548–554

Blair AC, Wolfe LM (2004) The evolution of an invasive plant: an experimental study with *Silene latifolia*. Ecology 85:3035–3042

Bossdorf O, Richards CL, Pigliucci M (2008) Epigenetics for ecologists. Ecol Lett 11:106–115

Bouchenak-Khelladi Y, Salamin N, Saolainen V et al (2008) Large multigene phylogenetic trees of the grasses (Poaceae): progress towards complete tribal and generic level sampling. Mol Phylogenet Evol 47:488–505

Bromilow C (1995) Problem plants of South Africa. Briza Press, Arcadia, South Africa

Brooks ML (1999a) Habitat invasibility and dominance by alien annual plants in the western Mojave Desert. Biol Invasions 1:325–337

Brooks ML (1999b) Alien annual grasses and fire in the Mojave Desert. Madrono 46:13–19

Brooks ML, Belnap J, Brown CS et al (2015) Exotic annual *Bromus* invasions - comparisons among species and ecoregions in the Western United States. In: Germino MJ, Chambers JC, Brown CS (eds) Exotic brome-grasses in arid and semiarid ecosystems of the Western US: causes, consequences, and management implications. Springer, New York, NY (Chapter 2)

Brown AHD (1990) Genetic characteristics of plant mating systems. In: Brown AHD, Glegg MT, Kahler AL et al (eds) Plant population genetics, breeding and genetic resources. Sinauer, Sunderland, MA, pp 145–162

Brown AHD, Burdon JJ (1987) Mating systems and colonizing success in plants. In: Gray AJ, Crawley MJ, Edwards PJ (eds) Colonization, succession and stability. The 26th symposium of the British Ecological Society held jointly with the Linnean Society of London. Blackwell, Oxford, pp 115–131

Brown AHD, Marshall DR (1981) Evolutionary changes accompanying colonization in plants. In: Scudder GG, Reveal JL (eds) Evolution today. Hunt Institute of Botanical Documents, Carnegie-Mellon University Press, Pittsburgh, PA, pp 351–363

Brown JS, Eckert CG (2004) Evolutionary increase in sexual and clonal reproductive capacity during biological invasion in an aquatic plant *Butomus umbellatus*. Am J Bot 92:495–502

Brunet J (1992) Sex allocation in hermaphroditic plants. Trends Ecol Evol 3:79–84

Burdon JJ, Marshall DR (1981) Biological control and the reproductive mode of weeds. J Appl Ecol 18:649–658

Campbell CS, Quinn JA, Cheplick GP et al (1983) Cleistogamy in grasses. Annu Rev Ecol Syst 14:411–441

Campbell DR (2000) Experimental tests of sex-allocation theory in plants. Trends Ecol Evol 15:227–232

Cano L, Escarre J, Fleck I et al (2008) Increased fitness and plasticity of an invasive species in its introduced range: a study using *Senecio pterophorus*. J Ecol 96:468–476

Catalan P, Kellogg EA, Olmstead RG (1997) Phylogeny of Poaceae subfamily Pooideae based on chloroplast *ndh*F gene sequences. Mol Phylogenet Evol 8:150–166

Chambers JC, Roundy BA, Blank RR et al (2007) What makes Great Basin sagebrush steppe ecosystems invasible by *Bromus tectorum*. Ecol Monogr 77:117–145

Charlesworth D, Charlesworth B (1981) Allocation of resources to male and female functions in hermaphrodites. Biol J Linn Soc 15:57–74

Clegg MT (1980) Measuring plant mating systems. BioScience 30:814–818

Colautti RI, MacIsaac HJ (2004) A neutral terminology to define "invasive" species. Divers Distrib 10:135–141

Colautti RI, Grigorovich IA, MacIsaac HJ (2006) Propagule pressure: a null model for biological invasions. Biol Invasions 8:1023–1037

Cox GW (2004) Alien species and evolution: the evolutionary ecology of exotic plants, animals, microbes, and interacting native species. Island Press, Washington, DC

D'Antonio CM, Vitousek PM (1992) Biological invasion by exotic grasses, the grass/fire cycle and global change. Annu Rev Ecol Syst 23:63–87

D'Antonio CM, Hughes RF, Tunison JT (2011) Long-term impacts of invasive grasses and subsequent fire in seasonally dry Hawaiian woodlands. Ecol Appl 21:1617–1628
Dlugosch KM, Parker IM (2007) Molecular and quantitative trait variation across the native range of the invasive species *Hypericum canariense*: evidence for ancient patterns of colonization via pre-adaptation? Mol Ecol 16:4269–4283
Dlugosch KM, Parker IM (2008) Founder events in species invasions: genetic variation, adaptive evolution, and the role of multiple introductions. Mol Ecol 17:431–449
Dudash MR, Fenster CB (2001) The role of breeding system and inbreeding depression in the maintenance of an outcrossing mating strategy in *Silene virginica* (Caryophyllaceae). Am J Bot 88:1953–1959
Durka W, Bossdorf O, Prati D et al (2005) Molecular evidence for multiple introductions of garlic mustard (*Alliaria petiolata*, Brassicaceae) to North America. Mol Ecol 14:1697–1706
Dyer AR, Rice KJ (1999) Effects of competition on resource availability and growth of a California bunchgrass. Ecology 80:2697–2710
Ekblom R, Galindo J (2011) Applications of next generation sequencing in molecular ecology of non-model organisms. Heredity 107:1–15
Eckert CG, Kalisz S, Geber MA et al (2010) Plant mating systems in a changing world. Trends Ecol Evol 25:35–43
Ellstrand NC, Elam DR (1993) Population genetic consequences of small population size: implications for plant conservation. Annu Rev Ecol Syst 24:217–242
Estoup A, Guillemaud T (2010) Reconstructing routes of invasion using genetic data: why, how and so what? Mol Ecol 19:4113–4130
Estoup A, Baird SJE, Ray N et al (2010) Combining genetic, historical and geographical data to reconstruct the dynamics of bioinvasions: application to the cane toad *Bufo marinus*. Mol Ecol Resour 10:886–901
Facon B, Genton BJ, Shykoff J et al (2006) A general eco-evolutionary framework for understanding bioinvasions. Trends Ecol Evol 21:130–135
Ficetola GF, Bonin A, Miaud C (2008) Population genetics reveals origin and number of founders in a biological invasion. Mol Ecol 17:773–782
Fisher AJ, DiTomasso JM, Gordon TR (2005) Intraspecific groups of *Claviceps purpurea* associated with grass species in Willapa Bay, Washington, and the prospects for biological control of invasive *Spartina alterniflora*. Biol Control 34:170–179
Fisher RA (1958) The genetical theory of natural selection. Dover, New York, NY
Fortune PM, Pourtau N, Viron N et al (2008) Molecular phylogeny and reticulate origins of the polyploidy *Bromus* species from section *Genea* (Poaceae). Am J Bot 95:454–464
Futuyma DJ (2005) Evolution. Sinauer, Sunderland, MA
Gaskin JF, Bon M-C, Cock MJW et al (2011) Applying molecular-based approaches to classical biological control of weeds. Biol Control 58:1–21
Gaskin JF, Schwarzlander M, Kinter CL et al (2013) Propagule pressure, genetic structure, and geographic origins of *Chondrilla juncea* (Asteraceae): an apomictic invader of three continents. Am J Bot 100:1871–1882
Gerlach JD, Rice KJ (2003) Testing life history correlates of invasiveness using congeneric plant species. Ecol Appl 13:167–179
Germino MJ, Belnap J, Stark JM et al (2015) Ecosystem impacts of exotic annual invaders in the genus *Bromus*. In: Germino MJ, Chambers JC, Brown CS (eds) Exotic brome-grasses in arid and semiarid ecosystems of the Western US: causes, consequences, and management implications. Springer, New York, NY (Chapter 3)
Goncalves da Silva A, Eberhard JR, Wright TF et al (2010) Genetic evidence for high propagule pressure and long-distance dispersal in monk parakeet (*Myiopsitta monachus*) invasive populations. Mol Ecol 19:3336–3350
Goolsby JA, De Barro PJ, Makinson JR et al (2006) Matching the origins of an invasive weed for selection of a herbivore haplotype for a biological control programme. Mol Ecol 15:287–297
Grevstad FS, Strong DR, Garcia-Rossi D et al (2003) Biological control of *Spartina alterniflora* in Willapa Bay, Washington using the planthopper *Prokelisia marginata*: agent specificity and early results. Biol Control 27:32–42

Harmon-Threatt AN, Burns JH, Shemyakina LA et al (2009) Breeding system and pollination ecology of introduced plants compared to their native relatives. Am J Bot 96:1544–1550
Hierro JL, Villarreal D, Eren O et al (2006) Disturbance facilitates invasion: the effects are stronger abroad than at home. Am Nat 168:144–156
Hobbs RJ, Huenneke LF (1992) Disturbance, diversity, and invasion: implications for conservation. Conserv Biol 6:324–337
Holm LG, Plucknett DL, Pancho JV et al (1977) The world's worst weeds: distribution and biology. University Press of Hawaii, Honolulu, HI
Hudson ME (2008) Sequencing breakthroughs for genomic ecology and evolutionary biology. Mol Ecol Res 8:3–17
Hufft RA, Zelikova TJ (2015) Ecological Genetics, Local Adaptation, and Phenotypic Plasticity in Bromus tectorum in the Context of a Changing Climate. In: Germino MJ, Chambers JC, Brown CS (eds) Exotic brome-grasses in arid and semiarid ecosystems of the Western US: causes, consequences, and management implications. Springer, New York, NY (Chapter 5)
Hulbert LC (1955) Ecological studies of *Bromus tectorum* and other annual bromegrasses. Ecol Monogr 25:181–213
Huttanus TD, Mack RN, Novak SJ (2011) Propagule pressure and introduction pathways of *Bromus tectorum* (cheatgrass; Poaceae) in the Central United States. Int J Plant Sci 172:783–794
Kao RH, Brown CS, Hufbauer RA (2008) High phenotypic and molecular variation in downy brome (*Bromus tectorum*). Invasive Plant Sci Manag 1:216–225
Keeley JE, Allen CD, Betancourt J et al (2006) A 21st century perspective on postfire seeding. J Forestry 104:103–110
Kellog EA (2001) Evolutionary history of grasses. Plant Physiol 125:1198–1205
Knaap PA (1996) Cheatgrass (*Bromus tectorum* L.) dominance in the Great Basin desert: history, persistence, and influence of human activities. Glob Environ Chang 6:37–52
Kolar CS, Lodge DM (2001) Progress in invasion biology: predicting invaders. Trends Ecol Evol 16:199–204
Kolbe JJ, Larson A, Losos JB (2007) Differential admixture shapes morphological variation among invasive populations of the lizard *Anolis sagrei*. Mol Ecol 16:1579–1591
Lavergne S, Molofsky J (2007) Increased genetic variation and evolutionary potential drive the success of an invasive grass. Proc Natl Acad Sci U S A 104:3883–3888
Lee CE (2002) Evolutionary genetics of invasive species. Trends Ecol Evol 17:386–391
Leger EA, Rice KJ (2003) Invasive California poppies (*Eschscholzia californica* Cham.) grow larger than native individuals under reduced competition. Ecol Lett 6:257–264
Leger EA, Espeland EK, Merrill KR et al (2009) Genetic variation and local adaptation at a cheatgrass (*Bromus tectorum*) invasion edge in western Nevada. Mol Ecol 18:4366–4379
Le Roux J, Wieczorek AM (2009) Molecular systematics and population genetics of biological invasions: towards a better understanding of invasive species management. Ann Appl Biol 154:1–17
Lockwood JL, Cassey P, Blackburn T (2005) The role of propagule pressure in explaining species invasions. Trends Ecol Evol 20:223–228
Lombaert E, Guillemaud T, Thomas CE et al (2011) Inferring the origin of populations introduced from a genetically structured native range by approximate Bayesian computation: case study of the invasive ladybird *Harmonia axyridis*. Mol Ecol 20:4654–4670
Mack RN (1981) Invasion of *Bromus tectorum* L. into western North America: an ecological chronicle. Agro-Ecosystems 7:145–165
Mack RN (1989) Temperate grasslands vulnerable to plant invasion: characteristics and consequences. In: Drake J (ed) Biological invasions: a global perspective. Wiley, New York, NY, pp 155–179
Mack RN (2000) Cultivation fosters plant naturalization by reducing environmental stochasticity. Biol Invasions 2:111–122
Mack RN (2005) Assessing biotic invasions in time and space: the second imperative. In: Mooney HA, Mack RN, McNeely JA et al (eds) Invasive alien species: a new synthesis. Island Press, Washington, DC, pp 179–208

Mack RN (2011) Fifty years of "Waging war on cheatgrass": research advances, while meaningful control languishes. In: Richardson DM (ed) Fifty years of invasion ecology: the legacy of Charles Elton. Wiley-Blackwell, Oxford, pp 253–265

Mack RN, Smith MC (2011) Invasive plants as catalysts for the spread of human parasites. NeoBiota 9:13–29

Mack RN, Simberloff D, Lonsdale WM et al (2000) Biotic invasions: causes, epidemiology, global consequences, and control. Ecol Appl 10:689–710

Maron JL, Montserrat V, Bommarco R et al (2004) Rapid evolution of an invasive plant. Ecol Monogr 74:261–280

McFadyen REC (1998) Biological control of weeds. Annu Rev Entomol 43:369–393

McKone MJ (1985) Reproductive biology of several bromegrasses (*Bromus*): breeding system, pattern of fruit maturation, and seed set. Am J Bot 72:1334–1339

McKone MJ (1987) Sex allocation and outcrossing rate: a test of theoretical predictions using bromegrasses (*Bromus*). Evolution 41:591–598

McKone MJ (1989) Intraspecific variation in pollen yield in bromegrass (Poaceae; *Bromus*). Am J Bot 76:231–237

Merrill KR, Meyer SE, Coleman CE (2012) Population genetic analysis of *Bromus tectorum* (Poaceae) indicates recent range expansion may be facilitated by specialist genotypes. Am J Bot 99:529–537

Meyer SE, Ghimire S, Decker S et al (2013) Ghost of outcrossing past in downy brome, an inbreeding annual grass. J Hered 104:476–490

Muhlenberg H (1793) Index *Florae Lancastriensis*. Trans Am Philos Soc Philadel 3:157–184

Muller-Scharer H, Schaffner U, Steinger T (2004) Evolution in invasive plants: implications for biological control. Trends Ecol Evol 19:417–422

Nadeau NJ, Jiggins CD (2010) A golden age of evolutionary genetics? Genomic studies of adaptation in natural populations. Trends Genet 26:484–492

Neuffer B, Hurka H (1999) Colonization history and introduction dynamics of *Capsella bursa-pastoris* (Brassicaceae) in North America: isozymes and quantitative traits. Mol Ecol 8:1667–1681

Nicotra AB, Atkin OIK, Bonser SP et al (2010) Plant phenotypic plasticity in a changing climate. Trends Plant Sci 15:684–692

Novak SJ (2007) The role of evolution in the invasion process. Proc Natl Acad Sci U S A 104:3671–3672

Novak SJ (2011) Geographic origins and introduction dynamics. In: Simberloff D, Rejmanek M (eds) Encyclopedia of biological invasions. University of California Press, Berkeley, CA, pp 273–280

Novak SJ, Mack RN (1993) Genetic variation in *Bromus tectorum* (Poaceae): comparison between native and introduced populations. Heredity 71:167–176

Novak SJ, Mack RN (2001) Tracing plant introduction and spread: genetic evidence from *Bromus tectorum* (Cheatgrass). BioScience 51:114–122

Novak SJ, Mack RN (2005) Genetic bottlenecks in alien plant species: influence of mating systems and introduction dynamics. In: Sax DF, Stachowicz JJ, Gaines SD (eds) Species invasions: insights into ecology, evolution and biogeography. Sinauer, Sunderland, MA, pp 201–228

Novak SJ, Rausch JH (2009) Use of field surveys, distributional data and genetic analyses to monitor alien species: *Taeniatherum caput-medusae* as an example of the approach. NeoBiota 8:169–182

Novak SJ, Sforza R (2008) Genetic analysis of native and introduced populations of *Taeniatherum caput-medusae* (Poaceae): implications for biological control. In: Julien MH, Sforza R, Bon M-C et al (eds) Proceedings of the XII international symposium on biological control of weeds. CAB International, Wallingford, pp 422–428

Novak SJ, Mack RN, Soltis DE (1991) Genetic variation in *Bromus tectorum* (Poaceae): population differentiation in its North American range. Am J Bot 78:1150–1161

Novak SJ, Mack RN, Soltis PS (1993) Genetic variation in *Bromus tectorum* (Poaceae): introduction dynamics in North America. Can J Bot 71:1441–1448

Parsons WT, Cuthbertson EG (1992) Noxious weeds of Australia. Inkata Press, Melbourne
Pawlak AR, Mack RN, Busch JW et al (2015) Invasion of *Bromus tectorum* (L.) into California and the American Southwest: rapid, multi-directional and genetically diverse. Biol Invasions 17(1):287–306
Pimentel D, Zuniga R, Morrison D (2005) Update on the environmental and economic costs associated with alien-invasive species in the United States. Ecol Econ 52:273–288
Prentis PJ, Wilson JRU, Dormontt EE et al (2008) Adaptive evolution in invasive species. Trends Ecol Evol 13:288–294
Prinzing A, Durka W, Klolz S et al (2002) Which species become aliens? Evol Ecol Res 4:385–405
Ramakrishnan AP, Meyer SE, Waters J et al (2004) Correlation between molecular markers and adaptively significant genetic variation in *Bromus tectorum*. (Poaceae), an inbreeding annual grass. Am J Bot 91:797–803
Ramakrishnan AP, Meyer SE, Fairbanks DJ et al (2006) Ecological significance of microsatellite variation in western North American populations of *Bromus tectorum*. Plant Species Biol 21:61–73
Reusch TB, Wood TE (2007) Molecular ecology of global change. Mol Ecol 16:3973–3992
Ricciardi A, Jones L, Kestrup ÅM et al (2011) Expanding the propagule pressure concept to understand the impact of biological invasions. In: Richardson DM (ed) Fifty years of invasion ecology: the legacy of Charles Elton. Wiley-Blackwell, Oxford, pp 225–235
Richards CL, Bossdorf O, Pigliucci M (2010) What role does heritable epigenetic variation play in phenotypic evolution? BioScience 60:232–237
Richards CL, Schrey AW, Pigliucci M (2012) Invasion of diverse habitats by few Japanese knotweed genotypes is correlated with epigenetic differentiation. Ecol Lett 15:1016–1025
Ritland K (2002) Extensions of models for the estimation of mating systems using independent loci. Heredity 88:221–228
Roderick GK, Navajas M (2003) Genes in new environment: genetics and evolution in biological control. Nat Genet 4:889–899
Ross KG, Shoemaker DD (2008) Estimation of the number of founders of an invasive pest insect population: the fire ant *Solenopsis invicta* in the USA. Proc R Soc B 275:2231–2240
Sakai AK, Allendorf FW, Holt JS et al (2001) The population biology of invasive species. Annu Rev Ecol Syst 32:305–332
Sala OE, Chapin FS III, Armesto JJ et al (2000) Global diversity scenarios for the year 2100. Science 287:1770–1774
Salamin N, Wuest RO, Lavergne S et al (2010) Assessing rapid evolution in a changing environment. Trends Ecol Evol 25:692–698
Sales F (1993) Taxonomy and nomenclature of *Bromus* sect. *Genea*. Edinb J Bot 50:1–31
Sales F (1994) Evolutionary tendencies in some annual species of *Bromus* (*Bromus* L. section *Genea* Dum. (Poaceae)). Bot J Linn Soc 115:197–210
Salo LF (2005) Red brome (*Bromus rubens* subsp. *madritensis*) in North America: possible modes for early introduction, subsequent spread. Biol Invasions 7:165–180
Schachner LJ, Mack RN, Novak SJ (2008) *Bromus tectorum* (Poaceae) in midcontinental United States: population genetic analysis of an ongoing invasion. Am J Bot 95:1584–1595
Schlotterer C (2004) The evolution of genetic markers—just a matter of fashion? Nat Rev Genet 5:63–69
Scott JW, Meyer SE, Merrill KR et al (2010) Local population differentiation in *Bromus tectorum* L. in relationship to habitat-specific selection regimes. Evol Ecol 24:1061–1080
Schierenbeck KA, Ellstrand NC (2009) Hybridization and the evolution of invasiveness of plants and other organisms. Biol Invasions 11:1093–1105
Schneider J, Doring E, Hilu KW et al (2009) Phylogenetic structure of the grass subfamily Pooideae based on comparison of plastid *matK* gene – 3′ *trnK* exon and nuclear ITS sequences. Taxon 58:405–424
Sexton JP, McKay JK, Sala A (2002) Plasticity and genetic diversity may allow saltcedar to invade cold climates in North America. Ecol Appl 12:1651–1660

Simberloff D (2009) The role of propagule pressure in biological invasions. Annu Rev Ecol Evol Syst 40:81–102
Slatkin M, Barton NH (1989) A comparison of three indirect methods for estimating average levels of gene flow. Evolution 43:1349–1368
Smith PM (1970) Taxonomy and nomenclature of the brome-grasses (*Bromus* L. s.l.). Notes from the Royal Botanical Garden Edinburgh 30: 361–375
Smith PM (1980) *Bromus* L. In: Tutin TG, Heywood VH, Burges NA et al (eds) Florae europaea, vol 5. Cambridge University Press, Cambridge, pp 182–189
Sork VL, Nason J, Campbell DR (1999) Landscape approaches to historical and contemporary gene flow in plants. Trends Ecol Evol 14:219–224
Sperry LJ, Belknap J, Evans RD (2006) *Bromus tectorum* invasion alters nitrogen dynamics in undisturbed arid grassland ecosystems. Ecology 87:603–615
Stebbins GL (1957) Self-fertilization and population variability in higher plants. Am Nat 91:337–354
Stewart G, Hull AC (1949) Cheatgrass (*Bromus tectorum* L.)—an ecological intruder in southern Idaho. Ecology 30:58–74
Stinchcombe JR, Hoekstra HE (2008) Combining population genomics and quantitative genetics: finding the genes underlying ecologically important traits. Heredity 100:158–170
Suarez AV, Tsutsui ND (2008) The evolutionary consequences of biological invasions. Mol Ecol 17:351–360
Sumners WH, Archibold OW (2007) Exotic plant species in the southern boreal forest of Saskatchewan. For Ecol Manag 251:156–163
Valliant MT, Mack RN, Novak SJ (2007) Introduction history and population genetics of the invasive grass *Bromus tectorum* (Poaceae) in Canada. Am J Bot 94:1156–1169
Van Kleunen M, Fischer M, Schmidt B (2000) Clonal integration in *Ranunculus reptans*: by-product or adaptation? J Evol Biol 13:237–248
Vergeer P, Rengelink R, Ouborg NJ et al (2003) Effects of population size and genetic variation on the response of *Succisa pratensis* to eutrophication and acidification. J Ecol 91:600–609
von Holle B, Simberloff D (2005) Ecological resistance to biological invasions overwhelmed by propagule pressure. Ecology 86:3212–3218
Wares JP, Hughes AR, Grosberg RK (2005) Mechanisms that drive evolutionary change: insights from species introductions and invasions. In: Sax DF, Stachowicz JJ, Gaines SD (eds) Species invasions: insights into ecology, evolution and biogeography. Sinauer, Sunderland, MA, pp 229–257
Wittenberg R, Cock MJW (2005) Best practices for the prevention and management of invasive alien species. In: Mooney HA, Mack RN, McNeely JA et al (eds) Invasive alien species, a new synthesis. Island Press, Washington, DC, pp 209–232
Wright S (1931) Evolution in Mendelian populations. Genetics 16:97–159
Xu C-Y, Julien MH, Fatemi M et al (2010) Phenotypic divergence during the invasion of *Phyla canescens* in Australia and France: evidence for selection-driven evolution. Ecol Lett 13:32–44

Chapter 5
Ecological Genetics, Local Adaptation, and Phenotypic Plasticity in *Bromus tectorum* in the Context of a Changing Climate

Rebecca A. Hufft and Tamara J. Zelikova

Abstract Effective management of invasive species spread requires understanding the potential of exotic species to colonize different habitat types. In the case of *Bromus tectorum*, colonization potential includes persisting in variable environments via phenotypic plasticity or via genetic variation. *Bromus tectorum* L. (cheatgrass or downy brome) is a highly invasive, self-pollinating, winter annual grass that was introduced to the intermountain region of North America around 1890 and expanded to its modern range within 40 years. Common garden studies have helped shed light on outcrossing frequency, microsite effects on establishment and growth, traits that could confer invasiveness, and variation in germination, morphology, and physiology. Here, we review the evidence for existing local adaptation and phenotypic plasticity in *B. tectorum* in its invaded range along with the potential for responses to climate change and discuss implications of both for its success as an invader and future management. All of these studies show that *B. tectorum* can tolerate a wide range of habitats as the result of genetic variation among populations, a range of locally adapted ecotypes, and phenotypic plasticity. The success of *B. tectorum* could be due to its ability to maintain fitness in both high-quality and marginal environments.

Keywords Adaptive evolution • Common garden • General-purpose genotype • Genetic variation • Phenotypic plasticity • Reciprocal transplant

---

R.A. Hufft (✉)
Denver Botanic Gardens, Denver, CO 80206, USA
e-mail: rebecca.hufft@botanicgardens.org

T.J. Zelikova
Department of Botany, University of Wyoming, Laramie, WY 82071, USA
e-mail: tzelikov@uwyo.edu

M.J. Germino et al. (eds.), *Exotic Brome-Grasses in Arid and Semiarid Ecosystems of the Western US*, Springer Series on Environmental Management,
DOI 10.1007/978-3-319-24930-8_5

5.1 Introduction

Ecological genetics, the study of the genetic basis for traits of ecological significance, such as those related to fitness, flowering time, and drought tolerance, provides valuable information to address ecological and evolutionary questions in natural systems. Heritable traits that confer high fitness can result in the adaptation of a population to its local environment, i.e., local adaptation (e.g., Turesson 1922; Clausen et al. 1941; Hiesey and Milner 1965; Gurevitch 1992; Dudley and Schmitt 1995; Cordell et al. 1998; Huey et al. 2000; Leimu and Fischer 2008). Understanding which fitness traits are advantageous in particular environments is especially relevant for invasive species and can help target management efforts to areas where invasive species might spread (Parker et al. 2003), especially in the face of climate change.

We know that plants can evolve over relatively short time scales to adapt to environmental conditions (Franks et al. 2007) and invasive species can do so even more quickly in their introduced range (Maron et al. 2004; Colautti and Barrett 2013). Multiple introductions are common for invasive species and can increase genetic variation and thus the potential for adaptive evolution (Dlugosch and Parker 2008). We would predict that invaders likely succeed in multiple environments if they are preadapted to local conditions, locally adapted via the creation of new genotypes through gene flow among independent introductions, or present general-purpose genotypes that survive through phenotypic or developmental plasticity (Parker et al. 2003). Effective management of the spread of invasive species requires differentiating among these options by quantifying the capacity for invasive species to have general-purpose genotypes (Baker 1965) with high levels of plasticity, assessing their ability to genetically adapt to new environments via in situ evolution of novel genotypes, or determining the prominence of preadapted genotypes (Parker et al. 2003). Understanding the relative roles of phenotypic plasticity and genetic adaptation in key life history traits of invasive species can help explain their success and help guide future management efforts.

Bromus tectorum L. (cheatgrass or downy brome) is a highly invasive, self-pollinating, winter annual grass that was introduced to the intermountain region of North America around 1890 and expanded to cover the intermountain West within 40 years, occupying much of the perennial shrublands in Washington, Idaho, Oregon, Nevada, Utah, and British Columbia (Mack 1981). This species is unique as one of the few invaders for which we have a good understanding of the dispersal history (Novak and Mack 1993, 2001; Novak et al. 1993). Historical and genetic data suggest multiple introductions into North America (Novak and Mack 2015). These multiple introductions have led to higher genetic diversity than would otherwise be expected in a predominantly selfing species (Novak and Mack 2015) and provide additional mechanisms underlying the success of its invasion (i.e., the potential to have preadapted genotypes and higher diversity on which evolution could act in situ to generate novel genotypes).

In its native range, *B. tectorum* succeeds in a variety of habitats, including the arid steppe region of Europe, the Middle East, and temperate and tropical parts of Asia, and this success in diverse habitats is thought to be due to high phenotypic plasticity (Fenesi et al. 2011). *Bromus tectorum* occurs in grasslands and shrublands on a range of soil types in North America, including salt desert, semidesert shrubland, shrub-steppe, grassland, low-elevation dry forest, and mid- and high-elevation mesic forests (Rice and Mack 1991a). While it can utilize the understory of forested areas, *B. tectorum* is often at low abundance (e.g., less than 5 % cover under *Cercocarpus ledifolius* Nutt., Leger et al. 2009), and forest understory is considered marginal habitat (Rice and Mack 1991a). A generalist ecological and evolutionary strategy, along with the genetic changes that may have occurred during and after its introduction, could be the reason for the ability of *B. tectorum* to invade a range of habitats (from Mojave Desert to intermountain montane and from open to understory vegetation), leading to its ultimate success as an invader in North America (Fig. 1.1 in Germino et al. 2015; Fig. 5.1).

There are two main approaches for studying genetic variation among populations and environments: (1) direct evaluation of neutral molecular genetic variation (with unknown ecological importance) and (2) measurement of genetic variation in quantitative traits (with known ecological function but unknown genetic basis) (Via 1990). Common garden and reciprocal transplant studies have been used for the past century to answer ecological and evolutionary questions, especially those related to local adaptation (Turesson 1922; Clausen et al. 1948). Common garden studies evaluate all genotypes in a common environment and through manipulations attempt to identify causal mechanisms for variation. Reciprocal transplant studies are essentially extensions of common garden studies that allow the incorporation of all the environmental variability in the home sites to test for local adaptation but do not necessarily inform a direct causal mechanism for adaptation. Reciprocal transplant studies can provide evidence of adaptive differentiation but must be combined with other approaches to identify causes of such variation, including genetic studies to determine if results are due to in situ evolution or pre-adapted genotypes, the role of multigenerational effects, and an examination of traits conferring adaptation. Common garden studies can be more logistically feasible than reciprocal transplant studies and can help identify putative causes of variation when experimental manipulations are used; however, they cannot provide direct evidence of adaptation, since not all genotypes are tested in their home environments, and environmental differences between the common garden location and the site of origin can lead to differences in performance rather than directly evaluating adaptation.

Common garden studies with *B. tectorum* have helped shed light on outcrossing frequency (Meyer et al. 2013), microsite effects on establishment and growth (Hoover and Germino 2012), possession of traits that could confer invasiveness (Fenesi et al. 2011), and variation in germination, morphology, and physiology (Rice and Mack 1991a; Kao et al. 2008; Bykova and Sage 2012). All these studies

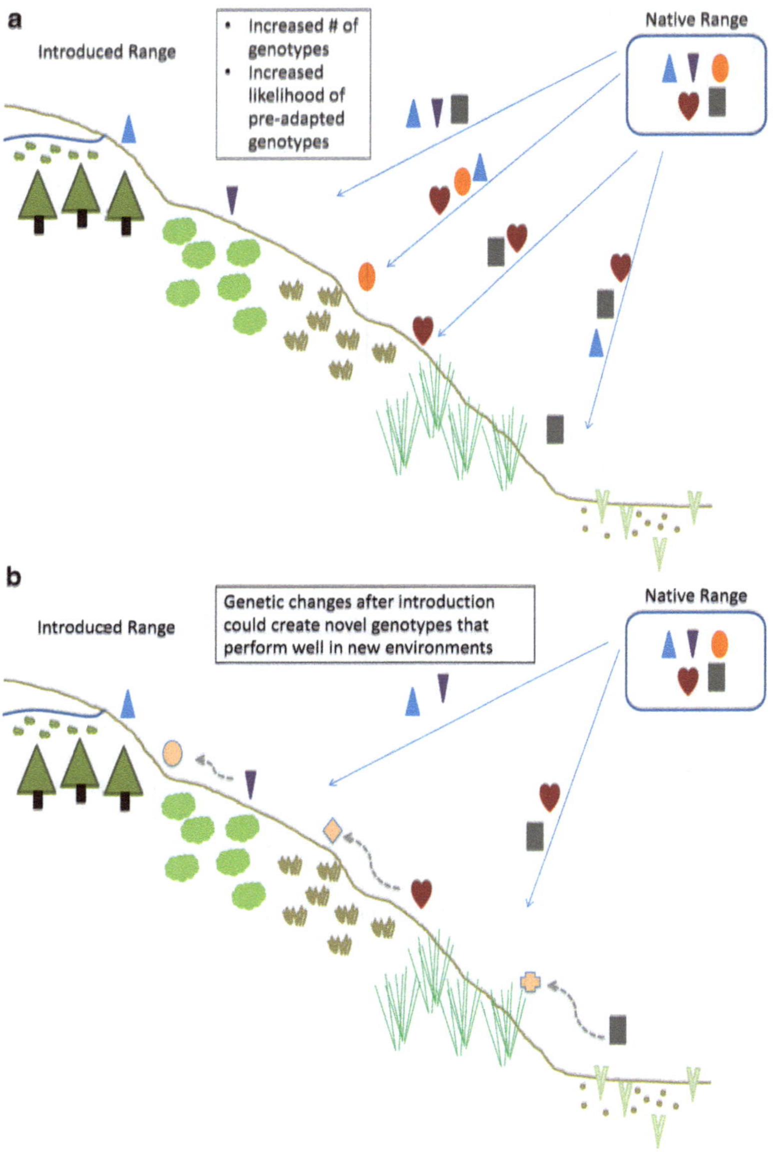

Fig. 5.1 Broad ecological tolerance in the introduced range can occur through three main processes described here. *Symbols* inside each box represent different genotypes. (**a**) Preadapted genotypes, (**b**) novel genotypes, and (**c**) phenotypic plasticity

show that *B. tectorum* can tolerate a wide range of environmental conditions as the result of genetic variation and locally adapted ecotypes. Here, we review the evidence for existing phenotypic plasticity and local adaptation in *B. tectorum* in its invaded range and discuss implications for future management in a changing climate.

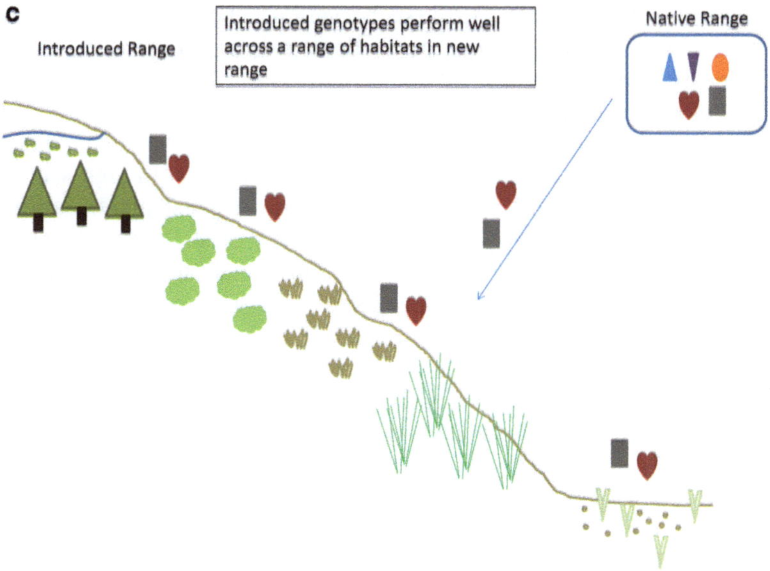

Fig. 5.1 (continued)

5.2 Evidence for Broad Environmental Tolerance

Bromus tectorum has a wide temperature tolerance that likely plays an important role in invasion success (Brooks et al. 2015), with the current northern range limit likely determined by its maximum cold tolerance (Salo 2005; Bykova and Sage 2012). *Bromus tectorum* could tolerate temperatures down to −31 °C in a greenhouse experiment and survived a sudden, severe fall freeze better than a congener (*Bromus rubens* L.) tested in the same experiment (Bykova and Sage 2012). In addition to surviving fall freezes, *B. tectorum* is able to take advantage of transient warm periods in autumn and spring with increased nitrate acquisition rates relative to native perennial grasses (Leffler et al. 2011) and greater root growth than natives at low temperatures. This ability to survive and thrive at low temperatures and take advantage of transient warm periods could give *B. tectorum* a competitive edge over native species in habitats characterized by cold winters. While *B. tectorum* can tolerate low temperature extremes, the optimal temperature range for *B. tectorum* growth is 10–20 °C, and it barely grows at or below 5 °C and above 40 °C (McCarlie et al. 2000). Thus, despite a low level of tolerance for extremes and fine-scale modeling efforts that predict an expansion of *B. tectorum* into colder habitats (West et al. 2015), a shift in temperature optima would be necessary for *B. tectorum* to continue to invade and succeed if climate extremes become the new norm as predicted (IPCC 2014).

In addition to the effects of temperature on growth and reproduction, *B. tectorum* is also responsive to other extreme growing conditions, with evidence for a genetic basis underlying the ability to tolerate such conditions. For example, there is genetic differentiation among *B. tectorum* populations growing in high and low salinity environments. A unique microsatellite genotype dominated a *B. tectorum* population in a high saline playa, but this genotype was absent in populations occurring on nonsaline soils (Scott et al. 2010). In a reciprocal transplant experiment, plants from the high salinity playa sites showed evidence of adaptation to high salinity, with larger size and seed production in high salinity treatments than individuals from less saline home sites in a greenhouse experiment (Scott et al. 2010). The same Scott et al. (2010) experiment also found that seeds from saline sites had higher germination and plant establishment rates than seeds from other sites. In addition, some seeds from the nonsaline habitat performed best at their site of origin, when compared to other reciprocal seeding transplant sites (Scott et al. 2010). Similar differences in plant performance and reproduction between home and away soils were found in a study in central Nevada (Haubensak et al. 2014). The genotypic variation and differential performance among habitats suggest local adaptation that may have resulted from selection for preadapted genotypes or the presence of novel genotypes. However, given low outcrossing rates, the creation of novel genotypes is less likely responsible for the presence of adapted genotypes on the high salinity playa sites (Scott et al. 2010).

At sites with established, high-density *B. tectorum* populations, generalist genotypes dominate. In contrast, at sites of recent expansions such as salt desert and warm desert habitats, unique genotypes dominate, suggesting they are specialist genotypes (Merrill et al. 2012). A more recent study confirms only genotypes specific to warm desert habitats (specialist genotypes) are found across a range of recently invaded warm and salt desert habitats in the Mojave Desert (Lara 2013). This suggests that while specialist genotypes may be important in the initial stages of the introduction and as populations get well established over time and reach high densities, generalist genotypes may be the force behind their long-term success.

Finally, although most *B. tectorum* seeds germinate within a year, some can remain in the seed bank for 3–5 years and germinate under more favorable environmental conditions. Variation in the timing of *B. tectorum* seed germination is under strong genetic control and likely contributes to its invasiveness (Meyer and Allen 1999). Greenhouse experiments show that the relationship between *B. tectorum* germination and temperature varies across populations (Meyer and Allen 1999), that seed dormancy is thermoregulated (Bair et al. 2006), and that the likelihood of seed dormancy appears to vary across *B. tectorum* populations (Rice and Dyer 2001; Kao et al. 2008). Comparisons of *B. tectorum* seed morphology among native and introduced populations show that North American seeds have thinner lemma and palea, morphological changes that help seeds germinate faster (Liu et al. 2013). Taken together, seed characteristics vary across *B. tectorum* populations, and this variation is genetically controlled and also influenced by the environment, with direct influence on germination timing and competitive success.

5.3 Evidence for Local Adaptation via Neutral Genetic Variation

With its annual, selfing life history, *B. tectorum* is expected to have very low levels of outcrossing and within-population genetic diversity. Indeed, outcrossing rates are low in introduced field populations in North America (Valliant et al. 2007) and even in experimentally created, high-diversity populations in a common garden study (Meyer et al. 2013). Despite low outcrossing rates, genetic diversity in its invaded range in North America seems higher than expected and sufficient to allow for adaptation into new environments (Table 5.1). Furthermore, large numbers of recombinant genotypes could be generated given the millions of plants per hectare in North America (Meyer and Leger 2010). This is largely because there is ample genetic variation within North American *B. tectorum* populations on which selection can act (Ramakrishnan et al. 2006; Ashley and Longland 2007; Valliant et al. 2007; Kao et al. 2008; Schachner et al. 2008; Huttanus et al. 2011; Avolio et al. 2012; Merrill et al. 2012). Because *B. tectorum* is a primarily selfing plant, genotypes with high fitness can be maintained, and low outcrossing rates (1 %) assure formation of novel genotypes (Meyer et al. 2013). Although some studies have concluded that the low outcrossing rates observed in *B. tectorum* are insufficient to introduce novel genotypes (0.58 %, Merrill et al. 2012), others have concluded that some populations do have outcrossing rates high enough to help create novel genotypes during range expansion into new habitats (1.62 %, Leger et al. 2009). In addition, genotypes can have variable outcrossing rates depending on the environment (Ashley and Longland 2007; Meyer et al. 2013), which means the rate of outcrossing could also change with continued range expansions and climate change. In order to better assess actual outcrossing rates and implications for the creation of novel genotypes and resultant species range expansion, a more sensitive molecular marker system, such as the previously developed SNP markers (Meyer et al. 2013), needs to be combined with a broader sample of individuals across the range of habitats (Meyer and Leger 2010).

Variation in genotypes among habitat types can indicate the presence of specialized ecotypes, and especially in *B. tectorum*, genetic markers correlate with and can be good surrogates for phenotypic traits (Ramakrishnan et al. 2004). In a microsatellite study using 21 western North American populations ranging from desert to montane from previous germination studies, genetic variation was correlated with habitat types (Meyer et al. 1997), indicating the presence of distinct *B. tectorum* ecotypes (Ramakrishnan et al. 2006). In addition, there was a genetic basis for seed germination traits that varied across populations (Ramakrishnan et al. 2006). Using two populations of *B. tectorum* in the Great Basin (one low-density population at high elevation and one high-density population at low elevation), Leger et al. (2009) used microsatellites to evaluate within-population genetic variation and a reciprocal transplant study to evaluate field performance. Within-population genetic and phenotypic variation differed between high- and low-elevation sites, suggesting a level of local adaptation to a particular set of environmental conditions that vary with elevation, namely, temperature and precipitation. This local adaptation could stem

Table 5.1 Results of genetic analyses of *Bromus tectorum*. (A) is the average number of alleles per locus, (FIS) is the inbreeding coefficient, and (H_T) is the total gene diversity

Study	Location	Marker	Sample size	Genotypes	A	FIS	H_T	Heterozygotes/%	Mean heterozygosity
Novak et al. (1991)	Western USA	Allozyme	1830		1.05	1	0.132	0	0
Bartlett et al. (2002)	Eastern USA	Allozyme	1248		1.01	1	0.075	0	0
Bartlett et al. (2002)	NV-CA	Allozyme	546		1.05	1	0.192	0	0
Bartlett et al. (2002)	Intermountain West	Allozyme	825		1.05	1	0.119	0	0
Bartlett et al. (2002)	British Colombia	Allozyme	329		1.08	1	0.261		0
Bartlett et al. (2002)	Europe	Allozyme	1242		1.02	0.98			0.0001
Bartlett et al. (2002)	Southwest Asia	Allozyme	484		1.04	0.96			0.0002
Valliant et al. (2007)	Canada	Allozyme	1892		1.04	0.9917	0.159	8/0.0042	
Schachner et al. (2008)	Midcontinent USA	Allozyme	1624	9	1.04	0.9778	0.228		0.0002
Huttanus et al. (2011)	Central USA	Allozyme	1841		1.05	0.9979	0.187	1/0.0005	
Ramakrishnan et al. (2006)	UT, ID, NV, CO	SSR	807	50	2.73	1		0	0
Ashley and Longland (2007)	NV	SSR	182	51		0.891–1	0.234	15/8.24	0.021
Kao et al. (2008)	CO	SSR	90	41	2.53	0.968–1		1/1.1	0.0002
Leger et al. (2009)	NV	SSR	185	29				3/1.62	
Scott et al. (2010)	UT	SSR	200	28				2/1	
Merrill et al. (2012)	Intermountain West	SSR	1884	131	10	0.977		11/0.58	0.0071
Meyer et al. (2013)	Intermountain West	SSR	2000			0.9851		15/0.0075	
Zelikova et al. (2013)	UT	SSR	36						0.257
Meyer et al. (2013)	Intermountain West	SNP	1429		1.70	0.97–0.99		119/8.32	0.00325

from the presence of preadapted genotypes or in situ evolution of novel genotypes, two non-mutually exclusive mechanisms that can lead to similar outcomes. The same study also found that while individuals from the high-density, low-elevation site were locally adapted as illustrated by higher germination rates and overall survival, there was still potential for adaptation at the invasion edge (high elevation) due to relatively high genetic diversity within those high-elevation sites and evidence of outcrossing. Such potential for further adaptation could mean that *B. tectorum* could continue to invade into higher elevations and additional habitats.

5.4 Evidence for Broad Environmental Tolerance via Phenotypic Plasticity

The success of species in a large range of environments suggests a high level of either genetic adaptation or phenotypic plasticity. Selection operates on phenotypic variation, and success in a broad range of environmental conditions can result from an increase in phenotypic plasticity (Bradshaw 1965). Environmental and habitat variation necessitates some level of physiological adaptation, and for *B. tectorum*, the most important physiological attributes that promote its invasiveness are (1) large temperature tolerance range, (2) early germination paired with rapid growth and high rates of water and nutrient uptake (Funk and Vitousek 2007; James et al. 2011), and (3) competitive ability in extreme environments (Chambers et al. 2015).

Invasive species generally show significantly more phenotypic plasticity than non-invasive species (Davidson et al. 2011). Because maternal effects can contribute to trans-generational plasticity (Dyer et al. 2010), the increase in phenotypic plasticity may be advantageous to invasive species (allowing them to perform well in newly invaded environments in the first generation) or may hinder local adaptation by impeding selection for locally adapted traits because phenotypes still reflect the former maternal environment (Moran and Alexander 2014 and references therein). In its native range, *B. tectorum* has exceptionally high phenotypic plasticity (Fenesi et al. 2011). Not only is this species-level plasticity beneficial to invasiveness, but it has been suggested that *B. tectorum* is even more plastic in its invasive range relative to its native range, including greater adaptive plasticity for freezing tolerance and more rapid seedling emergence in response to warm temperatures (Griffith et al. 2014), though more populations should be tested in the native range to confirm this finding.

Rice and Mack (1991b) evaluated *B. tectorum* phenotypic plasticity variation among six populations along a gradient in temperature and precipitation in habitats in its invaded range where populations were genetically differentiated (Rice and Mack 1991a). Across the range of habitats tested, phenotypic plasticity in flowering phenology and seed weight was very low, but aboveground biomass and seed number plasticity was high. Rice and Mack suggested that even when there is a genetic basis for phenological traits, *B. tectorum* fitness may still be driven by environmental factors given its high level of plasticity (Rice and Mack 1991b). High phenotypic plasticity in flowering, aboveground biomass, and seed production has been observed at these same sites over years that varied in temperature and precipitation, with plants producing few seeds in harsh years and many seeds under optimal climate conditions, attesting to the benefit of plasticity in seed production (Mack and Pyke 1983). While this is not unique to just *B. tectorum*, this does attest to the fact that although a genetic basis for many traits has been found, important fitness traits are still more likely controlled by the environment.

5.5 Implications of Local Adaptation and Phenotypic Plasticity for Invasion Success

Bromus tectorum is both phenotypically plastic and able to adapt to local conditions, traits that surely contribute to *B. tectorum* success in a wide range of habitats. This ability to maintain fitness advantages in both unfavorable and favorable environments is what Richards et al. (2006) call a Jack-and-master, a dual strategy that appears to also account for the success of another North American invader, *Tamarix ramosissima* Ledeb. (Sexton et al. 2002). Regardless of whether it results from the arrival of preadapted genotypes or in situ adaption, local adaptation may be associated with variation in phenotypic plasticity, such that populations with higher plasticity can take advantage of changing environmental conditions, while populations with lower plasticity have more fixed traits and thrive under stable conditions. In a field study monitoring *B. tectorum* phenology in response to late season moisture, mesic site plants had more phenological plasticity that allowed them to take advantage of the late season moisture, but xeric site plants had more fixed phenology and could not make use of the late season rains (Dyer et al. 2012). This is somewhat counterintuitive because xeric environments are broadly characterized by more variable precipitation regimes, and selection should favor traits that are compatible with high phenological plasticity (xeric sites, Dyer et al. 2012). However, not all traits would be expected to show increased phenological plasticity, and the timing of seed germination is generally less variable in more predictable environments (Meyer and Allen 1999).

In addition to germination, the phenology of growth and reproduction is under environmental and genetic regulation (Ball et al. 2004). In a common garden study, *B. tectorum* flowering times varied among habitat types, with shrub-steppe and semidesert shrubland ecotypes flowering the earliest and high-elevation mesic forest flowering the latest (Rice and Mack 1991a). The adaptive significance of such variation could be related to drought avoidance in semidesert shrubland and steppe habitats (Rice and Mack 1991a) but remains to be explored in other habitats. Vernalization (i.e., prechilling requirement for flowering) also varies by habitat type and appears to be adaptively significant, with warm desert populations flowering without vernalization and cooler foothills and montane sites flowering faster with increased vernalization (Meyer et al. 2004). The interaction between environmental and genetic control of phenology has implications for how *B. tectorum* will grow and reproduce under future climate conditions, with some ecotypes likely to expand, while others contract if genetic variation in traits related to phenology does not match the pace of changes in environmental conditions. Understanding the importance of phenological plasticity is especially important in light of changing climate, where higher plasticity may be associated with better ability to adapt to more unpredictable future climate.

5.6 Ecological and Evolutionary Adaptations to Climate Change

Plant evolutionary responses to climate change can be critical in determining species persistence, and this is especially true with invasive species that have higher phenotypic plasticity that can be adaptive (Davidson et al. 2011). As a result of both the potential introduction of preadapted genotypes and higher phenotypic plasticity, invasive species such as *B. tectorum* may be especially poised to show evolutionary responses to climate change, and their responses may be more rapid than those of native species (Clements and Ditommaso 2011). There are several lines of evidence to suggest that *B. tectorum*, along with other invasive grasses in the genus *Bromus*, should be capable of both ecological and evolutionary responses to climate change (Lee 2002), especially in light of its annual life cycle and primarily selfing reproduction that can maintain genotypes with high fitness (Meyer et al. 2013). As a result, *B. tectorum*'s responses to climate change could be especially rapid (Jump and Penuelas 2005; Leger et al. 2009). Furthermore, climate has a substantial influence on most aspects of *B. tectorum*'s life cycle, including seed vernalization (Meyer et al. 2004), germination (Meyer et al. 1997; Meyer and Allen 1999), establishment and growth (Bradford and Lauenroth 2006; Chambers et al. 2007), and flowering and senescence (Zelikova et al. 2013). Finally, both climate change and responses to those changes are expected to vary spatially (Bradley et al. 2009, 2015; Bradley 2010), influencing optimal *B. tectorum* growth conditions and pushing physiological limitations differentially across the invaded range. A clear understanding of which traits are under selection and how all these factors interact is necessary to model invasion potential under future climatic conditions.

Ecological responses to climate change are well documented, but there is little direct evidence for adaptation to climate change (Merila 2012), largely because documenting evolutionary responses is difficult and requires carefully designed experimental manipulations (Gienapp et al. 2008). In addition, traits under selection may be correlated or antagonistic to the direction of selection (Etterson and Shaw 2001). Finally, many other factors change along with climate, making it difficult to separate plastic and evolutionary responses. Climate change experiments are especially useful for understanding *B. tectorum* ecological and evolutionary responses to climate change, at least in part because experiments can cover *B. tectorum*'s entire life cycle. *Bromus tectorum* reproduces only by seed, facilitating measures of lifetime reproductive success (i.e., fitness) within experiments.

Ecological responses to climate change have been noted in both observational and experimental studies. Here, we summarize findings from climate manipulation experiments and discuss the evolutionary implications where possible. Although current knowledge of *B. tectorum* biology and spatial genetic variation across the western USA-invaded range is ample and growing (Ramakrishnan et al. 2006; Ashley and Longland 2007; Valliant et al. 2007; Kao et al. 2008; Schachner et al. 2008; Huttanus et al. 2011; Avolio et al. 2012; Merrill et al. 2012; Atkinson and Brown 2015; Novak and Mack 2015), few studies link genetic variation with adaptation to a changing climate, and no studies take that information into a distribution modeling framework. Thus, we present current knowledge to date and highlight knowledge gaps. In several instances, specific data on *B. tectorum* are not available. As a result, we present data from experiments on congeners and make predictions about how *B. tectorum* is likely to respond to similar stressors and environmental changes. *Bromus* species appear to respond strongly to experimental global change manipulations, including elevated CO_2, warming, altered precipitation, and N fertilization (as a proxy for N deposition) across a range of sites (discussed in detail below, see also Bradley et al. 2015; Belnap et al. 2015). Although a number of studies report ecological responses, genetic responses to climate change are rarely addressed. However, given the phenotypic variation and evidence that much of that variation can be adaptive, there is a clear potential for evolutionary responses (Clements and Ditommaso 2011) that have yet to be quantified (but see Zelikova et al. 2013; Grossman and Rice 2014). More broadly, evolutionary responses in other *Bromus* species can inform hypotheses of how *B. tectorum* might respond to elevated CO_2 in the future.

5.6.1 Elevated CO_2

Atmospheric CO_2 concentrations are predicted to reach 600 ppm by the end of the twenty-first century (IPCC 2014) and will present novel environmental conditions for plants that have not experienced such high atmospheric CO_2 concentrations for 20 million years (Pearson and Palmer 2000). Hundreds of studies have examined how elevated CO_2 affects plant development, but few studies have examined the influence of CO_2 as a selective agent, and none have examined potential evolutionary adaptation under elevated atmospheric CO_2 in *B. tectorum*. Elevated CO_2 generally increases plant water-use efficiency and, for C_3 plants, photosynthetic rates, which can translate to increases in biomass accumulation and reproduction and help facilitate biological invasions (Weltzin et al. 2003). This positive ecological effect of CO_2 enrichment has been shown for *B. tectorum* in low-elevation desert and shrubland sites and in greenhouse experiments with *B. tectorum* seeds collected from across an elevation gradient in northern Nevada (Ziska et al. 2005), for *Bromus erectus* Huds. and *B. rubens* in the Desert FACE site (Huxman et al. 1998; Smith et al. 2000; Nagel et al. 2004), and for *Bromus inermis* Leyss. in a Minnesota grassland (Lau et al. 2010; Steinger et al. 2000). In the Desert FACE experiment,

B. rubens plants produced more seeds under elevated CO_2 conditions, but the growth rate of the resultant seedlings was lower, and it appears that the increase in seed and plant C:N ratios under elevated CO_2 reduced the seed quality and seedling performance (Huxman et al. 1998). Given the generality of the positive CO_2 enrichment effect across *Bromus* species, future atmospheric CO_2 conditions are likely to favor *B. tectorum* in certain portions of its range. Nitrogen (N) limitation may constrain the stimulatory effects of CO_2, especially in N-limited systems (Larigauderie et al. 1988; Hungate et al. 1996), and water limitation may reduce effects of CO_2 during low precipitation years and during droughts (Nowak et al. 2004).

Although largely contingent on moisture availability and N status, the general ability of plants to respond to CO_2 enrichment can be both genotype and density dependent (Bazzaz et al. 1995; Ainsworth et al. 2008), and this variation in the strength of responses has evolutionary implications. In the Mojave Desert FACE experiment, Grossman and Rice (2014) examined the physiological responses of *B. rubens* to elevated CO_2 and found evidence of evolutionary adaptation. Specifically, they reported a reduction in phenotypic plasticity and a shift toward reduced stomatal conductance in plants grown under CO_2 enrichment (Grossman and Rice 2014). Reduced stomatal conductance can conserve water and can therefore be adaptive (Drake et al. 1997; Ainsworth and Long 2005), especially in water-limited systems such as the Mojave Desert, where water availability drives most measures of plant performance. In addition, the ability to respond to changes in atmospheric CO_2 concentrations can vary across genotypes within a population (Curtis et al. 1994).

5.6.2 Warming

Along with increasing atmospheric CO_2 concentrations, air temperatures and precipitation regimes are also changing, with ecological and evolutionary consequences for plants. There is consistent evidence that warming shifts plant phenology, with earlier emergence in spring that can provide individuals with access to limited resources (Verdu and Traveset 2005). There is also a link between the timing of life history events such as emergence, growth, and flowering and plant fitness (Ellwood et al. 2013), suggesting that changes in phenology can be under selection. In manipulation experiments on the Colorado Plateau, *B. tectorum* consistently responded to warming, with earlier growth and flowering, an overall longer growing season, and increased biomass and reproductive output, but only in years with ample spring precipitation (Zelikova et al. 2013). Similar phenology results were also observed in northern Utah's Wasatch Mountains, where warming also advanced spring phenology in *B. tectorum* (Compagnoni and Adler 2014).

The experimental warming-induced changes in *B. tectorum* biomass and reproductive output reported in Zelikova et al. (2013) also influenced offspring performance in follow-up greenhouse studies, leading to higher germination rates and lower mortality for plants from warmed seed sources. Shifts in phenology associated with warming were only evident in wetter years, however, and were dependent on soil water availability. The positive effects of warming were diminished in dry years (Zelikova et al. 2013; Compagnoni and Adler 2014). Coupled with other findings that *B. tectorum* growth and reproduction increased in wet years, there is also some indication that increasing temperature may select for particular genotypes (Zelikova et al. 2013), evidenced by the reduction in genotypic diversity under warming. Selection for warming-specific genotypes could potentially increase *B. tectorum* population growth in the future if the warming trajectory continues, though the relative size of the increase depends on concomitant change in effective soil moisture, which is a function of both precipitation and evapotranspiration.

Water availability mediates *B. tectorum* responses to other factors that influence plant establishment and success, including soil texture (Miller et al. 2006; Zelikova et al. 2013) and nutrient additions (Leffler et al. 2005). Plants can be limited either by water or combination of water and N (Concilio and Loik 2013), and the frequency of rain events early in the growing season has a disproportionally large enhancement of *B. tectorum* growth, fecundity, and abundance (Prevey et al. 2010; Concilio et al. 2013; Bradley et al. 2015), suggesting that future invasion will be especially sensitive to climate change. The ecological effects of warming and N deposition on *Bromus* have been well described, but the potential for adaptive and evolutionary responses has received little attention to date and presents a knowledge gap. Taken together, the results from field climate change experiments suggest that predicted increases in temperature, shifts from snow to rain in low to mid-elevations, and increasing snow inputs at higher elevations, particularly in northern latitudes, could facilitate the expansion of *B. tectorum* northward and upward in elevation. Since drought periods may have substantial influence on the ability of *B. tectorum* to reproduce, this expansion depends on the timing of rain events and the relative abundance of preadapted genotypes or potential for the creation of novel genotypes.

Similarly, decreases in precipitation and increases in temperature at low-elevation sites, particularly in southern latitudes, may decrease *B. tectorum* invasion potential in those populations that cannot adapt and instead favor *B. rubens* (Bradley et al. 2015). If *B. tectorum* ecological responses and fitness consequences to environmental change differ among genotypes, there is a potential for climate change to impose selection that favors certain genotypes with locally adapted traits or that confers increased phenotypic plasticity. There is some preliminary data to suggest the potential for adaptive evolution to occur in response to warming in short time frames. Microsatellite data collected on genotypes from a warming experiment on the Colorado Plateau in southeastern Utah are consistent with the hypothesis that over 5 years of experimental warming, selection has reduced genetic diversity and favored particular genotypes (Zelikova et al. 2013). For populations that are relatively continuous, gene flow from warming-adapted populations could bolster adaptation. For noncontiguous populations, adaptation may have to come from within. Taken together, future global change, including increasing atmospheric CO_2 concentrations, warming, and N deposition, is expected to favor *B. tectorum*, alleviating physiological limitations on early season growth as well as enhancing growth in N-limited habitats in those areas that are increasingly water limited due to decreases in precipitation or increases in evapotranspiration. Such responses have clear implications for managing *B. tectorum* and likely other exotic annual *Bromus*.

5.7 Management Implications

Evidence for local adaptation, along with high levels of phenotypic plasticity, makes *B. tectorum* a very serious threat for continued spread into novel habitats even under current climatic conditions. The risk that this species could spread or become more dominant in areas in which it already occurs depends on the direction of the change with increased risk in areas that receive more effective precipitation and decreased risk in areas that receive less (Bradley et al. 2015). There will likely be areas of expansion and contraction, with *B. tectorum* predicted to increase in areas that receive the majority of their yearly precipitation in winter and spring and decrease in areas that receive a high proportion of precipitation in summer (Bradley et al. 2009). Management strategies that limit gene flow among populations and limit additional introductions from the native range could reduce the adaptive potential of *B. tectorum* (Schachner et al. 2008), but these strategies are unlikely to prevent the evolution of novel invasive traits (Dlugosch and Parker 2008). Thus, strategies will need to also target the creation of novel genotypes and their expansion.

A comprehensive management strategy should include protection, prevention, restoration, monitoring, and adaptive management (Chambers et al. 2014) as well as building capacity for resistance (Fig. 1.1 in Germino et al. 2015). In addition to reducing habitat disturbance in relatively intact systems, the best strategy to control *B. tectorum* might be strengthening the competitive ability of native species. Research has shown that native grass species growing in areas invaded by *B. tectorum* can adapt to have traits important for restoration (Kulpa and Leger 2013) and that the evolution of native grass species in response to *B. tectorum* can happen rapidly (Rowe and Leger 2011). Successful management practices will have to take into consideration the roles of adaptation and plasticity and how they interact to affect the spread of regional ecotypes. For example, understanding the limits to phenological or physiological plasticity to environmental factors such as temperature could allow management to direct treatments at more appropriate times. In years with high winter and spring precipitation, increased resources to battle new and low-density invasions will be crucial to limit its continued spread. Above all, the highly successful strategies of both local adaptation and phenotypic plasticity require management strategies that focus on limiting disturbance and restoring landscapes with sufficient densities of native species such that *B. tectorum* cannot establish or compete. Managing impacts of the human footprint, including transportation and utility corridors, will be important in limiting the spread into new environments (Leu et al. 2008), which is especially important given its ability to thrive in such a broad range of environments. The best strategy might be focusing on the traits of the native species used in restoration that can increase resistance to and resilience from invasion (see Sect. 5.8).

5.8 Research Needs

The variation in physiological tolerance and the genetic basis for that variation help define potential areas of *B. tectorum* expansion and could inform future management efforts. Direct field observations and collections confirm that *B. tectorum* has expanded its range into more arid salt desert and warm desert communities (Meyer et al. 2001; Ramakrishnan et al. 2006). Although higher-elevation environments have historically been relatively free from invasion (Bradley and Mustard 2006), *B. tectorum* is increasing in abundance in these habitats as well (Chambers et al. 2007; Kao et al. 2008; Leger et al. 2009; Griffith and Loik 2010; Bromberg et al. 2011). *Bromus tectorum* is also actively expanding its range in South America in habitats that are similar to the Great Basin in the USA (Biganzoli et al. 2013). Ongoing expansion of *B. tectorum* into higher elevations, such as in the Rocky Mountains (Kao et al. 2008; Bromberg et al. 2011; West et al. 2015), and into drier areas, such as in the Mojave Desert (Lara 2013), suggests that we still do not know the full ecological limits of this species and highlights the importance of understanding the genetic mechanisms responsible for its success in new areas.

Understanding and increasing resilience to disturbance and resistance to invasion are key to successfully managing *B. tectorum* (Fig. 1.1 in Germino et al. 2015; Chambers et al. 2014). Although *B. tectorum* is one of the most well-studied invasive species, several ecological and evolutionary questions remain unaddressed. Expanding the existing molecular work on *B. tectorum* to include next-generation sequencing allows researchers to evaluate gene expression across ecotypes and in different habitats to better understand the interaction between plasticity and genetic adaptation (Meyer et al. 2013). Next-generation sequencing approaches are powerful tools that allow us to better assess the actual outcrossing rates and the implications for the creation of novel genotypes and resultant species range expansion, especially when applied across a broader sample of individuals and across the range of habitats (Meyer and Leger 2010).

In the face of climate change, there is a critical need to measure the rate and magnitude of species distributional shifts, especially for policy- and management-relevant species such as *B. tectorum*. We first need to understand ecological tolerances as they relate to future range shifts and adaptations to climate change. The degree to which evolution influences responses varies spatially and across ecotypes, but we lack sufficient information to quantify this variation and its spatiotemporal variance. Field and laboratory experimental studies that manipulate climate, namely, temperature and precipitation regimes, and focus on quantifying the potential for evolution can fill current knowledge gaps. We argue that it is absolutely essential to develop the best possible case studies that target environmental gradients that encompass the local range of the species and that have conservation-relevant outcomes and incorporate evolutionary responses within a modeling framework. *Bromus tectorum* represents a superlative species for such integration.

Incorporating the current knowledge of *Bromus* responses to climate change knowledge into future distribution modeling efforts and the potential for evolutionary responses is needed. Given our knowledge about historical and ongoing invasions,

B. tectorum presents an ideal model system for developing these modeling approaches (Parker et al. 2003). For example, developing models that incorporate the potential for *B. tectorum* to adapt to climate change in situ and extending such models across space to encompass the differential adaptation potential inherent across different ecotypes are a grand challenge. Such models will need to bring different kinds of data together, including data from studies to understand the physiological limits of adaptation across genotypes, common garden experiments with the same genotypes to link genotypic variation with phenotypic expression, and climate manipulation experiments that directly test the role of adaptation in responding to climate change under field conditions. Improving distribution modeling efforts is especially important when considering the allocation of limited management resources and focusing eradication efforts to areas where *B. tectorum* is expected to expand in the future, increasing the success of management efforts before *B. tectorum* further expands its range. In many instances, eradication or limiting range expansion may not be possible. Increasing predictive capacity and limiting impacts may be a better management option.

References

Ainsworth EA, Beier C, Calfapietra C et al (2008) Next generation of elevated CO_2 experiments with crops: a critical investment for feeding the future world. Plant Cell Environ 31(9):1317–1324

Ainsworth EA, Long SP (2005) What have we learned from 15 years of free-air CO_2 enrichment (FACE)? A meta-analytic review of the responses of photosynthesis, canopy. New Phytol 165(2):351–371

Ashley MC, Longland WS (2007) Microsatellite evidence of facultative outcrossing in cheatgrass (*Bromus tectorum*): implications for the evolution of invasiveness. Plant Species Biol 22(3):197–204

Atkinson SY, Brown CS (2015) Attributes that confer invasiveness and impacts across the large genus *Bromus* – lessons from the *Bromus* REEnet database. In: Germino MJ, Chambers JC, Brown CS (eds) Exotic brome-grasses in arid and semiarid ecosystems of the Western USA: causes, consequences, and management implications. Springer, New York, NY (Chapter 6)

Avolio ML, Beaulieu JM, Lo EYY et al (2012) Measuring genetic diversity in ecological studies. Plant Ecol 213(7):1105–1115

Bair NB, Meyer SE, Allen PS (2006) A hydrothermal after-ripening time model for seed dormancy loss in *Bromus tectorum* L. Seed Sci Res 16(1):17–28

Baker H (1965) Characteristics and modes of origin of weeds. In: Baker H, Stebbins GL (eds) The genetics of colonizing species. Academic, New York, NY, pp 147–172

Ball DA, Frost SM, Gitelman AI (2004) Predicting timing of downy brome (*Bromus tectorum*) seed production using growing degree days. Weed Sci 52(4):518–524

Bartlett E, Novak SJ, Mack RN (2002) Genetic variation in *Bromus tectorum* (Poaceae): differentiation in the eastern United States. Am J Bot 89(4):602–612

Bazzaz F, Jasieński M, Thomas S et al (1995) Microevolutionary responses in experimental populations of plants to CO_2-enriched environments: parallel results from two model systems. Proc Natl Acad Sci U S A 92(18):8161–8165

Belnap J, Stark JM, Rau BJ et al (2015) Soil moisture and biogeochemical factors influence the distribution of annual *Bromus* species. In: Germino MJ, Chambers JC, Brown CS (eds) Exotic brome-grasses in arid and semiarid ecosystems of the Western USA: causes, consequences, and management implications. Springer, New York, NY (Chapter 8)

Biganzoli F, Larsen C, Rolhauser AG (2013) Range expansion and potential distribution of the invasive grass *Bromus tectorum* in southern South America on the base of herbarium records. J Arid Environ 97:230–236

Bradford JB, Lauenroth WK (2006) Controls over invasion of *Bromus tectorum*: the importance of climate, soil, disturbance and seed availability. J Veg Sci 17(6):693–704

Bradley BA (2010) Assessing ecosystem threats from global and regional change: hierarchical modeling of risk to sagebrush ecosystems from climate change, land use and invasive species in Nevada, USA. Ecography 33(1):198–208

Bradley BA, Curtis CA, Chambers JC (2015) *Bromus* response to climate and projected changes with climate change. In: Germino MJ, Chambers JC, Brown CS (eds) Exotic brome-grasses in arid and semi-arid ecosystems of the Western US: causes, consequences and management implications. Springer, New York, NY (Chapter 9)

Bradley BA, Mustard JF (2006) Characterizing the landscape dynamics of an invasive plant and risk of invasion using remote sensing. Ecol Appl 16(3):1132–1147

Bradley BA, Oppenheimer M, Wilcove DS (2009) Climate change and plant invasions: restoration opportunities ahead? Glob Change Biol 15(6):1511–1521

Bradshaw AD (1965) Evolutionary significance of phenotypic plasticity in plants. Adv Genet 13(1):115–155

Bromberg JE, Kumar S, Brown CS et al (2011) Distributional changes and range predictions of downy brome (*Bromus tectorum*) in Rocky Mountain National Park. Invasive Plant Sci Manag 4(2):173–182

Brooks ML, Belnap J, Brown CS et al (2015) Exotic annual *Bromus* invasions – comparisons among species and ecoregions in the Western United States. In: Germino MJ, Chambers JC, Brown CS (eds) Exotic brome-grasses in arid and semiarid ecosystems of the Western USA: causes, consequences, and management implications. Springer, New York, NY (Chapter 2)

Bykova O, Sage RF (2012) Winter cold tolerance and the geographic range separation of *Bromus tectorum* and *Bromus rubens*, two severe invasive species in North America. Glob Change Biol 18(12):3654–3663

Chambers JC, Bradley BA, Brown CS et al (2014) Resilience to stress and disturbance, and resistance to *Bromus tectorum* L. invasion in cold desert shrublands of western North America. Ecosystems 17(2):360–375

Chambers JC, Germino MJ, Belnap J et al (2015) Plant community resistance to invasion by *Bromus* species – the roles of community attributes, *Bromus* interactions with plant communities, and *Bromus* traits. In: Germino MJ, Chambers JC, Brown CS (eds) Exotic brome-grasses in arid and semiarid ecosystems of the Western USA: causes, consequences, and management implications. Springer, New York, NY (Chapter 10)

Chambers JC, Roundy BA, Blank RR et al (2007) What makes Great Basin sagebrush ecosystems invasible by *Bromus tectorum*? Ecol Monogr 77(1):117–145

Clausen J, Keck DD, Hiesey WM (1941) Regional differentiation in plant species. Am Nat 75:231–250

Clausen J, Keck D, Hiesey W (1948) Experimental taxonomy. Carnegie Institute of Washington Year Book 47: 105–110

Clements DR, Ditommaso A (2011) Climate change and weed adaptation: can evolution of invasive plants lead to greater range expansion than forecasted? Weed Res 51(3):227–240

Colautti RI, Barrett SCH (2013) Rapid adaptation to climate facilitates range expansion of an invasive plant. Science 342(6156):364–366

Compagnoni A, Adler PB (2014) Warming, soil moisture, and loss of snow increase *Bromus tectorum*'s population growth rate. Elem Sci Anth 2(1):000020

Concilio AL, Loik ME (2013) Elevated nitrogen effects on *Bromus tectorum* dominance and native plant diversity in an arid montane ecosystem. Appl Veg Sci 16(4):598–609

Concilio AL, Loik ME, Belnap J (2013) Global change effects on *Bromus tectorum* L. (Poaceae) at its high-elevation range margin. Glob Change Biol 19(1):161–172

Cordell S, Goldstein G, Mueller-Dombois D et al (1998) Physiological and morphological variation in *Metrosideros polymorpha*, a dominant Hawaiian tree species, along an altitudinal gradient: the role of phenotypic plasticity. Oecologia 113(2):188–196

Curtis PS, Snow AA, Miller AS (1994) Genotype-specific effects of elevated CO_2 on fecundity in wild radish (*Raphanus raphanistrum*). Oecologia 97(1):100–105

Davidson AM, Jennions M, Nicotra AB (2011) Do invasive species show higher phenotypic plasticity than native species, and if so, is it adaptive? A meta-analysis. Ecol Lett 14(4):419–431

Dlugosch KM, Parker IM (2008) Founding events in species invasions: genetic variation, adaptive evolution, and the role of multiple introductions. Mol Ecol 17(1):431–449

Drake BG, GonzalezMeler MA, Long SP (1997) More efficient plants: a consequence of rising atmospheric CO_2? Annu Rev Plant Physiol Plant Mol Biol 48:609–639

Dudley SA, Schmitt J (1995) Genetic differentiation in morphological responses to simulated foliage shade between populations of *Impatiens capensis* from open and woodland sites. Funct Ecol 9(4):655–666

Dyer AR, Brown CS, Espeland EK et al (2010) The role of adaptive trans-generational plasticity in biological invasions of plants. Evol Appl 3(2):179–192

Dyer AR, Hardison JL, Rice KJ (2012) Phenology constrains opportunistic growth response in *Bromus tectorum* L. Plant Ecol 213(1):103–112

Ellwood ER, Temple SA, Primack RB et al (2013) Record-breaking early flowering in the eastern United States. PLoS One 8(1), e53788

Etterson JR, Shaw RG (2001) Constraint to adaptive evolution in response to global warming. Science 294(5540):151–154

Fenesi A, Redei T, Botta-Dukat Z (2011) Hard traits of three Bromus species in their source area explain their current invasive success. Acta Oecol 37(5):441–448

Franks SJ, Sim S, Weis AE (2007) Rapid evolution of flowering time by an annual plant in response to a climate fluctuation. Proc Natl Acad Sci U S A 104(4):1278–1282

Funk JL, Vitousek PM (2007) Resource-use efficiency and plant invasion in low-resource systems. Nature 446(7139):1079–1081

Germino MJ, Chambers JC, Brown CS (2015) Introduction: exotic annual *Bromus* in the Western USA. In: Germino MJ, Chambers JC, Brown CS (eds) Exotic brome-grasses in arid and semiarid ecosystems of the Western USA: causes, consequences, and management implications. Springer, New York, NY (Chapter 1)

Gienapp P, Teplitsky C, Alho JS et al (2008) Climate change and evolution: disentangling environmental and genetic responses. Mol Ecol 17(1):167–178

Griffith AB, Andonian K, Weiss CP et al (2014) Variation in phenotypic plasticity for native and invasive populations of *Bromus tectorum*. Biol Invasions 16(12):2627–2638

Griffith AB, Loik ME (2010) Effects of climate and snow depth on *Bromus tectorum* population dynamics at high elevation. Oecologia 164(3):821–832

Grossman JD, Rice KJ (2014) Contemporary evolution of an invasive grass in response to elevated atmospheric CO_2 at a Mojave Desert FACE site. Ecol Lett 17(6):710–716

Gurevitch J (1992) Differences in photosynthetic rate in populations of *Achillea lanulosa* from 2 altitudes. Funct Ecol 6(5):568–574

Haubensak KA, D'Antonio CM, Embry S et al (2014) A comparison of *Bromus tectorum* growth and mycorrhizal colonization in salt desert vs. sagebrush habitats. Rangel Ecol Manag 67(3):275–284

Hiesey WM, Milner HW (1965) Physiology of ecological races and species. Annu Rev Plant Physiol Plant Mol Biol 16(1):203–216

Hoover AN, Germino MJ (2012) A common-garden study of resource-island effects on a native and an exotic, annual grass after fire. Rangel Ecol Manag 65(2):160–170

Huey RB, Gilchrist GW, Carlson ML et al (2000) Rapid evolution of a geographic cline in size in an introduced fly. Science 287(5451):308–309

Hungate BA, Canadell J, Chapin FS (1996) Plant species mediate changes in soil microbial N in response to elevated CO_2. Ecology 77(8):2505–2515

Huttanus TD, Mack RN, Novak SJ (2011) Propagule pressure and introduction pathways of *Bromus tectorum* (Cheatgrass; Poaceae) in the central United States. Int J Plant Sci 172(6):783–794

Huxman TE, Hamerlynck EP, Jordan DN et al (1998) The effects of parental CO_2 environment on seed quality and subsequent seedling performance in *Bromus rubens*. Oecologia 114(2):202–208
Intergovernmental Panel on Climate Change (IPCC) (2013) Climate Change 2013: The physical science basis: working group I contribution to the fifth assessment report of the intergovernmental panel on climate change. Cambridge University Press, Cambridge, UK and New York
James JJ, Drenovsky RE, Monaco TA et al (2011) Managing soil nitrogen to restore annual grass-infested plant communities: effective strategy or incomplete framework? Ecol Appl 21(2):490–502
Jump AS, Penuelas J (2005) Running to stand still: adaptation and the response of plants to rapid climate change. Ecol Lett 8(9):1010–1020
Kao RH, Brown CS, Hufbauer RA (2008) High phenotypic and molecular variation in downy brome (*Bromus tectorum*). Invasive Plant Sci Manag 1(2):216–225
Kulpa SM, Leger EA (2013) Strong natural selection during plant restoration favors an unexpected suite of plant traits. Evol Appl 6(3):510–523
Lara D (2013) Population genetic structure of *Bromus tectorum* in the American Desert Southwest. Plant and Wildlife Sciences, Brigham Young University, Provo, UT
Larigauderie A, Hilbert DW, Oechel WC (1988) Effect of CO_2 enrichment and nitrogen availability on resource acquisition and resource allocation in a grass, *Bromus mollis*. Oecologia 77(4):544–549
Lau JA, Shaw RG, Reich PB et al (2010) Species interactions in a changing environment: elevated CO_2 alters the ecological and potential evolutionary consequences of competition. Evol Ecol Res 12(4):435–455
Lee CE (2002) Evolutionary genetics of invasive species. Trends Ecol Evol 17(8):386–391
Leffler AJ, Monaco TA, James JJ (2011) Nitrogen acquisition by annual and perennial grass seedlings: testing the roles of performance and plasticity to explain plant invasion. Plant Ecol 212(10):1601–1611
Leffler AJ, Peek MS, Ryel RJ et al (2005) Hydraulic redistribution through the root systems of senesced plants. Ecology 86(3):633–642
Leger EA, Espeland EK, Merrill KR et al (2009) Genetic variation and local adaptation at a cheatgrass (*Bromus tectorum*) invasion edge in western Nevada. Mol Ecol 18(21):4366–4379
Leimu R, Fischer M (2008) A meta-analysis of local adaptation in plants. PLoS One 3(12), e4010
Leu M, Hanser SE, Knick ST (2008) The human footprint in the west: a large-scale analysis of anthropogenic impacts. Ecol Appl 18(5):1119–1139
Liu Y, Zhang Y, Nowak RS et al (2013) Diaspore characteristics and ecological adaptation of *Bromus tectorum* L. from different distribution regions. J Arid Land 5(3):310–323
Mack RN (1981) Invasion of *Bromus tectorum* L. into western North America: an ecological chronicle. Agro-Ecosystems 7(2):145–165
Mack RN, Pyke DA (1983) The demography of *Bromus tectorum*: variation in time and space. J Ecol 71(1):69–93
Maron JL, Vila M, Bommarco R et al (2004) Rapid evolution of an invasive plant. Ecol Monogr 74(2):261–280
McCarlie VW, Hansen LD, Smith BN et al (2000) Respiratory and physiological characteristics in Subpopulations of great basin cheatgrass. In: McArthur ED, Fairbanks DJ (comp) Shrubland ecosystem genetics and biodiversity, Provo, UT. Gen Tech Rep RMRS-P-21. USDA, Forest Service, Rocky Mountain Research Station, Ogden, UT, pp 271–275
Merila J (2012) Evolution in response to climate change: in pursuit of the missing evidence. Bioessays 34(9):811–818
Merrill KR, Meyer SE, Coleman CE (2012) Population genetic analysis of *Bromus tectorum* (Poaceae) indicates recent range expansion may be facilitated by specialist genotypes. Am J Bot 99(3):529–537
Meyer SE, Allen PS (1999) Ecological genetics of seed germination regulation in *Bromus tectorum* L. I. Phenotypic variance among and within populations. Oecologia 120(1):27–34

Meyer SE, Allen PS, Beckstead J (1997) Seed germination regulation in *Bromus tectorum* (Poaceae) and its ecological significance. Oikos 78(3):475–485

Meyer SE, Ghimire S, Decker S et al (2013) The Ghost of outcrossing past in downy brome, an inbreeding annual grass. J Hered 104(4):476–490

Meyer SE, Leger EA (2010) Inbreeding, genetic variation, and invasiveness: the strange case of *Bromus tectorum*. Rangelands 32(1):6–11

Meyer SE, Nelson DL, Carlson SL (2004) Ecological genetics of vernalization response in *Bromus tectorum* L. (Poaceae). Ann Bot 93(6):653–663

Meyer SE, Nelson DL, Clement S (2001) Evidence for resistance polymorphism in the *Bromus tectorum* – *Ustilago bullata* pathosystem: implications for biocontrol. Can J Plant Pathol 23(1):19–27

Miller ME, Belnap J, Beatty SW et al (2006) Performance of *Bromus tectorum* L. in relation to soil properties, water additions, and chemical amendments in calcareous soils of southeastern Utah, USA. Plant Soil 288(1–2):1–18

Moran EV, Alexander JM (2014) Evolutionary responses to global change: lessons from invasive species. Ecol Lett 17(5):637–649

Nagel JM, Huxman TE, Griffin KL et al (2004) CO_2 enrichment reduces the energetic cost of biomass construction in an invasive desert grass. Ecology 85(1):100–106

Novak SJ, Mack RN (1993) Genetic variation in *Bromus tectorum* (Poaceae): comparison between native and introduced populations. Heredity 71:167–176

Novak SJ, Mack RN (2001) Tracing plant introduction and spread: genetic evidence from *Bromus tectorum* (Cheatgrass). Bioscience 51(2):114–122

Novak SJ, Mack RN (2015) Mating system, introduction and genetic diversity of *Bromus tectorum* in North America, the most notorious product of evolution within *Bromus* section Genea. In: Germino MJ, Chambers JC, Brown CS (eds) Exotic brome-grasses in arid and semi-arid ecosystems of the Western US: causes, consequences and management implications. Springer, New York, NY (Chapter 4)

Novak SJ, Mack RN, Soltis PS (1991) Genetic variation in *Bromus tectorum* (Poaceae): population differentiation in its North American range. Am J Bot 78(8):1150–1161

Novak SJ, Mack RN, Soltis PS (1993) Genetic variation in *Bromus tectorum* (Poaceae): introduction dynamics in North America. Can J Plant Pathol 71(11):1441–1448

Nowak RS, Ellsworth DS, Smith SD (2004) Functional responses of plants to elevated atmospheric CO_2: do photosynthetic and productivity data from FACE experiments support early predictions? New Phytol 162(2):253–280

Parker IM, Rodriguez J, Loik ME (2003) An evolutionary approach to understanding the biology of invasions: local adaptation and general-purpose genotypes in the weed *Verbascum thapsus*. Conserv Biol 17(1):59–72

Pearson PN, Palmer MR (2000) Atmospheric carbon dioxide concentrations over the past 60 million years. Nature 406(6797):695–699

Prevéy JS, Germino MJ, Huntly NJ, Inouye RS (2010) Exotic plants increase and native plants decrease with loss of foundation species in sagebrush steppe. Plant Ecol 207:39–51

Ramakrishnan AP, Meyer SE, Fairbanks DJ et al (2006) Ecological significance of microsatellite variation in western North American populations of *Bromus tectorum*. Plant Species Biol 21(2):61–73

Ramakrishnan AP, Meyer SE, Waters J et al (2004) Correlation between molecular markers and adaptively significant genetic variation in *Bromus tectorum* (Poaceae) an inbreeding annual grass. Am J Bot 91(6):797–803

Rice KJ, Dyer AR (2001) Seed aging, delayed germination and reduced competitive ability in *Bromus tectorum*. Plant Ecol 155(2):237–243

Rice KJ, Mack RN (1991a) Ecological genetics of *Bromus tectorum* I. A hierarchical analysis of phenotypic variation. Oecologia 88(1):77–83

Rice KJ, Mack RN (1991b) Ecological genetics of *Bromus tectorum* II. Intraspecific variation in phenotypic plasticity. Oecologia 88(1):84–90

Richards CL, Bossdorf O, Muth NZ et al (2006) Jack of all trades, master of some? On the role of phenotypic plasticity in plant invasions. Ecol Lett 9(8):981–993
Rowe CLJ, Leger EA (2011) Competitive seedlings and inherited traits: a test of rapid evolution of *Elymus multisetus* (big squirreltail) in response to cheatgrass invasion. Evol Appl 4(3):485–498
Salo LF (2005) Red brome (*Bromus rubens* subsp *madritensis*) in North America: possible modes for early introductions, subsequent spread. Biol Invasions 7(2):165–180
Schachner LJ, Mack RN, Novak SJ (2008) *Bromus tectorum* (Poaceae) in mid-continental United States: population genetic analysis of an ongoing invasion. Am J Bot 95(12):1584–1595
Scott JW, Meyer SE, Merrill KR et al (2010) Local population differentiation in *Bromus tectorum* L. in relation to habitat-specific selection regimes. Evol Ecol 24(5):1061–1080
Sexton JP, McKay JK, Sala A (2002) Plasticity and genetic diversity may allow saltcedar to invade cold climates in North America. Ecol Appl 12(6):1652–1660
Smith SD, Huxman TE, Zitzer SF et al (2000) Elevated CO_2 increases productivity and invasive species success in an arid ecosystem. Nature 408(6808):79–82
Steinger T, Gall R, Schmid B (2000) Maternal and direct effects of elevated CO_2 on seed provisioning, germination and seedling growth in *Bromus erectus*. Oecologia 123(4):475–480
Turesson G (1922) The species and the variety as ecological units. Hereditas 3(1):100–113
Valliant MT, Mack RN, Novak SJ (2007) Introduction history and population genetics of the invasive grass *Bromus tectorum* (Poaceae) in Canada. Am J Bot 94(7):1156–1169
Verdu M, Traveset A (2005) Early emergence enhances plant fitness: a phylogenetically controlled meta-analysis. Ecology 86(6):1385–1394
Via S (1990) Ecological genetics and host adaptation in herbivorous insects: the experimental study of evolution in natural and agricultural systems. Ann Rev Entomol 35:421–446
Weltzin JF, Belote RT, Sanders NJ (2003) Biological invaders in a greenhouse world: will elevated CO_2 fuel plant invasions? Front Ecol Environ 1(3):146–153
West AM, Kumar S, Wakie T et al (2015) Using high-resolution future climate scenarios to forecast *Bromus tectorum* invasion in Rocky Mountain National Park. PLoS One 10(2)
Zelikova TJ, Hufbauer RA, Reed SC et al (2013) Eco-evolutionary responses of *Bromus tectorum* to climate change: implications for biological invasions. Ecol Evol 3(5):1374–1387
Ziska LH, Reeves JB, Blank B (2005) The impact of recent increases in atmospheric CO_2 on biomass production and vegetative retention of Cheatgrass (*Bromus tectorum*): implications for fire disturbance. Glob Change Biol 11(8):1325–1332

Chapter 6
Attributes That Confer Invasiveness and Impacts Across the Large Genus *Bromus*: Lessons from the *Bromus* REEnet Database

Sheryl Y. Atkinson and Cynthia S. Brown

Abstract *Bromus* (L.) species are cool-season grasses of temperate regions and tropical high elevations. Some species in the genus *Bromus* have been widely introduced into new areas of the globe and are invasive in the Western United States, while others occur only in their native ranges. We developed a database with information about traits of *Bromus* species and their interactions with biotic and abiotic features of their environments. Using the collected data, we looked for correlations among wide introduction, weediness, a suite of traits including taxonomic section, year, life span, seed awn length, average seed mass, polyploidy, human use and cultivar availability, and climate factors. Annual *Bromus* species were often destructive crop weeds, ruderal weeds, and environmental (natural habitat) weeds. Long awn length was associated with wide introduction and weediness in annual *Bromus* grasses. Perennial *Bromus* grasses generally remained confined to their native regions unless they were polyploid species cultivated for hay, forage, and revegetation, and few were invasive. Invasiveness in *Bromus* species was associated with the ability to grow at high and low temperature and precipitation levels and with human activities. Most research focuses on highly invasive species such as *Bromus tectorum* L. (downy brome or cheatgrass) and cultivated species such as *Bromus inermis* Leyss. (smooth brome), while information about most other species is more limited. Information about *Bromus* species in a central location facilitates comparisons among species and provides data that can be used for modeling, prediction, management, and control of *Bromus* grass invasions.

Keywords Taxonomy • Plant traits • Introduction • Invasiveness • Human use • Species distribution

---

S.Y. Atkinson (✉) • C.S. Brown
Department of Bioagricultural Sciences and Pest Management, Colorado State University, Fort Collins, CO 80523-1177, USA
e-mail: syatkinson@comcast.net; cynthia.s.brown@colostate.edu

M.J. Germino et al. (eds.), *Exotic Brome-Grasses in Arid and Semiarid Ecosystems of the Western US*, Springer Series on Environmental Management,
DOI 10.1007/978-3-319-24930-8_6

6.1 Introduction

The grass family (Poaceae) contains a number of species that are of exceptional importance to humans, including wheat, rice, corn, barley, rye, millet, sugarcane, and grasses that provide forage for domesticated animals. Unfortunately, some grass species that have been introduced by humans into new regions have become destructive both to natural environments and to human agricultural systems. Pyšek (1998) listed the Poaceae family in the top four families that provide the most species listed as alien in world floras and in the top six families that have the highest proportion of invasive species relative to total species. He suggested that highly evolved inflorescences, successful seed dispersal mechanisms, and ability to grow in and adapt to diverse habitats all contribute to invasiveness in grasses.

The grasses in genus *Bromus* include a number of weedy species that have impacted croplands, rangelands, and natural habitats and that have spread far outside of their native ranges. There are also many *Bromus* grasses that are not considered weedy and that have limited distributions. Western North America has had exceptionally severe problems with *Bromus* invasions. Two species that are highly invasive in North America are *Bromus tectorum* L. (cheatgrass or downy brome) and *Bromus rubens* L. (syn. *Bromus madritensis* ssp. *rubens* (L.) Husnot, red brome). *B. tectorum* is a weed of winter cereal crops. It is also an aggressive invader in natural habitats, especially in the Great Basin region where it is common in sagebrush ecosystems (Concilio et al. 2013). *B. rubens* has become a serious invader in the Mojave Desert and Sonoran region and is broadly distributed in lower elevations in the Great Basin (Salo 2004, 2005; Salo et al. 2005; Brooks and Berry 2006).

A somewhat elusive goal of invasion science is to find a fast and easy way to determine which species are likely to become invaders and when invasions are likely to occur. Unfortunately, it is increasingly clear that there is no easy way to identify species with a potential for invasiveness or to recognize the early stages of invasions. Researchers have identified a number of factors associated with invasiveness, but few of these are consistent across a broad range of plant taxa and types. According to the National Research Council (2002, quoted by Powell 2004), there are "no known broad scientific principles or reliable procedures for identifying the invasive potential of plants, plant pests, or biological control agents in new geographical ranges." Kolar and Lodge (2002) commented that "characteristics that apply generally to all taxonomic groups and in all ecosystems" do not exist and that characteristics important in the early stages of an invasion may differ from those that facilitate later invasion stages. Lodge (1993) suggested that prediction requires "focused studies at more specific levels of taxonomy or habitats." Information about invasiveness in specific plant families, genera, and species could better identify factors that can help to predict whether a given species is likely to become invasive (Perrins et al. 1992; Burns 2004; Simberloff 2005; Pyšek and Richardson 2007; Rejmánek et al 2005). Data on plant behavior, growth requirements, and ecological interactions at the genus and species levels can also help managers evaluate the likelihood of invasion by a given type of plant in a specific area or habitat and to better manage existing invasions.

Several studies have investigated invasiveness in annual *Bromus* grasses. Hulbert (1955) studied ten species that have been introduced into the Western United States and suggested that winter hardiness, ability to germinate at lower temperatures, rapid growth, spring maturation, and high seed output all help to make *B. tectorum* a successful invader in semiarid areas. A more recent study of genetic and demographic characteristics of annual *Bromus* species (Roy 1990; Roy et al. 1991) found a positive relationship between the number of climactic zones in the native range and the number of regions with a Mediterranean climate occupied worldwide but did not find other clear differences between invasive and noninvasive species.

Databases for taxa that have high numbers of invasive species and that contain information about characteristics of interest can be used in conjunction with prediction and modeling tools to contribute to a better understanding of invasiveness and to improve prevention and management of invasions. We developed and populated a database of information about grass species in the genus *Bromus* to investigate the use of databases for research into plant invasions and to develop an understanding of characters associated with invasiveness in *Bromus* grasses. The database was developed in conjunction with a *Bromus* Research, Education, and Extension (REEnet) project to coordinate networking on the distribution and abundance, impacts, and management of exotic and invasive *Bromus* species in the Western United States. It contains information about range and distribution, habitat, plant traits, environmental interactions, weediness and invasiveness, human uses, and management for over 150 species (http://greatbasin.wr.usgs.gov/GBRMP/bromus/bromus.html).

Grasses in the genus *Bromus* belong to the subfamily Pooideae and the tribe Bromeae. There are two other genera in the Bromeae tribe, *Boissiera* (one species) and *Littledalea* (four species) (Clayton and Renvoize 1986). The number of species in genus *Bromus* is sometimes listed at over 400, but about 150 species are currently considered to be valid (Soderstrom and Beaman 1968 cited by Saarela et al. 2007; Clayton and Renvoize 1986). *Bromus* grasses use the C3 photosynthetic pathway, grow in temperate regions and at high elevations in the tropics, and are closely related to important cereal grasses including wheat, rye, oats, and barley (Grass Phylogeny Working Group 2000, 2001). They differ from most other grasses in having simple rounded starch grains rather than compound starch grains (Clayton and Renvoize 1986).

Taxonomists have divided the grasses of the genus *Bromus* into sections (Smith 1970) or subgenera (Stebbins 1981). Tsvelev (1984) considered the differences between *Bromus* subgroups to be so large that he separated the genus into separate genera (Tsvelev and Fedorov 1984). All three classifications are used today, typically with *Bromus* species separated into six sections, subgenera, or genera (Table 6.1). The section names will be used here. Grasses in three sections, *Bromus* (a section in genus *Bromus*), *Genea*, and *Nevskiella*, are annuals. Those in section *Bromopsis* are perennials with one exception. The grasses in section *Ceratochloa* and *Neobromus* tend to have somewhat intermediate life cycles and may be variously described as annual, biennial, short-lived perennial, or perennial.

In the fall of 2012 and the spring of 2013, we used the database to investigate patterns of introduction and weediness in order to develop and examine hypotheses

Table 6.1 Taxonomy of the tribe Bromeae

Genus	Methods of subdividing *Bromus* grasses	Life span	No. species	Ploidy	Native distribution
Genus *Bromus*	Genus *Bromus* Linnaeus Section *Bromus* Subgenus *Bromus*	Annual, occ. biennial	ca. 30	Diploid, polyploid	Asia, Europe, Africa
	Section *Triniusia* (Steudel) Nevski	Annual	2–3	Diploid	Asia, Europe
	Genus *Bromopsis* (Dumortier) Fourreau Section *Bromopsis* Dumortier or *Pnigma* Dumortier or *Festucoides* Cosson & Durieu Subgenus *Festucaria* Link	Perennial, 1 annual	70–90	Diploid, polyploid	Asia, Europe, Africa, Australia, North America, South America
	Section *Sinobromus* Keng	Perennial	5		Asia
	Genus *Ceratochloa* Beauvois Section *Ceratochloa* (Beauv.) Grisebach Subgenus *Ceratochloa* (Beauv.) Hackel	Annual, biennial, Short-lived perennial, perennial	ca. 20	Polyploid	North America, South America
	Genus *Anisantha* (C. Koch) Section *Genea* Dumortier Subgenus *Stenobromus* Hackel	Annual	5–8	Diploid, polyploid	Europe, North Africa, and Western Asia, especially around the Mediterranean
	Genus *Nevskiella* Kreczetovich & Vvedensky Section *Nevskiella* (Krecz. & Vved.) Tournay Subgenus *Nevskiella* (Krecz. & Vved.) Krecz. & Vved.	Annual	1	Diploid	Western and Central Asia
	Genus *Trisetobromus* Nevski Section *Neobromus* (Shear) Hitchcock Subgenus *Neobromus* Shear	Annual (or perennial)	2	Polyploid	South America (Chile)
Genus *Boissiera* Hochstetter ex Steudel		Annual	1	Diploid, polyploid	Eastern Mediterranean, Asia
Genus *Littledalea* Hemsley		Perennial	4	?	Western China, Central Asia

This table shows three systems of taxonomy that are commonly in use today. The *left column* shows three genera: *Bromus*, *Boissiera*, and *Littledalea*. The genus *Bromus* is usually subdivided into six groups, shown in the second column (Smith 1970; Stebbins 1981; Tsvelev 1981; Clayton et al. 2006 onwards; Saarela et al. 2007). Tsvelev (1981) broke the genus *Bromus* into six genera. Smith (1970) subdivided it into sections and Stebbins (1981) used subgenera. Additional sections proposed by other authors are sometimes used (Scholz 1998; Liu et al. 2006) about invasiveness in *Bromus* grasses. We evaluated the data in the database to identify factors for which information for many species could be found. We then developed hypotheses about associations between these factors and invasiveness based on the literature (Table 6.2).

6.2 Methods

6.2.1 Creation and Population of the Database

We created the database of information about *Bromus* species using Microsoft Access 2007 and populated it with information from many sources including regional floras and manuals, online databases, papers in peer-reviewed journals, government documents, dissertations, and theses. Over three hundred references are cited in the database, and these are associated with species. The bibliography section includes many more references. Basic searches by scientific name and country are available. Advanced search option allows users to find and compare species with specific traits or other properties and to search for references.

For analyses, we used the data available in the *Bromus* database. We used the species listed in the Kew Gardens GrassBase database of global grass species (Clayton et al. 2006+) and the Integrated Taxonomic Information System (ITIS) database as valid species for data analyses. In some cases, valid species differed among these two databases and the USDA PLANTS database.

6.2.2 Statistical Analysis

For the statistical analyses, we defined four dependent variables: wide introduction, crop weed, ruderal weed, and environmental weed. To define these variables, we separated commonly used definitions of invasiveness into two parts: (1) wide introduction and (2) negative impact. To define wide introduction, we developed a 1–5 scale based on the number of biogeographic realms occupied by each species—Palearctic, Nearctic, Afrotropic, Neotropic, Indo-Malay, Australasia, Oceania, and Antarctic (Udvardy 1975; Olson and Dinerstein 2002). These biogeographic realms, which "represent the unique flora and fauna of the earth's continents and ocean basins" (World Wildlife Fund 2014 http://worldwildlife.org/biomes) differ slightly from continents. North Africa is included in the Palearctic realm along with Europe and northern Asia. This puts all of the Mediterranean region into a single realm which maps better to typical *Bromus* native ranges. Species were assigned a distribution score of 1–3 (*not widely introduced*) if they are confined to their native range or to a single biogeographic realm outside of their native range and 4 or 5 (*widely introduced*) if they are found in two (4) or more (5) biogeographic realms outside of their native range. To identify species that are considered to have a *negative impact*,

Table 6.2 Hypotheses evaluated in this chapter, based on factors that are commonly listed as contributing to invasiveness in research papers by Baker (1974), Roy (1990), Rejmánek (2000), Rejmánek et al. (2005), Pyšek and Richardson (2007), Whitney and Gabler (2008), and Firn et al. (2011)

	Factor	Hypothesis	Basis/rationale	Independent variables	Comments based on analyses
1	Taxa	Invasiveness is correlated with taxonomic section, with species in sections *Genea* and *Bromus* more likely to be invasive	This is often stated in literature	Section (*Genea*, *Bromus*, *Ceratochloa*, *Bromopsis*)	Species in section *Genea* and *Bromus* were the most widely introduced and weedy, and species in section *Bromopsis* were the least
2	Life span	Shorter life spans are correlated with invasiveness	Annual grasses have replaced perennial grasses in parts of the western United States	Life span (short, medium, long)	Shorter life span was correlated with both wide introduction and weediness
3	Year species named and described	Species named and described early are more likely to be invasive than those described later	Species common in their native ranges are likely to be described early and to be widely introduced and weedy	Year	Species named and described earlier were more likely to be widely introduced and weedy
4	Effective seed distribution	Long lemma awns are correlated with invasiveness	Long lemma awns aid in dispersal by vertebrate animals, help to protect the seed from predation, and aid in seed burial	Maximum awn length	Longer maximum awn length was correlated with wide introduction and weediness
5	Effective seed distribution	Low mean seed mass is correlated with invasiveness	Low mass seeds may be created in large numbers and more widely dispersed	Mean seed mass	Lower mean seed mass was marginally correlated with crop weediness. There were no significant correlations with wide introduction or other weed categories

6	Ploidy level	Polyploid chromosome counts are correlated with invasiveness	Polyploidy may result in more robust plants and provides greater opportunity for genetic variation	Ploidy level (diploid, polyploid, both)	Polyploid species and species with both diploid and polyploid individuals were more likely to be widely introduced and to be crop and ruderal weeds. This pattern was much weaker for annual species than for perennials
7	Human use	Use by humans for forage, hay, and revegetation is correlated with invasiveness	Species are introduced into new areas and grown in large quantities under conditions which encourage establishment	Human use (hay/forage/revegetation, no hay/forage/revegetation)	Use for hay/forage/revegetation was correlated with wide introduction and with weediness
8	Cultivar availability	Availability of cultivars is correlated with invasiveness	Cultivar development may promote increased genetic variation and allow plants to grow in more areas	Cultivar availability (cultivars, no cultivars)	Cultivar availability (having cultivars) was correlated with wide introduction and weediness
9	Climate	Ability to grow at both low and high temperatures is correlated with invasiveness	Ability to adapt to extreme environmental conditions may provide competitive advantages	Minimum temperature, Maximum temperature	Lower minimum temperature and higher maximum temperature were correlated with wide introduction and weediness
10	Climate	Ability to grow at a wide range of temperatures is correlated with invasiveness	Ability to adapt to extreme environmental conditions may provide competitive advantages	Temperature range	Wider temperature range was correlated with wide introduction and weediness

(continued)

Table 6.2 (continued)

	Factor	Hypothesis	Basis/rationale	Independent variables	Comments based on analyses
11	Climate	Ability to grow in arid regions is correlated with invasiveness	Ability to adapt to extreme environmental conditions may provide competitive advantages	Minimum precipitation, Maximum precipitation	Lower minimum precipitation and higher maximum precipitation were correlated with wide introduction and weediness
12	Climate	Ability to adapt to varying precipitation levels is associated with invasiveness	Ability to adapt to extreme environmental conditions may provide competitive advantages	Precipitation range	Wider precipitation range was correlated with wide introduction and weediness

Variables used for analyses and comments are provided for each hypothesis we looked for species that are listed as a weed in one or more of three categories: crop weed (weeds of crops), ruderal weed (weeds of waste and disturbed areas), and environmental weed (weeds of natural and seminatural habitats). We defined crop weed values as *crop weed* or *not a crop weed*, ruderal weed values as *ruderal weed* or *not a ruderal weed*, and environmental weed values as *environmental weed* or *not an environmental weed*. Species classified both as *widely introduced* and as an *environmental weed* were considered to be *invasive*.

We used 14 independent variables in the statistical analyses. These are section, life span, year, maximum awn length, mean seed mass, ploidy level, human use, cultivar availability, minimum temperature, maximum temperature, temperature range, minimum precipitation, maximum precipitation, and precipitation range (Table 6.2).

For Section, we used values *Genea*, *Bromus*, *Ceratochloa*, and *Bromopsis*. The three species in small sections *Neobromus* and *Nevskiella* were eliminated along with *Bromus andringitrensis* (Camus) about which little is known (Camus 1956). We used three values for life span: *short* for annuals, *medium* for biennials and short-lived perennials, and *long* for perennials. Species sometimes described as annual and sometimes as perennial were classified as having a medium life span. Year is the year in which a species was described and named. Information for this variable came from the Missouri Botanic Garden Tropicos database (2013).

Seeds of *Bromus* grasses have lemma awns of various lengths from zero to over 40 mm. For maximum awn length, we used the high value of the awn length range in the species descriptions in the GrassBase database (Clayton et al. 2006+). Seed mass data were found for 64 species and entered into the database as the mass of 1000 seeds in grams. Seed mass values came from a variety of sources, including the USDA ARS Germplasm Resource Information Network (GRIN) database (USDA National Plant Information System 2011) and the Kew Gardens Seed Information Database SID (Royal Botanical Gardens Kew 2008). Seed mass values from all sources were averaged for each species to determine mean mass in mg.

For ploidy level, species were categorized as *diploid*, *polyploid*, or *both*, based on evaluation of chromosome count and ploidy level data and also on information in the literature. These categories are non-overlapping. Species were categorized as diploid if all or nearly all records of ploidy level found were 2N (having 14 chromosomes) and if the literature indicated that the species is normally considered to be diploid. Other species were categorized as polyploid if all or nearly all records of ploidy level found were 4N or higher and if the literature indicated that the species is normally considered to be polyploid. All *Ceratochloa* species were categorized as polyploid, as this is a characteristic of the section.

We defined human use with values *forage/hay/revegetation* and *no forage/hay/revegetation* based on information about *Bromus* species that are currently used for these purposes or have been used in the past. Other human uses of *Bromus* species were not included in the analyses. We defined cultivar availability as *cultivars* and *no cultivars* based on whether or not we found information stating that cultivars had been developed (named varieties intentionally bred or selected for cultivation).

All temperature and precipitation data used in the analyses were based on climate data collected from 1961 to 1990 that were associated with locations listed in species collection records. The online GrassPortal database (Osborne et al. 2011) was used to access these data. GrassPortal is a collaborative project that allows species data from Kew Gardens GrassBase to be combined with phylogenetic data developed at the University of Lausanne, Lausanne, France, collection records available in the Global Biodiversity Information Facility (GBIF 2013), and average climate data compiled by the University of East Anglia Climatic Research Unit (Osborne et al. 2011). Appendix 6.1 contains a list of data sets and data providers for collection records accessed through GrassPortal and GBIF. We defined the following variables based on data available in GrassPortal: minimum temperature (mean annual minimum temperature), maximum temperature (mean annual maximum temperature), temperature range (maximum temperature minus minimum temperature), minimum precipitation (minimum annual mean precipitation), maximum precipitation (maximum annual mean precipitation), and precipitation range (maximum precipitation minus minimum precipitation).

We ran a series of univariate linear logistic regressions with SAS 9.2 using Procedure Logistic to analyze relationships between the dependent variables and the independent variables. Each of these regressions used one dependent variable and one independent variable. The number of species used for each univariate regression depended on available data.

We ran four logistic multiple regressions (each with one of the four dependent variables and multiple independent variables) for the 54 species for which multifactor data were available. For these analyses, we used Procedure Logistic with a backward stepwise elimination and the following independent variables: year, life span, mean seed mass, maximum awn length, ploidy level, human use, cultivar availability, minimum temperature, maximum temperature, minimum precipitation, and maximum precipitation. We then repeated the four multiple regressions substituting temperature range and precipitation ranges for the minimum and maximum temperature and precipitation variables. When we evaluated the results of the logistic multiple regressions, we considered factors to be significantly correlated if p-values were less than 0.05.

6.2.3 Histograms

We created histograms showing species counts for section, year, maximum awn length, mean seed mass, ploidy level, human use, cultivar availability, and climate variables. All species were included in the histograms, not just those included in the statistical analyses. For some of these histograms, we separated species into three groups: sections *Bromus*, *Genea*, and *Nevskiella* (Eastern Hemisphere species with mostly annual life spans), sections *Ceratochloa* and *Neobromus* (polyploid Western Hemisphere species with annual to perennial life spans), and section *Bromopsis* (with perennial life spans).

6.3 Results and Discussion

6.3.1 Summary of Results

Based on the data we collected in the database, we categorized 24 species in the genus *Bromus* as widely introduced, 35 species as crop weeds, 37 species as ruderal weeds, and 23 species as environmental weeds. These categories are not exclusive; 15 species were categorized as widely introduced and were in all three weed categories and 16 species were categorized as both widely introduced and environmental weeds, meeting the criteria used here to be considered invasive.

Table 6.2 includes a summary of the analyses results for each hypothesis. Tables 6.3 and 6.4 show the number of species used, Wald Chi-Square values, *p*-values, and odds ratios generated by SAS for the logistic regressions. The univariate logistic regressions showed most of the independent variables with the exception of mean seed mass to be significantly correlated with wide introduction and with all three categories of weediness. Polyploidy was significantly correlated with wide introduction and weediness for crop and ruderal weeds, but not for environmental weeds.

The species used for the logistic multiple regression were those species for which data for multiple independent variables were available. This species set is more biased toward species that are invasive, weedy, or cultivated by humans. These results showed significant correlations between early year and weediness. Long maximum awn length and low minimum precipitation were significantly correlated with weediness for crop and ruderal weeds, and wide temperature range was significantly correlated with weediness for ruderal weeds. Also, in this data set, low mean seed mass was weakly correlated with weediness for crop weeds.

6.3.2 Taxonomy and Life Span (Hypotheses 1 and 2)

We hypothesized that the short-lived species in sections *Genea* and *Bromus* would be more widely introduced and weedy than long-lived species (Hypotheses 1 and 2, Table 6.2). The results of the univariate analysis supported these hypotheses (Table 6.3). In the genus *Bromus*, taxonomy and life span are closely related. Species in sections *Bromus*, *Genea*, and *Nevskiella* are mostly annuals. A few are biennials. All species in section *Bromopsis* are perennials with the single exception of *Bromus texensis* (Shear) Hitchc. (Texas brome). The species in sections *Ceratochloa* and *Neobromus* tend to fall somewhere in between, with life spans ranging from annual to perennial (Fig. 6.1).

Stebbins (1981) believed that the genus *Bromus* evolved in Eurasia, along with grassland ungulates including wild cattle, bison, and sheep and that sections *Bromopsis* (subgenus *Festucaria*), *Ceratochloa*, and *Neobromus* differentiated during the Pliocene. He suggested that *Bromus* grasses in sections *Ceratochloa* and

Table 6.3 Results of univariate logistic regressions

Independent variable	No.	Values	Is widely introduced			Is a crop weed			Is a ruderal weed			Is an environmental weed		
			Wald Chi Sq.	Pr> Chi Sq.	Odds ratio point est.	Wald Chi Sq.	Pr> Chi Sq.	Odds ratio point est.	Wald Chi Sq.	Pr> Chi Sq.	Odds ratio point est.	Wald Chi Sq.	Pr> Chi Sq.	Odds ratio point est.
Section	147		22.0715	<0.0001		40.069	<0.0001		38.9054	<0.0001		21.2555	<0.0001	
		Bromus vs. *Bromopsis*			21.250 (4.386–102.951)			36.037 (10.627–124.044)			28.699 (9.104–90.466)			10.5 (2.633–41.868)
		Ceratochloa vs. *Bromopsis*			15.179 (2.679–86.006)			3.890 (0.794–10.069)			4.373 (1.051–18.189)			7. 467 (1.516–36.783)
		Genea vs. *Bromopsis*			127.5 (15.187–>999.999)			62.239 (9.414–411.481)			49.198 (7.834–308.977)			84 (11.696–603.298)
		Bromus vs. *Ceratochloa*			1.400 (0.400–4.894)			9.333 (2.250–38.711)			6.562 (1.768–24.353)			1.406 (0.367–5.386)
		Bromus vs. *Genea*			0.167 (0.029–0.965)			0.583 (0.101–3.358)			0.583 (0.101–3.358)			0.125 (0.021–0.737)
		Ceratochloa vs. *Genea*			0.119 (0.018–0.795)			0.063 (0.008–0.471)			0.089 (0.013–0.621)			0.089 (0.013–0.621)
Early year	147		25.6954	<0.0001	1.042 (1.026–1.059)	30.1648	<0.0001	0.971 (0.961–0.982)	31.0694	<0.0001	0.971 (0.961–0.981)	24.5799	<0.0001	0.965 (0.952–0.979)

Life span	147		22.827	<0.0001		38.9695	<0.0001		37.5581	<0.0001		20.5218	<0.0001	
		Medium vs. long			31.333 (5.778–169.899)			71.250 (10.793–470.342)			39.428 (6.681–232.696)			31.330 (5.778–169.885)
		Short vs. long			10.123 (3.337–30.706)			35.625 (10.970–116.357)			17.781 (6.596–47.937			8.057 (2.615–24.827)
Low mean seed mass	64		0.0056	0.9404	0.993 (0.815–1.208)	2.9032	0.0884	0.831 (0.672–1.028)	1.0911	0.2962	0.899 (0.737–1.1097)	0.0877	0.7671	0.969 (0.785–1.195)
High max. awn length	147		11.5794	0.0007	1.123 (1.051–1.201)	12.6954	0.0004	1.138 (1.060–1.222	12.3416	0.0004	1.134 (1.057–1.216)	10.6148	0.0011	1.110 (1.042–1.182)
Ploidy	92		8.3911	0.0151		8.5254	0.0141		7.5350	0.0231		3.0221	0.2207	
		Both vs. diploid			8.666 (1.948–38.562)			8.214 (1.955–34.510)			7.200 (1.753–29.567)			3.611 (0.792–16.472)
		Both vs. polyploid			4.000 (1.153–13.876)			5.000 (1.355–18.450)			3.000 (0.869–10.363)			1.496 (0.422–5.300)
		Polyploid vs. diploid			2.167 (0.628–7.479)			1.643 0.582–4.635)			2.400 (0.824–6.987			2.414 (0.706–8.261)
Human use Used for hay/forage/revegetation	147		15.2639	<0.0001	6.427 (2.527–16.345)	11.2623	0.0008	4.133 (1.805–9.468)	12.2860	0.0005	4.322 (1.908–9.835)	14.2934	0.0002	6.406 (2.446–16.777)

(continued)

Table 6.3 (continued)

Independent variable	No.	Values	Is widely introduced			Is a crop weed			Is a ruderal weed			Is an environmental weed		
			Wald Chi Sq.	Pr> Chi Sq.	Odds ratio point est.	Wald Chi Sq.	Pr> Chi Sq.	Odds ratio point est.	Wald Chi Sq.	Pr> Chi Sq.	Odds ratio point est.	Wald Chi Sq.	Pr> Chi Sq.	Odds ratio point est.
Cultivars available	147		21.0028	<0.0001	16.856 (5.036–56.414)	10.5678	0.0012	6.491 (2.092–19.693)	9.5889	0.0020	5.833 (1.910–17.810)	18.8994	<0.0001	13.731 (4.215–44.731)
Low min. temp.	110		11.5505	0.0007	0.872 (0.805–0.943)	4.4480	0.0349	0.935 (0.878–0.995)	5.0512	0.0246	0.931 (0.876–0.991)	7.0060	0.0081	0.903 (0.836–0.974)
High max. temp.	110		13.4449	0.0002	1.267 (1.116–1.437)	9.0555	0.0026	1.143 (1.048–1.246)	10.1639	0.0014	1.151 (1.056–1.256)	10.3829	0.0013	1.220 (1.081–1.377)
Wide temp range	110		20.9695	<0.0001	1.178 (1.098–1.264)	12.2717	0.0005	1.099 (1.040–1.148)	13.7853	<0.0002	1.099 (1.046–1.155)	15.1755	<0.0001	1.133 (1.064–1.206)
Low min. precip.	110		15.2055	<0.0001	0.991 (0.987–0.996)	16.6878	0.0001	0.993 (0.999–0.997)	10.2624	0.0014	0.996 (0.994–0.999)	12.7326	0.0004	0.992 (0.988–0.996)
High max. precip.	110		15.3495	<0.0001	1.001 (1.000–1.001)	7.6914	0.0055	1.001 (1.000–1.001)	10.7369	0.0011	1.001 (1.000–1.001)	9.7984	.0017	1.001 (1.000–1.001)
Wide precip. range	110		21.2073	<0.0001	1.001 (1.001–1.002)	14.6014	0.0001	1.001 (1.000–1.001)	16.8083	0.0001	1.001 (1.000–1.001)	15.5000	<0.0001	1.001 (1.000–1.001)

Odds ratio point estimates are means across species with the Wald confidence intervals in parentheses. Correlations were considered significant if the *p*-value (Pr>Chi Sq.) was less than 0.05

Table 6.4 Results of two multiple logistic regressions with a backward stepwise elimination conducted for each of the four dependent variables

	Is widely introduced			Is a crop weed			Is a ruderal weed			Is an environmental weed		
	Wald Chi Sq.	Pr> Chi Sq.	Odds ratio point est.	Wald Chi Sq.	Pr> Chi Sq.	Odds ratio point est.	Wald Chi Sq.	Pr > Chi Sq.	Odds ratio point est.	Wald Chi Sq.	Pr> Chi Sq.	Odds ratio point est.
High max. awn-length				5.5659	0.0183	1.194 (1.030–1.384)						
Early year							8.9630	0.0028	0.971 (0.952–0.990)	13.6144	0.0002	0.963 (0.944–0.983)
Low mean seed mass				4.1653	0.0413	0.738 (0.551–0.988)						
Low min. precip.				4.7191	0.0298	0.996 (0.992–1.000)	5.6404	0.0176	0.933 (0.987–0.999)			
High max. awn length							4.6317	0.03514	0.0915 (4.6317–0.0314)			
Early year				12.7009	0.0004	0.974 (0.959–0.988)	5.9404	0.0148	0.976 (0.956–0.995)	13.6144	0.0002	0.963 (0.944–0.983)
Wide temp. range							4.4334	0.0352	1.130 (1.008–1.256)			

The same independent variables were used as those used in the univariate regressions with the exception of section. In one set of multiple regressions, minimum temperature, maximum temperature, minimum precipitation, and maximum precipitation were used. In the other, temperature range and precipitation ranges were used. Fifty-four species were used for the multiple regressions and correlations were considered significant if the *p*-value (Pr>Chi Sq.) was less than 0.05

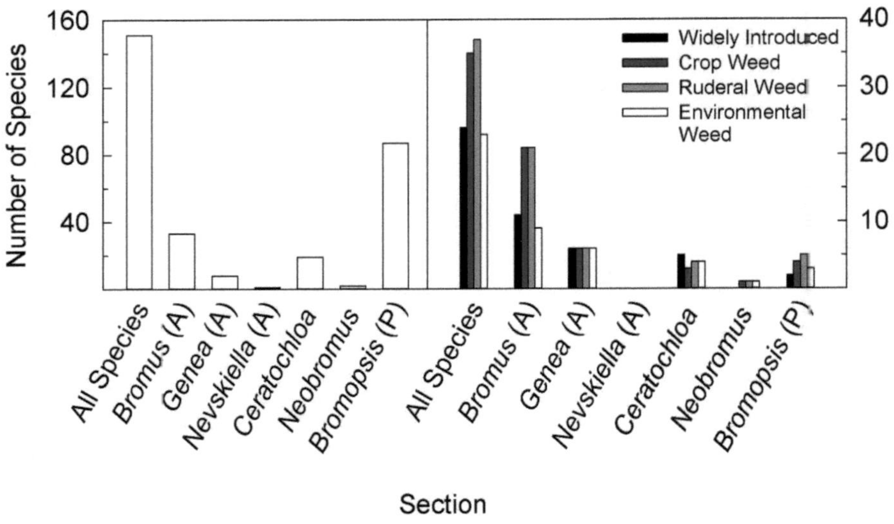

Fig. 6.1 Number of species in each taxonomic section of the *Bromus* genus for all introduction and impact categories combined (*left panel*) or separated (*right panel*). Species may be both widely introduced and weedy and may be included in multiple weed categories. "A" and "P" are used to show sections that are (almost) all annual or perennial. Note the different *y*-axis scales for the *left* (total) and *right* (by category) panels

Neobromus reached North America and eventually South America by the end of the Pliocene, with diploid and tetraploid *Ceratochloa* and *Neobromus* species eventually becoming extinct and only New World species remaining extant. Grasses in section *Bromopsis* also spread to Africa and the Americas during the Pliocene. The species in sections *Bromus* and *Genea* (subgenera *Bromus* and *Stenobromus*) probably developed from different early Central Asian *Bromopsis* species during the Pleistocene in conjunction with human activities. They also spread into Europe during the Pleistocene and are considered archaeophytes (introduced before about 1500 AD) (Stebbins 1981; Preston et al. 2004).

Annual species in sections *Genea* and *Bromus* are species that have long been associated with human habitation and have adapted to grow and reproduce effectively in disturbed areas and in association with crops (Stebbins 1981). They are native to the Eastern Hemisphere, but many have been introduced into the Americas. The evolutionary association of the species in these sections with grazing and agriculture suggests that they have developed adaptations that are likely to make them highly competitive weeds in areas heavily used by humans. The *Genea* species, especially, have seeds with long awns and other features that facilitate distribution by grazing animals. This section has only eight species, but six are widely naturalized and are destructive crop weeds, ruderal weeds, and environmental weeds (Andersson et al. 2002; Kleeman and Gill 2006; Fortune et al. 2008; Williams et al. 2011). Section *Genea* includes the seriously invasive species *B. tectorum* and *B. rubens.*

Many species in section *Bromus* are widely introduced and weedy, with some species falling into all three weed categories. Out of 32 species in section *Bromus*, at least 18 species are widely introduced, 11 (or more) are crop weeds, 11 (or more) are ruderal weeds, and at least eight are environmental weeds. *Bromus hordeaceus* L. (soft brome) is one of the annual grasses that have replaced native bunchgrass species in much of California (Howard 1998; Hobbs et al. 2007; Orrock and Hoisington-López 2009). *Bromus japonicus* Houtt. (Japanese brome), a serious crop weed in the North Central United States, is also invading natural environments in northern mixed-prairie communities (Whisenant 1990; O'Connor et al. 1991; Haferkamp et al. 1997). Some species in this section are associated with specific crops. *Bromus secalinus* L. (rye brome) is a primarily a weed of cereal crops (Tsvelev and Fedorov 1984; Cowbrough et al. 2007). *Bromus interruptus* (Druce) grew with sainfoin, and *Bromus grossus* (Desf. ex DC.) was associated with spelt wheat. Both are probably extinct in the wild (Ainouche and Bayer 1997; Rich and Lockton 2002; Bilz 2011; Gigot 2011).

Genea and *Bromus* crop weeds reduce yields and increase weed control costs. Many countries have restrictions limiting imports of commodities that contain their seeds (Cowbrough et al. 2007; Walters 2011). Changes in agricultural methods may be reducing populations of some species. *B. secalinus* is reported to be decreasing in abundance in many areas, probably because of improved seed-cleaning techniques and other control methods (Darbyshire 2003; Luneva 2003–2009; Cowbrough et al. 2007; Stace et al. 2005).

The perennial section *Bromopsis* contains far fewer weeds and invaders. Out of 89 species in the section *Bromopsis*, only *B. inermis* Leyss. (smooth brome) and *Bromus erectus* Huds. (erect brome) are invasive. Taxonomists have suggested that species in this section fall into two groups. The first group is a mostly polyploid Eurasian group with large anthers, small chromosomes, and a rhizomatous or densely tufted growth pattern. Grasses in this group often cross-pollinate. This group includes *B. inermis*, *Bromus pumpellianus* Scribn. (Pumpelly's brome), *B. erectus*, *Bromus riparius* Rehmann (meadow brome), and possibly South American species *Bromus auleticus* Trin. ex Nees. The second group contains mostly self-pollinating American species with small anthers and large chromosomes. These are less densely tufted and non-rhizomatous. Some Eurasian species of *Bromopsis*, including *Bromus ramosus* Huds. (hairy brome) and *Bromus benekenii* (Lange) Trimen, are more similar to the North American species (Stebbins 1981; Armstrong 1983; Saarela 2001; Saarela et al. 2007; Sutkowska and Mitka 2008). The species in the Eurasian group have traits that suggest they may pose more risk than the species in the American group. Rhizomes facilitate spread of plants introduced into new environments, and rhizomatous species like *B. inermis* may form monocultures (Otfinowski et al. 2007; Otfinowski and Kenkel 2008; Sinkins and Otfinowski 2012). The denser growth patterns of grasses in the Eurasian group also make them useful for hay and for revegetation and soil stabilization.

The species in section *Ceratachloa* and *Neobromus* include annual, biennial, and perennial species and are somewhat more likely to be widely introduced and weedy than the species in section *Bromopsis*, but less so than the species in sections *Bromus* and *Genea.* They may behave as short-lived crop weeds and ruderal weeds under some circumstances. The *Ceratochloa* species of most concern, *Bromus catharticus*

Vahl (rescue brome), a valuable forage species with many cultivars, has become an invasive environmental weed in some areas where it has been introduced for human use (Halvorson and Guertin 2003; Randall 2007; Wu et al. 2009).

6.3.3 Year (Hypothesis 3)

We included year of taxonomic recognition (naming) in analyses after noticing that a number of invasive species had been named and described by Linnaeus. Hypothesis 3 (Table 6.2) says that species that were named earlier are more likely to be invasive than species that were named later. Both the simple and multiple regressions showed significant correlations between early species names and both wide introduction and weediness (Tables 6.3 and 6.4). We believe that species that were named early are those that were common, distinctive, and accessible to botanists during the eighteenth and nineteenth centuries. Many native European *Bromus* species that are invasive were named before 1800 (Fig. 6.2). Most widely introduced or weedy non-European species were named before 1850. One interesting exception is the annual European weed *Bromus lepidus* Holmb. (slender soft brome). *B. lepidus*, which was described in 1924, is interfertile with *B. hordeaceus* but has smaller seeds similar in size to those of *Fescue pratensis* Huds. (meadow fescue). It is believed to have been introduced in Britain as a contaminant of commercial lawn and turf seeds (Smith 1968). Recent species that are spreading into new areas could be of concern, as they may have adaptations that help them to propagate and grow in human-dominated environments,

6.3.4 Seed Awns and Seed Mass (Hypotheses 4 and 5)

One feature of the *Bromus* grasses that stands out is the presence of short to long subapical lemma awns. Hypothesis 4 states that species with long-awned seeds are more likely to be invasive. Our analyses showed long awns to be significantly correlated with both wide introductions and weediness (Tables 6.3 and 6.4).

Genea species have medium to long awns, while those of species in other sections are short to medium (Fig. 6.3). Long awns can aid in seed dissemination by sticking in fur and wool and can discourage grazing. Some *Genea* have stiff awns that can injure livestock and wildlife. *Bromus tectorum* is a useful forage grass early in the season, but the dried awns can damage the mouths of grazing animals (Reid et al. 2008; Germino et al. 2015). *Bromus diandrus* Roth (ripgut brome) has long stiff awns, which can penetrate the skin, eyes, intestines, and feet of livestock (Murrumbidgee Catchment Management Authority 2008). Awns may also help orient seeds correctly during burial and promote germination in upper soil layers (Peart 1984; Humphreys et al. 2011). However, awnless seeds may become buried deeper in soil, remain dormant longer, and show better fire survival (Peart 1984; Humphreys et al. 2011).

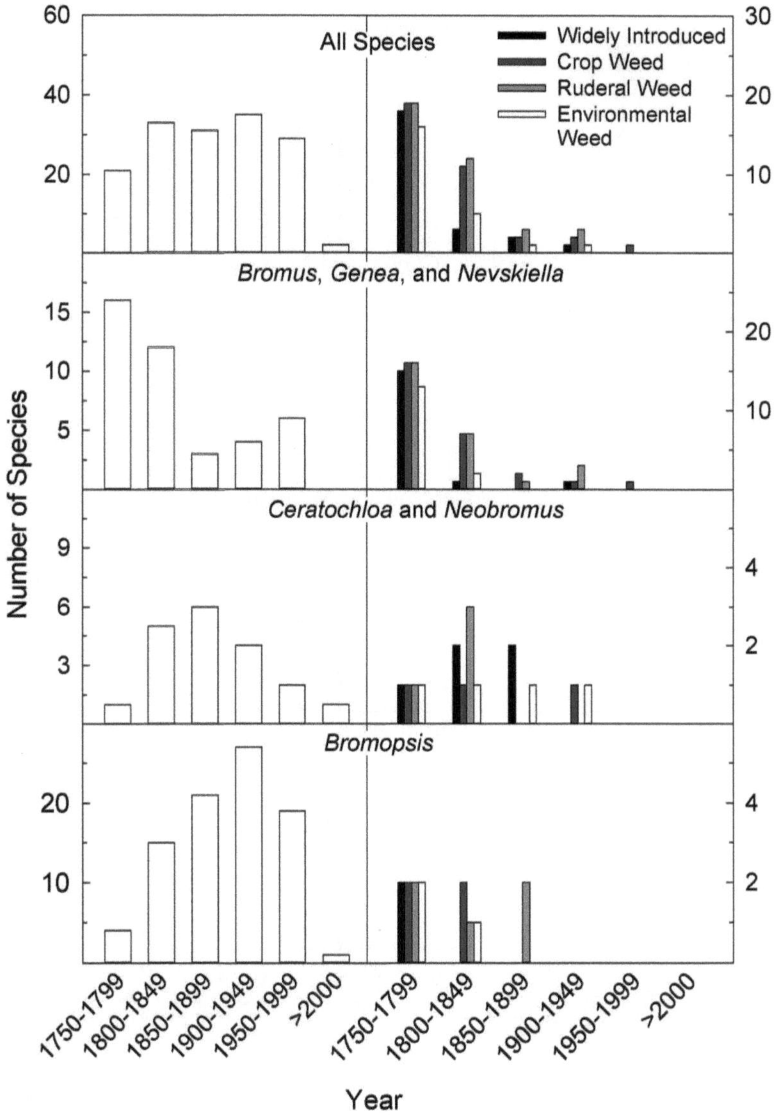

Fig. 6.2 Frequency distribution of species' year of naming (i.e., recognition) within each taxonomic section of the *Bromus* genus, for introduction or impact categories combined (*left panels*) or separated (*right panels*). Species may be both widely introduced and weedy and may be included in multiple weed categories

Hypothesis 5 (Table 6.2) states that species with low mean seed mass are more likely to be invasive. Our analyses did not support this, although there may be a weak association between light seeds and crop weed species (Tables 6.3 and 6.4). Annual species in sections *Genea* and *Bromus* have seeds that range from light to heavy. Most of the perennial *Bromopsis* species have light to medium seeds,

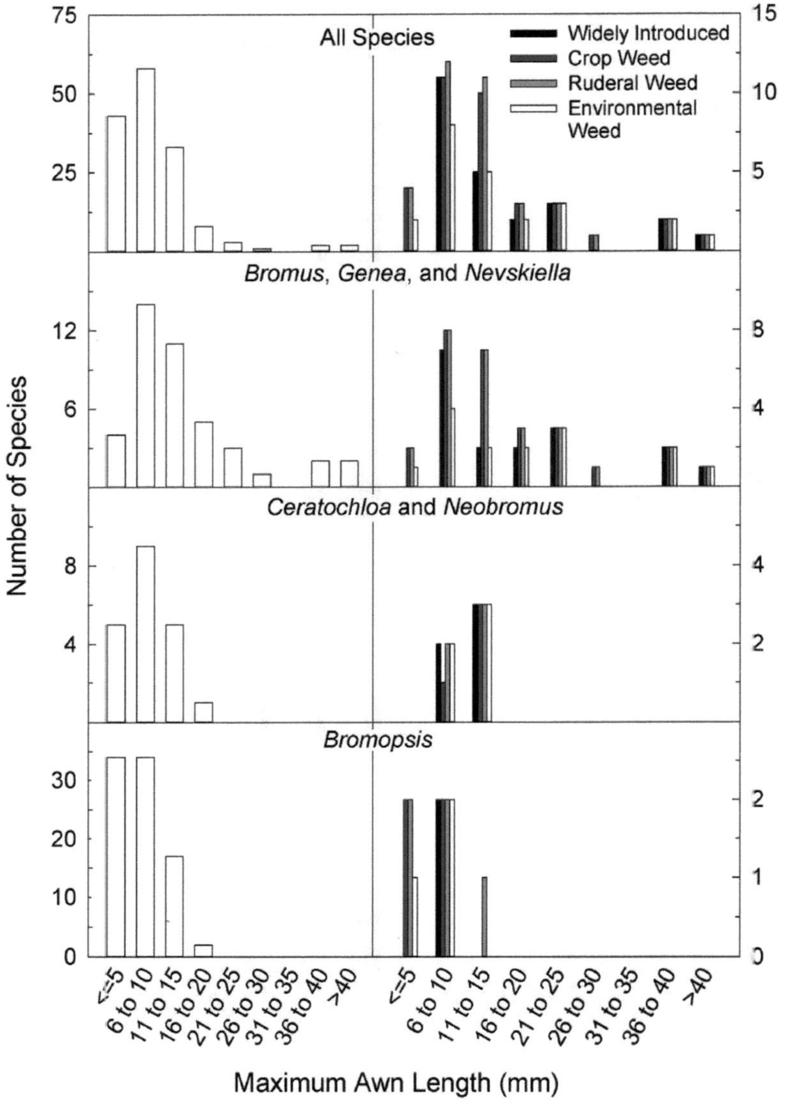

Fig. 6.3 Frequency distribution of species' maximum awn lengths by taxonomic section in the *Bromus* genus, for introduction or impact categories combined (*left panels*) or separated (*right panels*). Species may be both widely introduced and weedy and may be included in multiple weed categories

although a few have very heavy seeds, and most *Ceratochloa* species have heavy seeds (Fig. 6.4).

Some researchers have related low seed mass to high seed production, better dispersal, soil persistence, high initial germinability, and shorter required chilling period (Rejmánek 1996; Hamilton et al. 2005). In a study of nonnative species in Australia,

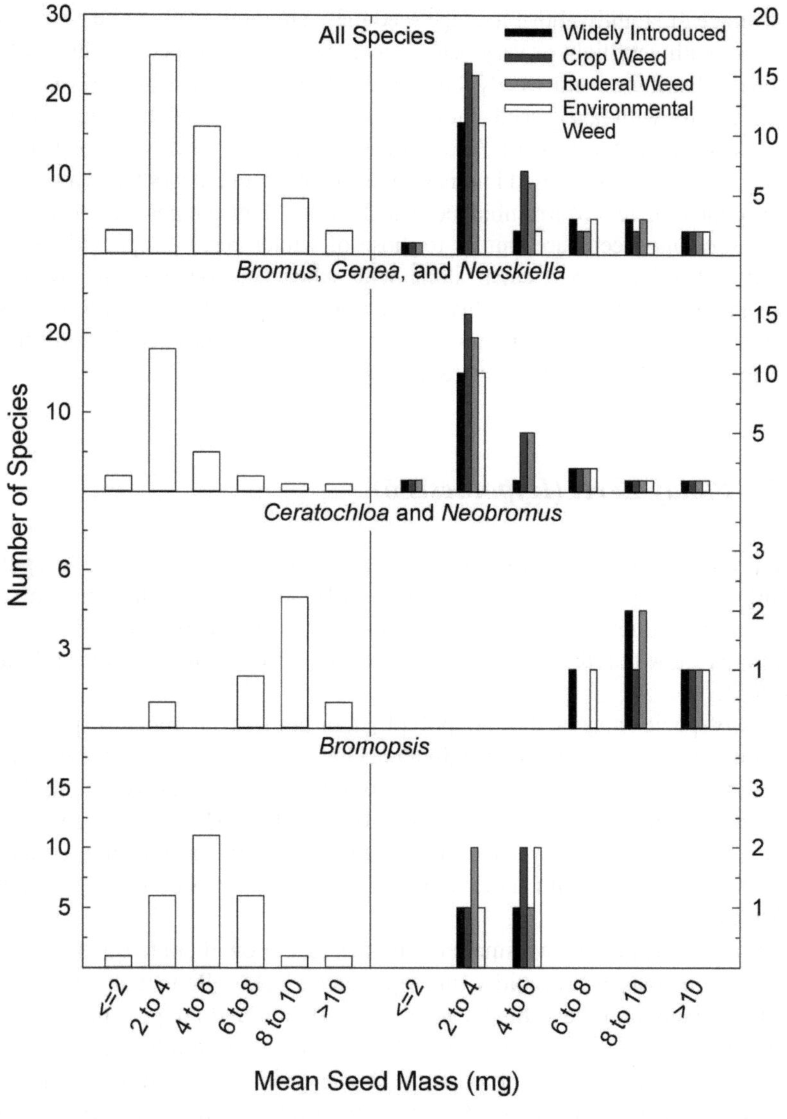

Fig. 6.4 Frequency distribution of species' mean seed masses by taxonomic section in the *Bromus* genus, for introduction or impact categories combined (*left panels*) or separated (*right panels*). Species may be both widely introduced and weedy and may be included in multiple weed categories

Hamilton et al. (2005) suggested that introduced plants with low seed mass are likely to be successful invaders in areas disturbed by humans. However, Moles and Westoby (2006) found that larger and heavier seeds may be more likely to survive in soil and that species with high seed mass have a higher seedling survival rate. Schmidt et al. (2012) studied North American native plants and found that species having seeds with near-average or slightly above-average seed mass are more likely to be pests (weeds) than those with very light or very heavy seeds. Some annual species with light seeds are widely distributed and weedy, but some with heavier seeds are also. *Genea* and *Bromus* species with heavy seeds may take advantage of animal and human activities for seed dissemination. *Bromus diandrus* has rather heavy seeds with long stiff awns and may have been transported into new areas by livestock, in wool, and as a seed and forage contaminate (Murrumbidgee Catchment Management Authority 2008). *Bromus secalinus* seeds are similar to those of winter cereal crops and were often harvested, distributed, and eaten along with wheat and other grains by neolithic Europeans (Behre 2008). The highly invasive species *B. tectorum* and *B. rubens* both have fairly light seeds and long awns. This combination may facilitate dispersal in rangelands and other open habitats (Sales 1994).

6.3.5 Ploidy Level (Hypothesis 6)

We hypothesized that polyploid species are more likely to be invasive (Hypothesis 6, Table 6.2). Our analyses showed that polyploidy is significantly correlated with wide introduction and with weediness for crop and ruderal weeds, but the histograms show that this pattern is much stronger in perennial species (Table 6.3) (Fig. 6.5).

Bromus species may be diploid, polyploid, or both. In a number of species, some individuals are diploid and other individuals have varying levels of polyploidy. The species in section *Ceratochloa* are all polyploid with ploidy levels ranging from 6× to 12×. South American species are generally hexaploid. Most North American species are octaploid and may have developed the higher chromosome counts as a result of hybridization with species in section *Bromopsis* (Stebbins 1981; Stebbins and Tobgy 1944).

Diploid species may have smaller genomes that allow them to quickly complete mitosis and meiosis, grow rapidly, and reproduce quickly (Bennett and Smith 1972). Polyploid plants may have higher levels of variation and adapt better to new and fluctuating environmental conditions. In addition, polyploidy may affect plant physiology and morphology, producing slower growth rates, more robust plants, larger flowers and seeds, delayed or prolonged reproduction, greater stress tolerance, and better winter survival (Rejmánek 1996; Monty et al. 2010; te Beest et al. 2012). This suggests that polyploidy could be especially valuable for perennial species. Stebbins (1956) commented that diploid species are likely to be better adapted to their original habitats, but polyploidy allows plants to better deal with novel environmental conditions, especially when combined with hybridization. Roy (1990) suggested that the absence of a relationship between ploidy level and invasiveness in annual *Bromus* grasses might be related to the recent development of both polyploidization and invasions.

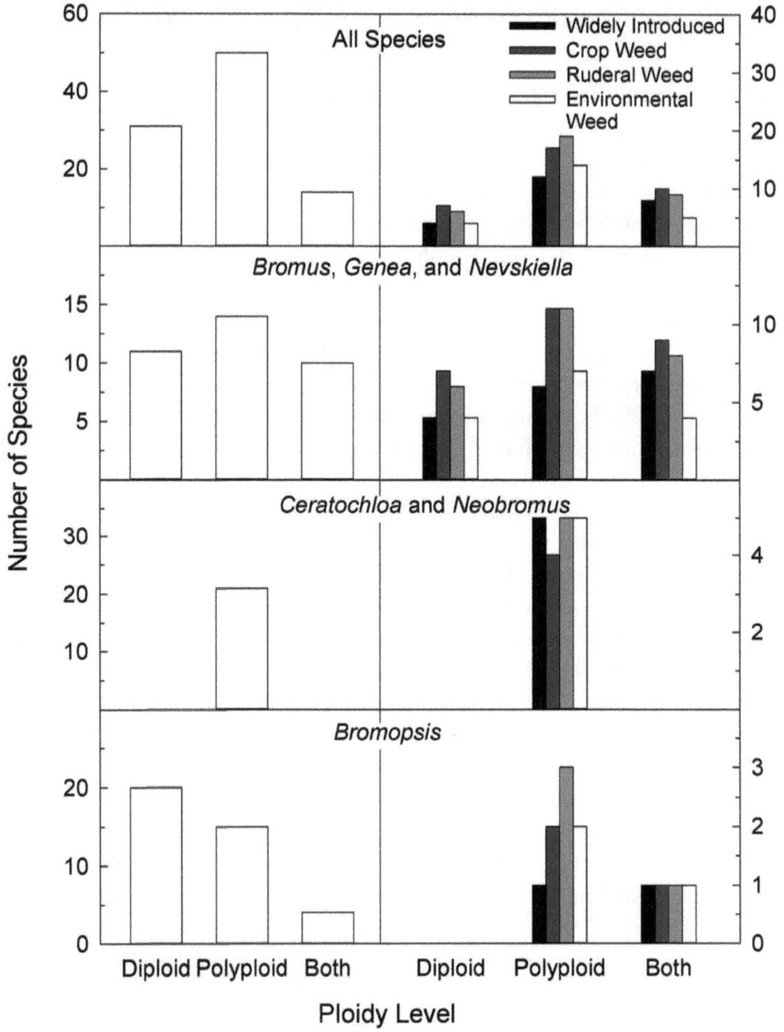

Fig. 6.5 Number of species that are diploid or polyploid in each taxonomic section of the *Bromus* genus, for all introduction or impact categories combined (*left panels*) or separated (*right panels*). Species may be both widely introduced and weedy and may be included in multiple weed categories

In perennial section *Bromopsis*, widely introduced and weedy species are either polyploid or both diploid and polyploid. The association of polyploidy with wide introduction and weediness in the longer-lived sections *Bromopsis* and *Ceratochloa* may also be related to extensive use of polyploid species for hay, forage, and revegetation and for development of cultivars.

6.3.6 Human Use and Cultivar Availability (Hypotheses 7 and 8)

Hypotheses 7 and 8 (Table 6.2) state that species that are used for hay, forage, and revegetation are more likely to be invasive than those that are not and that development of named cultivars is associated with invasiveness. The logistic regressions showed human use and cultivar availability to be significantly correlated with both wide introduction and weediness (Table 6.3).

Sections *Bromopsis* and *Ceratochloa* contain species that are widely used for hay, forage, and revegetation (Fig. 6.6). *Bromus inermis*, in section *Bromopsis*, is widely cultivated and has been used extensively for hay and forage and for revegetation of roadsides, ditches, mine tailings, old logging trails, overgrazed rangeland, and other disturbed areas in North America (Sampson 1913; Weintraub 1953; Otfinowski et al. 2007). Many cultivars are available. These are typically grouped into northern meadow types based on strains introduced from relatively wet temperate regions in Eastern Europe and Russia, southern steppe types based on strains probably introduced from Hungary, and intermediate types (Williams et al. 2011; Smoliak et al. 1990). Cultivar "Polar" is a hybrid of *B. inermis* and *B. pumpellianus* (Walsh 1994). *Bromus pumpellianus* is native across both North America and Asia and is sometimes considered a subspecies of *B. inermis*. For this reason, *B. inermis* is often listed as native in North America. However, *B. inermis* (*Bromus inermis* ssp. *inermis)* was introduced from Eurasia in the 1880s and has interbred with and often replaced *B. pumpellianus* (Elliot 1949). *Bromus inermis* has invaded natural grasslands, shrublands, and forests in western North America. It often forms monocultures and can exclude other species (Otfinowski et al. 2007; Sumners and Archibold 2007; Dillemuth et al. 2009; United States Geological Survey 2013).

Bromus erectus, a Eurasian *Bromopsis* species, is used for pasture and hay in Europe (Jongepierova et al. 2007). It now grows in disturbed areas in eastern North America and is listed as a crop and environmental weed in Australia (Barkworth et al. 2007; Randall 2007). Two other species in section *Bromopsis*, the closely related (or conspecific) *B. riparius* and *Bromus biebersteinii* Roem. & Schult. (meadow brome), have also been introduced into North America for hay and forage and are only occasionally reported as being weedy (Lass and Prather 2007; Williams et al. 2011). *Bromus riparius* is sometimes used as a less aggressive alternative to *B. inermis*, and the two species hybridize (Williams et al. 2011). Other species in section *Bromopsis*, including *Bromus leptoclados* Nees in Africa and *B. auleticus* in South America, are used for hay or forage, but have not been widely introduced into new regions (Stewart 1996; Iannone et al. 2012).

Cultivars of South American *Ceratochloa* species *B. catharticus* (rescue brome) are used for forage and hay in North America, Australia, New Zealand, and Europe (Stewart 1996; Williams et al. 2011). The taxonomy of *B. catharticus* is difficult, and a number of *Ceratochloa* grasses formerly listed as species are now considered to be conspecific with *B. catharticus*. Grasses in the *B. catharticus* complex have

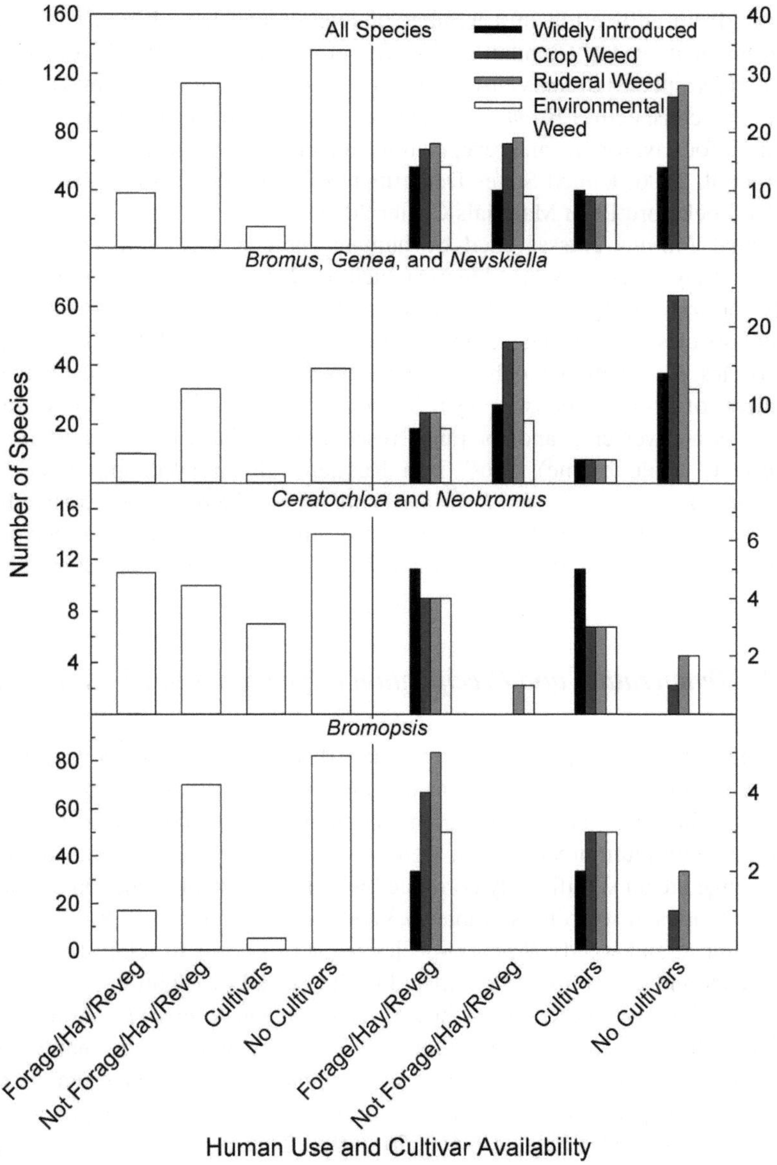

Fig. 6.6 Number of species in different categories of human use or cultivar availability in each section of the *Bromus* genus, for introduction or impact categories combined (*left panels*) or separated (*right panels*). Species may be both widely introduced and weedy and may be included in multiple weed categories

escaped from cultivation in a number of regions and have become ruderal and environmental weeds (Randall 2007; Wu et al. 2009). North American native *Ceratochloa* species *Bromus carinatus* Hook. & Arn. (California brome) and closely related (or conspecific) *Bromus marginatus* Nees ex Steud. (mountain brome) are also used for hay, forage, and revegetation, and cultivars are available (Darris 2007; Tilley et al. 2006; United States Department of Agriculture Natural Conservation Service Lockeford Plant Materials Center 2012).

Annual *Bromus* grasses used by humans include species that have already invaded many parts of North America such as *B. tectorum*, which is used extensively for early season forage (Upadhyaya et al. 1986). A few cultivars of annual *Bromus* species have been developed, but are less widely used than those of perennial species. *B. rubens* "Panoche" was developed in California for soil stabilization. A cultivar of *Bromus mollis* L. (syn. *B. hordeaceus*), also developed in California, is used as a cover crop and for range reseeding and burn rehabilitation. *Bromus arvensis* L. (field brome) "Dos" is a Russian cultivar (Williams et al. 2011). *B. rubens*, *B. hordeaceus*, and *B. arvensis* are all considered to be invasive both in North America and in other parts of the world, although *B. arvensis* has become less common in some parts of Europe (Ainouche and Bayer 1997; Stace et al. 2005).

6.3.7 Temperature and Precipitation (Hypotheses 9, 10, 11 and 12)

We hypothesized that species that are widely distributed and weedy would be able to grow in both hot and cold regions and also in arid regions. Our results showed that low minimum temperature, high maximum temperature, wide temperature range, low minimum precipitation, high maximum precipitation, and wide precipitation range are all significantly correlated with both wide introduction and with all three categories of weediness (Tables 6.3 and 6.4) (Figs. 6.7 and 6.8).

Introduced and weedy *Bromus* species tended to occur in relatively warm and both wetter and drier climates compared to the overall distribution of *Bromus* species. Some invasive species, including *B. tectorum* and *B. inermis*, are also able to survive cold temperatures. A number of *Bromus* grasses are native to the Mediterranean region where rainfall occurs primarily in the winter and are winter annuals that can germinate in the fall, survive the winter, and resume growth in early spring (Jackson 1985; Sales 1994; Salo 2005). This type of life cycle can give annual *Bromus* species a competitive advantage over native species in western North America (DeFalco et al. 2003; Norton et al. 2007). Winter annuals often produce seed and die in early to mid summer. Their dried foliage then provides fuel for fires (Brooks et al. 2015; Germino et al. 2015). Two highly invasive species, *B. tectorum* and *B. rubens*, are associated with an increase in fire frequency that has seriously damaged natural habitats (Melgoza et al. 1990; Melgoza and Nowak 1991; Zouhar 2003; Salo 2004, 2005; Brooks 2012). Some *Bromus* species grow at high elevations in South America and Central Asia. These species tend to have restricted ranges, and relatively little information about them is available.

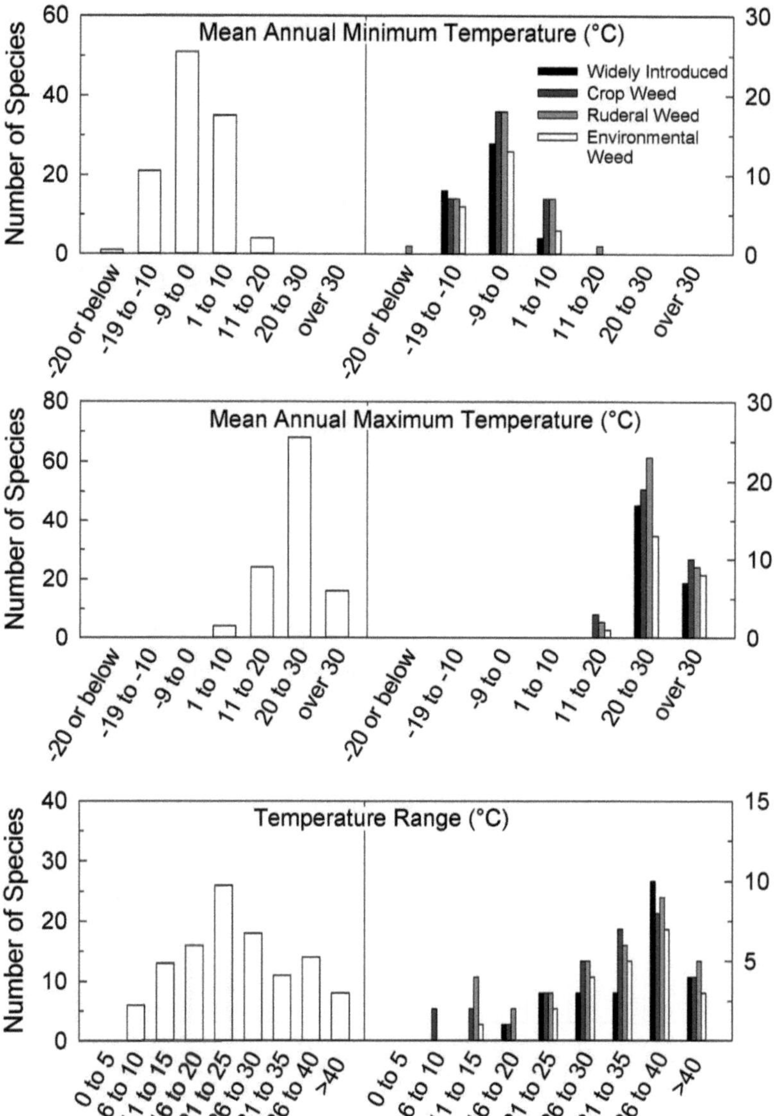

Fig. 6.7 Frequency distribution of minimum, maximum, and range (maximum–minimum) of mean annual temperatures for all species in the *Bromus* genus combined (*left panel*) or separated by introduction or impact category (*right panel*). Species may be both widely introduced and weedy and may be included in multiple weed categories

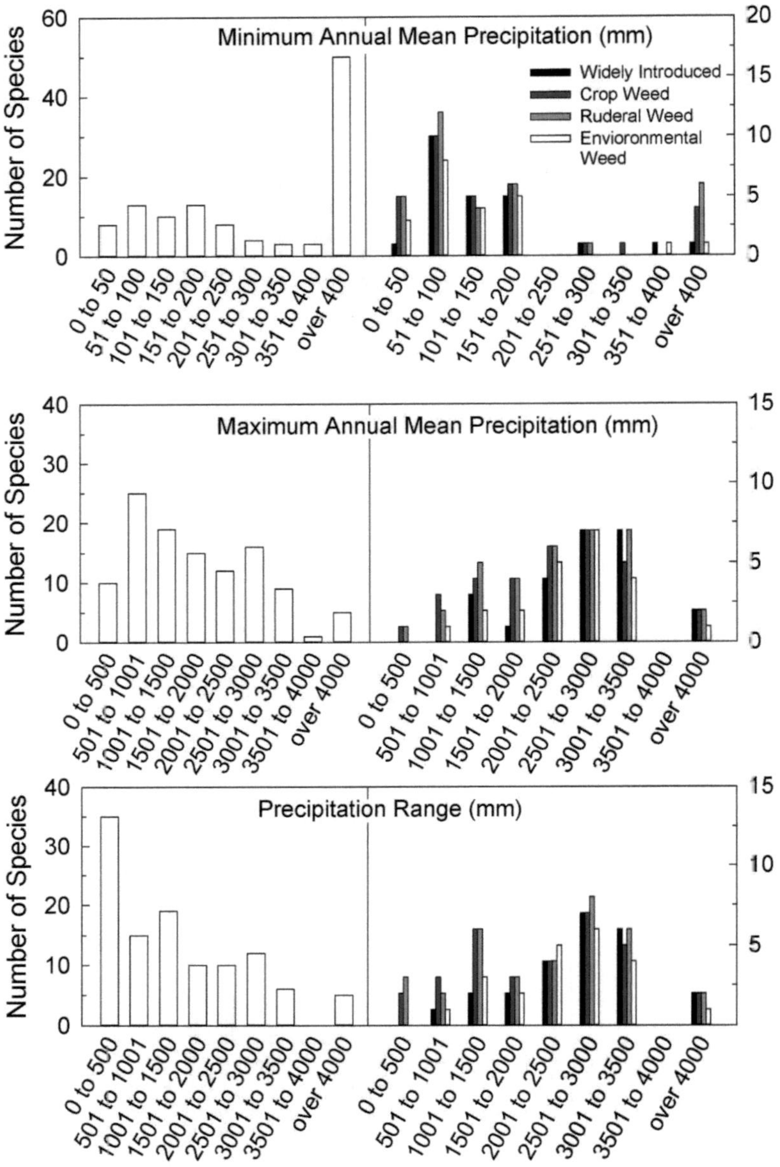

Fig. 6.8 Frequency distribution of minimum, maximum, and range (maximum–minimum) of annual mean precipitation for all species in the *Bromus* genus combined (*left panel*) or separated by introduction or impact category (*right panel*). Species may be both widely introduced and weedy and may be included in multiple weed categories

6.4 Management Implications

A number of *Bromus* grasses that have been introduced into the Western United States have become natural area invaders, crop weeds, and ruderal weeds (Table 6.5). This suggests that the introduction of additional species of *Bromus* grasses should be carefully evaluated and possibly regulated or prohibited. Monitoring and management of *Bromus* species in the Western United States may help to limit new invasions and prevent or reduce damage to additional habitat.

Cultivars of introduced species should be used in ways that minimize movement into natural habitats. Today land managers often prefer to use native *Bromus* species for revegetation and restoration of natural habitats (Bugg et al. 1997; May et al. 1999; Tilley et al. 2006; Wood et al. 2015). Use of introduced species like *B. inermis* for revegetation in natural and seminatural areas should be avoided if possible. Substitution of less aggressive species and cultivars for hay and forage should also be considered.

6.5 Research Needs

In the past, the time and labor required to collect large amounts of data have been barriers to the use of comparison studies with large numbers of species. Today, the increasing availability of large collections of data on the Internet and the development of tools and data collections that integrate data from multiple sources make analyses of large datasets easier. Using the *Bromus* database, we were able to show that species traits such as annual life span, long awns, and polyploidy are correlated with invasiveness. In most cases, the correlations that we saw supported existing hypotheses about the relationships between invasiveness and plant traits. Unfortunately, many types of data that would be useful for comparison studies are hard to find, especially for less common and noninvasive species. For example, information on soil type, pH, and nutrient requirements is often vague or unavailable for many species.

Much of the research on annual *Bromus* species has focused on the highly invasive *Genea* species *B. tectorum.* There is less information available for other weedy and invasive annual species. Research that improves our understanding of the factors that have limited their distribution and evaluates their invasive potential can contribute to prediction and prevention of future invasions. For example, five of the seven species other than *B. tectorum* in section *Genea* are present in the United States and are crop, ruderal, and environmental weeds. The other two species, *B. fasciculatus* and *B. sericeus*, seem to be confined to their native Eurasian ranges and may be adapted to specific types of environments (Sales 1994; Acedo and Llamas 2001; Oja 2002; Fortune et al. 2008). Little information about these two species is available. *Bromus sericeus* is another species that is little studied but could have serious negative impacts if it is introduced because it is very similar to *B. tectorum* and is sometimes considered to be conspecific (*Bromus tectorum* ssp. *lucidus* Sales) (Sales 1991). Many species in the annual section *Bromus* are weeds

Table 6.5 Species of concern in the Western United States

Species categorized as invasive and present in the Western United States		Some other species of possible concern for the Western United States		
Section	Species	Section	Species	Comment
Bromus	*Bromus arvensis*	*Bromus*	*Bromus alopecurus*	Introduced in California and other parts of the United States and weedy
Ceratochloa	*Bromus catharticus*	*Bromus*	*Bromus arenarius*	Introduced into the Western United States and weedy; may be a hybrid between grasses in sections *Bromus* and *Genea*
Ceratochloa	*Bromus cebadilla*	*Neobromus*	*Bromus berteroanus* (syn. *Bromus trinii)*	Introduced into the Western United States and weedy
Genea	*Bromus diandrus*	*Bromus*	*Bromus briziformis*	Introduced into the Western United States and weedy
Bromopsis	*Bromus erectus*	*Bromus*	*Bromus commutatus*	Introduced into the Western United States and weedy
Bromus	*Bromus hordeaceus*	*Bromus*	*Bromus danthoniae*	Weedy elsewhere
Bromopsis	*Bromus inermis*	*Bromus*	*Bromus lepidus*	Introduced into the Eastern United States and weedy; may be distributed as a contaminant in grass seed mixes used for revegetation
Bromus	*Bromus japonicus*	*Bromus*	*Bromus pectinatus*	Crop and ruderal weed in Africa; may be a hybrid between grasses in sections *Bromus* and *Genea*
Bromus	*Bromus lanceolatus*	*Bromus*	*Bromus scoparius*	Introduced into the Western United States and weedy
Genea	*Bromus madritensis*	*Genea*	*Bromus sericeus*	Similar to highly invasive *Bromus tectorum*; adapted to arid habitats
Bromus	*Bromus racemosus*			
Genea	*Bromus rigidus*			
Genea	*Bromus rubens*			
Bromus	*Bromus squarrosus*			
Genea	*Bromus sterilis*			
Genea	*Bromus tectorum*			

Species categorized as invasive are those introduced to two or more biogeographic realms outside of their native range and also as an environmental weed. Other species may also be of management concern (Smith 1968; Wilcox 1986; Sales 1991; Taa et al. 2004; Oja 2007; Saarela et al. 2007; United States Department of Agriculture NRCS 2011; Angelo and Boufford 2013) in their native ranges that have been introduced into North America. Thus, their ecology in their native ranges forecasts their behavior in an introduced range. Additional information on invasive species such as *B. hordeaceus*, *B. rubens*, and *B. japonicus* would inform their management. Data for other less invasive annual species in section *Bromus* is limited, but could shed light on reasons for lack of invasiveness and invasiveness alike.

Research on perennial *Bromus* grasses has focused on breeding and cultivation of species used for hay and forage. More research about the relationship between invasion and human use of perennial *Bromus* species for animal feed and revegetation could help to encourage the development of varieties that are less likely to escape from cultivation and to damage natural habitats and could also contribute to a better understanding of invasion. The history of *Bromus inermis* in North America is an important case study.

Research that focuses on climatic factors is becoming increasingly important. Changes in temperature and rainfall patterns could affect the current ranges of introduced *Bromus* grasses. Some invasive species such as *B. tectorum* and *B. arvensis* may become less common in parts of their current ranges, but could expand into new areas. Other *Bromus* species could also move into new areas or become more serious invaders (Zelikova et al. 2013; Sherry et al. 2012; Bradley and Wilcove 2009; Hobbs et al. 2007).

A number of patterns revealed by our analyses could be studied in greater depth. For example, our analyses showed that species that were described and named early are more likely to be widely distributed and weedy than species that were named and described more recently. If this pattern also occurs in other taxa and to better understand the underlying factors, it could be used to predict and prevent future invasions. We also showed that polyploidy is associated with weediness, especially for perennial *Bromus*. Exploration of this pattern could reveal important mechanisms of plant invasions.

Finally, while many *Bromus* grasses are invasive, some are rare and others are likely to be threatened by changes in agricultural and land management practices, habitat loss, and climate change. A number of species grow in limited areas, sometimes at high elevations or in other areas with extreme environmental conditions. Information about these grasses is difficult to find and in some cases their current status is unknown.

References

Acedo C, Llamas F (2001) Variation of micromorphological characters of lemma and palea in the genus *Bromus* (Poaceae). Ann Bot Fenn 38:1–14

Ainouche ML, Bayer RJ (1997) On the origins of the tetraploid *Bromus* species (section *Bromus*, Poaceae): insights from internal transcribed spacer sequences of nuclear ribosomal DMA. Genome 40:730–743

Andersson L, Milberg P, Schütz W, Steinmetz O (2002) Germination characteristics and emergence time of annual *Bromus* species of differing weediness in Sweden. Weed Res 42:135–147

Angelo R, Boufford DE (2013) Atlas of the flora of New England. http://neatlas.org/index.html. Accessed 23 Jun 2013

Armstrong KC (1983) The relationship between some Eurasian and American species of *Bromus* section Pnigma as determined by the karyotypes of some F1 hybrids. Can J Bot 61:700–707

Baker HG (1974) The evolution of weeds. Annu Rev Ecol Syst 5:1–24

Barkworth ME, Anderton LK, Capels KM, Long S, Piep MB (2007) Manual of grasses for North America. Intermountain Herbarium and Utah State University Press, Logan, Utah

Behre K-E (2008) Collected seeds and fruits from herbs as prehistoric food. Veg Hist Archaeobot 17:65–73

Bennett MD, Smith JB (1972) The effects of polyploidy on meiotic duration and pollen development in cereal anthers. Proc R Soc Lond B Biol 181:81–10

Bilz M (2011) *Bromus interruptus*. IUCN 2012. IUCN red list of threatened species. Version 2012.1. http: http://www.iucnredlist.org. Accessed 23 Aug 2012

Bradley BA, Wilcove DS (2009) When invasive plants disappear: transformative restoration possibilities in the western United States resulting from climate change. Restor Ecol 17:715–721

Brooks ML (2012) Effects of high fire frequency in creosote bush scrub vegetation of the Mojave Desert. Int J Wildland Fire 21:61–68

Brooks ML, Belnap J, Brown CS et al (2015) Exotic annual *Bromus* invasions – comparisons among species and ecoregions in the Western United States. In: Germino MJ, Chambers JC, Brown CS (eds) Exotic brome-grasses in arid and semiarid ecosystems of the Western USA: causes, consequences, and management implications. Springer, New York, NY (Chapter 2)

Brooks ML, Berry KH (2006) Dominance and environmental correlates of alien annual plants in the Mojave Desert, USA. J Arid Environ 57:100–124

Bugg RL, Brown CS, Anderson JH (1997) Restoring native perennial grasses to rural roadsides in the Sacramental Valley of California: establishment and evaluation. Restor Ecol 5:214–228

Burns JH (2004) A comparison of invasive and non-invasive dayflowers (Commelinaceae) across environmental nutrient and water gradients. Divers Distrib 10:387–397

Camus A (1956) Graminées nouvelles des genres *Craspedorhachis*, *Agrostis*, et *Bromus*, Notulae Systematica, Herbier du Museum de Paris 15: 134–137

Clayton WD, Renvoize SA (1986) Genera Graminum – Grasses of the World. Royal Botanic Gardens, Kew

Clayton WD, Vorontsova MS, Harman KT et al (2006+) GrassBase – the online grass flora – The Royal Botanic Gardens at Kew. http://www.kew.org/data/grasses-db.html. Accessed Mar 2012

Concilio AL, Loik ME, Belhamp J (2013) Global change effects on *Bromus tectorum* L. (Poaceae) at its high-elevation range margin. Glob Change Biol 19:161–172

Cowbrough M, Darbyshire S, Sikkema P et al (2007) Weed Profile: Chess, *Bromus secalinus*. Ministry of Agriculture, Food, and Rural Affairs. Ontario, Canada. http://www.omafra.gov.on.ca/english/crops/facts/info_chess.htm. Accessed 29 Oct 2012

Darbyshire SJ (2003) Inventory of Canadian weeds. http://publications.gc.ca/collections/Collection/A42-100-2003E.pdf. Agriculture and Agri-Food Canada

Darris D (2007) California Brome – *Bromus carinatus* Hook. & Arn. – Plant Symbol-BRCA5. United States Department of Agriculture National Resources Conservation Service. http://plants.usda.gov/factsheet/pdf/fs_brca5.pdf. Accessed 10 Aug 2010

DeFalco LA, Bryla DR, Smith-Longozo V, Nowak RS (2003) Are Mojave Desert annual species equal? Resource acquisition and allocation for the invasive grass *Bromus madritensis* subsp. *rubens* (Poaceae) and two native species. Am J Bot 90:1045–1053

Dillemuth FP, Rietschier EA, Cronin JT (2009) Patch dynamics of a native grass in relation to the spread of invasive smooth brome (*Bromus inermis*). Biol Invasions 11:1381–1391

Elliot FC (1949) *Bromus inermis* and *Bromus pumpellianus* in North America. Evolution 3:142–149

Firn J, Moore JL, MacDougall AS et al (2011) Abundance of introduced species at home predicts abundance away in herbaceous communities. Ecol Lett 14:274–291

Fortune PM, Portau N, Viron N et al (2008) Molecular phylogeny and reticulate origins of the polyploid *Bromus* species from section *Genea* (Poaceae). Am J Bot 95:454–464

GBIF (2013) Global biodiversity information facility. http://www.gbif.org. Accessed 1 Jan 1–28 Feb 2013 through Osborne et al. (2011) GrassPortal

Germino MJ, Belnap J, Stark JM et al (2015) Ecosystem impacts of exotic annual invaders in the genus bromus. In: Germino MJ, Chambers JC, Brown CS (eds) Exotic brome-grasses in arid and semiarid ecosystems of the Western USA: causes, consequences, and management implications. Springer, New York, NY (Chapter 3)

Gigot G (2011) *Bromus bromoideus*. IUCN 2012. IUCN red list of threatened species. Version 2012.1. http://www.iucnredlist.org. Accessed 23 Aug 2012

Grass Phylogeny Working Group (2000) A phylogeny of the grass family (Poaceae), as inferred from eight character sets. In: Jacobs SWL, Everett J (eds) Grasses: systematics and evolution. Csiro Publishing, Collingwood, Victoria, pp 3–7

Grass Phylogeny Working Group (2001) Phylogeny and subfamilial classification of the grasses (Poaceae). Ann Mo Bot Gard 88:373–457

Haferkamp MR, Heitschmidt RK, Karl MG (1997) Influence of Japanese brome on western wheatgrass yield. J Range Manag 50:44–50

Halvorson WL, Guertin P (2003) USGS weeds in the west project: status of introduced plants in southern Arizona parks – factsheet for: *Bromus* L. spp. US Geological Survey, Southwest Biological Science Center, Sonoran Desert Field Station, University of Arizona, Tucson, AZ

Hamilton MA, Murray BR, Cadotte MW, Hose GC, Baker AC, Harris CJ, Licari D (2005) Life-history correlates of plant invasiveness at regional and continental scales. Ecol Lett 8:1066–1074

Hobbs RJ, Yates S, Mooney HA (2007) Long-term data reveal complex dynamics in grassland in relation to climate and disturbance. Ecol Monogr 77:545–568

Howard JL (1998) *Bromus hordeaceus*. In: US Department of Agriculture, Forest Service, Rocky Mountain Research Station, Fire Sciences Laboratory, Fire Effects Information System. http://www.fs.fed.us/database/feis/

Hulbert LC (1955) Ecological studies *of Bromus tectorum* and other annual bromegrasses. Ecol Monogr 25:181–213

Humphreys AM, Antonelli A, Pirie MD et al (2011) Ecology and evolution of the diaspore burial syndrome. Evolution 65:1163–1180

Iannone LJ, Pinget AD, Nagabhyru P et al (2012) Beneficial effects of *Neotyphodium tembladerae* and *Neotyphodium pampeanum* on a wild forage grass. Grass Forage Sci 67:382–390

Integrated Taxonomic Information System (ITIS). http://www.itis.gov. Accessed Mar 2012

Jackson LE (1985) Ecological origins of California's Mediterranean grasses. J Biogeogr 12:349–361

Jongepierova I, Mitchley J, Tzanopoulos J (2007) A field experiment to recreate species rich hay meadows using regional seed mixtures. Biol Conserv 139:297–305

Kleeman SG, Gill GS (2006) Differences in the distribution and seed germination behavior of populations of *Bromus rigidus* and *Bromus diandrus* in South Australia: adaptations to habitat and implications for weed management. Aust J Agric Res 57:213–219

Kolar C, Lodge DM (2002) Ecological predictions and risk assessment for alien fishes in North America. Science 298:1233–1236

Lass L, Prather (2007) A scientific evaluation for noxious and invasive weeds of the highway 95 construction project between Uniontown Cutoff and Moscow. Aquilavision, Inc. http://www.nortwestmedia.net/us95/pdf-final/Final-Weed-Report.pdf. Accessed 12 Aug 2012

Liu L, Zhu G, Ammann KH (2006) Flora of China – *Bromus* Linnaeus, Sp. Pl: 76. 1753. http://www.efloras.org/florataxon.aspx?flora_id=2&taxon_id=104664

Lodge DM (1993) Species invasions and deletions: community effects and responses to climate and habitat change. In: Kareiva PM, Kingsolver JG, Huey RB (eds) Biotic interactions and global change. Sinauer, Sunderland, MA, pp 367–401

Luneva NN (2003–2009) *Bromus secalinus* L. – Rye brome. Interactive agricultural ecological Atlas of Russia and neighboring countries. http://www.agroatlas.ru/en/content/weeds/Bromus_secalinus. Accessed 21 Aug 2012

May KW, Willms WD, Stout DG, Coulman B, Fairey NA, Hall JW (1999) Seed yield of three Bromus species native to western Canada. Can J Plant Sci 79:551–555

Melgoza G, Nowak RS, Tausch RJ (1990) Soil water exploitation after fire: competition between *Bromus tectorum* (Cheatgrass) and two native species. Oecologia 83:1–7

Melgoza G, Nowak RS (1991) Competition between cheatgrass and two native species after fire: implications from observations and measurements of root distribution. J Range Manag 44:27–33

Missouri Botanic Garden (2013) Tropicos.http://www.tropicos.org. Accessed April–June 2013

Moles A, Westoby M (2006) Seed size and plant strategy across the whole life cycle. Oikos 113:92–105

Monty A, Maurice S, Mahy G (2010) Phenotypic traits variation among native diploid, native tetraploid and invasive tetraploid *Senecio inaequidens* DC. (Asteraceae). Biotechnol Agron Soc Environ 14:627–632

Murrumbidgee Catchment Management Authority (2008) Best management practices for dryland cropping systems – Great Brome (*Bromus diandrus*). http://www.murrumbidgee.cma.nsw.gov.au/downloads/info_sheets/dryland_cropping/Great_Brome_Revised.pdf. Accessed 3 Aug 2010

National Research Council (2002) Predicting invasions of nonindigenous plants and plant pests. National Academy Press, Washington, DC, p 9

Norton JB, Monaco TA, Norton U (2007) Mediterranean annual grasses in western North America: kids in a candy store. Plant Soil 298:1–5

O'Connor BJ, Paquette SP, Gusta LV (1991) A comparison of the freezing tolerance of downy brome, Japanese brome and Norstar winter wheat. Can J Plant Sci 71:565–569

Oja T (2002) *Bromus fasciculatus* Presl – a third diploid progenitor of *Bromus* section *Genea* allopolyploids (Poaceae). Hereditas 137:113–118

Oja T (2007) Preliminary isozyme evidence on the hybrid origin and diploid progenitors of *Bromus pectinatus* (Poaceae). Aliso 23:468–471

Olson DM, Dinerstein E (2002) The global 200: priority ecoregions for global conservation. Ann Mo Bot Gard 89:199–224

Orrock J, Hoisington-López JL (2009) Mortality of exotic and native seeds in invaded and uninvaded habitats. Acta Oecol 35:758–762

Osborne CP, Visser V, Chapman S et al (2011) GrassPortal: an online ecological and evolutionary data facility. http://www.grassportal.org. Accessed 1 Jan–28 Feb 2013

Otfinowski R, Kenkel NC (2008) Clonal integration facilitates the proliferation of smooth brome clones invading northern fescue prairies. Plant Ecol 199:235–242

Otfinowski R, Kenkel NC, Catling PM (2007) The biology of Canadian weeds. 134. *Bromus inermis* Leyss. Can J Plant Sci 87:183–198

Peart MH (1984) The effects of morphology, orientation and position of grass diaspores on seedling survival. J Ecol 72:437–453

Perrins J, Williamson M, Fitter A (1992) Do annual weeds have predictable characters. Acta Oecol 13:517–533

Powell MR (2004) Risk assessment for invasive plant species. Weed Technol 18:1305–1308

Preston CD, Pearman DA, Dines TD (2004) New atlas of the British and Irish flora. Oxford University Press, Oxford, NY

Pyšek P (1998) Is there a taxonomic pattern to plant invasions? Oikos 82:282–294

Pyšek P, Richardson DM (2007) Traits associated with invasiveness in alien plants, where do we stand? In: Nentwig W (ed) Biological invasions, vol 193, Ecological Studies. Springer, Berlin, pp 97–126

Randall RP, CRC for Australian Weed Management (2007) The introduced flora of Australia and its weed status. http://www.agric.wa.gov.au/objtwr/imported_assets/content/hort/intro_flora_australia.pdf. Accessed 23 Aug 2012

Reid CR, Goodrich S, Brown JE (2008) Cheatgrass and red brome; the history and biology of two invaders. In: Kitchen SG, Pendleton RL, Monaco TA et al (eds) Shrublands under fire: disturbance and recovery in a changing world, 6–8 June 2006. Gen Tech Rep RMRS-P-52. USDA, Forest Service, Rocky Mountain Research Station, Cedar City, UT, pp 27–32

Rejmánek M (1996) A theory of seed plant invasiveness: the first sketch. Biol Conserv 78:171–181

Rejmánek M (2000) Invasive plants: approaches and predictions. Austral Ecol 25:497–506

Rejmánek M, Richardson DM, Pyšek P (2005) Plant invasions and invasibility of plant communities. In: van der Maurel E (ed) Vegetation ecology. Blackwell Science, Oxford, pp 332–355

Rich TCG, Lockton AJ (2002) *Bromus interruptus* (Hack.) Druce (Poaceae) – an extinct English endemic. Watsonia 24:69–80

Roy J (1990) In search of the characteristics of plant invaders. In: Di Castri F, Hansen AJ, Debussche M (eds) Biological invasions in Europe and the Mediterranean Basin. Kluwer Academic, Dordrecht, pp 336–352

Roy J, Navas ML, Sonié L (1991) Invasion by annual brome grasses: a case study challenging the homoclime approach to invasions. In: Groves RH, Di Castri F (eds) Biogeography of Mediterranean invasions. Cambridge University Press, Cambridge, pp 207–224

Royal Botanic Gardens Kew (2008) Seed Information Database (SID) v. 7.1. http://data.kew.org/sid. Accessed 26 Jan–17 Feb 2011

Saarela JM (2001) Molecular systematic studies in the commelind monocots. Dissertation, The University of British Columbia, NR26789

Saarela JM, Peterson PM, Keane RM, Cayouette J, Graham SW (2007) Molecular phylogenetics of *Bromus* (Poaceae, Pooideae) based on chloroplast and nuclear DNA sequence data. Aliso 23:450–467

Sales F (1991) A reassessment of *Bromus tectorum*: a computer assessment, synaptospermy and chorispermy. Flora et Vegetatio Mundi 9:29–41

Sales F (1994) Evolutionary tendencies in some annual species of *Bromus* (*Bromus* L. sect. *Genea* Dum. (Poaceae)). Bot J Linn Soc 115:197–210

Salo LF (2004) Population dynamics of red brome (*Bromus madritensis* subsp. *rubens*): times for concern, opportunities for management. J Arid Environ 57:291–296

Salo LF (2005) Red Brome (*Bromus rubens* subsp. *madritensis*) in North America: possible modes for early introductions, subsequent spread. Biol Invasions 7:165–180

Salo LF, Mcpherson GR, Williams DG (2005) Am Midl Nat 153:95–109

Sampson AW (1913) The reseeding of depleted grazing lands to cultivated forage plants. Bulletin of the USDA, No. 4

Scholz H (1998) Notes on *Bromus danthoniae* and relatives (Gramineae). Willdenowia 28:143–150

Schmidt JP, Stephens PR, Drake JM (2012) Two sides of the same coin? Rare and pest plants native to the United States and Canada. Ecol Appl 22:1512–1525

Sherry RA, Arnone JA III, Johnson DW et al (2012) Carry over from previous year environmental conditions alters dominance hierarchy in a prairie plant community. J Plant Ecol 5:134–146

Simberloff D (2005) Non-native species do threaten the natural environment. J Agric Environ Ethic 18:595–607

Sinkins PA, Otfinowski R (2012) Invasion or retreat? The fate of exotic invaders on the northern prairies, 40 years after cattle grazing. Plant Ecol 213:1251–1262

Smith P (1968) The *Bromus mollis* aggregate in Britain. Watsonia 6:327–344

Smith P (1970) Taxonomy and nomenclature of the brome grasses. Notes R Bot Gard Edinb 30:361–376

Smoliak S, Ditterline RL, Scheetz JD et al (1990) Smooth brome (*Bromus inermis*). In: Montana interagency plant materials handbook for forage production, conservation, reclamation, and wildlife (EB69). http://animalrangeextension.montana.edu/articles/forage/species/grasses/Smoothbromegrass.htm. Accessed 12 Jul 2012

Soderstrom TR, Beaman JH (1968) The genus *Bromus* (Gramineae) in Mexico and Central America. In: Publication of the Museum, Michigan State University, Biological series 3, pp 465–520

Stace C, van der Meijden, de Kort I (eds) (2005) Rye Brome. Interactive flora of Northwest Europe. http://wbd.etibioinformatics.nl/bis/flora.php?selected=beschrijving&menuentry=soorten&id=4598. Accessed 21 Aug 2012

Stebbins GL (1956) Cytogenetics and evolution of the grass family. Am J Bot 43:890–905

Stebbins GL (1981) Chromosomes and evolution in the genus *Bromus* (Gramineae). Bot Jahrb Syst Bot 102:359–379

Stebbins GL, Tobgy HA (1944) The cytogenetics of hybrids in *Bromus*. I. Hybrids within the section *Ceratochloa*. Am J Bot 31:1–11

Stewart AV (1996) Potential value of some *Bromus* species of the section *Ceratochloa*. N Z J Agric Res 39:611–618

Sumners WH, Archibold OW (2007) Exotic plant species in the southern boreal forest of Saskatchewan. Forest Ecol Manag 251:156–163

Sutkowska A, Mitka J (2008) RAPD analysis points to old world *Bromus* species as ancestral to new world subgen. *Festucaria*. Acta Biol Ser Bot 50:117–125

Taa A, Tanner D, Bennie ATP (2004) Effects of stubble management, tillage and cropping sequence on wheat production in the south-eastern highlands of Ethiopia. Soil Till Res 75:69–82

te Beest M, Le Roux JJ, Richardson DM et al (2012) The more the better? The role of polyploidy in facilitating plant invasions. Ann Bot 109:19–45

Tilley DJ, Ogle D, St. John L et al (2006) Plant Guide – Mountain Brome – *Bromus marginatus* Nees ex Steudel. USDA National Resources Conservation Service. http://plants.usda.gov/plantguide/pdf/pg_brma4.pdf. Accessed 25 Jan 2011

Tsvelev NN, Fedorov AnA (ed) (1984) Grasses of the Soviet Union Part 1. Russian Translation Series 8. AA Balkema, Rotterdam

Udvardy MDF (1975) A classification of the biogeographical provinces of the world. IUCN Occasional Paper No. 18. Morges, Switzerland

United States Department of Agriculture ARS (2011) National plant germplasm system. http://www.ars-grin.gov/npgs. Accessed 10 Oct–14 Nov 2011

United States Department of Agriculture Natural Conservation Service Lockeford Plant Materials Center (2012) 'Cucamonga' California Brome. Lockeford, California. http://www.nrcs.usda.gov/Internet/FSE_PLANTMATERIALS/publications/capmcrb11063.pdf. Accessed 16 Jul 2015

United States Department of Agriculture NRCS (2011) The PLANTS Database. National Plant Data Team, Greensboro, NC 27401-4901 USA. http://plants.usda.gov. Accessed 10 Oct–14 Nov 2011

United States Geological Survey (2013) An assessment of exotic plant species of the rocky mountains: *Bromus inermis* Leyss. (*Bromopsis inermis*) Smooth Brome (Poaceae – Graminae). http://www.npwrc.usgs.gov/resource/plants/explant/brominer.htm. Accessed 13 May 2014

Upadhyaya MK, Turkington R, McIlvride D (1986) The biology of Canadian weeds. 75. – *Bromus tectorum* L. Can J Plant Sci 66:689–709

Walters TW (2011) A resource for California Central Valley table grapes. Identification Technology Program, CPHST, PPQ, APHIS, USDA. Fort Collins, CO. http://itp.lucidcentral.org/id/table-grape/resource. Accessed 31 Oct 2012

Walsh RA (1994) *Bromus pumpellianus*. In: U.S. Department of Agriculture, Forest Service

Weintraub FC (1953) Grasses Introduced into the United States – Forest Service Agricultural Handbook No. 58

Whisenant SG (1990) Postfire population dynamics of Bromus japonicus. Am Midl Nat 123:301–308

Whitney KD, Gabler CA (2008) Rapid evolution in introduced species, 'invasive traits' and recipient communities: challenges for predicting invasion potential. Divers Distrib 14:569–580

Wilcox DH (1986) Biology and control of *Bromus pectinatus* Thunb. Dissertation, The University of Manitoba, Canada

Williams WM, Stewart AV, Williamson ML (2011) *Bromus*. In: Kole C (ed) Wild crop relatives: genomic and breeding resources, millets and grains. Springer, Berlin, pp 15–29 (Chapter 2)

Wood TE, Doherty K, Padgett W (2015) Development of native plant materials for restoration and rehabilitation of Colorado plateau ecosystems. Nat Area J 35:134–149

World Wildlife Fund, Inc (WWF) (2014) Ecoregions. http://worldwildlife.org/biomes. Accessed 19 Apr 2014

Wu S-H, Tsai JK, Sun HT et al (2009) Patterns of plant invasions in the preserves and recreation areas of Shei-Pa National Park in Taiwan. Bot Stud 50:217–227

Zelikova TJ, Hufbauer RA, Reed SC et al (2013) Co-evolutionary responses of *Bromus tectorum* to climate change: implications for biological invasions. Ecol Evol 3:1374–1387

Zouhar K (2003) *Bromus tectorum.* In: Fire Effects Information System. USDA, Forest Service, Rocky Mountain Research Station, Fire Sciences Laboratory. http://www.fs.fed.us/database/feis. Accessed 1 Nov 2013

Chapter 7
Community Ecology of Fungal Pathogens on *Bromus tectorum*

Susan E. Meyer, Julie Beckstead, and JanaLynn Pearce

Abstract *Bromus tectorum* L. (cheatgrass or downy brome) presents a rich resource for soil microorganisms because of its abundant production of biomass, seeds, and surface litter. Many of these organisms are opportunistic saprophytes, but several fungal species regularly found in *B. tectorum* stands function as facultative or obligate pathogens. These organisms interact dynamically with abiotic factors such as interannual variation in weather, with other soil microorganisms, with their hosts, and with each other to create spatially and temporally varying patterns of endemic or epidemic disease. Five principal soilborne pathogens, *Ustilago bullata* Berk. (head smut pathogen), *Tilletia bromi* (Brockm.) Nannf. (chestnut bunt pathogen), *Pyrenophora semeniperda* (Brittlebank & Adams) Shoemaker (black fingers of death pathogen), *Fusarium* Link sp. n. (*Fusarium* seed rot pathogen), and a new species in the Rutstroemiaceae (bleach blonde syndrome pathogen), are known to have sometimes major impacts on *B. tectorum* seed bank dynamics, seedling emergence, and seed production. These pathogens exhibit niche specialization, so that they are rarely in direct competition. They sometimes interact to increase the total impact on *B. tectorum* stand structure, which can result in stand failure or "die-off." Die-offs represent areas where *B. tectorum* has been controlled by natural processes, suggesting that these areas might be suitable targets for restoration. Naturally occurring fungal pathogens that can have a strong negative impact on *B. tectorum* success have also been considered as candidate organisms for *B. tectorum* biocontrol using an augmentative mycoherbicidal strategy.

S.E. Meyer (✉)
US Department of Agriculture, Forest Service, Rocky Mountain Research Station, Provo, UT 84606, USA
e-mail: smeyer@fs.fed.us

J. Beckstead
Department of Biology, Gonzaga University, 502 E. Boone Ave, Spokane, WA 99258, USA
e-mail: beckstead@gonzaga.edu

J. Pearce
Department of Plant and Wildlife Sciences, Brigham Young University, Provo, UT 84606, USA

M.J. Germino et al. (eds.), *Exotic Brome-Grasses in Arid and Semiarid Ecosystems of the Western US*, Springer Series on Environmental Management,
DOI 10.1007/978-3-319-24930-8_7

Keywords Biocontrol • Epidemic disease • Stand failure • Seed bank dynamics • Soilborne pathogen

7.1 Introduction

Monocultures of exotic annuals in the *Bromus* genus represent a valuable resource for many pathogens, which in turn can have strong impacts on stand dynamics, although stands are remarkably well buffered from these impacts in most years. Most of the information we have on stand dynamics and almost all the information on the effects of fungal pathogens come from work with *B. tectorum*, and this information forms the basis for most of the discussion that follows. What we know about other annual bromes largely confirms that they have similar life histories and seed bank dynamics. For *Bromus tectorum* L. (downy brome or cheatgrass), very high seed production, efficient seed dispersal, and the ability to form short-persistent (1–2 years) seed banks contribute to the buffering of disease impacts (Meyer et al. 2007a; Smith et al. 2008). Productivity varies dramatically from year to year due to variation in precipitation, but years without seed production are rare. The seeds of *B. tectorum* would seem to present a major resource for granivores, but native rodents do not prefer them, and even harvester ants and birds apparently consume only a small fraction of the crop (Connolly et al. 2014). There is no indication of secondary metabolites produced by the plant or its endophytes that would deter granivores, although this has not been specifically examined. Fall seed bank densities are generally similar to seed production estimates, suggesting that few seeds are removed. Densities range from 5000 to as high as 50,000 seeds m^{-2}, which is approximately equivalent to a grain yield of up to 1500 kg ha^{-1} (Meyer et al. 2007a; Smith et al. 2008). However, this surfeit of seeds does not go to waste. A diverse community of soilborne pathogens is ready to take advantage of this resource as soon as the first rains arrive following dispersal. *Bromus tectorum* seeds are dormant in early summer at dispersal and lose dormancy through dry afterripening under dry summer conditions (Christensen et al. 1996). They are poised to germinate rapidly with the first substantial rains of autumn, but if rainfall is intermittent or insufficient to trigger complete germination, or if precipitation arrives late in the fall, ungerminated seeds may enter secondary dormancy (Allen et al. 2010). These dormant seeds comprise the persistent seed bank. Seedling emergence can take place any time from late summer through early spring depending on precipitation patterns. Regardless of emergence time, the plants bolt and set seed in spring, that is, they exhibit a facultative winter annual life cycle. The seeds, more accurately referred to as florets containing single caryopses, are quite large, weighing an average of 3 mg. As with most grasses, the primary storage compound is starch.

Plant ecologists and range managers working in *B. tectorum*-dominated systems have long been aware of the presence of fungal pathogens in *B. tectorum* stands and have speculated about their importance in regulating stand dynamics. Piemeisel (1938) was among the first to document *B. tectorum* successional processes. He observed that stands of overwintered plants sometimes suffered high mortality in the very early spring, a phenomenon he called "winterkill." He suggested that perhaps *Microdochium nivale* (Fr.) Samuels & Hallett, then called *Fusarium nivale* (Fr.) Sorauer (pink snow mold), might be the causal organism but also considered that abiotic (weather) factors could be responsible. Klemmedson and Smith (1964) reported that pink snow mold was a common pathogen on *B. tectorum* in the inland Pacific Northwest and also that many other pathogens had been reported from this host (Sprague 1953 in Klemmedson and Smith 1964). They also specifically reported on the occurrence of head smut epidemics caused by *Ustilago bullata* (head smut pathogen) that resulted in smutting of >95 % of the individuals and consequent succession to perennial plants.

Piemeisel (1951) also described a phenomenon that he called cyclic succession in *B. tectorum* monocultures on abandoned cropland in southern Idaho. He observed that as *B. tectorum* stands became more and more dense in the years following initial establishment, they ultimately reached a "degenerate" state in which seed production was prevented and stand loss ensued. *Bromus tectorum* would then reestablish on the newly vacant site at low density, and the cycle would repeat itself. He credited this effect to increasing intraspecific competition, but it seems plausible that plant pathogens associated with the heavy litter and crowded conditions of "degenerate" stands could have played a role. This process is very similar to the "die-off" or stand failure in *B. tectorum* monocultures documented in recent years (Baughman and Meyer 2013; Meyer et al. 2014a). The die-off phenomenon is therefore not new, even though it has only recently come to the attention of land managers. The term "die-off" refers to the complete lack of a current-year stand on a site previously occupied by a *B. tectorum* monoculture, in other words, establishment failure (Baughman and Meyer 2013). It is usually the result of mass mortality of germinating seeds or preemergent seedlings, though loss of the persistent seed bank or failure of seed bank replenishment through lack of seed production the previous year can also be involved. All of the processes involved with die-offs can potentially be mediated through fungal pathogens.

The objective of our research with *B. tectorum* fungal pathogens is to understand how they interact with abiotic factors, with the soil microbial community, with their hosts, and with each other to impact stand dynamics in *B. tectorum* and specifically to cause stand failure. We hope to use this knowledge as a restoration tool to create *B. tectorum* stand failure, either through in situ manipulation of the factors controlling disease levels or through inoculum augmentation. The objective would be to temporarily reduce *B. tectorum* competition in the context of restoration seeding. The more recent information presented here is based largely on our own research. We present our data and current understanding of these pathogens in a historical sequence, with the best known species presented first, followed by less well-studied and more recently discovered organisms.

7.2 Principal Pathogens on *Bromus tectorum*

7.2.1 Ustilago bullata *(Head Smut Pathogen)*

The head smut disease of *B. tectorum*, caused by the cosmopolitan basidiomycete pathogen *U. bullata*, is the *B. tectorum* disease most familiar to land managers, ranchers, and recreationists. It is macroscopically visible and conspicuous late in the spring as smutted flowering heads on diseased plants that are interspersed in *B. tectorum* stands among healthy plants in seed (Fig. 7.1a). Under epidemic

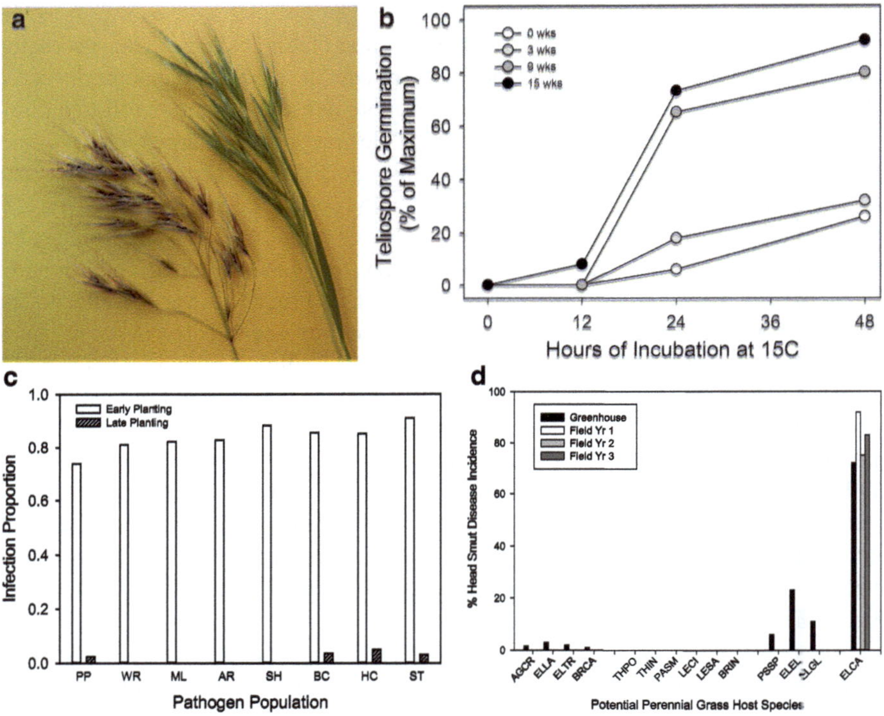

Fig. 7.1 (**a**) Normal *Bromus tectorum* inflorescence (*right*) and smutted inflorescence infected with *Ustilago bullata* (*left*). (**b**) Dormancy loss during 15 weeks in dry storage at 30 °C for a representative *U. bullata* teliospore collection as measured by germination time courses during incubation at 15 °C on PDA for 48 h (each data point represents proportion of teliospores germinated out of 100 examined, corrected for viability; the data points are independent. The experiment was repeated in time for multiple teliospore collections with similar results). (**c**) Field disease incidence following inoculation with teliospores of eight *U. bullata* populations and planting either in early fall or late fall (from Boguena et al. 2007). (**d**) Disease incidence on 12 perennial grass species after inoculation with pathogen strains from *B. tectorum*. (Group 1: AGCR *Agropyron cristatum*, ELLA *Elymus lanceolatus*, ELTR *Elymus trachycaulus*, BRCA *Bromus carinatus*. Group 2: THPO *Thinopyrum ponticum*, THIN *Thinopyrum intermedium*, PASM *Pascopyrum smithii*, LECI *Leymus cinereus*, LESA *Leymus salinus*, BRIN *Bromus inermis*. Group 3: PSSP *Pseudoroegneria spicata*, ELEL *Elymus elymoides*, ELGL *Elymus glaucus*. Group 4: ELCA *Elymus canadensis*)

conditions, its black, soot-like spores can collect as a visible powder on the boots and pant-legs of people who walk through a heavily diseased stand.

7.2.1.1 *Ustilago bullata* Life Cycle

The head smut pathogen is an obligate biotroph, which means it must grow inside the tissues of a living host plant in order to complete its life cycle (Fischer and Holton 1957). This cycle includes spore dispersal, germination, infection of emerging coleoptiles (grass seedling leaves) of the host plant, and systemic growth inside the host. This is followed by preemption of the flowering physiology of the host, so that the florets, instead of containing seeds, are modified into smut bullae for the production and delivery of teliospores (pathogen propagules). Many people assume that the head smut pathogen attacks the inflorescence directly, but this is not the case. The head smut pathogen is not a flower-infecting pathogen but instead is a systemic, seedling-infecting fungus that lives inside vegetative tissues until the plant flowers. Conditions during seed germination and very early seedling growth are therefore the major environmental determinants of disease levels each year.

Teliospores are the dispersal stage of the pathogen. They are released after rainfall causes the bullae to expand and rupture. Most are probably dispersed in raindrops onto the soil immediately adjacent to the smutted plant, though they can also be dispersed by wind or animals. The spores are usually dormant at the time of dispersal. They lose dormancy over the summer under hot, dry conditions in a pattern that parallels dormancy loss in seeds of the host (Fig. 7.1b). This coevolved pattern ensures that smut spores and seeds will germinate synchronously. The spores as well as the seeds are highly germinable in early autumn when temperatures are still warm. If seed germination-triggering rains arrive early, disease levels can be very high. This pathogen has minimal ability to infect at cooler temperatures. If autumn rains arrive late, disease levels will generally be low even if inoculum levels were high (Fig. 7.1c; Boguena et al. 2007). The window of infection is narrow, from the time the coleoptile (seedling leaf) begins to emerge from the seed until the seedling tissue becomes hardened and resistant to penetration. The pathogen can increase its chances of infection during this narrow window by producing free-living, yeast-like sporidia. Individual teliospores are dikaryotic (i.e., they contain two nuclei per spore). The teliospores undergo nuclear fusion, meiosis, and germination, after which a period of asexual multiplication begins. In this way one teliospore can potentially produce hundreds of the sporidial cells that can take part in mating. Two sporidial cells of opposite mating type must fuse to form the infection hypha (mycelial strand), which again contains two nuclei per cell. The infection hypha is not capable of free-living growth. It must encounter a susceptible host coleoptile and penetrate before it exhausts its limited resources.

Smut teliospores do not form persistent spore banks, so that the occurrence of disease is dependent on the presence in the seed bank of spores produced the previous season or, in the longer term, on aerial or seedborne spore dispersal. In years following *B. tectorum* stand failure, which prevents production and dispersal of smut teliospores along with seeds, the incidence of head smut disease on plants that establish from the persistent seed bank is essentially zero.

Once safely inside a plant, the fungus almost always survives to reproduce. The mycelium lives in the crown of the plant from the time of hyphal penetration until bolting commences in mid-spring, then follows the elongating flowering shoots upward and takes over the flowering physiology of the plant, preventing seed formation. The teliospores mature in the resulting bullae, completing the life cycle of the fungus. Plant pathogens with this mode of attack have been called sterilizing or castrating fungi, as opposed to pathogens that cause outright plant death. For an annual plant like *B. tectorum*, the outcome is essentially the same, namely, termination of the host life cycle.

7.2.1.2 Host Range

Ustilago bullata has one of the widest host ranges of any smut fungus, infecting several genera of cool-season annual grass weeds and forage grasses, including *Bromus*, *Agropyron*, and *Elymus* (Fischer 1940; Kreizinger et al. 1947; Meiners and Fischer 1953). The pathogen exhibits a high degree of host specialization, in spite of being an apparent generalist. Specific pathogen races are virulent on only a small subset of the host range of the species as a whole, most often on a single species and sometimes only on specific genotypes within a species.

We have examined the host range of *U. bullata* strains isolated from *B. tectorum* as part of an evaluation of nontarget effects associated with the possible use of this organism for *B. tectorum* biocontrol. We tested six strains at very high inoculum loads on 14 species of perennial grasses (32 plants per species per strain) in the genera *Agropyron*, *Pseudoroegneria*, *Thinopyrum*, *Pascopyrum*, *Leymus*, *Elymus*, and *Bromus* in a series of greenhouse and field inoculation trials. *Bromus tectorum* controls were 100 % diseased in every test. Pathogen strains showed similar patterns of disease and were combined for data presentation.

Nontarget species generally fell into four susceptibility categories (Fig. 7.1d). Ten species were clearly nonhosts with respect to pathogen strains from *B. tectorum*, exhibiting very low disease levels the first year in the greenhouse (Group 1) or, for those that failed to flower in the greenhouse the first year, during three subsequent years after out-planting (Group 2). These included the four introduced forage grasses in the test as well as the native species *Leymus cinereus* (Scribn. & Merr.) Á. Löve (basin wildrye), *Leymus salinus* (M.E. Jones) Á. Löve (salina wildrye), *Pascopyrum smithii* (Rydb.) Á. Löve (western wheatgrass), *Elymus lanceolatus* (Scribn. & J.G. Sm.) Gould (thickspike wheatgrass), *Elymus trachycaulus* (Link) Gould ex Shinners (slender wheatgrass), and *Bromus carinatus* Hook. & Arn (mountain brome). Three native species, *Elymus glaucus* Buckley (blue wildrye), *Elymus elymoides* (Raf.) Swezey (squirreltail), and *Pseudoroegneria spicata* (Pursh) Á. Löve (bluebunch wheatgrass), showed low to moderate levels of disease (6–23 %) the first time they flowered in the greenhouse, but when these smutted individuals were outplanted, they flowered normally and did not exhibit the disease in three subsequent years (Group 3). These species were consistently able to outgrow the disease and are probably not true hosts for strains from *B. tectorum*. Only one perennial species, the native *Elymus canadensis* L. (Canada wildrye; Group 4), showed disease at moderate to high levels (70–90 %) for multiple years following inoculation.

The fact that head smut pathogen strains from *B. tectorum* are largely avirulent on native grasses makes it unlikely that these strains originated in the North American range but instead arrived along with *B. tectorum* seeds from the native Eurasian range. Thus, while *B. tectorum*-infecting strains are very common and widely distributed in western North America, they are probably not truly native, even though *U. bullata* clearly includes native strains on native grasses (Fischer 1940).

We also examined susceptibility of *B. tectorum* and two other introduced annual bromes to several head smut races from different brome species in an unreplicated pilot inoculation experiment with 12 plants per treatment combination. *Bromus tectorum* was quite susceptible to a smut race from a conspecific host population (82 % disease incidence) and somewhat susceptible to races from the exotic annual brome species *Bromus arvensis* L. (field or Japanese brome; syn. *B. japonicus* Thunb.; 58 % disease incidence) and *Bromus sterilis* L. (poverty brome; 30 % disease incidence). It was completely resistant to smut populations from *Bromus diandrus* Roth (ripgut brome) and *Bromus rubens* L. (red brome). *Bromus diandrus* and *B. rubens* were each completely resistant to smut populations from the other four annual bromes but susceptible to smut populations from conspecific hosts (*B. diandrus*, 100 % disease incidence; *B. rubens*, 91 % disease incidence). All three annual species tested were also completely resistant to a smut population from the native perennial *B. carinatus*. This study provides preliminary evidence for race-specific resistance against head smut races from other *Bromus* species in all three introduced annual brome species tested.

7.2.1.3 *Ustilago bullata* Distribution, Epidemiology, and Genetics

Head smut disease is ubiquitous and common throughout the western North American range of *B. tectorum*, and it is almost always possible to find smutted plants in a population. Epidemic levels of disease (>30 % smutted tillers) are encountered sporadically, usually but not always at more mesic sites with reliable fall precipitation and early *B. tectorum* emergence. We carried out a five-state survey of *B. tectorum* diseases at 32 sites in 2005, using a point-intercept method with 40 placements of a ten-pin sampling frame in each population. We recorded a mean head smut disease incidence of 12 % (range 1–69 %). A similar survey at 45 sites in 2006 yielded a mean head smut disease incidence of 16 % (range 0–51 %).

We followed the course of a head smut epidemic at a disturbed sagebrush steppe site in the foothills above Boise, Idaho, over a 4-year period (1999–2003; Meyer et al. 2010a). At the height of this epidemic, over 95 % of the *B. tectorum* population was smutted. The epidemic resulted in near extinction of *B. tectorum* over several hectares and consequently local extinction of the head smut pathogen as well. The site became dominated by *Poa bulbosa* L. (bulbous bluegrass), which was a minor constituent of the vegetation at the beginning of the epidemic.

Head smut epidemics are a result of a complex interplay of factors that make their occurrence difficult to predict. These include inoculum loads resulting from previous-year disease, weather patterns, and the genetic composition of both host and pathogen populations, which can sometimes respond dramatically to short-term selection (Meyer et al. 2010a). In addition to host specialization at the species level, *U. bullata* is also divided into a complex series of pathogen races both among and within populations on *B. tectorum* (Meyer et al. 2001, 2005, 2010a). The patterns of virulence in the pathogen and corresponding resistance or susceptibility in the host generally follow the gene-for-gene model that has been demonstrated in many plant–pathogen interactions that exhibit race-specific resistance (Crute et al. 1997). In a gene-for-gene system, the gene product of a single locus in a pathogen race can be recognized by a complementary gene product of a single locus (a resistance locus) in a host resistance phenotype, resulting in host resistance to this particular pathogen race. If the pathogen has an allele at that locus that does not make the elicitor gene product, or if the host lacks the allele at the complementary resistance locus enabling recognition of the elicitor, then the host exhibits susceptibility to this race. The smuts and bunts are among the few systemic pathogens for which race-specific host resistance has been documented (Crute et al. 1997).

While several resistance loci have been identified in *B. tectorum* that have corresponding avirulence loci in the head smut pathogen, most *B. tectorum* lineages are susceptible to most head smut races, making the function of resistance genes difficult to discern (Meyer et al. 2001, 2005, 2010a). A notable exception is the dominant *B. tectorum* genotype in the Mojave Desert population at Potosi Pass in southern Nevada, which is completely resistant to all head smut pathogen races from Great Basin populations. It is attacked by a unique co-occurring race of the pathogen (Meyer et al. 2005). This unique Mojave Desert race can infect Great Basin *B. tectorum* lineages in greenhouse inoculation trials but has not been found in populations north of the Mojave Desert (Meyer et al. 2005).

7.2.2 Tilletia bromi *(Chestnut Bunt Pathogen)*

Tilletia bromi (chestnut bunt pathogen), like the head smut pathogen, is a basidiomycete seedling-infecting, systemic smut fungus that prevents seed set in infected plants (Duran and Fischer 1961). It is not as common as *U. bullata* and occurs over a much narrower range of environmental conditions. It also produces symptoms that are much less conspicuous than the smutting caused by *U. bullata*, and its presence in *B. tectorum* populations usually goes unnoticed (Fig. 7.2a). However, epidemic levels of chestnut bunt disease can occur and can have a strong negative impact on seed production.

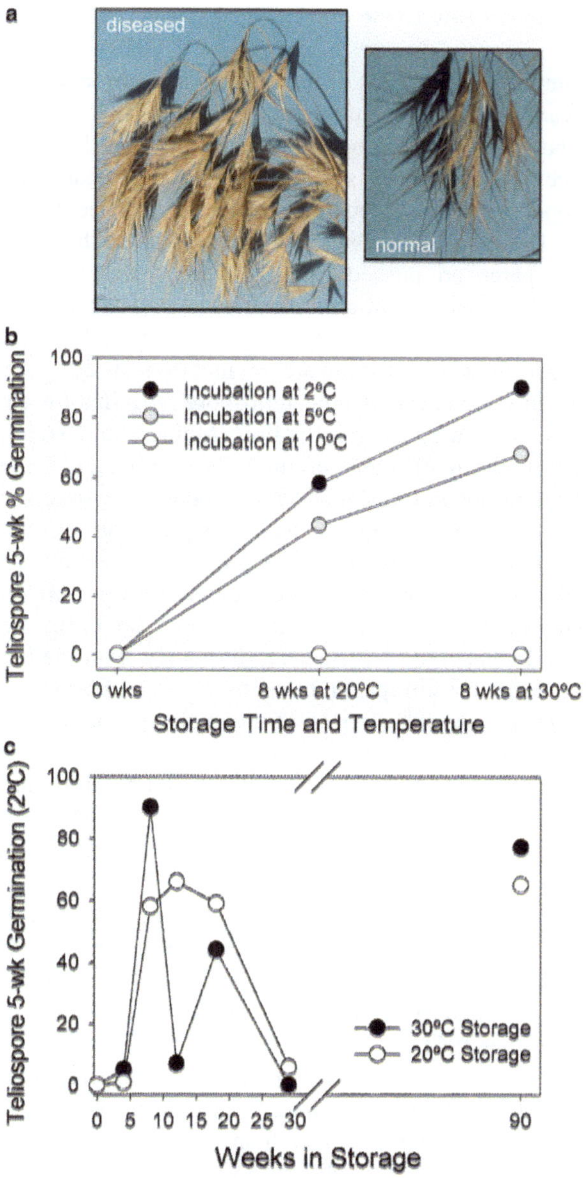

Fig. 7.2 (**a**) Normal *Bromus tectorum* inflorescence (*right*) and diseased inflorescence infected with *Tilletia bromi* (*left*), (**b**) Final germination percentages after 5 weeks on PDA for a representative *T. bromi* teliospore collection tested at three incubation temperatures when recently harvested and after 8 wks of dry storage at 20 and 30 °C, (**c**) Cyclic changes in *T. bromi* teliospore germination percentage at an incubation temperature of 2 °C during time in storage for 90 wks at constant temperatures of 20 and 30 °C. (No data are available for the period 30–90 weeks but the 90-wk data point is provided to show that at least one more cycle of dormancy release ensued. For (**b**) and (**c**), each data point represents proportion of teliospores germinated out of 100 examined, corrected for viability; the data points are independent. Each experiment was repeated in time for a different spore lot with similar results)

7.2.2.1 *Tilletia bromi* Life Cycle

As with head smut, the teliospores of this pathogen are produced in bullae, which are modified ovaries in the host inflorescence. In this case the bullae (sometimes referred to as "bunts") are also produced in florets within the spikelet that do not normally produce seeds, giving the spikelet a "chevron" appearance that is absent in normally developed spikelets. It requires a trained eye to see this difference in the field (Fig. 7.2a). These bullet-like bullae do not rupture on the plant, but remain in the litter after the plants are pushed over by winter storms, where they eventually disintegrate, releasing the chestnut-colored teliospores. The teliospores appear to be long-lived in the surface litter, forming a persistent spore bank. This is advantageous because conditions for infection are not met every year.

The requirements for chestnut bunt teliospore germination contrast strongly with those for head smut spore germination (Meiners and Waldher 1959). The spores do not germinate at all at temperatures above 5 °C, and they germinate best at the near-freezing temperature found under persistent winter snow cover (Fig. 7.2b). We have found that chestnut bunt teliospores are dormant at maturity within the bullae and initially lose dormancy during dry storage in much the same fashion as head smut spores, as long as the requirement for low temperature germination is met. The spores do not necessarily remain nondormant, however, but instead appear to be capable of reentering dormancy, even under constant temperature conditions in dry storage, and then to once again become nondormant in a cyclic pattern (Fig. 7.2c). The period of the cycle appears to be related to temperature, with storage at 30 °C giving a more rapid cycling pattern than storage at 20 °C. This ability to cycle between the dormant and nondormant states is probably related to the ability of the teliospores to form persistent spore banks, but much remains to be learned about this process and how it operates under field conditions.

As in the head smut pathogen, both the chestnut bunt vegetative mycelium inside the plant and the teliospores are dikaryotic, and nuclear fusion and meiosis take place prior to spore germination. This is immediately followed by mating to produce the dikaryotic secondary basidiospores that are the infective units in this fungus (Duran and Fischer 1961).

The ecology of *T. bromi* is very similar to that of its close relative *T. controversa*, the causal agent of dwarf bunt of winter wheat (Meiners 1958; Mathre 1996). Infection takes place in winter, underneath snow cover, after seedling emergence from the soil. The spores are not seedborne, and inoculating seeds directly does not result in disease. Instead, the spores must germinate on the surface of the litter and form secondary basidiospores, which must then intercept the seedling coleoptile after emergence. High levels of disease are confined to years when snow remains on the ground for extended periods.

Once the seedling has been penetrated and infected, fungal mycelium resides systemically in the seedling, then in the crown of the vegetative plant. It grows upward with the bolting flowering stalks in spring and preempts the flowering physiology of the plant, prevents seed set, and produces the "bunts" that contain pathogen teliospores.

7.2.2.2 *Tilletia bromi* Host Range

Tilletia bromi is a pathogen of worldwide distribution that infects members of the grass genera *Bromus*, *Festuca*, *Ventenata*, and some species of *Vulpia* (Castlebury et al. 2005). A closely related species, *Tilletia fusca* Ellis & Everh., is known to infect only two native North American *Vulpia* species (Boyd and Carris 1997, 1998). Within *T. bromi*, there are at least two major pathotypes that show strong host specialization and that may be distinct species. In Washington state, one pathotype infects *B. tectorum* while the other infects *B. arvensis*. Even in intermixed populations of the two hosts, the pathotypes are strongly genetically differentiated, indicating a high degree of host specialization (Pimentel et al. 2000). This finding supports earlier work on host specialization in this group of fungi (Hoffmann and Meiners 1971). It is not known whether pathogen races from *B. tectorum* can infect closely related species that are also known hosts for this pathogen, e.g., *B. sterilis*, or whether race-specific resistance against this pathogen occurs within *B. tectorum*, as we have demonstrated for *U. bullata*. Detailed work on host range in this group has been largely precluded by the technical difficulties associated with experimental inoculation trials.

7.2.2.3 *Tilletia bromi* Distribution and Epidemiology

The chestnut bunt pathogen is widely distributed on *B. tectorum* throughout the Intermountain Region, but its occurrence is sporadic. Many populations contain no sign of this organism, and only occasionally is it detected at epidemic levels. In *B. tectorum* disease surveys in 2005 and 2006, described earlier for head smut disease, chestnut bunt disease incidence averaged 8.3 and 6.0 %, respectively, at 32 and 45 survey sites. In 2005, the disease was epidemic (>20 % incidence) at five sites, four of which were in upper foothill or montane environments likely to have winter snow in most years. Similarly, in 2006, the disease was epidemic at six sites, all of which were in the upper foothill or montane zone. Conversely, we never found any sign of the disease at six Mojave Desert sites. These findings support the earlier conclusion that snow cover in winter is essential for the development of even moderate levels of chestnut bunt disease. But because the spores are likely long-lived in soil, even a single successful infection year can leave a legacy of spores that can cause occasional bunted plants even in suboptimal environments. The fact that this pathogen has no obvious means of spore dispersal also makes it likely that the disease is absent in many environments favorable for its development. This could account for its apparently sporadic occurrence even in montane environments.

7.2.3 *Pyrenophora semeniperda (Black Fingers of Death Pathogen)*

Pyrenophora semeniperda is a well-known generalist ascomycete seed pathogen found throughout the temperate regions of the world (Medd et al. 2003; Stewart et al. 2009). It was dubbed "black fingers of death" because of its conspicuous black, fingerlike fruiting structures (stromata) that protrude from the surface of killed seeds (Fig. 7.3a). It has been regarded as only a weak pathogen that causes little or no damage to cereal crops, and it received little study until Richard Medd and colleagues in Australia initiated studies of the potential of this organism for grass weed biocontrol in wheat (Medd et al. 2003; Campbell and Medd 2003; Medd and Campbell 2005). An exception was the early study by Kreitlow and Bleak (1964). They studied the natural occurrence of the disease on native and introduced grasses at wildland sites in northern Utah using bait seed experiments and also performed greenhouse studies of host susceptibility.

7.2.3.1 *Pyrenophora semeniperda* Life Cycle

The asexual state of this fungus is by far the most frequently encountered state, although sexual structures (perithecia) have been found on a few seeds from *B. tectorum* field seed bank samples. In contrast, seed bank samples may contain literally hundreds of killed seeds with the protruding stromata that produce asexual spores (conidia; Meyer et al. 2007a). There have been conflicting reports on the life cycle of this organism. Campbell and Medd (2003), working with wheat seeds, found that direct conidial inoculation of mature seeds resulted in infection but that the seedlings easily outgrew the fungus and suffered no long-term consequences. They concluded that the fungus must infect during flowering and be internally seedborne in order to cause seed death.

We determined early in our study of this organism that seed germination rate was the key factor in determining whether or not *P. semeniperda* infection would result in seed mortality, a phenomenon we called the "race for survival" (Beckstead et al. 2007). *Bromus tectorum* seed germination rate is a function of dormancy status. When mature nondormant seeds are inoculated with the pathogen, most escape through very rapid germination. When mature seeds are inoculated in the dormant state, most are killed. We determined that infection levels are as high on nondormant seeds as on dormant seeds and that pathogen development can take place on nondormant seeds that have successfully germinated and produced seedlings. Campbell and Medd (2003) worked with the rapidly germinating seeds of wheat and found that the pathogen could cause infection, but not mortality. Because they did not test the pathogen on slow-germinating seeds, they concluded that it had limited ability to kill mature seeds.

Medd and Campbell (2005) were able to infect developing seeds in the inflorescence of annual grass weeds such as ripgut brome (*B. diandrus*) with *P. semeniperda* using

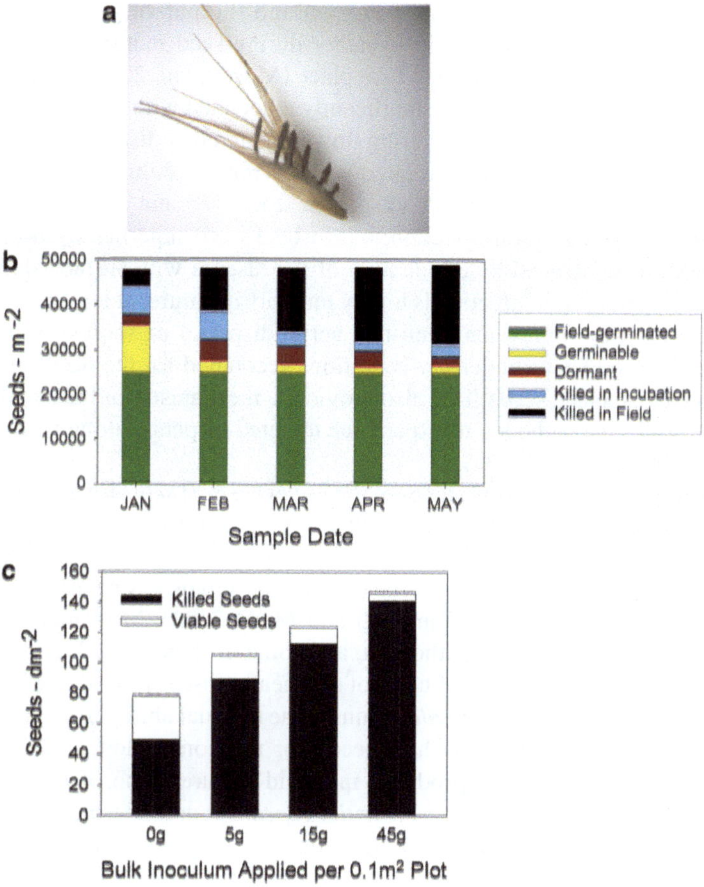

Fig. 7.3 (**a**) *Bromus tectorum* seed killed by *Pyrenophora semeniperda*, showing the protruding stromata that engendered the moniker 'black fingers of death', (**b**) *Bromus tectorum* seed density in different categories as measured monthly during a field seed bank study in 2005–2006 at the Whiterocks exclosure in Skull Valley, Utah. (For each date 20 seed bank samples were collected and processed as described for the cheatgrass disease survey), (**c**) Densities of viable and killed seeds in the seed bank at Haven Flats on the Hanford Reach National Monument in spring 2010 after application of *P. semeniperda* inoculum at three levels the previous fall ($n = 10$ for controls and $n = 40$ for inoculated plots, i.e., mean of four pathogen strains at each inoculum level. Data were obtained from seed bank samples collected from each plot as described earlier)

an aerial inoculum spray method. Wallace (1959) also succeeded in producing infection and subsequent mortality as mature seeds in wheat and oats using this method. In both studies, extended dew periods were required in order for successful floret infection to occur, calling into question whether this process would ever occur naturally in the semiarid environments where these annual grass weeds are a problem. We approached this question indirectly by examining *P. semeniperda* disease levels on undispersed seeds of *B. tectorum* and the native grass *E. elymoides* and examining the correlation with weather during seed maturation, when floret infection on the plant would have to take place (Meyer et al. 2008a). We found that the highest levels of disease were significantly associated with the driest conditions rather than the wettest conditions during flowering and that disease levels on undispersed seeds were positively correlated with current inoculum levels in the soil. We also found that the conidia on undispersed seeds were not deeply seedborne, as would be the case for floral infection, but instead were superficially borne on the floret bracts, as evidenced by elimination of the disease with surface sterilization. We concluded that floral infection is highly unlikely in nature, at least in ecosystems where *B. tectorum* is prevalent, and that aerial dispersal of spores from the seed bank into the seed heads under dry conditions accounted for the occurrence of the disease on undispersed seeds. This also provides a mechanism for the conidia of this otherwise soilborne pathogen to experience targeted dispersal along with the seeds of its host (Meyer et al. 2008a).

Both the asexual and sexual stages of the life cycle of *P. semeniperda* are reported to occur exclusively on seeds (Shoemaker 1966; Paul 1969). This pathogen has been reported to cause a foliar disease called ring spot on young wheat plants, but this disease is of no economic importance, and the fungus has never been reported to sporulate on living leaves (Campbell and Medd 2003). Most members of the genus *Pyrenophora* are foliar pathogens, and some of them, e.g., *P. tritici-repentis* (Died.) Drechsler, causal agent of tan spot of wheat, cause serious damage on cereal crops. It appears that *P. semeniperda* retains some residual ability to cause leaf spots even though it is dependent on host seeds for the completion of its life cycle. Interestingly, *P. semeniperda* produces in liquid culture some of the same toxic compounds (spirocyclic lactams) as *P. tritici-repentis*, but it does not produce these compounds in solid culture on seeds, suggesting that they are not necessary for seed pathogenesis (Masi et al. 2014b).

We have some evidence that *P. semeniperda* can occur as an endophyte in *B. tectorum* plants, presumably by growing into the seedling following infection of a rapidly germinating seed (Beckstead et al. 2012). There is no evidence that the disease is vertically transmitted through seeds, however. The infected litter could potentially act as an inoculum source in early summer soon after production, but it loses its effectiveness prior to contact with the seed bank.

7.2.3.2 *Pyrenophora semeniperda* Host Range

As mentioned earlier, *P. semeniperda* has a very wide host range among the grasses, with >36 genera reported as hosts (Medd et al. 2003). The species has also been reported occasionally from dicot seeds. Most reports have been made in the context of laboratory tests of seed quality. Prior to our work, there were no published reports of this organism in soil seed banks, explaining why it was regarded as relatively uncommon.

Generalist pathogens can exhibit differential success on different hosts, resulting in complex host range patterns. Several factors operate to reduce realized host range relative to potential host range, particularly under field conditions. We explored factors influencing host range of *P. semeniperda* by first measuring potential host range in laboratory experiments at high inoculum loads with 26 grass species, including the primary host *Bromus tectorum*, and developing models to predict susceptibility and tolerance based on host traits, including germination speed, seed hardness, seed size, and phylogenetic relationships (Beckstead et al. 2014). Susceptibility was defined by the level of infection whether seeds survived or were killed, whereas tolerance was defined by the ability of infected seeds to survive. All species tested were at least somewhat susceptible to the pathogen at high inoculum loads, but both infection and mortality varied widely. Species more closely related to the original host (*B. tectorum*) were more susceptible to infection, whereas species with slower seed germination were less tolerant and therefore more likely to suffer mortality. We also examined the effect of inoculum load on host susceptibility and tolerance to *P. semeniperda* in laboratory experiments (Beckstead et al. 2014). Both infection and mortality were sharply reduced as inoculum load was reduced. Intermediate loads had major negative impacts on dormant *B. tectorum* seeds but generally minimal effects on native grass species.

We also searched for this pathogen in the seed banks of co-occurring native grasses and determined that *P. semeniperda* rarely exploits the seeds of native hosts under field conditions (Beckstead et al. 2010, 2014). This marked reduction in realized host range relative to potential host range suggests that laboratory host range studies are potentially a poor predictor of either the current or possible future realized host range for wildland plant pathogens. Subsequent theoretical and field experimental studies on this pathosystem have supported the conclusion that *P. semeniperda* poses low risk to native grass species even when they are planted directly into seed beds with high inoculum loads (Mordecai 2013; Meyer et al. 2014b).

7.2.3.3 *Pyrenophora semeniperda* Distribution and Epidemiology

Medd et al. (2003) reported that *P. semeniperda* was definitely known from Australia, North and South America, and South Africa, with one report from Egypt, and Stewart et al. (2009) found it in Turkey and Greece. It is probable that seed bank studies in drier temperate regions of the world where annual bromes are important members of the flora would reveal a wider distribution.

We examined the distribution of *P. semeniperda* in the Intermountain Region as part of the *B. tectorum* disease survey described earlier. Disease incidence was measured as density of *P. semeniperda*-killed seeds in the soil seed bank. Ten spring seed bank samples were collected at each site. Killed seeds with visible stromata were counted, and apparently viable seeds were allowed to lose dormancy, then incubated and scored as viable, nonviable due to other causes, or killed by *P. semeniperda* (Meyer et al. 2007a). The mean killed seed density was similar each year:

ca. 3500 killed seeds m^2 in 2005 and 3900 killed seeds m^{-2} in 2006, with values ranging from 0 to as high as 20,000 killed seeds m^{-2}. In the 2005 data set, there was a significant trend for a larger proportion of potential carryover seeds to be killed at sites with higher seed densities in the potential carryover seed bank, namely, at drier sites (Meyer et al. 2007b). This pathogen is thus more important at sites with less reliable autumn rainfall and a higher probability that seeds will enter secondary dormancy and become part of the potential carryover seed bank. In essence, when large numbers of seeds fail to germinate in the first germination-triggering storms and subsequently become secondarily dormant, most of these seeds are killed by the pathogen. At more mesic sites, where the potential carryover seed bank is small, the pathogen is present only at low levels, and most of the small number of seeds that remain ungerminated can escape mortality. Even though the pathogen can sporulate on germinated seeds that go on to form seedlings, its fitness is clearly increased by causing seed mortality.

Seed bank studies with more frequent sampling dates were carried out in 2005–2006 at the Whiterocks study site in Skull Valley, Utah, the location of many of our published studies on this pathosystem (e.g., Beckstead et al. 2007, 2012; Meyer et al. 2007a, 2014b), permitting us to examine these patterns in more detail (Fig. 7.3b). The fall of 2005 was extremely dry, so that the first germination-triggering rainfall event took place during a warm period just before New Year's Day. Approximately half of the 48,000 seeds m^{-2} in the seed bank germinated during this storm and about half of the remaining seeds were still germinable. These remaining seeds rapidly entered dormancy under winter conditions and became prey to attack by *P. semeniperda*. By the end of spring, the pathogen had killed 76 % of the potential carryover seed bank in the field, and another 12 % were likely already infected, as they developed pathogen stromata in subsequent incubation, for a total of 88 % mortality of the potential carryover seed bank and 42 % mortality of the previous-year seed crop.

The demographic consequences of the high 2005–2006 seed mortality for *B. tectorum* were very likely negligible. Germinated *B. tectorum* seeds can successfully establish a stand and produce a new crop of seeds regardless of the impact of *P. semeniperda* on ungerminated seeds. It is only following years of stand failure that *P. semeniperda* becomes potentially important to *B. tectorum* demographics, because in those years, the stand must reestablish from the in situ carryover seed bank, and the density of viable seeds remaining in the seed bank becomes a major factor limiting stand density. In effect, the carryover seed bank only serves as an insurance policy in the event of stand failure, and in most years, stand failure does not occur. The pathogen exploits excess seed production but leaves the *B. tectorum* population largely unharmed and able to produce large quantities of seeds to support pathogen success in subsequent years. *Bromus tectorum* would likely form much larger carryover seed banks in the absence of *P. semeniperda* in the dry environments that favor seed bank carryover, but in spite of this, destruction by the pathogen of a major fraction of the seed crop each year poses little threat to *B. tectorum* persistence.

Most studies support the idea that the main target of this pathogen is dormant seeds, but we have also encountered pathogen strains that can kill fast-germinating, nondormant seeds (Meyer et al. 2010b). We first thought that faster-growing pathogen strains would be more likely to kill fast-germinating seeds, but in fact the opposite proved to be true. The strains that caused the highest mortality on fast-germinating seeds were the slowest-growing strains. This apparent contradiction could be due to the high cost of producing toxins that could quickly disable a germinating seed. This fungus produces large quantities of cytochalasin B, a toxin that prevents cell division following mitosis (Evidente et al. 2002), making this toxin a likely candidate. We measured cytochalasin B production in a series of pathogen strains with different growth rates and obtained a significant negative correlation between cytochalasin B production and mycelial growth rate (Masi et al. 2013). This resource trade-off between mycelial growth and toxin production was later demonstrated more conclusively (Meyer et al. 2015).

Field evidence for pathogen-caused nondormant seed mortality comes from inoculum addition experiments in which the density of killed plus viable seeds in the carryover seed bank was much increased with inoculum addition, as well as the proportion of seeds killed (Fig. 7.3c). This implies that the pathogen at augmented inoculum loads killed seeds that would otherwise not have carried over, i.e., nondormant seeds.

Another mechanism that could explain how *P. semeniperda* kills nondormant seeds is through water stress associated with intermittent small autumn precipitation events. Mortality of nondormant seeds was greatly increased if they were first incubated postinoculation at water potentials that suppressed radicle emergence but permitted pathogen development prior to incubation in free water (Finch et al. 2013). This could explain the mortality of seeds that should otherwise have been fast-germinating and able to escape in field seed banks.

7.2.3.4 *Pyrenophora semeniperda* Genetics

We recently published a genome assembly for *P. semeniperda*, opening the door for comparative genomic studies with other species of *Pyrenophora* for which sequenced genomes are available (Soliai et al. 2014). This could be especially helpful in elucidating the evolutionary origin and function of the phytotoxins that it produces (Evidente et al. 2002; Masi et al. 2013, 2014a, b). We also sequenced the ITS region (internal transcribed spacer sequence from ribosomal DNA) of a total of 417 strains from 20 of the pathogen populations in the *B. tectorum* disease survey described earlier (Boose et al. 2011). Genetic analysis revealed high diversity, with 12 different ITS haplotypes, but very little population structure. Most of the variation (>80 %) was accounted for by within-population variance. There was weak but significant differentiation between northern (Washington and Idaho) and southern (Utah and Colorado) population groups, and the northern group had significantly higher gene diversity than the southern group. Overall, these results suggest that the

P. semeniperda populations on *B. tectorum* did not originate from local populations of native grass hosts. The relationship between genetic and geographic distance was only weakly supported ($r=0.146$, $P=0.053$). It seems more likely that the populations on *B. tectorum* traveled as seedborne inoculum from the Eurasian range and also accompanied *B. tectorum* during its subsequent expansion throughout the West. The strains isolated from seeds collected in Turkey and Greece belonged to ITS haplotypes also found in Intermountain populations, supporting a Eurasian origin for the pathogen, at least for populations on *B. tectorum* (Stewart et al. 2009). The pathogen is so rare in native seed banks that we have been unable to unequivocally identify strains originating from native grasses. It is possible that all the strains in North America were introduced along with their exotic annual grass hosts.

7.2.4 Fusarium *sp. n.* (Fusarium *Seed Rot Pathogen)*

As mentioned earlier, stand failure is a widespread phenomenon in *B. tectorum* dominated ecosystems that is commonly referred to as "die-off" (Baughman and Meyer 2013). As part of studies to understand the causes of stand failure, we planted *B. tectorum* bait seeds into die-off soils in both field and greenhouse experiments and isolated putative causal organisms from killed seeds (Meyer et al. 2014a). We detected an array of fungal organisms that could potentially be seed or preemergent seedling pathogens, but *Fusarium* was by far the most commonly isolated. We therefore initiated studies to investigate whether *Fusarium* could be a die-off causal organism (Meyer et al. 2014a).

Members of the ascomycete genus *Fusarium* are ubiquitous in soils worldwide and include many important pathogens of cultivated plants, particularly vegetables and winter cereal crops (Nelson et al. 1981). The occurrence of *Fusarium* species in natural ecosystems is also frequently reported, but its role in the microbial ecology of these ecosystems is much less well documented (e.g., Walsh et al. 2010).

Molecular-genetic characterization of isolates from *B. tectorum* die-off soils using both ITS and TEF (translation elongation factor) sequence data determined that they belonged to the *Fusarium tricinctum* species complex (O'Donnell et al. 2013). The strains in our study represent one or more undescribed biological species within this complex, but a more exhaustive multilocus molecular-genetic analysis will be required to clarify their status.

Species of *Fusarium* have been reported to be pathogenic on seeds or newly germinated seedlings and to cause diseases referred to as "seed rot" that can result in emergence failure (Slykhuis 1947) or, in the case of the seeds of plant-parasitic plants, host penetration failure (Sauerborn et al. 1996; Muller-Stover et al. 2009). We have demonstrated conclusively that *Fusarium* strains isolated from diseased *B. tectorum* seeds are pathogenic on *B. tectorum* seeds and are capable of causing sometimes high mortality, especially under conditions of intermittent water stress (Meyer et al. 2014a). However, the role of *Fusarium* in *B. tectorum* stand failure in the field has not yet been conclusively demonstrated.

7.2.4.1 *Fusarium* Seed Rot Life Cycle

The asexual life cycle of *Fusarium* begins with conidia (asexually produced spores; Fig. 7.4a). Two types of conidia, macroconidia and microconidia, are normally produced, although some of the strains from *B. tectorum* rarely if ever produce

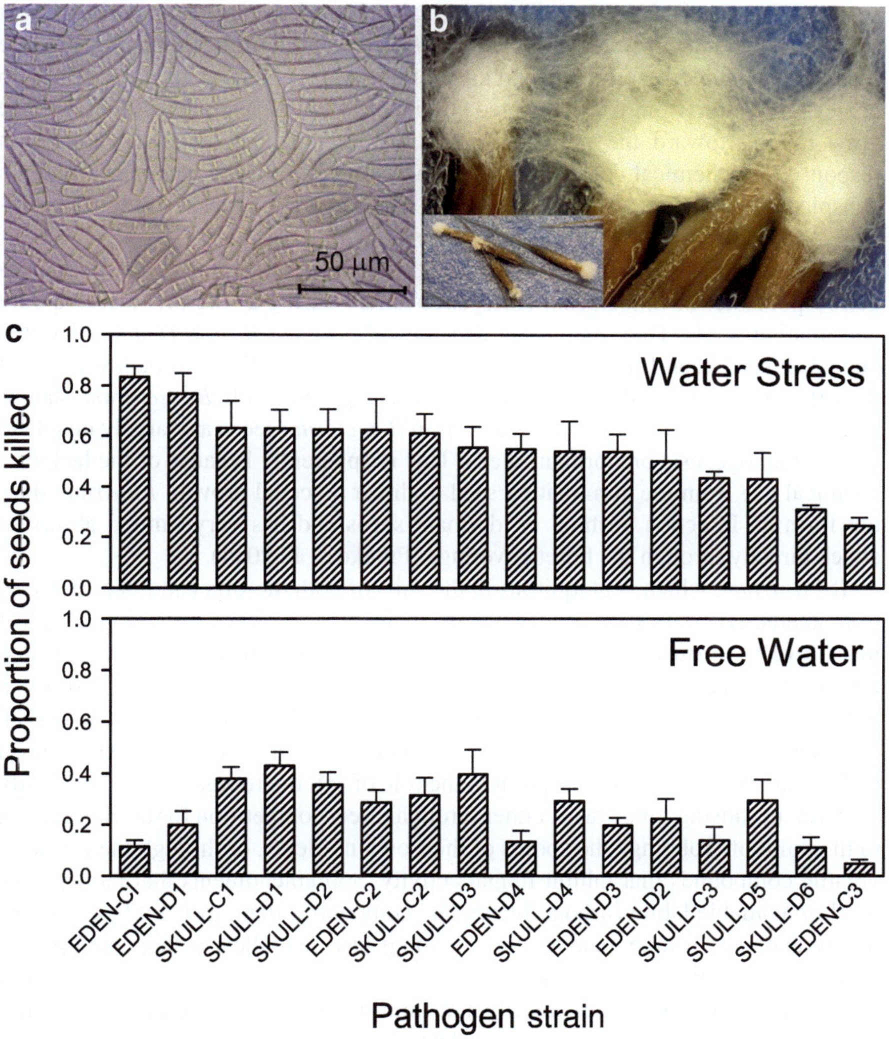

Fig. 7.4 (**a**) Macroconidia produced by a *Fusarium* strain from a die-off soil. (**b**) The development of abundant mycelium on killed *Bromus tectorum* seeds that had been inoculated with *Fusarium* macroconidia, incubated for 7 days at −1.5 MPa, and then transferred to free water (inset shows the development of infection cushions or "white tufts" over the point of incipient radicle emergence during incubation at low water potential). (**c**) *Bromus tectorum* seed mortality caused by 16 strains of *Fusarium* when inoculated seeds were incubated for 7 days at −1.5 MPa prior to transfer to water (*upper panel*) and when inoculated seeds were placed directly into free water (*lower panel*) at 25 °C (From Meyer et al. 2014a)

macroconidia and some sporulate only weakly in culture. It is likely that most of these strains also produce chlamydospores, which are thick-walled resting spores that can persist in the soil for longer periods of time, but we have not yet observed these in culture (Leslie and Summerell 2006).

We have recently elucidated the asexual life cycle of a *Fusarium* strain from *B. tectorum* as it is expressed during pathogenesis on seeds in the laboratory using scanning electron microscopy (Franke et al. 2014). Inoculated nondormant host seeds were held under water stress (−1.5 MPa) to retard germination and to provide *Fusarium*, which can germinate and grow at this water potential, the opportunity to achieve infection. Conidia germinated within a few hours, and the resulting hyphae grew rapidly toward the point of impending radicle emergence, apparently in response to a chemical cue produced during germination. The pathogen formed a conspicuous infection cushion within 48 h, and penetration and seed mortality followed soon after transfer to free water (Franke et al. 2014).

Baughman and Meyer (2013) produced circumstantial evidence that the pathogen responsible for *B. tectorum* emergence failure during a "die-off" affected only germinating seeds. They found that densities of dormant *B. tectorum* seeds in the persistent seed bank following a die-off were the same in the seed banks of recent die-off areas and in adjacent areas that had supported full *B. tectorum* stands. Interestingly, *Fusarium* strains isolated from *B. tectorum* seeds are largely unable to initiate pathogenesis on dormant seeds. This is apparently because of the lack of a chemical cue from the germinating seed to direct mycelial growth. *Fusarium* does not form an infection cushion on dormant seeds and has very limited ability to attack directly through the floret coverings (Franke et al. 2014).

If *Fusarium*, which is ubiquitous in the soils of both die-offs and intact *B. tectorum* stands, is a die-off causal organism, then stand recovery following a die-off must involve *Fusarium* suppression. There is evidence from many studies that fungal spore germination and hyphal growth can often be suppressed in field soil, a phenomenon referred to as "fungistasis" (Lockwood 1977; Garbeva et al. 2011). This suppression is usually alleviated in autoclaved soil, indicating that it has a biological cause. Many studies point to the role of soil microorganisms, specifically bacteria, in causing fungistasis, either through direct competition for nutrients, even to the point of "robbing" the spores of their own nutrients, or through the action of volatile compounds that inhibit fungal activity. Soil amendments that increase the level of available labile carbon, the organic compounds that most soil heterotrophs use as an energy source, tend to alleviate fungistasis and allow pathogenic fungi to resume activity in the soil (Bonanomi et al. 2013). These soil amendments could either make labile carbon temporarily non-limiting, or they could provide the pathogen with the energy to produce its own defensive compounds (Garbeva et al. 2011). Studies on the role of fungistasis in mediating *B. tectorum* stand failure and recovery have been initiated.

Fusarium species are capable of producing a host of secondary metabolites, many of which have phytotoxic, mycotoxic, or antibiotic activity (O'Donnell et al. 2013). It has been recently demonstrated in a very interesting study that *Fusarium* toxin production is upregulated in the presence of common soil bacteria such as *Bacillus subtilis* (Ola et al. 2013). In these experiments, toxin production was increased by an order of magnitude in co-culture with this bacterium. This strongly suggests that the fungus makes this toxic compound at least partly as a defense response against microorganisms. The role of toxic secondary metabolites produced by *Fusarium* in alleviating fungistasis in the soil and thus in regulating the cycle of disease represented by periodic *B. tectorum* stand failure is a topic for further investigation.

7.2.4.2 *Fusarium* Seed Rot Host Range

Fusarium species generally exhibit wide host range, but many species are made up of series of *formae speciales* that are highly host specific (e.g., *F. solani*; O'Donnell 2000). Strains isolated from *B. tectorum*-infested soils have caused mortality on seeds of the native perennial grasses *Pseudoroegneria spicata* and *E. elymoides* as well as *B. tectorum* in both field and laboratory studies (Meyer et al. 2014b). It is apparent that these strains are not strictly host specific and can cause disease on multiple cool-season grass species. There is still much to be learned about genetic variation in this group and its role in host specificity. The strains from *B. tectorum* soils may represent a series of closely related species with somewhat different host ranges.

7.2.4.3 *Fusarium* Seed Rot Distribution and Epidemiology

In surveys using bait seed experiments at *B. tectorum*-infested sites in northern Nevada, western Utah, and central Washington, all *Fusarium* strains identified to date belong to the *F. tricinctum* species group (O'Donnell et al. 2013; Meyer et al. 2014a). This suggests that it is widely distributed and common in the Western United States and is likely the primary *Fusarium* taxon present in *B. tectorum* monocultures throughout the Intermountain Region.

In a laboratory pathogenicity test with 16 strains isolated from killed *B. tectorum* seeds, wide variation in virulence was observed (Fig. 7.4c, Meyer et al. 2014a). Mortality was generally higher (27–83 %) when nondormant seeds were held at low water potential during the initial stages of infection as described earlier. However, some strains caused mortality of >40 % even in free water, while the least virulent strain caused <10 % mortality under this condition. As these strains are likely capable of living saprophytically in the soil, high virulence may not be a prerequisite for long-term survival, but it is likely that strains capable of causing seed mortality under field conditions have a selective advantage in terms of resources available for sporulation. There may be a continuum in this group such that some strains are almost exclusively saprophytes whereas others exhibit a strongly pathogenic life history strategy.

7.2.5 *Rutstroemiaceae sp. n. (Bleach Blonde Pathogen)*

A newly discovered ascomycete pathogen that may be quite important on *B. tectorum* has been identified using molecular-genetic techniques as a relative of *Sclerotinia homoeocarpa* Bennett, the dollar spot pathogen of turfgrass (Franke et al. 2014). We first observed the symptoms of the disease caused by this pathogen many years ago but were uncertain whether the syndrome was caused by abiotic stress or pathogen activity.

Unlike the pathogens described earlier, this organism apparently infects the crowns of already established seedlings or young plants. The plants survive to bolting, but the spikelets in the inflorescences abort prior to the completion of seed filling, leaving the plants sterile. These sterile plants are stunted in comparison with healthy plants, and they turn straw colored while healthy plants are still at the green or purple stages of ripening. These wispy, straw-colored heads with unfilled seeds are symptoms of the bleach blonde syndrome (Fig. 7.5a).

We thought this syndrome might be caused by some pathogen that causes similar symptoms on winter cereals, e.g., *Gaeumannomyces graminis* (Sacc.) Arx & Olivier (the causal agent of take-all disease) or *Fusarium culmorum* (W.G. Sm.) Sacc. (the causal agent of dryland foot rot). These pathogens produce clear disease signs on the lower stems and roots, whereas bleach blonde plants have no readily visible disease signs. Isolations from the crowns of diseased plants consistently yielded an organism that grouped within the family Rutstroemiaceae based on its ITS sequence but was not a perfect match for any known species, indicating that it likely represents an undescribed taxon. More intensive molecular-genetic work will be necessary to typify and name this new organism.

We have demonstrated in greenhouse pathogenicity tests that the organism isolated from the crowns of diseased *B. tectorum* plants in the field is definitely the causal organism responsible for the bleach blonde syndrome (J. Pearce, unpublished data). Disease incidence in the pathogenicity test was 18 %. The pathogen was readily re-isolated from diseased plants, which had significantly lower total biomass, floret number per tiller, and mass per floret than healthy plants (Fig. 7.5b). They closely resembled bleach blonde-affected plants observed in the field.

7.2.5.1 Bleach Blonde Pathogen Life Cycle

Members of the family Rutstroemiaceae generally do not produce asexual spores but instead produce asexual resting structures called sclerotia or stromata. They are potentially capable of sexual reproduction, but the fleshy cup mushrooms that are formed are seen much more often in species of mesic environments. The bleach blonde pathogen is soilborne and apparently infects the host plants after seedlings are established, either through the roots or directly into the crown. There seems to be no movement of the fungus within the bolting flowering stalk, and unlike the smuts and bunts, it does not produce spores or other reproductive structures in the

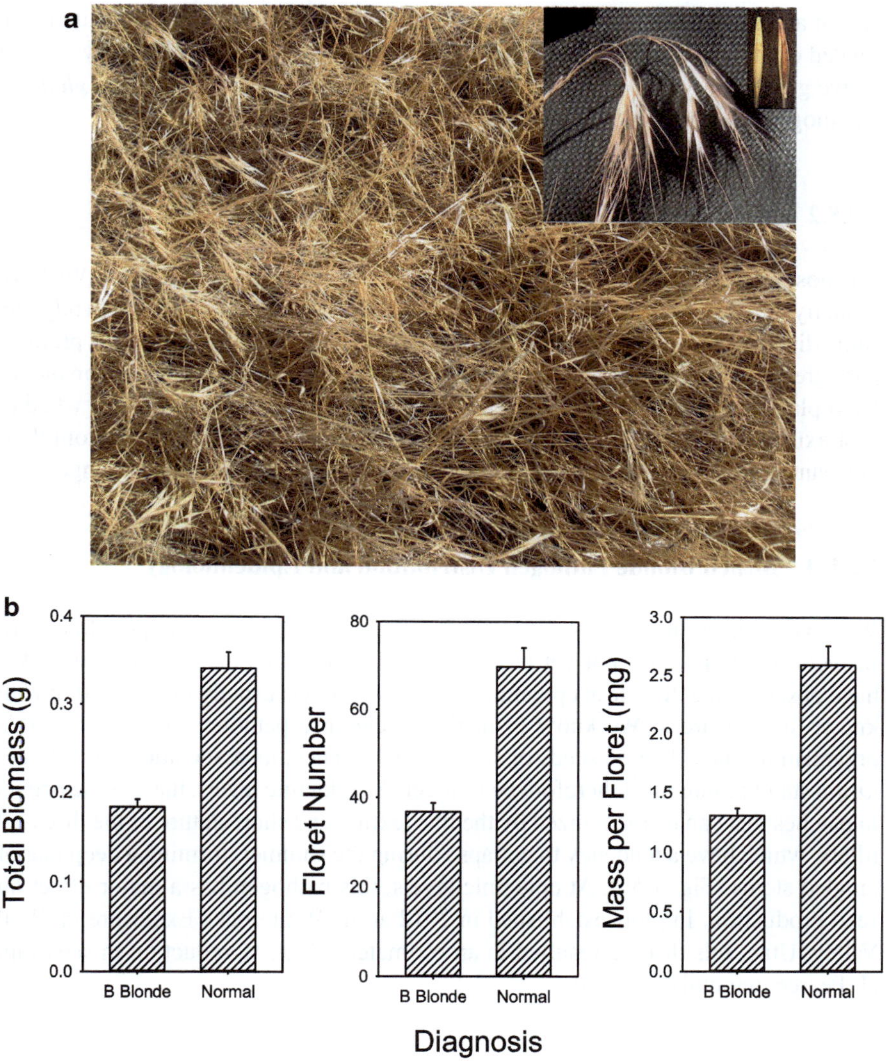

Fig. 7.5 (**a**) A patch of *Bromus tectorum* monoculture that suffered epidemic disease caused by the bleach blonde pathogen (Rutstroemiaceae sp. n.). Inset shows a close-up of a sterile inflorescence from a diseased plant and a comparison of a sterile floret (*left*) with a normal filled floret (*right*). (**b**) Effect of disease caused by the bleach blonde pathogen on whole plant biomass, number of florets per tiller, and mean mass per floret in a greenhouse pathogenicity test. (Florets with a mean mass of <1.5 mg are nonviable)

inflorescences. These inflorescences abort because disease development at the base of the plant apparently blocks the vascular tissue and causes water stress in the flowering shoots, in a manner similar to a vascular wilt disease. If surface-sterilized stem bases of diseased plants are incubated in the laboratory, the irregularly shaped black stromata characteristic of the family Rutstroemiaceae are produced within the crown and stem tissue. Similar stromatal structures are also produced in culture. In related organisms, the stromata can persist in the soil for many years. They resume active growth only in the presence of host root exudates (e.g., *Stromatinia gladioli*, a pathogen of cultivated gladiolus; Jeves and Coley-Smith 1980).

7.2.5.2 Bleach Blonde Syndrome Pathogen Host Range

The host range of this newly discovered pathogen is completely unknown. We have casually observed individuals of *B. rubens* and *B. diandrus* with apparently the same disease syndrome, but these plants were not critically examined. In a preliminary greenhouse pathogenicity test, we observed no bleach blonde disease on inoculated plants of *B. arvensis*. We are currently engaged in tests to determine whether root exudates of different potential hosts can release pathogen stromata from their dormant state, as a next step in examining bleach blonde pathogen host range.

7.2.5.3 Bleach Blonde Pathogen Distribution and Epidemiology

We have detected individuals with bleach blonde syndrome in many populations of *B. tectorum*, but we do not yet have any quantitative data on its distribution. We have observed the disease at epidemic levels a few times, usually in association with known die-off areas. We know from the greenhouse pathogenicity test that this organism has no effect on seeds or seedling emergence; its association with *B. tectorum* stand failure must therefore be indirect. At epidemic levels, the disease occurs in patches that can be recognized by the fine texture and short stature of the diseased tillers, which have a tendency to collapse during the summer, forming a deep mat of tangled stems (Fig. 7.5b). At epidemic levels, this pathogen has a major effect on seed production. In plots established in 2012 at the Whiterocks Exclosure in Skull Valley, Utah, this disease resulted in an estimated 60–80 % reduction in seed rain (J. Pearce, unpublished data).

7.3 Community Ecology of Pathogens on *Bromus tectorum*

The five principal pathogens on *B. tectorum* described above exhibit host relationships and infection phenologies that tend to minimize their competitive interactions. Each pathogen exhibits niche specialization by attacking at a specific stage of the *B. tectorum* life cycle (Fig. 7.6). *Pyrenophora semeniperda* usually attacks dormant seeds, either seeds in primary dormancy (if there are summer rains) or secondarily dormant seeds in the carryover seed bank. *Fusarium* usually attacks nondormant seeds in the germinable autumn seed bank. *Ustilago bullata* infects the coleoptiles of newly germinated seeds under warm autumn conditions, while *T. bromi* infects the coleoptiles of emerged seedlings under cold winter conditions. Finally, the

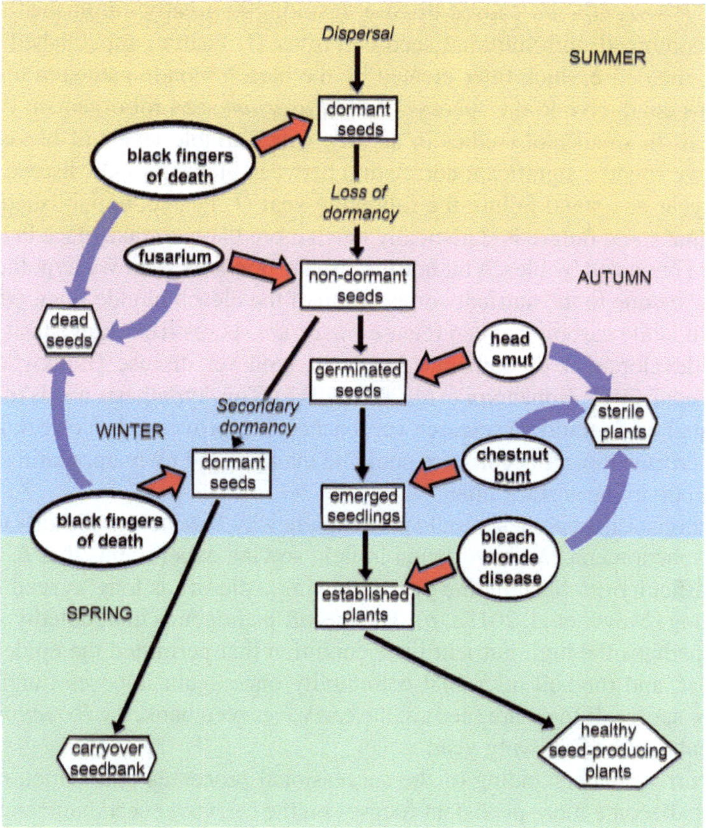

Fig. 7.6 Schematic diagram showing the relationship of each of the five pathogens discussed in the text to the stage of the *B. tectorum* life cycle when infection takes place, the season of infection, and the outcome of disease (seed death or sterility). *Black arrows* represent transitions between different *B. tectorum* life stages (*square boxes*); *red arrows* indicate life stage impacted by each of the five pathogens (*oval boxes*); *purple arrows* indicate pathogen impacts on final outcomes (*hexagonal boxes*). Pathogens are *Pyrenophora semeniperda* (black fingers of death), *Fusarium* sp. (*Fusarium* seed rot disease), *Ustilago bullata* (head smut disease), *Tilletia bromi* (chestnut bunt disease), and Rutstroemiaceae sp. n. (bleach blonde syndrome)

bleach blonde pathogen attacks the crowns of juvenile established plants. These pathogens thus clearly exhibit niche differentiation with regard to their mode of utilization of the *B. tectorum* host. There may be preemptive competition, in that host individuals killed at earlier stages in the life cycle are not available as prey for pathogens that operate at later stages. But as different weather scenarios favor infection at different life stages, opportunities for high levels of disease for each pathogen tend to be separated either in time (among years) or space (different habitats).

We do have some evidence that these pathogens can sometimes interact synergistically to increase the negative impact on *B. tectorum* stand dynamics. Specifically, epidemic levels of the bleach blonde syndrome can be associated with very high levels of *P. semeniperda*-caused disease, resulting in greatly diminished seed production combined with minimal seed carryover (J. Pearce, unpublished data). In addition, the dense, thick litter created by the bleach blonde pathogen may create conditions conducive to the success of the *Fusarium* seed rot organism the subsequent year. In small plot studies in an area with variable levels of bleach blonde disease, we found a significant correlation between bleach blonde disease levels in the first year and stand failure the following year (J. Pearce, unpublished data). It appears that stand failure is statistically much more likely to take place in areas that have been impacted by bleach blonde disease the previous year. We hypothesize that this could be due to the nutrient composition of the bleach blonde litter, which may be high in labile carbon that can release *Fusarium* spores from fungistasis and permit the development of epidemic levels of seed rot disease (Lockwood 1977; Garbeva et al. 2011; Bonanomi et al. 2007, 2013). This hypothesis needs to be rigorously tested using multiple research approaches. If it proves to be correct, it implies that *B. tectorum* die-off occurrence could be manipulated by manipulating levels of labile carbon in the surface litter.

The fungistasis hypothesis could also explain why *B. tectorum* die-offs tend to be transient phenomena. We have found in field sowing experiments that *B. tectorum* has no difficulty establishing the year following a die-off as long as seed supply is not limiting (Meyer et al. 2013a, b). The die-off pathogen is undoubtedly still present, but perhaps the high-nutrient litter condition that permitted the epidemic does not persist, and the soil microbial community once again imposes fungistasis on *Fusarium* spores. If there are seeds in the carryover seed bank, the *B. tectorum* stand can reestablish the following year.

Our current understanding of the successional processes that sometimes cause die-offs to become more persistent focuses on the carryover seed bank and the supporting role of *P. semeniperda*. Litter dynamics once again appear to be key to this process (Beckstead et al. 2012). Without an adequate carryover seed bank to establish a stand, lack of *B. tectorum* cover the second year can result in litter loss (Smith et al. 2008). Die-off soils that lose their litter cover are often colonized by dicot weeds that are adapted to colonize bare soil, namely, *Salsola tragus* L. (prickly Russian thistle), *Sisymbrium altissimum* L. (tall tumblemustard), *Bassia scoparia* (L.) A.J. Scott Show (burningbush or ironweed), and *Ceratocephala testiculata* (Crantz) Roth (bur buttercup or curveseed butterwort). These species in turn may create litter that can promote recolonization by *B. tectorum*, but this process may take several years. The rate of *B. tectorum* recovery may also depend on whether the dicot weeds were present in the seed bank or must disperse in. Most of these are "tumbleweeds" that are effectively dispersed into the openings created by die-offs. The size of the die-off area may also be a factor in recovery rate because of increased dispersal distance for *B. tectorum* in larger die-offs.

In the course of our investigations of *B. tectorum* stand failure, we have encountered several other soilborne fungal pathogens whose impacts are still not known. Many of these, such as *Alternaria* spp., were only weakly pathogenic in laboratory tests and were subject to complete suppression in the presence of *Fusarium* (Pearce and Beckstead, unpublished data). The exception was *Epicoccum nigrum*, which caused *B. tectorum* seed mortality under water stress in the laboratory at levels comparable to moderately pathogenic *Fusarium* strains (Poh and Saunders, unpublished data). It also suppressed *Fusarium* in co-inoculations. *Epicoccum* has also been isolated from killed seedlings and juvenile *B. tectorum* plants from the field and may be another important player in the die-off phenomenon. We have too little information to speculate further on what its role might be.

7.4 Management Implications

The main lesson to be learned from studies of *B. tectorum* disease organisms is that these pathogens exist in dynamic equilibrium with *B. tectorum* and likely have a long evolutionary history on this host. While catastrophic stand failure can occur, it is much more common for diseases to persist long term at endemic levels that fluctuate as a function of yearly weather patterns and population levels of other microorganisms, including other pathogens, in the soil. A "good" pathogen does not drive its host to local extinction because that often implies its own local extinction as well. But even when this happens, the high seed dispersal capability of *B. tectorum* means that reinvasion is just a matter of time. The pathogens find a way to reinvade as well, either on seeds or via wind or animal dispersal, or to persist in the soil.

The management goal in *B. tectorum*-infested rangelands is ultimately to replace these annual grass monocultures with perennial plant communities, preferably native communities, that offer higher resource value in terms of soil and watershed protection, biodiversity, carbon sequestration, and forage production. The study of *B. tectorum* disease epidemiology offers two possible approaches to achieve this goal. The first is to use artificially produced inoculum of a pathogen in a short-term mycoherbicidal strategy to temporarily knock down *B. tectorum* populations in the context of restoration seeding (Meyer et al. 2008b). The idea is to create disease epidemics that can provide a window for native seedlings to establish. This approach has been investigated for two of the pathogens discussed above, namely, *U. bullata* and *P. semeniperda*. For *U. bullata*, the problem of the narrow infection window and the requirement for warm temperatures during coleoptile infection would preclude its successful use in most of the environments where *B. tectorum* is a problem. For *P. semeniperda*, we have sometimes achieved complete mortality of the carryover seed bank with inoculum augmentation, but this biocontrol method must be combined with some other method for control of the current-year stand in order to be effective, greatly limiting its usefulness.

The second approach for using *B. tectorum* disease as a management tool is more promising. The idea is to take advantage of naturally occurring stand failure as an opportunity for restoration seeding or to manipulate conditions in the field to cause die-offs that can then be used as restoration opportunities. This would eliminate the need for inoculum augmentation, which is the most difficult and controversial component of mycoherbicidal biocontrol. We have already learned that die-offs are usually transient phenomena followed by a period with much reduced disease potential.

We have also learned that die-offs can create seed bed conditions conducive to the emergence and early establishment of native grasses (Baughman 2014). By developing a more detailed mechanistic understanding of the causes and consequences of die-offs, we may be able to develop a strategy for the successful restoration of *B. tectorum*-infested rangelands.

7.5 Research Needs

While we have made a great deal of progress in understanding the diseases that can affect *B. tectorum* stand dynamics in the last 15 years, much remains to be accomplished before these diseases can be successfully exploited in a management context. Specifically, we need to undertake an integrated research program aimed at understanding the full complexity of the *B. tectorum* seed bank microbial community, including direct and indirect interactions of each pathogen with other pathogens, with other microorganisms, and with microenvironmental factors such as litter dynamics and composition, all of which can affect disease incidence as well as post-die-off successional trajectories. This includes detailed studies of the systematics, ecology, and genetics of the pathogens involved, as mentioned in earlier sections. A more landscape-level approach that examines *B. tectorum* stand dynamics in space and time, especially as affected by epidemic disease and its interactions with interannual weather variation, would also be a valuable addition to the research program.

References

Allen PS, Meyer SE, Foote K (2010) Induction and release of secondary dormancy under field conditions in *Bromus tectorum*. In: Pendleton R, Meyer SE, Schultz B (eds) Seed Ecology III – The third international society for seed science meeting on seeds and the environment – "Seeds and Change", 20–24 June 2010, Salt Lake City, UT. USDA, Forest Service, Rocky Mountain Research Station, Albuquerque, NM, pp 2–3

Baughman OW (2014) Will native plants succeed where exotic invaders fail? Cheatgrass die-off as an opportunity for restoration in the Great Basin, USA. MS thesis, University of Nevada, Reno, NV

Baughman OW, Meyer SE (2013) Is *Pyrenophora semeniperda* the cause of downy brome (*Bromus tectorum*) die-offs? Invasive Plant Sci Manag 6:105–111

Beckstead J, Meyer SE, Molder CJ et al (2007) A race for survival: can *Bromus tectorum* seeds escape *Pyrenophora semeniperda*-caused mortality by germinating quickly? Ann Bot 99:907–914

Beckstead J, Meyer SE, Connolly BM et al (2010) Cheatgrass facilitates spillover of a seed bank pathogen onto native grass species. J Ecol 98:168–177

Beckstead J, Miller LE, Connolly BM (2012) Direct and indirect effects of plant litter on a seed–pathogen interaction in *Bromus tectorum* seed banks. Seed Sci Res 22:135–144

Beckstead J, Meyer SE, Reinhart K et al (2014) Factors affecting host range in a generalist seed pathogen of semi-arid shrublands. Plant Ecol 215:427–440

Boguena T, Meyer SE, Nelson DL (2007) Low temperature during infection limits *Ustilago bullata* disease incidence on *Bromus tectorum*. Biocontrol Sci Tech 17:33–52

Bonanomi G, Antignani V, Pane C et al (2007) Suppression of soilborne fungal diseases with organic amendments. J Plant Pathol 89:311–324

Bonanomi G, Gaglione SA, Incerti G et al (2013) Biochemical quality of organic amendments affects soil fungistasis. Appl Soil Ecol 72:135–142

Boose D, Harrison S, Clement S et al (2011) Population genetic structure of the seed pathogen *Pyrenophora semeniperda* on *Bromus tectorum* in western North America. Mycologia 103:85–93

Boyd ML, Carris LM (1997) Molecular relationships among varieties of the *Tilletia fusca* (*T. bromi*) complex and related species. Mycol Res 101:269–277

Boyd ML, Carris LM (1998) Evidence supporting the separation of the *Vulpia*-and *Bromus-infecting* isolates in the *Tilletia fusca* (*T. bromi*) complex. Mycologia 90:1031–1039

Campbell MA, Medd RW (2003) Leaf, floret and seed infection of wheat by *Pyrenophora semeniperda*. Plant Pathol 52:437–447

Castlebury LA, Carris LM, Vánky K (2005) Phylogenetic analysis of *Tilletia* and allied genera in order Tilletiales (Ustilaginomycetes; Exobasidiomycetidae) based on large subunit nuclear rDNA sequences. Mycologia 97:888–900

Christensen M, Meyer SE, Allen PS (1996) A hydrothermal time model of seed after-ripening in *Bromus tectorum* L. Seed Sci Res 6:155–164

Connolly BM, Pearson DE, Mack RN (2014) Granivory of invasive, naturalized, and native plants in communities differentially susceptible to invasion. Ecology 95:1759–1769

Crute IR, Holub EB, Burdon JJ (eds) (1997) The gene-for-gene relationship in plant – parasite interactions. CAB International, Oxon

Duran R, Fischer GW (1961) The genus *Tilletia*. Washington State University, Pullman, WA

Evidente A, Andolfi A, Vurro M et al (2002) Cytochalasins Z1, Z2 and Z3, three 24-oxa[14] cytochalasans produced by *Pyrenophora semeniperda*. Phytochemistry 60:45-53

Finch H, Allen PS, Meyer SE (2013) Environmental factors influencing Pyrenophora semeniperda-caused seed mortality in *Bromus tectorum*. Seed Sci Res 23:57–66

Fischer GW (1940) Host specialization in the head smut of grasses, *Ustilago bullata*. Phytopathology 30:991–1017

Fischer GW, Holton CS (1957) Biology and control of the smut fungi. Ronald Press, New York, NY

Franke J, Meyer SE, Geary B (2014) Bleach blonde syndrome, a new disease of *Bromus tectorum* implicated in cheatgrass die-offs. Botanical Society of America 2014 Annual Meeting, Boise ID. Abstract 446.

Franke J, Geary B, Meyer SE (2014) Identification of the infection route of a *Fusarium* seed pathogen into non-dormant *Bromus tectorum* seeds. Phytopathology 104:1306-1313

Garbeva P, Hol WH, Termorshuizen AJ et al (2011) Fungistasis and general soil biostasis–a new synthesis. Soil Biol Biochem 43:469–477

Hoffmann JA, Meiners JP (1971) Host specialization in the complex species, *Tilletia fusca*. Phytopathology 61:225–227

Jeves TM, Coley-Smith JR (1980) Germination of sclerotia of *Stromatinia gladioli*. Trans Br Mycol Soc 74:13–18

Klemmedson JO, Smith JG (1964) Cheatgrass (*Bromus tectorum* L.). Bot Rev 30:226–262

Kreitlow KW, Bleak AT (1964) *Podosporiella verticillata*, a soilborne pathogen of some western Gramineae. Phytopathology 54:353–357

Kreizinger EJ, Fischer GW, Law AG (1947) Reactions of mountain brome and Canada wild-rye strains to head smut (*Ustilago bullata*). J Agric Res 75:105–111

Leslie JF, Summerell BA (2006) The *Fusarium* laboratory manual. Blackwell, Ames, IA

Lockwood JL (1977) Fungistasis in soils. Biol Rev 52:1–43

Masi M, Evidente A, Meyer S et al (2013) Effect of strain and cultural conditions on the production of cytochalasin B by the potential mycoherbicide *Pyrenophora semeniperda* (Pleosporaceae, Pleosporales). Biocontrol Sci Tech 24:53–64

Masi M, Meyer S, Cimmino A et al (2014a) Pyrenophoric acid, a phytotoxic sesquiterpenoid penta-2,4-dienoic acid produced by *Pyrenophora semeniperda*, a potential mycoherbicide for biocontrol of annual bromes. J Nat Prod 77:925–930
Masi M, Meyer S, Clement S et al (2014b) Spirostaphylotrichin W, a spirocyclic γ-lactam isolated from liquid culture of *Pyrenophora semeniperda*, a potential mycoherbicide for cheatgrass (*Bromus tectorum*) biocontrol. Tetrahedron 70:1497–1501
Mathre DE (1996) Dwarf bunt: politics, identification, and biology. Ann Rev Phytopathol 34:67–85
Medd RW, Campbell MA (2005) Grass seed infection following inundation with *Pyrenophora semeniperda*. Biocontrol Sci Tech 15:21–36
Medd RW, Murray GM, Pickering DI (2003) Review of the epidemiology and economic importance of *Pyrenophora semeniperda*. Australas Plant Pathol 32:539–550
Meiners JP (1958) Studies on the biology of *Tilletia bromi-tectorum*. Phytopathology 48:211–216
Meiners JP, Fischer GW (1953) Further studies on host specialization in the head smut of grasses, *Ustilago bullata*. Phytopathology 43:200–203
Meiners JP, Waldher JT (1959) Factors affecting spore germination of twelve species of *Tilletia* from cereals and grasses. Phytopathology 49:724–728
Meyer SE, Nelson DL, Clement S (2001) Evidence for resistance polymorphism in the *Bromus tectorum-Ustilago bullata* pathosystem: implications for biocontrol. Can J Plant Pathol 23:19–27
Meyer SE, Nelson DL, Clement S et al (2005) Genetic variation in *Ustilago bullata*: molecular genetic markers and virulence on *Bromus tectorum* host lines. Int J Plant Sci 166:105–115
Meyer SE, Quinney D, Nelson DL et al (2007a) Impact of the pathogen *Pyrenophora semeniperda* on *Bromus tectorum* seedbank dynamics in North American cold deserts. Weed Res 47:54–62
Meyer SE, Smith DC, Beckstead J (2007b) *Bromus tectorum* seed banks: impact of the pathogen *Pyrenophora semeniperda*. In: Turner S, Merritt D, Clarke S et al (eds) Seed ecology II. Conference proceedings. The second international meeting on seeds and the environment, Perth, Australia, 9–13 September 2007. Kings Park and Botanic Garden, Perth, Australia
Meyer SE, Beckstead J, Allen PS et al (2008a) A seed bank pathogen causes seedborne disease: *Pyrenophora semeniperda* on undispersed grass seeds in western North America. Can J Plant Pathol 30:525–533
Meyer SE, Nelson DL, Clement S et al (2008b) Cheatgrass (*Bromus tectorum*) biocontrol using indigenous fungal pathogens. In: Kitchen SG, Pendleton RL, Monaco TA et al (comps) Proceedings—Shrublands under fire: disturbance and recovery in a changing world, 6–8 June 2006. Cedar City, UT. Gen Tech Rep RMRS-P-52. USDA, Forest Service, Rocky Mountain Research Station, Fort Collins, CO, pp 61–67
Meyer SE, Nelson D, Clement S et al (2010a) Ecological genetics of the *Bromus tectorum* (Poaceae)–*Ustilago bullata* (Ustilaginaceae) pathosystem: a role for frequency-dependent selection? Am J Bot 97:1304–1312
Meyer SE, Stewart TE, Clement S (2010b) The quick and the deadly: growth vs. virulence in a seed bank pathogen. New Phytol 187:209–216
Meyer SE, Leger EA, Beckstead J et al (2013a) Understanding the causes and consequences of cheatgrass die-offs in the Great Basin. First Summary Report: August 2011–December 2012. January 10, 2013. Unpublished report on file at the Idaho State Office of the USDI Bureau of Land Management, Boise, ID
Meyer SE, Leger EA, Beckstead J et al (2013b). Understanding the causes and consequences of cheatgrass die-offs in the Great Basin. Second Summary Report: FY2013. November 18, 2013. Unpublished report on file at the Idaho State Office of the USDI, Bureau of Land Management, Boise, ID
Meyer SE, Franke J, Baughman OW et al (2014a) Does *Fusarium*-caused seed mortality contribute to *Bromus tectorum* stand failure in the Great Basin? Weed Res 54(5):511–519
Meyer SE, Merrill K, Allen P et al. (2014b) Indirect effects of an invasive annual grass on perennial grass seed fates. Oecologia 174:1401-1413

Meyer SE, Masi M, Clement S et al (2015) Mycelial growth rate and toxin production in the seed pathogen *Pyrenophora semeniperda*: resource trade-offs and temporally varying selection. Plant Pathol DOI: 10.1111/ppa.12377

Mordecai EA (2013) Despite spillover, a shared pathogen promotes native plant persistence in a cheatgrass-invaded grassland. Ecology 94:2744–2753

Muller-Stover D, Kohlschmid E, Sauerborn J (2009) A novel strain of *Fusarium oxysporum* from Germany and its potential for biocontrol of *Orobanche ramosa*. Weed Res 2:175–182

Nelson PE, Toussoun TA, Cook RJ (1981) *Fusarium*: diseases, biology and taxonomy. Pennsylvania State University Press, University Park, PA

O'Donnell K (2000) Molecular phylogeny of the *Nectria haematococca-Fusarium solani* species complex. Mycologia 92:919–938

O'Donnell K, Rooney AP, Proctor R et al (2013) Phylogenetic analyses of *RPB1* and *RPB2* support a middle Cretaceous origin for a clade comprising all agriculturally and medically important fusaria. Fungal Genet Biol 52:20–31

Ola AR, Thomy D, Lai D et al (2013) Inducing secondary metabolite production by the endophytic fungus *Fusarium tricinctum* through co-culture with *Bacillus subtilis*. J Nat Prod 76:2094–2099

Paul AR (1969) The production of *Pyrenophora semeniperda* in culture. Trans Br Mycol Soc 52:373–379

Piemeisel RL (1938) Changes in weedy plant cover on cleared sagebrush land and their probable causes. USDA Technical Bulletin 654

Piemeisel RL (1951) Causes affecting change and rate of change in a vegetation of annuals in southern Idaho. Ecology 32:53–72

Pimentel G, Peever TL, Carris LM (2000) Genetic variation among natural populations of *Tilletia controversa* and *T. bromi*. Phytopathology 90:376–383

Sauerborn J, Dorr I, Abbasher A et al (1996) Electron microscopic analysis of the penetration process of *Fusarium nygamai*, a hyperparasite of *Striga hermonthica*. Biol Control 7:53–59

Shoemaker RA (1966) A pleomorphic parasite of cereal seeds, *Pyrenophora semeniperda*. Can J Bot 44:1451–1456

Slykhuis JT (1947) Studies on *Fusarium culmorum* blight of crested wheat and brome grass seedlings. Can J Res 25:155–180

Smith DC, Meyer SE, Anderson VJ (2008) Factors affecting *Bromus tectorum* seed bank carryover in western Utah. Rangel Ecol Manag 61:430–436

Soliai MM, Meyer SE, Udall JA et al (2014) De novo genome assembly of the fungal plant pathogen *Pyrenophora semeniperda*. PLoS One 9:1–11

Stewart T, Meyer SE, Allen PS (2009) First report of *Pyrenophora semeniperda* in Turkey and Greece. Plant Dis Rep 93:1351

Wallace HAH (1959) A rare seed-borne disease of wheat caused by *Podosporiella verticillata*. Can J Bot 37:509–515

Walsh J, Laurence MH, Liew EC et al (2010) *Fusarium*: two endophytic novel species from tropical grasses of northern Australia. Fungal Divers 44:149–159

Part III
Understanding Environmental Controls and *Bromus* Distribution (Invasibility of Landscapes by *Bromus*)

Chapter 8
Soil Moisture and Biogeochemical Factors Influence the Distribution of Annual *Bromus* Species

Jayne Belnap, John M. Stark, Benjamin M. Rau, Edith B. Allen, and Susan Phillips

Abstract Abiotic factors have a strong influence on where annual *Bromus* species are found. At the large regional scale, temperature and precipitation extremes determine the boundaries of *Bromus* occurrence. At the more local scale, soil characteristics and climate influence distribution, cover, and performance. In hot, dry, summer-rainfall-dominated deserts (Sonoran, Chihuahuan), little or no *Bromus* is found, likely due to timing or amount of soil moisture relative to *Bromus* phenology. In hot, winter-rainfall-dominated deserts (parts of the Mojave Desert), *Bromus rubens* is widespread and correlated with high phosphorus availability. It also responds positively to additions of nitrogen alone or with phosphorus. On the Colorado Plateau, with higher soil moisture availability, factors limiting *Bromus tectorum* populations vary with life stage: phosphorus and water limit germination, potassium and the potassium/magnesium ratio affect winter performance, and water and potassium/

J. Belnap (✉)
US Geological Survey, Southwest Biological Science Center, Moab, UT 84532, USA
e-mail: jayne_belnap@usgs.gov

J.M. Stark
Department of Biology, and the Ecology Center, Utah State University, Logan, UT 84322-5305, USA
e-mail: jstark@biology.usu.edu

B.M. Rau
US Department of Agriculture, Forest Service, Southern Research Station, Aiken, SC 29803, USA
e-mail: benjaminmrau@fs.fed.us

E.B. Allen
Department of Botany and Plant Sciences, University of California, Riverside, Riverside, CA, USA
e-mail: edith.allen@ucr.edu

S. Phillips
US Geological Survey, Forest and Rangeland Ecosystem Center, Corvallis, OR 97330-6169, USA
e-mail: sue_phillips@usgs.gov

M.J. Germino et al. (eds.), *Exotic Brome-Grasses in Arid and Semiarid Ecosystems of the Western US*, Springer Series on Environmental Management,
DOI 10.1007/978-3-319-24930-8_8

magnesium affect spring performance. Controlling nutrients also change with elevation. In cooler deserts with winter precipitation (Great Basin, Columbia Plateau) and thus even greater soil moisture availability, *B. tectorum* populations are controlled by nitrogen, phosphorus, or potassium. Experimental nitrogen additions stimulate *Bromus* performance. The reason for different nutrients limiting in dissimilar climatic regions is not known, but it is likely that site conditions such as soil texture (as it affects water and nutrient availability), organic matter, and/or chemistry interact in a manner that regulates nutrient availability and limitations. Under future drier, hotter conditions, *Bromus* distribution is likely to change due to changes in the interaction between moisture and nutrient availability.

Keywords Climate • Geomorphology • Nitrogen • Nutrients • Phosphorus • Soils

8.1 Introduction

Despite much research, we still know little about what makes arid and semiarid (hereafter referred to as dryland) ecosystems susceptible to invasion by exotic annual grasses such as *Bromus tectorum* (L.) and *Bromus rubens* (L) (Fig. 8.1). Because annual grasses are often associated with soil surface disturbance, this is

Fig. 8.1 Unless the invasion is a result of fire, annual grass invasions in the drier parts of the Western USA often occur in distinct patches (as indicated by *arrows*) as can be seen in this photo (taken adjacent to the Great Salt Lake on lacustrine sediments from Lake Bonneville)

thought to be the factor most responsible for such invasions (Hobbs and Huenneke 1992). However, invasive annual grasses are also able to establish in relatively undisturbed communities yet not always occur in disturbed sites (Tausch et al. 1994; Belnap and Phillips 2001), indicating there are other factors influencing ecosystem invasibility.

In this chapter, we explore the potential role of soil biogeochemical factors in controlling the distribution of *B. tectorum* and *B. rubens* (*B. madritensis* ssp. *rubens*) in the mid- and lower-elevation semiarid and arid lands of the Western USA. We review the literature and other data on how water and nutrient availability, as influenced by climate, controls the spread and range extent of exotic annual *Bromus* (*Bromus* hereafter). We examine studies correlating *B. tectorum* and *B. rubens* distribution in five regions where these plants occur in the Western USA: the Chihuahuan Desert, the Mojave Desert, the Colorado Plateau Desert, the Great Basin/Columbia Plateau Deserts, and the California coastal sage scrub. All these regions have different climatic regimes, which likely influence the ability of these species to establish and thrive (Table 8.1). The Chihuahuan is a hot desert with the majority of rainfall occurring during summer, whereas the Mojave Desert is a hot desert with predominantly winter precipitation. The Colorado Plateau and Great Basin/Columbia Plateau Deserts are both much cooler. The Colorado Plateau receives both summer and winter precipitation, whereas the Great Basin/Columbia Plateau regions receive almost exclusively winter precipitation. We end the chapter with a hypothesis on how soils and climate may interact to limit cover of *Bromus* throughout the low elevations of the Western USA.

8.2 Soil Nutrient Availability and Water in Dryland Settings

Plants require nitrogen (N), phosphorus (P), potassium (K), magnesium (Mg), sulfur (S), and calcium (Ca) in large amounts, which are thus considered macronutrients, whereas essential nutrients needed in small quantities (e.g., manganese [Mn], zinc [Zn], copper [Cu], and iron [Fe]) are called micronutrients (Marschner 1995).

Table 8.1 Climate regimes of the different regions of Western US semiarid and arid lands

	Mean annual temperature (C)	Mean annual precipitation (mm)	Timing of most precipitation
Columbia Plateau	4–14	230–380	Spring and fall
Colorado Plateau	5–17	150–400	Summer and winter/early spring
Great Basin	6–11	150–300	Winter
California coastal sage scrub	17–22	260–300	Late fall to early spring
Mojave	17–33	130–160	Winter
Chihuahuan	19–24	150–400	Summer

Years of research on plant nutrition has demonstrated that annual plants, including annual *Bromus*, require higher levels of available soil nutrients on a shorter timescale than perennial plants, because annuals require all of the nutrients necessary to complete their life cycle within a single season, whereas perennial plants can store and recycle nutrients in plant tissue for use during successive years (Epstein 1961). Because many soils in dryland regions contain low levels of many essential nutrients, they may be less or not invasible by annual *Bromus* plants compared to other sites with higher levels of soil nutrients.

Dryland soils generally have low levels of N. As N is often limiting to plants, most studies have focused on N as the nutrient likely to be most limiting for *Bromus*. Higher N in both lab and field settings (e.g., postfire or removal of perennials) has generally elicited a positive response in *Bromus* (see below; Chambers et al. 2007; Rau et al. 2014). Nitrogen can be measured as total or available pools (e.g., nitrate, ammonium), and the relationship with *Bromus* has been shown to be positive and negative regardless of the form measured (see text below). Desert soils can also have low levels of total P, or the P present can be biotically unavailable to plants as it readily forms insoluble precipitates with calcium, often found in great abundance in desert soils as calcium or magnesium carbonates ($CaCO_3$, $MgCO_3$, respectively; we refer to $CaCO_3$ and $MgCO_3$ collectively in terms of their acid-neutralizing potential or ANP). As P is an essential macronutrient, low P availability can also be highly limiting to desert plants (Schlesinger et al. 1989; Parker 1995). Many fewer studies have addressed the role of P in *Bromus* compared to N. Micronutrients can also be very important in dryland soils; for instance, Mn can alleviate salt stress in plants (Krishnamurti and Huang 1988), yet almost no studies have addressed the role of micronutrients in *Bromus* invasions. Interactions among nutrients can also be critical in determining their bioavailability. Higher K can increase N uptake (Dibb and Thompson 1985) and is also linked to increased Mn availability (Krishnamurti and Huang 1988). These nutrient interactions may be especially important to annual plants. For example, Scott and Billings (1964) observed that soils with high K/Mg ratios were dominated by annual plants, whereas soils with low K/Mg ratios were dominated by perennials.

Soil texture has several ways in which it regulates water and nutrient availability that can influence establishment and growth of annual *Bromus* (Miller et al. 2006a, b). First, it affects soil moisture availability. When rain events are large, infiltration can be greater in sandy soils than fine-textured soils, as incoming water drains down and away from the surface, thus out of the evaporative zone, whereas in finer-textured soils, water is held closer to the surface and thus evaporates more readily (Sala et al. 1988). This increased evaporative loss of water from fine-textured soils can also concentrate salts at the soil surface, which can increase plant water stress. Decomposition and nutrient transformations that increase soil nutrient bioavailability require moisture. Nutrient uptake by plants can only occur when soils are moist (Leffler and Ryel 2012). Water is required by plants and very high temperatures that result in high evapotranspiration rates and thus low soil moisture limit where plants grow, especially annuals with shallow root systems. Precipitation timing also interacts with soil texture to determine soil moisture, as rain falling at high summer temperatures evaporates more quickly than rain falling during cooler winter temperatures.

On the other hand, finer-textured soils are usually more fertile than sandy soils, as nutrients adsorbed to the fine particles are prevented from leaching downward out of plant root zones. In addition, finer-textured soils tend to accumulate greater concentrations of soil organic matter (SOM) (Nichols 1984), which acts as a reservoir for plant nutrients.

Soils form from weathered parent material present *in situ* (i.e., bedrock) and include materials deposited by wind or water. Soil characteristics (e.g., texture, nutrients) thus depend to a large degree on parent material type and degree of weathering. Soils from parent materials with high nutrient content may initially have similar nutrient levels; however, over time, many nutrients may leach downward into the subsoil, where they precipitate as insoluble and plant-unavailable forms, or they may be transported downslope by erosional processes. Longer weathering times result in finer-textured soils. Soils farther from mountain sources also tend to be finer textured than upslope soils because the finer, lighter particles stay suspended longer in water and are transported farther downslope. Therefore, depositional zones, such as depressions and the base of hillslopes, have finer, deeper, more fertile soils than upslope soils. Because *B. tectorum* and *B. rubens* are annual plants, we would expect them to favor more fertile sites. Thus, we would predict these species to be more successful on geomorphic units where soils are derived from parent materials with more nutrients, those that weather to a finer texture, and/or those that occur in depositional settings (e.g., downslope, depressions). Slope aspect can also affect distribution patterns, although there is little data from which to draw conclusions among regions.

The interactions among parent materials and soil formation and geomorphic processes (e.g., landslides, overland flow, aeolian [wind-blown] deposition) create a mosaic of unique geomorphic units, highly variable in space and often in time (McAuliffe and McDonald 1995; Hamerlynck et al. 2002). As these units determine plant distribution, they create a mosaic of vegetation communities as well (Webb et al. 1988). However, even within the framework of local to regional settings, the occurrence of *Bromus* can be highly heterogeneous at various scales. The legacy of previous vegetation can result in patchy distributions of SOM (e.g., islands of fertility) that persist long after the vegetation has changed. Dryland soils have inherently low SOM contents and thus are low in nutrients that are tightly associated with SOM (e.g., N, S, and to a lesser extent P); however, it is at this low end of the nutrient availability spectrum where plants are most responsive to changes in nutrients and where heterogeneity is most strongly expressed in terms of its effect on plant growth (Stark 1994). The high soil moisture and nutrient requirements of annual plants may restrict them to microhabitats where these resources are more abundant, such as the depositional zones mentioned above or under shrub canopies (or where shrub canopies previously existed), especially in hotter and drier regions (Abella et al. 2011). Even when annual grasses are able to invade the interspaces between the native perennials, these invaded patches can be directly adjacent to seemingly similar, but uninvaded, interspace areas (Fig. 8.1). This patchwork pattern is often repeated across a given landscape. When highly localized, these small patches of *Bromus* suggest that these invasions are not always controlled by climate, seed availability, herbivory, or soil disturbance. Instead, these patterns support the hypothesis that invasion at the local scale is most likely controlled by microclimate and/or soil characteristics. Larger-scale invaded patches can be a result of disturbance and/or removal of native competitors through processes such as fire or heavy grazing. There are exceptions to this patchy distribution pattern where large areas are covered by *Bromus*, such as areas covered by highly fertile loess, lacustrine soils (e.g., Snake River Plain in southern Idaho, areas of northern Nevada, Lake Bonneville sediments in Utah, Mancos Shale on the Colorado Plateau) or invasions following chronic (e.g., excessive grazing) or acute (e.g., fire) disturbance. In these areas, the cover of annual grasses often is, or has the potential to be, quite high and homogenous.

8.3 Studies on the Potential Soil Controls on *B. rubens* and *B. tectorum* Distribution

There are only a handful of studies examining how soil characteristics influence *B. rubens* and *B. tectorum* invasion into low-elevation dryland ecosystems, except after major disturbance (e.g., fire, plowing). The majority of studies examining *Bromus* distribution are correlative, not mechanistic, and thus, it is important to recognize that unless soil chemistry was analyzed prior to invasion, these studies cannot determine whether the particular soil characteristics were present before the invasion and caused the invasion or if they were due to plant-soil feedbacks by *Bromus* (e.g., Germino et al. 2015). In addition, many soil characteristics are frequently correlated with each other. For example, changes in pH change the solubilities of multiple nutrients simultaneously; SOM, water, and nutrient availability are frequently associated with higher silt and clay contents. Therefore, it is not possible to conclusively identify the actual limiting factor based on plant-soil correlations. Nevertheless, we will discuss the soil factors most frequently associated with the presence of *Bromus* in an attempt to identify factors that may regulate its distribution. Below, we divide this discussion into five climatic regions: Chihuahuan Desert, Mojave Desert, Colorado Plateau Desert, Great Basin Desert, and California coastal sage scrub.

8.3.1 *Chihuahuan and Mojave Deserts*

Chihuahuan Desert An extensive field survey (Soil Interactions with *Bromus* [SIB] study; Belnap et al., unpublished data) found no invasive exotic annual-grass patches (of any species, including *Bromus*) on any soil type. This was despite the fact that total vegetation cover and the major soil nutrients of interest at these sites were similar to those found in the other deserts (Table 8.1).

Mojave Desert There are several studies from the Mojave Desert that examine the correlation between soil characteristics and *B. rubens* cover. The SIB study compared soil (0–10 cm in depth) in randomly selected patches invaded by *B. rubens* with adjacent uninvaded patches (Fig. 8.2) at 172 sites across the eastern Mojave Desert on multiple parent material types (Table 8.2). When all sites were combined, multiple regression analysis showed a significant positive relationship between *B. rubens* cover and the ratio of bicarbonate-extractable P (P_{be}) to the acid-neutralizing potential (which binds P, making it bio-unavailable) (P_{be}/ANP) ($R^2=0.36$) (for all results reported from this study, significance was defined as $P<0.05$). A more resolved model resulted when only sites with >4 % *B. rubens* cover were considered (P_{be}/ANP = R^2 of 0.83). These results indicate that overall, *B. rubens* cover may be limited by P availability in this desert. Because soil factors affecting *B. rubens* distribution could vary with available precipitation, this dataset was also analyzed by elevation classes of 200–500, 500–950, 950–1100, and 1100–1775 m height above sea level (Table 8.3). For 200–500 m, extractable calcium (Ca_{ex}), Zn_{ex}, and Na_{ex} were negatively correlated, and Mn_{ex} was positively related, to *B. rubens* cover (total $R^2=0.73$). *Bromus rubens* cover at 500–950 m was positively related to P_{be}/ANP and K_{ex}/Mg_{ex} and negatively related to copper (Cu_{ex}; total $R^2=0.62$). The 950–1100 m sites showed a very weak positive relationship between *B. rubens* cover, and Mn_{ex} ($R^2=0.08$). *Bromus* cover at sites above 1100 m had a positive relationship with both P_{be}/ANP and silt (total $R^2=0.41$).

Three other studies in the Mojave Desert addressed the relationship between soil characteristics and invasive *B. rubens*. The first study examined the effects of N fertilization on response of *B. rubens* and native forbs at Joshua Tree National Park (Allen et al. 2009, unpublished data). N fertilization experiments were done at two pinyon-juniper woodlands with lower and higher levels of anthropogenic N deposition, 6 and 12 kg N ha^{-1} $year^{-1}$ (Tonnesen et al. 2007; Fenn et al. 2010, unpublished data). The site with low N deposition had initial low cover and low seed bank density of *B. rubens*, and the site with high deposition had higher cover of *B. rubens* with high seed bank density (Allen et al. 2009; Schneider and Allen 2012). Plots were fertilized with 0, 5, or 30 kg N ha^{-1} as NH_4NO_3 each fall 2002–2004 and percent cover assessed in spring 2005. Experimental N fertilization promoted increased cover of brome grasses at the high N deposition site with a subsequent decrease in native forb cover (Fig. 8.2). However, grass cover was both low and highly variably at the low deposition site and grasses did not respond significantly. In contrast, native forb cover was initially high at the low deposition site, and native forbs respond to N fertilizer. This implies that native forbs are able to respond to N fertilizer when there is reduced competition from exotic grasses (Allen et al. 2009, unpublished data). However, *B. rubens* had a sufficiently large seed bank even at the lower N site and could eventually increase with high inputs of N (Schneider and Allen 2012). Thus, increasing N deposition appears to make these ecosystems more invasible by brome grasses. These invasions (Fig. 8.2, Allen et al. 2009) occurred in soils with moderate resin-extractable P (P_{re}) concentrations (~3–10 mg kg^{-1}) and pH of 6.5–8 (Rao and Allen 2010). In two additional Mojave Desert studies, Brooks (1999) showed that *B. rubens* cover was higher in lower micro-topographic positions where total N, P, and water were higher than elevated hummocks, but this

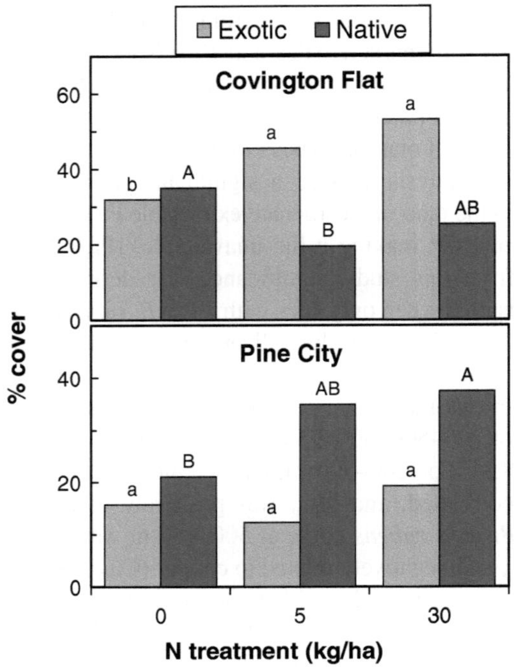

Fig. 8.2 Cover of exotic grass *B. rubens* and native forbs (~25 species) in 2005 at two sites in Joshua Tree NP Covington Flat has relatively higher anthropogenic N deposition with ~12 kg N ha^{-1} $year^{-1}$ and Pine City has ~6 kg N ha^{-1} $year^{-1}$ (Tonnesen et al. 2007; Fenn et al. 2010, unpublished data). Plots were fertilized with 0, 5, or 30 kg N ha^{-1} as NH_4NO_3 each fall 2002–2004 and percent cover of herbaceous vegetation assessed in spring 2005 (redrawn from Allen et al. 2009, unpublished data). Different letters above columns indicate significant differences within exotic or native species

study did not distinguish among these three soil factors. However, an N fertilization experiment confirmed that N was limiting *B. rubens* productivity (Brooks 2003).

8.3.2 Colorado Plateau Desert

The interaction of *Bromus* with soil factors for the Colorado Plateau region has only been addressed by the SIB study. This study sampled 195 sites for soil chemical characteristics and *B. tectorum* cover at small (80 ha), intermediate (8000 ha), and large scales (80,000 ha). At the 80 ha scale, three uninvaded and three *B. tectorum*-*invaded* areas were randomly selected, and within each area, a block of 30 plots were randomly placed for sampling. Multiple regression showed a significant correlation between *Bromus* cover and higher soil K_{ex}, K_{ex}/Mg_{ex}, K_{ex}/Ca_{ex}, and soil CEC, with the strongest relationship being a positive correlation with K_{ex}/Mg_{ex} ($R^2 = 0.80$; Table 8.3). Soil nutrients were measured at this site before the invasion

Table 8.2 Ranges of nutrients, nutrient ratios, and annual and perennial plant cover in the different deserts sampled

	Mojave ($n=172$)		Colorado Plateau ($n=195$)		Great Basin ($n=30$)		Chihuahuan ($n=14$)	
	Minimum	Maximum	Minimum	Maximum	Minimum	Maximum	Minimum	Maximum
Sand (%)	42	95	22	93	13	77	33	70
Very coarse sand (%)	0	49	0	14	0	10	0	20
Coarse sand (%)	0	36	0	21	0	21	1	22
Medium sand (%)	2	58	0	39	0	34	1	60
Fine sand (%)	8	51	4	72	1	34	5	47
Very fine sand (%)	4	35	3	42	5	25	3	29
Clay (%)	4	28	4	51	12	40	14	35
Silt (%)	1	40	0	52	11	59	21	40
pH	6.4	8.0	7.0	9.0	4.9	8.1	7	8.0
ANP (%)	<1	27	0	63	0	28	<1	35
Ca_{ex}	500	5796	1012	26,880	960	4842	2384	54,620
Cu	0.12	1.7	0.05	5.2	0.88	3.7	0.01	1.5
Fe	0.9	14	1	63	3.8	84	0.4	3.1
Mg_{ex}	29	692	31	820	120	849	10	7272
Mn	<1	14	<1	32	6	113	<1	10
Total N	1	1463	39	6994	369	2551	33	1427
Na_{ex}	0.1	1013	0.2	912	52	980	26	30,180
K_{ex}	30	1053	26	887	270	1724	<1	1554
K_{be}	40	529	38	816	125	1130	10	1004
P_{be}	0.8	27	0.9	37	11	60	0.1	26

(continued)

Table 8.2 (continued)

	Mojave ($n=172$)		Colorado Plateau ($n=195$)		Great Basin ($n=30$)		Chihuahuan ($n=14$)	
	Minimum	Maximum	Minimum	Maximum	Minimum	Maximum	Minimum	Maximum
Zn	0.12	0.81	0.06	3.5	0.4	8.2	0.02	1.6
K_{be}/Mg	0.64	6.2	0.13	4.6	0.46	4.3	0.04	5.2
P/ANP	<1	166	<1	391	<1	528	<1	1.5
Perennial plant % cover	0	57	0	80	12	80	14	50
Annual plant % cover	0	48	0	50	0	75	0	0

Subscripts indicate the extracting agent. Phosphorus (P_{be}) and K (K_{be}) were extracted with $NaHCO_3$, and exchangeable cations (Na_{ex}, Ca_{ex}, K_{ex}, and Mg_{ex}) were extracted with $NH_4C_2H_3O_2$ buffered to pH 8.5 to match the soils. Micronutrients (Cu_{ex}, Fe_{ex}, Zn_{ex}, and Mn_{ex}) were extracted with diethylenetriaminepentaacetic acid (DTPA). P/ANP is the ratio of P_{be} to the acid-neutralizing capacity of the soil (ANP). Concentrations of elements are expressed as mg element/kg dry soil. Subscripts av = available; ex = exchangeable; be = bicarbonate extractable. Total N was measured using the Kjeldahl method

Table 8.3 Results of stepwise regression models for Mojave, Colorado Plateau, and Great Basin sites with percent cover of exotic annual grass as the dependent variable

	Predictor	R^2	Correlation sign
Mojave Desert ($n=172$)			
All sites combined	P/ANP	0.36	+
>4 % annual grass ($n=37$)	P/ANP	0.83	+
Elevation 200–500 m ($n=24$)	Ca_{ex}	0.26	–
	Mn	0.46	+
	Zn	0.64	–
	Na_{ex}	0.73	–
Elevation 500–950 m ($n=85$)	P/ANP	0.47	+
	K/Mg	0.55	+
	Cu	0.62	–
Elevation 950–1100 m ($n=31$)	Mn	0.08	+
Elevation 1100–1775 m ($n=32$)	P/ANP	0.25	+
	Silt	0.41	+
Colorado Plateau ($n=195$)			
All sites combined	K/Mg	0.20	+
Elevation 1200–1400 m ($n=22$)	Soil depth	0.69	+
	K/Mg	0.94	+
	P	0.99	+
Elevation 1400–1585 m ($n=124$)	P/ANP	0.29	+
	Soil depth	0.41	+
Elevation 1585–2160 m ($n=49$)	P/Ca[a]	0.63	+
	Fine sand	0.73	–
	Very fine sand	0.78	+
Great Basin ($n=30$)			
All sites combined	K_{av}	0.24	+
	Survey clay	0.35	–
	Mn	0.48	+
Elevation 865–1400 m ($n=14$)	Very fine sand	0.46	+
	Mg	0.61	+
	Survey clay	0.80	–
Elevation 1400–2159 m ($n=16$)	Mn	0.35	+
	ANP	0.61	+
	Survey clay	0.76	–

The reported R^2 values are cumulative. "Survey clay" refers to data obtained from Natural Resources Conservation Service soil surveys, rather than from analyses of collected soil. Subscripts av = available; ex = exchangeable

[a]This ratio was used in place of P/ANP, as we did not have ANP values for these sites

occurred (Kleiner and Harper 1977) and at that time, K_{ex}, K_{ex}/Mg_{ex}, and soil CEC were higher in the plots that were invaded; thus, the correlation between invasion and soil chemistry was not due solely to post-invasion plant-soil feedbacks. In addition, soil chemistry has been measured in invaded and uninvaded plots since the invasion and shows no consistent directional changes in total soil N, nitrate (NO_3), ammonium (NH_4^+), P_{be} K_{ex}, K_{be}, Ca_{ex}, or micronutrients (Belnap et al. 2005, 2006; Schaeffer et al. 2012), further supporting the idea that soil chemistry differences existed pre-invasion.

At the intermediate scale (8000 ha), soil maps were used to identify different map units, and within each unit, a sample point was randomly selected. If this point was uninvaded, then the nearest invaded site (if present within 200 m) was also sampled; if invaded, then the nearest uninvaded site (if present within 200 m) was sampled as well. Similarly to the 80 ha area, sites dominated by *B. tectorum* had higher K_{ex}/Mg_{ex} than uninvaded sites. At the larger (80,000 ha) scale, the same approach was adopted but over the larger area. Again, K_{ex}/Mg_{ex} was significantly positively related to *B. tectorum* cover, but the relationship was weaker than at the small scale ($R^2=0.20$ vs. 0.80, respectively, Table 8.3). When sites from the largest region were divided into elevation classes, *B. tectorum* cover at sites with an elevation of 1200–1400 m was positively correlated with soil depth, K_{ex}/Mg_{ex} and P_{re} (total $R^2=0.99$; Table 8.3). As seen in the Mojave, *B. tectorum* cover at sites of intermediate elevation (1400–1585 m) was positively correlated with P_{be}/ANP and soil depth (total $R^2=0.41$). *Bromus tectorum* cover at 1585–2160 m sites was correlated with P_{re}/Ca_{ex} (+), fine sand (−), and very fine sand fractions (+, total $R^2=0.78$).

8.3.3 *Great Basin/Intermountain Region and Columbia Plateau*

Despite a great deal of research on *B. tectorum* in the Great Basin/Intermountain-Columbia Plateau regions, few studies have used a field approach to examine how a suite of soil characteristics may control this species. Belnap's SIB survey visited 30 sites in the Great Basin. Multiple regression analysis found that *B. tectorum* cover was positively correlated with K_{ex} and Mn_{ex} and negatively correlated with soil survey clay data (obtained from the Natural Resources Conservation Service soil surveys rather than analyzed soil; NRCS 2004) (total $R^2=0.48$; Table 8.3). When divided into elevation groups, *B. tectorum* cover at lower elevations (865–1400 m) was positively related to very fine sand content and Mg_{ex} and negatively related to soil survey clay (total $R^2=0.80$). At higher elevations (1400–2159 m), *B. tectorum* cover was positively related to Mn_{ex}, ANP, and survey clay (total $R^2=0.76$; Table 8.3).

Rau et al. (2014) sampled 72 plots at four sites in Utah, Idaho, Oregon, and Washington. Soils were sandy or silt loams, all supporting intact sagebrush communities with a *B. tectorum* understory that ranged from 0 to 50 % areal cover. Across all plots, P_{re} was the factor most strongly (and positively) correlated with *B. tectorum* cover pre-disturbance ($R^2=0.57$). Measurements immediately after severe experimental disturbances (fire, mowing) showed P increased across all treatments and became less important than water and soil texture. Bashkin et al. (2003) also found a landscape-level correlation between *B. tectorum* biomass and P_{re}. Chambers et al. (2007) found no relationship between *B. tectorum* biomass and soil N in experimentally seeded plots but did find that soil water was limiting to *B. tectorum* establishment, biomass, and seed production at low-elevation sites. Blank et al. (2013) found that areas invaded by *B. tectorum* had significantly more phosphatase activity and more solution-phase ortho-P than uninvaded areas. Neither study found a relationship with N.

In contrast, Jones et al. (2015) did find a correlation between soil N and B. *tectorum* biomass and seed production. Stark and Norton (2015) used a common garden study to examine growth of *B. tectorum* in intact soil cylinders (23 cm diam. × 45 cm deep) collected from high-elevation (2020 m) field plots in a sagebrush-steppe ecosystem western Colorado. The soil cylinders were collected from field plots planted to different plant communities 24 years earlier (Stevenson et al. 2000). In the field, without supplemental water, *B. tectorum* biomass was lowest when grown on soil that had been scraped and planted to sagebrush (*Artemisia tridentata* Nutt. ssp. *wyomingensis* Beetle & Young) and native perennial grasses 24 years earlier. Biomass was moderate when grown on soil scraped and seeded to *B. tectorum* 24 years earlier and was highest on soil collected from mature, undisturbed *A. tridentata* communities. Analyses of plant tissue concentrations showed that N availability regulated *B. tectorum* production across the three soil types. Other studies in the Great Basin and elsewhere have also shown that sites invaded by *B. tectorum* have greater soil N and N availability than adjacent uninvaded sites (Booth et al. 2003b; Norton et al. 2004; Hooker et al. 2008; Adair and Burke 2010). In SE Oregon, Bansal et al. (2014) found a strong interaction of time of sampling and community type (invaded, mixed, native) and total resin-extractable N, resin NH_4, and resin NO_3^-. Many of these studies assume that *B. tectorum* created the greater N availability following invasion; however, as discussed earlier, with these correlative studies it is difficult to know whether *B. tectorum* invaded sites that already had greater N availability or it created the greater soil N post-invasion. In at least one study, the additional soil N in the invaded site was far in excess of ecosystem N input rates during the time since invasion by *B. tectorum* (e.g., Blank 2008), and we can reasonably assume that the differences in soil N existed prior to invasion (see also Germino et al. 2015). In this case, greater N availability may have promoted invasion. In other cases, the answer is not as obvious, and in fact, *B. tectorum* may have both invaded because of greater soil N and subsequently contributed to greater N availability.

8.3.4 Coastal Sage Scrub

The impacts of anthropogenic N deposition were assessed on coastal sage scrub vegetation in western Riverside County, CA, USA. This vegetation type has been extensively invaded by annual *Bromus* species. Greenhouse experiments indicated

Table 8.4 Abundance (% cover) and richness (species/3 ha) of plant groups along a nitrogen deposition gradient in western Riverside County, CA

Site	Exotic grass (% cover)	Native forb (% cover)	Shrub (% cover)	Native forbs no. of species per 3 ha	Soil N (mg kg^{-1})	N dep (kg ha^{-1} $year^{-1}$)
Jurupa Hills	63.5	4	2.2	16	37.7	19.6
Box Springs	69.2	18.5	2.4	31	32.6	14.7
Botanic Garden	36.0	25.4	0.2	20	28.9	13.4
Lake Perris	0.5	26.1	2.8	30	20.3	11.1
Mott Reserve	6.7	14.3	11.2	37	30.6	11.1
Lopez Canyon	11.1	19.6	19.3	67	9.6	9.0
Tucalota Hills	1.5	35.7	35	50	10.5	8.7

Sites along an urban to rural gradient in southern CA. Soil N is extractable nitrate plus ammonium. N deposition is from the CMAQ model. Table adapted from Fenn et al. (2010)

B. rubens would respond to atmospheric N deposition (Padgett and Allen 1999; Yoshida and Allen 2001, 2004). Along an urban to rural gradient, N deposition ranged from 19.6 to 8.6 kg N ha^{-1} $year^{-1}$, based on the Community Multiscale Air Quality (CMAQ) (Tonnesen et al. 2007) model (Table 8.4), whereas other variables (soils, elevation, and rainfall) were all similar. Extractable soil N (NO_3+NH_4) ranged from 39 mg N kg^{-1} at the urban end of the gradient to 10 mg kg^{-1} at the rural end. Exotic grass cover was positively correlated with soil N and atmospheric N deposition, while cover of native shrubs and forbs was negatively correlated with N ($p<0.001$). Exotic grass cover was dominated by *B. rubens* (up to 7.7 % cover) and *Bromus diandrus* Roth (ripgut brome) (up to 61 %) with <1 % cover of *Bromus hordeaceus* L. (soft brome) and *B. tectorum*. Exotic grass cover generally increased with higher N deposition (Table 8.4).

8.4 Manipulative Studies

8.4.1 Nitrogen

Multiple greenhouse and field studies have examined how N additions influence *Bromus* performance. In the field where *B. tectorum* already exists, N fertilization can increase *Bromus* production, especially in areas or years of higher precipitation and cooler temperatures. Kay (1966) showed that application of ammonium nitrate (NH_4NO_3) to field plots in northeastern California increased production in all but the driest years. Similarly, Brooks (2003) and Rao and Allen (2010) found that in the Mojave Desert, addition of NH_4NO_3 increased production of *B. rubens*, with the greatest increases occurring during years with the highest precipitation. Interestingly, Gundale et al. (2008) examined nutrient limitations to *B. tectorum* growth in a greenhouse study and found that N was limiting to *B. tectorum* growth in soil collected from beneath the canopy of ponderosa pine trees, but P was limiting to *B. tectorum* growth in soil collected from nearby bunchgrass patches. In another greenhouse study, Monaco et al. (2003) found that while N fertilization stimulated *B. tectorum* production, it was primarily NO_3 that produced the greatest response in shoot biomass. A number of studies have also found that lowering N availability by adding labile C (such as sucrose) to soils to promote N immobilization has the effect of lowering *B. tectorum* production: in a Wyoming big sagebrush community (McLendon and Redente 1991), in a shortgrass steppe (Paschke et al. 2000), and in soil from a sagebrush community in northern Utah (Monaco et al. 2003). Most studies suggest that *B. tectorum* production is favored by high N availability.

Many fertilization studies have shown N additions stimulate *Bromus* production while often increasing perennials as well. For example, on the Columbia Plateau, James (2008) showed that N fertilization increased both *B. tectorum* and perennial grass production. More importantly, James et al. (2011) showed that while low N availability reduced *B. tectorum* production, *B. tectorum* still outproduced the native perennial grass seedlings even at low N availability. Because *B. tectorum* appears to outperform native perennial grass seedlings under both high and low N availabilities, they proposed that manipulation of N availability may not be an effective restoration practice for shifting the competitive balance in favor of native grasses during seedling establishment. Outcomes are life-stage dependent, and in intact ecosystems, N uptake in mature, native perennial grasses is strongly coupled with N availability, and the mature, perennial grasses that dominate these systems may keep N and other resources at such low concentrations that *Bromus* seedlings are not competitive. However, any disturbance that decouples this relationship by eliminating established grasses results in an ecosystem highly vulnerable to invasion (Chambers et al. 2007).

In addition, other studies have shown that manipulating N availability does shift community composition where *B. tectorum* has already become established because of stronger relative responses of *B. tectorum* to N availability. Studies in a relatively warm and dry *A. tridentata* ssp. *wyomingensis* community in western Colorado (McLendon and Redente 1991, 1992) and in a shortgrass steppe in central Colorado (Paschke et al. 2000) found that addition of N fertilizer to mixed stands of annual and perennial vegetation stimulated production by *B. tectorum* to a much greater degree than perennials and shifted the community composition in favor of *B. tectorum* dominance. While N additions frequently increase *B. tectorum* production, the effect on native perennials appears to be highly dependent on the plant species and life stage and the characteristics of the soils and climate. For example, in a meta-analysis examining studies from sagebrush, shortgrass steppe, tallgrass prairie, mixed grass prairie, and coastal California grasslands, Blumenthal et al. (2003) found that of nine studies using C addition to reduce N availability, seven showed suppression of weeds while not affecting native perennials, whereas two studies showed no effect on weeds or natives.

While a number of studies have examined the effect of N availability on production of *B. tectorum* and perennial grasses in semiarid and arid lands, very few have examined the effect of N availability (or other nutrients) on the initial establishment of *B. tectorum* into existing stands of perennial grasses. In a study in the shortgrass steppe of central Colorado, Paschke et al. (2000) reported that lowering N availability by adding sucrose reduced recruitment of weedy annuals into established perennial grass communities. Similarly, Beckstead and Augspurger (2004) found that in western Utah, addition of sucrose to field plots with high *B. tectorum* densities reduced *B. tectorum* density and biomass during the establishment and active growth phases; however, in low-density *B. tectorum* stands, N addition had no effect on establishment density or overall biomass. More future research should focus on the effect of nutrients on this aspect of *B. tectorum* invasion.

There is also the intriguing possibility that while soil N may influence annual-grass production directly, it also may do so via interactions with other nutrients. For example, studies have shown that N addition stimulates soil phosphatase concentrations and activity, thus likely increasing available P (Phuyal et al. 2008; Collins et al. 2008). As *B. tectorum* produces and secretes phosphatases (Bolton et al. 1993), a stimulation of *B. tectorum* by N could enhance its phosphatase production as well.

8.4.2 Phosphorus and Others Nutrients

Multiple studies have experimentally addressed the effect of nutrients other than N on *Bromus* performance. Manganese (Mn) can stimulate the growth of *B. tectorum* and other annual grasses (Bildusas et al. 1986; Cramer and Nowak 1992; Miller et al. 2006a, b). With burning, soil Mn doubled in low P/ANP soils and increased almost four times in high P/ANP soils (Belnap, unpublished data). Rau et al. (2008) found burned soils contained 21 kg Mn ha^{-1} compared to 12 kg Mn ha^{-1} in unburned soils, which may at least partially explain *Bromus* success in invading burned areas, as burning does not always elevate other nutrients such as N or P. Potassium has been found to stimulate *B. tectorum* growth in the greenhouse (Howell 1998; Morrison 1999) and in the field on the Colorado Plateau (Miller et al. 2006a, b). Using manipulative experiments and correlations, Miller et al. (2006a, b) found that in the field, the limiting soil factors for *B. tectorum* can vary with different life stages: P and water limited germination in the fall and K_{ex} and K_{ex}/Mg_{ex} were positively correlated with winter performance. A combined analysis of winter–spring performance showed that growth was negatively correlated with ANP and $CaCO_3$ (both negatively affect P_{be} availability) and Mg_{ex} and positively correlated with P_{be}, silt, clay, and dust content (in addition to the Mn_{ex} and K_{ex}, mentioned above). In greenhouse experiments, addition of P has been shown to stimulate production of *B. tectorum* biomass (e.g., Dakheel et al. 1993; Blackshaw et al. 2004; Cherwin et al. 2009), but, in some cases, not as much as N (Dakheel et al. 1993). Belnap and Sherrod (2009) showed in a greenhouse experiment that *B. tectorum* germination was suppressed when P availability was reduced. DeLucia et al. (1989) found that *B. tectorum* biomass was reduced by over 90 % when growing in P-limited soils. Schlesinger et al. (1989) in the Great Basin Desert and Parker (1995) the Sonoran Desert reported P was limiting to plants in desert settings. In contrast, McLendon and Redente (1991) saw no effect on *B. tectorum* when field plots in western Colorado were amended with P.

8.4.3 Soil Moisture

Soil moisture has an important influence on *Bromus* performance directly, and it also can influence *Bromus* success indirectly through its effect on the availability of soil nutrients. Soil water and nutrient pools are not independent of each other, because plants must obtain soil nutrients in solution (Barber 1995; Leffler and Ryel 2012 and references therein). To date there have been very few studies which examine the interaction between soil water, nutrients, and *Bromus* growth, despite the importance of understanding these interactions. The greatest productivity of *B. rubens* occurred under the combination of high water and N in the Mojave Desert in a garden plot and also in experimentally fertilized natural vegetation that was measured during a series of dry to wet years (Rao and Allen 2010). Using intact monoliths obtained from a *B. tectorum*-infested site on the Columbia Plateau, Link et al. (1995) determined that the addition of water or N alone had no influence on *B. tectorum* growth, but when the two were added together, biomass production and leaf area were nearly twice that of any other treatment (control, water alone, or N alone), indicating that water and N were co-limiting. On the Colorado Plateau, Miller et al. (2006a) reported that water additions alone significantly increased establishment and plant density of *B. tectorum*. In contrast, Beckstead and Augspurger (2004), Link et al. (1995), and Cline and Rickard (1973) all reported that at Great Basin sites, water additions had no influence on *B. tectorum*; however, Beckstead and Augspurger did show N additions alone significantly increased *B. tectorum* biomass and plant density. In addition to these direct water and nutrient manipulations, several studies have shown that removal of competing vegetation significantly increases *B. tectorum* performance, predominantly due to increased moisture availability (Melgoza et al. 1990; Dodd et al. 1998; Beckstead and Augspurger 2004; Chambers et al. 2007). However, given that many areas have *Bromus* spp. occurring in patches directly adjacent to uninvaded patches, water alone is likely not the sole driver in many situations. Lastly, nutrients such as NO_3^- and sulfate move to the root largely by mass flow, and uptake is thus less constrained by soil water content than more diffusion-limited nutrients such as P and NH_4^+. Thus, soil moisture conditions may constrain nutrient uptake by plants, with P and NH_4^+ possibly more limiting than NO_3^- or S in dry desert soils, despite being present in sufficient amounts.

8.5 Discussion

8.5.1 How Soil Texture and Biochemistry Can Limit *Bromus*

Based mostly on experimental additions of N to field plots and greenhouse plants, and increases in available N following disturbance or management treatments, elevated soil N has long been generally believed to be the most important soil factor controlling the invasion and success of annual exotic plants (Smaytohlgren et al. 1999; Ehrenfeld 2003). In our literature review, we found that the influence of N depend on which climatic region is being considered. Under hotter and drier soil temperature and moisture regimes, extensive field surveys (172 sites in the Mojave Desert, 195 sites on the Colorado Plateau) and several experimental studies showed that nutrients other than N were the most correlated with *Bromus* occurrence. However, N additions did stimulate *B. rubens* performance in greenhouse pots, in field plots, and along two N gradients in the Mojave Desert (Padgett and Allen 1999; Yoshida and Allen 2001, 2004; Allen et al. 2009). This may have been due to the soil pH at these sites, which was relatively acidic (Table 8.2) compared to most dryland soils, which are generally >7.5. Low pH can result in less co-limitation by other nutrients than generally occurs in more alkaline soils.

Many field and greenhouse studies in the wetter and cooler Great Basin/Intermountain region showed a correlation between soil N and an increase in *B. tectorum* production, especially during wetter years. However, other greenhouse and field studies found the strongest correlation with *B. tectorum* occurrence or performance to be with K, P, or Mn (e.g., DeLucia et al. 1989; Miller et al. 2006a, b). As nutrient availability can vary widely with pH, soil moisture, and other factors, these results may not be conflicting but instead reflect unmeasured site conditions (e.g., soil pH, depth, moisture) or sampling time (e.g., wet versus dry years or life cycle stage). Unfortunately, despite the amount of data we have on factors correlated with *Bromus* occurrence and production, we have far less information on nutrients correlated with its initial germination and establishment phases.

The idea that P may limit *Bromus* invasion success is fairly recent and, from our review, may be an important driver in the hotter regions (thus having lower soil moisture) such as the eastern Mojave and Colorado Plateau Deserts and also sometimes in cooler regions (thus having higher soil moisture), such as the Great Basin. This might be due to the low availability of P in highly alkaline soils due to several interrelated factors: (1) carbonate ($CaCO_3$, $MgCO_3$) can complex with P, reducing its availability; thus, P availability can change with factors that alter the interaction of these compounds and P; (2) marginally soluble calcium phosphate compounds can precipitate out of solution; and (3) high levels of Ca and HCO_3 can reduce or prevent the dissolution of carbonate and Ca-P compounds, thus reducing amounts of available P via common ion effects: the neutralization of acids secreted by plants, mycorrhizal symbionts, other rhizosphere microbes, and/or root-respired CO_2, all of which would otherwise increase P solubility (Barber 1995). Other factors that are also likely to contribute to lower P availability in these soils is the low level of soil organic matter, which can be the dominant source of plant available P when soil physiochemical processes tie up P (Stevenson 1986) and low initial concentrations of P in the soil parent material, which in these regions is often aeolian material derived from sandstone.

Climate may also affect the bioavailability of P via changes in $CaCO_3$ solubility (as postulated by Miller et al. 2006a, b). In high-pH desert soils, most P is bound to CaCO3, rendering it unavailable for plant uptake. The generation of H_2CO_3 facilitates carbonate dissolution and thus the transition of solid-phase P to solution-phase P (Jungk and Claassen 1997). The rate of H_2CO_3 formation is partially controlled by soil water content and the solubility of CO_2 in that water (Krauskopf and Bird 1995). The solubility of CO_2 in water, like that of other gases, is greater at cold than warm temperatures, and thus a theoretical maximum in H_2CO_3 production should occur when soils are cool and wet. Therefore, release of $CaCO_3$-bound P should also be highest under cool and wet soil conditions. Root growth can also contribute respiratory CO_2, facilitating the release and therefore acquisition of carbonate-bound nutrients, and *B. tectorum* shows relatively high levels of winter root growth when soils are cold and moist (Harris 1967). Therefore, under the above scenario, winter would be the time when P would be most available in dryland soils. This is supported by several studies. Results from in situ resin bags at Colorado Plateau sites indicate an increase in available P during the cold, moist conditions found in winter (Miller et al. 2006a, b). *Bromus tectorum* growth rates were greatest at these sites in winter and were positively related with P/Ca and inversely related with ANP. Lajtha and Schlesinger (1988) also found that in situ resin bag P concentrations peaked in cool winter conditions in the Chihuahuan Desert. Magid and Nielsen (1992) showed that laboratory extractions done at 4 °C recover significantly more P than those done at 25 °C. In general, lower diffusion rates at decreased temperatures may partially counteract the increase in available soil P. However, the situation may be different in calcareous soils, as diffusion rates may actually increase with decreasing temperatures (but above freezing) (Jungk and Claassen 1997). Increased solubility of P at lower temperatures may also play a large role. In addition, *Bromus* establishes in the fall and is active through winter in many areas. During this time, the roots can actively uptake P when soil temperatures and soil water content are high enough to allow efficient transport of P to the roots.

Increased soil moisture due to climate factors (higher precipitation and lower temperatures) may also impact biotically mediated ways that free bound P. Soil fungi and plants, including *Bromus*, secrete phosphatases which can make bound P available, and increased soil moisture, both in space and time, can stimulate this production. As mentioned above, root-respired CO_2 can acidify the root environment and thus increase levels of available P.

Other cations, including micronutrients, may influence invasions as well (e.g., Blank et al. 2007). In contrast to N and P, much less work has been done with K as a limiting nutrient in dryland soils. Plant species differ in their K uptake (Gray et al. 1953), and uptake is positively and highly related with plant root cation-exchange capacity (CEC) ($r=0.78$; Crooke and Knight (1962). Root CECs vary widely among plant functional types: annuals have higher root CECs than perennials, and grasses have higher CECs than herbaceous and woody species (Crooke and Knight 1962; Scott and Billings 1964; Woodward et al. 1984). Therefore, annual grasses should generally have the highest CECs of all plants. Possibly due to high root CECs that are found in *Bromus* (Belnap, unpublished data), they can have higher tissue concentrations of K than adjacent native perennial plants (Blank et al. 2002), which may also indicate they have a higher requirement for K than the native grasses (Tilman 1982). Tilman et al. (1999) also reported K to be limiting in the field for the perennial *Taraxacum officinale*. Traditionally, agriculture has regarded exchangeable K values of 140 ppm to be the minimum required for adequate plant growth (Marschner 1995; Leigh and Storey 1991). Using this standard, most of the soils sampled in the regional field surveys (Belnap, unpublished data) were K deficient.

Potassium interacts with other cations and micronutrients, which may also explain its importance to annual grasses. Osmoregulation in plants is mediated by K (Mäser et al. 2002; Wang et al. 2002). The presence of high Na can be toxic to many plants, and there are multiple studies showing that K ameliorates Na toxicity in plants (Mäser et al. 2002) and other organisms such as bacteria (Kraegeloh and Kunte 2002). The preferential transport of K over Na is especially pronounced in actively photosynthesizing tissues such as young leaves and developing seeds (Wang et al. 2002). The extent to which plants utilize K to avoid Na stress varies among species (Mäser et al. 2002). In addition, K has been implicated in plant avoidance of water stress (Xu et al. 2002). Multiple studies support the observation that high levels of Mg and Ca can restrict plant uptake of K in both the laboratory and the field (Epstein 1961; Sinanis et al. 2003). Additions of K can also enhance Mn availability to plants, which may be very important in understanding the distribution of exotic annual grasses (see below; Krishnamurti and Huang 1988). Crooke and Knight (1962) and Scott and Billings (1964) were the first to note that soils with high K_{ex}/Mg_{ex} ratios were dominated by annual plants. This finding was followed up by Harner and Harper (1973), Pederson and Harper (1979), and Woodward et al. (1984) to explain patterns of plant distribution in the arid Western USA.

Micronutrients are seldom considered in studies of native plant distribution and/or performance. However, studies in the Western USA have found that micronutrients are important in predicting the distribution of disparate organisms such as soil lichens, annual grasses, and perennial shrubs (Cramer and Nowak 1992; Bowker et al. 2005). There is very little information on how micronutrients may influence plant distribution patterns in dryland soils characterized by high pH, low SOM, and high $CaCO_3$ levels, as almost all studies have been conducted in agricultural settings. That said, it is known that the bioavailability of Zn, Mn, and Fe is very low in alkaline soils that are commonly found in dryland regions and can thus be limiting to plants under these conditions (Clark and Baligar 2000). Although these nutrients are only required in very small amounts, they are essential for plant growth. These micronutrients also form oxides that bind P, reducing its availability in soils. In the field surveys reported here, Mn_{ex} was often found to be positively related to *Bromus* cover, indicating there are places where this micronutrient may be limiting. Greenhouse studies have found that Mn_{ex} can stimulate the growth of annual grasses (Bildusas et al. 1986; Cramer and Nowak 1992). Perhaps analogously, soil lichens are associated with soils containing higher Mn (Blank et al. 2001; Bowker et al. 2005).

Soil depth and texture were also important in characterizing invaded sites in all the deserts. At some sites in the Great Basin, deep loamy (fine sand to silt) soils with high water-holding capacity appear best able to retain a mix of *B. tectorum* and perennial grasses, in contrast to shallow coarse sandy soils that can have greater abundance of *B. tectorum* (e.g., Rau et al. 2014). This may be because the deep loams have and hold resources longer throughout the growing season than shallower coarse soils. Consequently, although *B. tectorum* uses much of the surface soil water to complete its life cycle, sufficient amounts are left, especially in subsoils (Ryel et al.2010) for native herbaceous perennials to persist. Conversely, shallow sandy soils with low water-holding capacity may be more prone to *B. tectorum* dominance, as this plant can more fully utilize what soil moisture is present, depleting the resources that natives need (Melgoza et al. 1990). Thus, when sites experience several water-limited growing seasons, repeated herbivory, or disturbance such as fire, native perennial herbaceous species may experience sufficient stress that they can no longer produce adequate carbohydrate stores to survive, leaving the site open to *B. tectorum* dominance. Once *B. tectorum* is dominant on these sites, native seedlings cannot compete even when adequate resources are available (Booth et al. 2003a; Monaco et al. 2003). Surprisingly, though, an opposite phenomenon appears to occur on the Colorado Plateau, where *B. tectorum* only dominates deeper but finer-textured (fine sandy loam to clay loam) soils rather than shallow coarser sandy soils. The reason for this contrast to the Great Basin is not fully understood. It is possible that *B. tectorum* cannot invade the coarser shallow soils of the Colorado Plateau because of their very low fertility and instead require finer-textured soils where nutrients are sufficient to support a high cover of annual plants. In contrast, the Great Basin and Columbia Plateau soils are generally more fertile and thus competition with other plants and water-holding capacity may be the main drivers behind *B. tectorum* dominance.

8.5.2 Why Were Different Nutrients Limiting in Different Regions and What Does This Mean for Predicting Annual Grass Distribution in Western US Semiarid and Arid Lands?

Figure 8.3 presents a conceptual model of how different nutrients may limit *Bromus* occurrence among different dryland regions of the Western USA, depending on climate (Belnap 2011). This hypothesis is based on the supposition that climate can alter the availability of soil nutrients, especially those made bio-unavailable by soil carbonates (e.g., P and micronutrients). This model suggests that in regions where such nutrients are limiting to *Bromus*, climatic factors can determine whether this genus is able to invade and persist. This model proposes that the ratio of the average

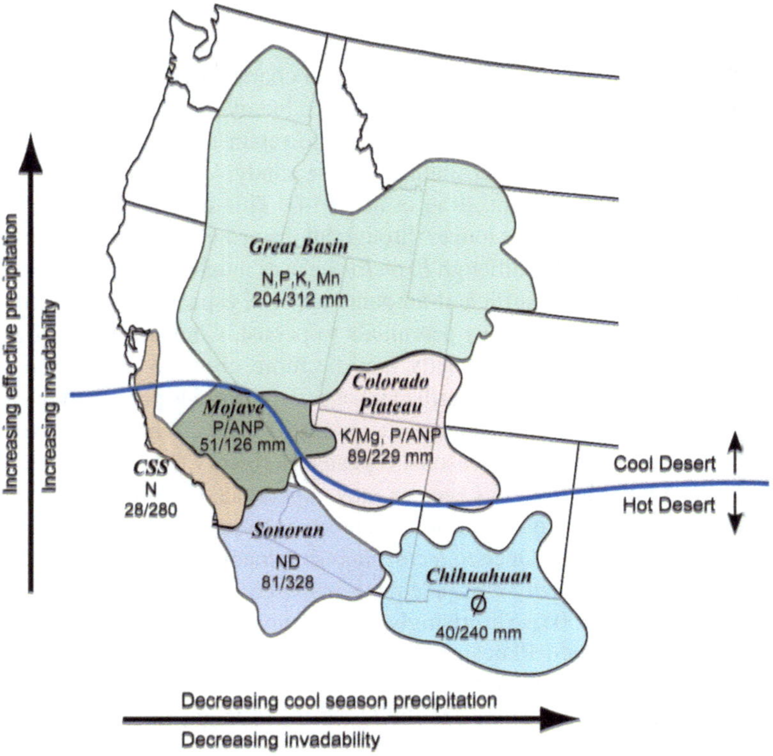

Fig. 8.3 A hypothesized relationship between climate and soil factors (nutrients and water) limiting *Bromus* cover in the different deserts of the Western USA. Limiting nutrients are listed in suggested order of importance within regional boundaries. The numbers within the regional boundaries indicate the average annual amount of precipitation (mm) when mean daily air temperatures are below 10 °C relative to total precipitation. We suggest that as this ratio and the total amount of cool season precipitation increase, so do conditions that allow the conversion of bio-unavailable phosphorus (P) into bioavailable P. As P becomes less limiting, other nutrients such as N nitrogen (N), potassium (K), manganese (Mn), and water become more important to *Bromus* establishment. We do not have data on the soil factors constraining invasive annual grasses in the Sonoran desert. Based on this model, P is expected to be most limiting to *Bromus* in hotter and drier regions

amount of precipitation that falls when air temperatures are below 10 °C (winter) relative to total precipitation (winter/total precipitation; W/T), as well as the total amount of cool season precipitation, indicates the frequency with which conditions occur that allow the conversion of bio-unavailable P into bioavailable P, as discussed in the P section above. The desert regions discussed occur along a gradient of a W/T ratio and total winter precipitation. The Chihuahuan Desert occurs at one end of the gradient, as it has the lowest total winter precipitation, the lowest proportion of winter rain, and the highest winter temperatures of the deserts we sampled. Under these conditions, it is expected that P is seldom freed from carbonates and thus available P in the soils remains extremely low most of the year. The Mojave has the next lowest winter precipitation and W/T ratio and the next highest winter temperatures, followed by the Colorado Plateau. The Great Basin/Columbia Plateau is at the other end of the gradient, with the highest winter precipitation, the highest W/T ratio, and the lowest winter temperatures of all deserts sampled. Therefore, we suggest that P is likely the most limiting at the Chihuahuan Desert end of the gradient, becoming less so in the Mojave, even less on the Colorado Plateau, and finally the least limiting in the Great Basin. Studies in this review generally support this pattern, as invasive annual grasses are less common (or even absent, as in the SIB data) where winter precipitation is low and temperatures high (Chihuahuan Desert), and thus P is theoretically the least available, despite adequate levels of all soil nutrients in the soil at these sites. At slightly higher winter precipitation and cooler winter temperatures (the Mojave Desert), annual grasses can still be most limited by factors associated with P availability. As total winter precipitation increases and temperatures decrease (going northward to the Colorado Plateau), factors associated with P availability may become less limiting and other factors (K and water availability) become important, although P availability is still part of the regression model. At the Great Basin sites, factors influencing N, P, K, and water availability become most important. Soils of the Mediterranean climate region of southern California should show the least P limitation of all, with its cool, wet winter. Soil organic matter also increases along this gradient from dry, hot desert regions to the cooler wetter semiarid lands and is important in providing nutrients to plants. Therefore, SOM levels are important in understanding nutrient limitations. However, there are exceptions to this model at the site or landscape level. For instance, P is highly available on soils with a pH of 6.5 or less (Marschner 1995). Such soils can be found in many areas of the Mojave Desert (Rao and Allen 2010) and the California sage scrub habitat (Padgett and Allen 1999). Further research is needed to verify this hypothesis and to refine its application to various landscapes.

The role of N, P, or other nutrients in promoting *B. tectorum* and *B. rubens* invasion and dominance is also likely to depend on the perennial plants involved (e.g., grasses versus shrubs) and the role of fire. For example, for fire to be carried through vegetation, there must be sufficient fuel loads. In some communities, sufficient fuels are only achieved during very wet years; however, high nutrient availability may also stimulate *B. tectorum* and *B. rubens* production such that even under moderately wet years, there is sufficient grass biomass to carry fire. This is undoubtedly a major contributor to the well-known annual-grass fire cycle that has allowed huge areas in the Western USA to be converted to monocultures of *Bromus*. For example, fire probability was modeled to increase above 3.9 kg ha^{-1} $year^{-1}$ (*Bromus* and *Schismus*) because of increased fine fuel production under moderate precipitation in the Mojave Desert (Rao et al. 2010). Balch et al. (2012) found cheatgrass dominates ~6 % of the Great Basin, with 13 % of these areas burning from 1980 to 2009, double that of other vegetation types. A historic record of 582 fires in the Mojave Desert was compared with N deposition rates and showed overall more area burned than expected when N deposition rates exceeded 7 kg ha^{-1} $year^{-1}$ (Rao et al. 2014). Fires driven by annual grasses in the Mojave Desert reduce both native shrub recovery and native forb diversity, resulting in persistent dominance by exotic grasses (Steers and Allen 2011, 2012). If fire occurs in a stand of perennial grass, it may be of little consequence. But if the fire occurs in a stand of fire-intolerant shrubs such as *A. tridentata*, postfire increases in N availability and the loss of *A. tridentata* could make lower precipitation areas open to invasion by *Bromus*.

This broad pattern of relationships, however, will have limits and exceptions when applied to particular soil types and across large areas. For instance, P availability appears limiting at higher-elevation sites in all regions in the SIB study. This seems to be contrary to the P-limitation model presented above and may show one of its limitations. However, it may also be that temperatures at the higher elevations are often below freezing, at which point the dissolution of the bonds between P and other elements would not occur. Soil texture and depth will also modify the applicability of this model at the site level as well. In addition, the water requirement of *Bromus* at various points in its life cycle likely plays a major role in determining the degree to which it can dominate a site. For instance, although *B. tectorum* is found in nearly every US state and southern Canadian provinces west of Quebec, it has only become dominant over large landscapes in semiarid regions dominated by wet winter-spring and dry summer precipitation regimes (Pyke and Novak 1994; Bradley 2009; Mack 2011). This may be because the most favorable conditions for germination and establishment within this region occur where spring precipitation is relatively abundant and temperatures are relatively warm (10–20 °C), but not hot (Roundy et al. 2007; Bradley 2009). In contrast, on the Colorado Plateau and hotter deserts, *Bromus* is generally dominant in relatively small patches.

8.6 Management Implications

The current state of knowledge suggests that there are critical interactions that occur among climate regimes, soil texture, nutrient availability, and soil surface disturbance which determine *Bromus* success in different settings. Even with our currently incomplete understanding, managers can utilize soil and climate maps to make initial assessments of the settings (parent material, soils, geomorphic settings, climate) more or less resistant and/or resilient to *Bromus* invasion. In addition, this understanding can aid in prioritizing sites for restoration or protection. Advancement of the understanding of *Bromus*-soil relationships is also needed to aid land managers in better predicting where *Bromus* can invade under future conditions as well as the resultant effects on native communities.

8.7 Research Needs

Additional research is needed to further refine our knowledge on how various factors affect *Bromus* establishment, such as (1) the less studied nutrients (e.g., Mg, K, micronutrients); (2) the presence of different native species, as they can utilize soil nutrients differently among each other; (3) the distribution of different soil acid-neutralizing capacities because it is not often measured in soils; and (4) how the interactions of these soil factors are expected to change among regions as climates fluctuate and type, intensity, and timing of surface disturbances change. The latter is increasingly important with future moisture and temperature changes, and if our hypothesis is correct, large changes in *Bromus* distributions could occur. In addition, soils in many areas of the Western USA are unmapped or maps are inadequate for this purpose. Therefore, more effort is needed in this arena as well.

Acknowledgments Funding for JB was provided by USGS Ecosystems and the Climate and Land Use Program and for EA, the National Park Service PMIS No. 72123 and National Science Foundation DEB 04-21530. Patti Haggerty helped with graphics. Any use of trade names is for descriptive purposes only and does not imply endorsement by the US government.

References

Abella SR, Craig DJ, Chiquoine LP et al (2011) Relationships of native desert plants with red brome (*Bromus rubens*): toward identifying invasion-reducing species. Invasive Plant Sci Manag 4:115–124

Adair EC, Burke IC (2010) Plant phenology and life span influence soil pool dynamics: *Bromus tectorum* invasion of perennial C3-C4 grass communities. Plant Soil 335:255–269

Allen EB, Rao LE, Steers RJ et al (2009) Impacts of atmospheric nitrogen deposition on vegetation and soils in Joshua Tree National Park. In: Webb RH, Fenstermaker LF, Heaton JS et al (eds) The Mojave desert: ecosystem processes and sustainability. University of Nevada Press, Las Vegas, NV, pp 78–100

Balch JK, Bradley BA, D'Antonio CM et al (2012) Introduced annual grass increases regional fire activity across the arid western USA (1980–2009). Glob Chang Biol 19:173–183

Bansal S, Sheley RL, Blank B et al (2014) Plant litter effects on soil nutrient availability and vegetation dynamics: changes that occur when annual grasses invade shrub-steppe communities. Plant Ecol 215:367–378

Barber SA (1995) Soil nutrient bioavailability. Wiley, New York, NY

Bashkin M, Stohlgren TJ, Otsuki Y et al (2003) Soil characteristics and plant exotic species invasions in the Grand Staircase—Escalante National Monument, Utah, USA. Appl Soil Ecol 22:67–77

Beckstead J, Augspurger CK (2004) An experimental test of resistance to *B. tectorum* invasion: limiting resources at different life stages. Biol Invasions 6:417–432

Belnap J (2011) Biological phosphorus cycling in dryland regions. In: Bünemann EK (ed) Phosphorus in action: biological processes in soil phosphorus cycling, vol 26. Springer, Berlin, pp 371–406

Belnap J, Phillips SJ (2001) Soil biota in an ungrazed grassland: response to annual grass (*Bromus tectorum)* invasion. Ecol Appl 11:1261–1275

Belnap J, Sherrod SK (2009) Soil amendment effects on the exotic annual grass *Bromus tectorum* L. and facilitation of its growth by the native perennial grass *Hilaria jamesii* (Torr.) Benth. In: Van der Valk AG (ed) Herbaceous plant ecology. Springer, Netherlands, pp 345–357

Belnap J, Phillips SL, Sherrod S et al (2005) Soil biota can change after exotic plant invasion: does this affect ecosystem processes? Ecology 86:3007–3017

Belnap J, Phillips SL, Troxler T (2006) Soil lichen and moss cover and species richness can be highly dynamic: the effects of invasion by the annual exotic grass *Bromus tectorum* and the effects of climate on biological soil crusts. Appl Soil Ecol 32:63–76

Bildusas IJ, Pfleger FL, Stewart EL et al (1986) Response of *Bromus inermis* inoculated with *Glomus fasciculatum* to potassium fertilization and drought stress. Plant Soil 95:441–444

Blackshaw RE, Molnar LJ, Janzen HH (2004) Nitrogen fertilizer timing and application method affect weed growth and competition with spring wheat. Weed Sci 52:614–622
Blank RR (2008) Biogeochemistry of plant invasion: a case study with downy brome (*Bromus tectorum*). Invasive Plant Sci Manag 1:226–238
Blank RR, Allen FL, Young JA (2001) Physical and chemical comparisons between microphytic and non-microphytic soil seedbeds. J Range Manag Arch 54:194–196
Blank RR, Qualls RG, Young JA (2002) *Lepidium latifolium*: plant nutrient competition-soil interactions. Biol Fertil Soils 35:458–464
Blank RR, Chambers JC, Roundy B et al (2007) Nutrient availability in rangeland soils: influence of prescribed burning, herbaceous vegetation removal, overseeding with *Bromus tectorum*, season, and elevation. Rangel Ecol Manag 60:644–655
Blank RR, Morgan T, Clements CD, Mackey BE (2013) *Bromus tectorum* L. invasion: changes in soil properties and rates of bioturbation. Soil Sci 178:281–290
Blumenthal DM, Jordan NR, Russelle MP (2003) Soil carbon addition controls weeds and facilitates prairie restoration. Ecol Appl 13:605–615
Bolton H Jr, Smith JL, Link SO (1993) Soil microbial biomass and activity of a disturbed and undisturbed shrub-steppe ecosystem. Soil Biol Biochem 25:545–552
Booth MS, Caldwell MM, Stark JM (2003a) Overlapping resource use in three Great Basin species: implications for community invisibility and vegetation dynamics. J Ecol 91:36–48
Booth MS, Stark JM, Caldwell MM (2003b) Inorganic N turnover and availability in annual- and perennial-dominated soils in a northern Utah shrub-steppe ecosystem. Biogeochemistry 66:311–330
Bowker MA, Belnap J, Davidson DW et al (2005) Evidence for micronutrient limitation of biological soil crusts: importance to arid-lands restoration. Ecol Appl 15:1941–1951
Bradley BA (2009) Regional analysis of the impacts of climate change on *B. tectorum* invasion shows potential risk and opportunity. Glob Chang Biol 15:196–208
Brooks ML (1999) Habitat invasibility and dominance by alien annual plants in the western Mojave Desert. Biol Invasions 1:325–337
Brooks ML (2003) Effects of increased soil nitrogen on the dominance of alien annual plants in the Mojave Desert. J Appl Ecol 40:344–353
Chambers JC, Roundy BA, Blank RR et al (2007) What makes Great Basin sagebrush ecosystems invasible by *Bromus tectorum*? Ecol Monogr 77:117–145
Cherwin K, Seastedt TR, Suding KN (2009) Effects of nutrient manipulations and grass removal on cover, species composition, and invasibility of a novel grassland in Colorado. Restor Ecol 17:818–826
Clark RB, Baligar VC (2000) Acidic and alkaline soil constraints on plant mineral nutrition. In: Wilkinson RE (ed) Plant-environment interactions. Marcel Dekker, New York, NY, pp 133–177
Cline JF, Rickard WH (1973) Herbage yields in relation to soil water and assimilated nitrogen. J Range Manag 26:296–298
Collins SL, Sinsabaugh RL, Crenshaw C (2008) Pulse dynamics and microbial processes in arid-land ecosystems. J Ecol 96:413–420
Cramer GR, Nowak RS (1992) Supplemental manganese improves the relative growth, net assimilation and photosynthetic rates of salt-stressed barley. Physiol Plant 84:600–605
Crooke MW, Knight AH (1962) An evaluation of published data on the mineral composition of plants in the light of the cation-exchange capacities of their roots. Soil Sci 93:365–373
Dakheel AJ, Radosevich SR, Barbour MG (1993) Effect of nitrogen and phosphorus on growth and interference between *Bromus tectorum* and *Taeniatherum asperum*. Weed Res 33:415–422
DeLucia EH, Schlesinger WH, Billings WD (1989) Edaphic limitations to growth and photosynthesis in Sierran and Great Basin vegetation. Oecologia 78:184–190
Dibb DW, Thompson WR Jr (1985) Interaction of potassium with other nutrients. In: Munson RD (ed) Potassium in agriculture. American Society of Agronomy, Crop Science Society of America, Soil Science Society of America, Madison, pp 515–533

Dodd MB, Lauenroth WK, Welker JM (1998) Differential water resource use by herbaceous and woody plant life-forms in a shortgrass steppe community. Oecologia 117:504–512
Ehrenfeld JG (2003) Effects of exotic plant invasions on soil nutrient cycling processes. Ecosystems 6:503–523
Epstein E (1961) The essential role of calcium in selective cation transport by plant cells. Plant Physiol 36:437–444
Fenn ME, Allen EB, Weiss SB et al (2010) Nitrogen critical loads and management alternatives for N-impacted ecosystems in California. J Environ Manag 91:2404–2423
Germino MJ, Belnap J, Stark JM et al (2015) Ecosystem impacts of exotic annual invaders in the genus *Bromus*. In: Germino MJ, Chambers JC, Brown CS (eds) Exotic brome-grasses in arid and semiarid ecosystems of the Western US: causes, consequences, and management implications. Springer, New York, NY (Chapter 3)
Gray B, Drake M, Colby WG (1953) Potassium competition in grass-legume associations as a function of root cation exchange capacity. Soil Sci Soc Am Proc 17:235–239
Gundale MJ, Sutherland S, DeLuca TH (2008) Fire, native species, and soil resource interactions influence the spatio-temporal invasion pattern of *Bromus tectorum*. Ecography 31:201–210
Hamerlynck EP, McAuliffe JR, McDonald EV et al (2002) Ecological responses of two Mojave Desert shrubs to soil horizon development and soil water dynamics. Ecology 83:768–779
Harner RF, Harper KT (1973) Mineral composition of grassland species of the eastern Great Basin in relation to stand productivity. Can J Bot 51:2037–2046
Harris GA (1967) Some competitive relationships between *Agropyron spicatum* and *Bromus tectorum*. Ecol Monogr 37:89–111
Hobbs RJ, Huenneke LF (1992) Disturbance, diversity, and invasion: implications for conservation. Conserv Biol 6:324–337
Hooker TD, Stark JM, Norton U et al (2008) Distribution of ecosystem C and N within contrasting vegetation types in a semiarid rangeland in the Great Basin, USA. Biogeochemistry 90:291–308
Howell W (1998) Germination and establishment of *Bromus tectorum* L. in relation to cation exchange capacity, seedbed, litter, soil cover and water. MA thesis, Prescott College
James JJ (2008) Effect of soil nitrogen stress on the relative growth rate of annual and perennial grasses in the Intermountain West. Plant Soil 310:201–210
James JJ, Drenovsky RE, Monaco TA et al (2011) Managing soil nitrogen to restore annual grass-infested plant communities: effective strategy or incomplete framework? Ecol Appl 21:490–502
Jones R, Chambers JC, Johnson DW et al (2015) Effect of repeated burning on plant and soil carbon and nitrogen in cheatgrass (Bromus tectorum) dominated ecosystems. Plant Soil 386:47–64
Jungk A, Claassen N (1997) Ion diffusion in the soil-root system. Adv Agron 61:53–110
Kay BL (1966) Fertilization of cheatgrass ranges in California. J Range Manag 19:217–220
Kleiner EF, Harper KT (1977) Soil properties in relation to cryptogamic groundcover in Canyonlands National Park. J Range Manag 30:202–205
Kraegeloh A, Kunte HJ (2002) Novel insights into the role of potassium for osmoregulation in *Halomonas elongata*. Extremophiles 6:453–462
Krauskopf KB, Bird DK (1995) Introduction to geochemistry. McGraw Hill, New York, NY
Krishnamurti GSR, Huang PM (1988) Kinetics of manganese released from selected manganese oxide minerals as influenced by potassium chloride. Soil Sci 146:326–334
Lajtha K, Schlesinger WH (1988) The biogeochemistry of phosphorus cycling and phosphorus availability along a desert soil chronosequence. Ecology 69:24–39
Leffler AJ, Ryel RJ (2012) Resource pool dynamics: conditions that regulate species interactions and dominance. In: Monaco TA, Sheley RL (eds) Invasive plant ecology and management: linking processes to practice. CAB International, Oxfordshire, pp 57–78
Leigh RA, Storey R (1991) Nutrient compartmentation in cells and its relevance to the nutrition of the whole plant. In: Porter JR, Lawlor DW (eds) Plant growth: interactions with nutrition and the environment. Cambridge University Press, Cambridge, pp 33–54

Link SO, Bolton H Jr, Thiede ME et al (1995) Responses of downy brome to nitrogen and water. J Range Manag 48:290–297

Mack RN (2011) Fifty years of 'waging war on cheatgrass': research advances, while meaningful control languishes. In: Richardson DM (ed) Fifty years of invasion ecology: the legacy of Charles Elton. Wiley-Blackwell, Oxford, pp 253–265

Magid J, Nielsen NE (1992) Seasonal variation in organic and inorganic phosphorus fractions of temperate-climate sandy soils. Plant Soil 144:155–165

Marschner H (1995) The mineral nutrition of plants. Academic, London

Mäser P, Gierth M, Schroeder JI (2002) Molecular mechanisms of potassium and sodium uptake in plants. Plant Soil 247:43–54

McAuliffe JR, McDonald EV (1995) A piedmont landscape in the eastern Mojave Desert: examples of linkages between biotic and physical components. San Bernardino Cty Mus Assoc Q 42:53–63

McLendon T, Redente EF (1991) Nitrogen and phosphorus effects on secondary succession dynamics on a semi-arid sagebrush site. Ecology 72:2016–2024

McLendon T, Redente EF (1992) Effects of nitrogen limitation on species replacement dynamics during early secondary succession on a semiarid sagebrush site. Oecologia 91:312–317

Melgoza G, Nowak RS, Tausch RJ (1990) Soil water exploitation after fire: competition between *Bromus tectorum* (B. tectorum) and two native species. Oecologia 83:7–13

Miller ME, Belnap J, Beatty SW et al (2006a) Effects of water additions, chemical amendments, and plants on in situ measures of nutrient bioavailability in calcareous soils of southeastern Utah, USA. Plant Soil 288:19–29

Miller ME, Belnap J, Beatty SW et al (2006b) Performance of *Bromus tectorum* L. in relation to soil properties, water additions, and chemical amendments in calcareous soils of southeastern Utah, USA. Plant Soil 288:1–18

Monaco TA, Johnson DA, Norton JM et al (2003) Contrasting responses of intermountain west grasses to soil nitrogen. J Range Manag 56:282–290

Morrison R (1999) Potassium as a limiting nutrient for germination and production of *B. tectorum* (*Bromus tectorum*) in the Canyonlands National Park, Utah. Honor's thesis, University of Denver

Nichols JD (1984) Relation of organic-carbon to soil properties and climate in the southern Great Plains. Soil Sci Soc Am J 48:1382–1384

Norton JB, Monaco TA, Norton JM et al (2004) Soil morphology and organic matter dynamics under cheatgrass and sagebrush-steppe plant communities. J Arid Environ 57:445–466

NRCS (2004) Natural resources conservation service, USDA, list of soil surveys by state. http://soils.usda.gov/survey/printed_surveys/. Cited 21 April 2004

Padgett PE, Allen EB (1999) Differential responses to nitrogen fertilization in native shrubs and exotic annuals common to Mediterranean coastal sage scrub of California. Plant Ecol 144:93–101

Parker KC (1995) Effects of complex geomorphic history on soil and vegetation patterns on arid alluvial fans. J Arid Environ 30:19–39

Paschke MW, McLendon T, Redente EF (2000) Nitrogen availability and old-field succession in a shortgrass steppe. Ecosystems 3:144–158

Pederson JC, Harper KT (1979) Chemical composition of some important plants of southeastern Utah summer ranges related to mule deer reproductions. Great Basin Nat 39:122–128

Phuyal M, Artz RR, Sheppard L et al (2008) Long-term nitrogen deposition increases phosphorus limitation of bryophytes in an ombrotrophic bog. Plant Ecol 196:111–121

Pyke DA, Novak SJ (1994) *B. tectorum* demography-establishment attributes, recruitment, ecotypes, and genetic variability. In: Monsen SB, Kitchen SG, (comps) Ecology and management of annual rangelands, Boise, ID, May 1992. General Technical Report INT-GTR-313. USDA, Forest Service, Intermountain Research Station, Ogden, UT, pp 12–21

Rao LE, Allen EB (2010) Combined effects of precipitation and nitrogen deposition on native and invasive winter annual production in California deserts. Oecologia 62:1035–1046

Rao LE, Allen EB, Meixner T (2010) Risk-based determination of critical nitrogen deposition loads for fire spread in southern California deserts. Ecol Appl 20:1320–1335

Rao LE, Matchett JR, Brooks ML et al (2014) Relationships between annual plant productivity, nitrogen deposition and fire size in low elevation California desert scrub. Int J Wildland Fire 24:48–58

Rau BM, Chambers JC, Blank RR et al (2008) Prescribed fire, soil, and plants: burn effects and interactions in the central Great Basin. Rangel Ecol Manag 61:169–181

Rau BM, Chambers JC, Pyke DA et al (2014) Soil resources influence vegetation and response to fire and fire-surrogate treatments in sagebrush-steppe ecosystems. Rangel Ecol Manag 67:506–521

Roundy BA, Hardegree SP, Chambers JC et al (2007) Prediction of *B. tectorum* field germination potential using wet thermal accumulation. Rangel Ecol Manag 60:613–623

Ryel R, Leffler AJ, Ivans C, Peek MS, Caldwell MM (2010) Functional differences in water-use patterns of contrasting life forms in Great Basin steppelands. Vadose Zone J 9:548–560

Sala OE, Parton WJ, Joyce LA et al (1988) Primary production of the central grassland region of the United States. Ecology 69:40–45

Schaeffer SM, Ziegler SE, Belnap J et al (2012) Effects of *Bromus tectorum* invasion on microbial carbon and nitrogen cycling in two adjacent undisturbed arid grassland communities. Biogeochemistry 111:427–441

Schlesinger WH, DeLucia EH, Billings WD (1989) Nutrient-use efficiency of woody plants on contrasting soils in the western Great Basin, Nevada. Ecology 70:105–113

Schneider HE, Allen EB (2012) Effects of elevated nitrogen and exotic plant invasion on soil seed bank composition in Joshua Tree National Park. Plant Ecol 213:1277–1287

Scott D, Billings WD (1964) Effects of environmental factors on standing crop and productivity of an alpine tundra. Ecol Monogr 34:243–270

Sinanis C, Keramidas VZ, Sakellariadis S (2003) Thermodynamics of potassium-magnesium exchange in two alfisols of northern Greece. Commun Soil Sci Plant Anal 34:439–456

Stark JM (1994) Causes of soil nutrient heterogeneity at different scales. In: Caldwell MM, Pearcy R (eds) Exploitation of environmental heterogeneity by plants: ecophysiological processes above and below ground. Wiley, New York, NY, pp 255–284

Stark JM, Norton JM (2015) The invasive annual cheatgrass increases nitrogen availability in 24-year-old replicated field plots. Oecologia 177:799–809

Steers RJ, Allen EB (2011) Fire effects on perennial vegetation in the western Colorado Desert, USA. Fire Ecol 7:59–74

Steers RJ, Allen EB (2012) Impact of recurrent fire on annual plants: a case study from the western edge of the Colorado Desert. Madrono 59:14–24

Stevenson FJ (1986) Cycles of soil. Wiley, New York, NY

Stevenson BA, McLendon T, Redente EF (2000) Effects of soil fumigation and seeding regimes on secondary succession in a semiarid shrubland. Arid Soil Res Rehabil 14:87–99

Stohlgren TJ, Binkley D, Chong GW et al (1999) Exotic plant species invade hot spots of native plant diversity. Ecol Monogr 69:25–46

Tausch RJ, Svejcar T, Burkhardt JW (1994) Patterns of annual grass dominance on Anaho Island: implications for Great Basin vegetation management. In: Monsen SB, Kitchen SG (comps) Ecology and management of annual rangelands, Boise, ID, May 1992. General Technical Report INT-GTR-313. USDA, Forest Service, Intermountain Research Station, Ogden, UT, pp 120–125

Tilman D (1982) Resource competition and community structure. Monographs in population biology. Princeton University Press, Princeton, NJ

Tilman EA, Tilman D, Crawley MJ et al (1999) Biological weed control via nutrient competition: potassium limitation of dandelions. Ecol Appl 9:103–111

Tonnesen G, Wang Z, Omary M et al (2007) Assessment of nitrogen deposition: modeling and habitat assessment. California Energy Commission, PIER Energy-Related Environmental Research CEC-500-2005-032. http://www.energy.ca.gov/2006publications/CEC-500-2006-032/CEC-500-2006-032.pdf

Wang S, Zheng W, Ren J et al (2002) Selectivity of various types of salt-resistant plants for K^+ over Na^+. J Arid Environ 52:457–472

Webb RH, Steiger JW, Newman EB (1988) The response of vegetation to disturbance in Death Valley National Monument, California. US Geological Survey Bulletin 1793. US Government Printing Office, Denver, CO

Woodward RA, Harper KT, Tiedemann AR (1984) An ecological consideration of the significance of cation-exchange capacity of roots of some Utah range plants. Plant Soil 79:169–180

Xu S, An L, Feng H et al (2002) The seasonal effects of water stress on *Ammopiptanthus mongolicus* in a desert environment. J Arid Environ 51:437–447

Yoshida LC, Allen EB (2001) Response to ammonium and nitrate by a mycorrhizal annual invasive grass and a native shrub in southern California. Am J Bot 88:1430–1436

Yoshida LC, Allen EB (2004) ^{15}N uptake by mycorrhizal *Artemisia californica* and the invasive *Bromus madritensis* of a N-eutrophied shrubland. Biol Fertil Soils 39:243–248

Chapter 9
Bromus Response to Climate and Projected Changes with Climate Change

Bethany A. Bradley, Caroline A. Curtis, and Jeanne C. Chambers

Abstract A prominent goal of invasive plant management is to prevent or reduce the spread of invasive species into uninvaded landscapes and regions. Monitoring and control efforts often rely on scientific knowledge of suitable habitat for the invasive species. However, rising temperatures and altered precipitation projected with climate change are likely to shift the geographic range of that suitable habitat. Here, we review experimental and modeling studies of climatic limits to exotic annual *Bromus* (*Bromus* hereafter) distribution in the Intermountain West in the context of projections of future climate change. We update empirical models of range shifts to test whether *Bromus rubens* L. (red brome) is likely to expand into ranges that become less suitable for *Bromus tectorum* L. (cheatgrass or downy brome). Warming temperatures are likely to create an advantage for *Bromus* species throughout much of the Intermountain West, potentially enhancing invasion into formerly resistant ecosystems if native species mortality increases with warming and drought. *Bromus rubens* is likely to expand into areas of the Southern Great Basin and Colorado Plateau as warmer winters reduce range constraints caused by cold intolerance. However, a primary limitation to exotic annual *Bromus* invasion and expansion is growing season precipitation. Projections for precipitation change are uncertain, but increased precipitation during periods critical for exotic annual *Bromus* germination and growth is forecast for the Northern Great Basin. Increased *Bromus* reproduction and biomass may exacerbate *B. tectorum* invasion and associated fire risk, especially if coupled with longer fire seasons and more extreme

B.A. Bradley (✉)
Department of Environmental Conservation, University of Massachusetts,
Amherst, MA 01003, USA
e-mail: bbradley@eco.umass.edu

C.A. Curtis
Graduate Program in Organismic and Evolutionary Biology, University of Massachusetts,
Amherst, MA 01003, USA
e-mail: cacurtis@eco.umass.edu

J.C. Chambers
USDA Forest Service, Rocky Mountain Research Station,
Reno, NV 89512, USA
e-mail: jchambers@fs.fed.us

M.J. Germino et al. (eds.), *Exotic Brome-Grasses in Arid and Semiarid Ecosystems of the Western US*, Springer Series on Environmental Management,
DOI 10.1007/978-3-319-24930-8_9

fire weather. Managers should anticipate both shifts in the overall distribution of *Bromus* species, as well as changes in relative abundance within its existing range.

Keywords Biogeography • Climate change • Experiments • Range limits • Species distribution model

9.1 Introduction

Both native and invasive plants in the Western USA will be affected by changing climate conditions caused by anthropogenic greenhouse gas emissions. However, different physiological tolerances to temperature and precipitation of individual species (see Brooks et al. 2015) will create opportunities for some species but disadvantages for others. Warmer temperatures are leading to species distribution shifts poleward and upward in elevation—a trend that has been observed for thousands of species globally (e.g., Parmesan and Yohe 2003; Root et al. 2003; Chen et al. 2011). For plants, which are relatively dispersal limited (compared to animals), warmer temperatures tend to lead to an earlier spring and longer growing season (Root et al. 2003; Morisette et al. 2009). A longer growing season may give an advantage to some plant species that are able to initiate growth and reproduce as soon as resources are available, e.g., many annual species and species with high phenotypic plasticity in phenology (Willis et al. 2008). Plants that are able to quickly disperse large numbers of propagules into newly suitable habitat are likely to be favored in a changing climate (Johnston 2011). Exotic annual *Bromus* (*Bromus* hereafter) species have highly plastic responses to resource availability (Hulbert 1955; Monaco et al. 2003; James et al. 2011), high reproductive capacity (Mack and Pyke 1983), and propagules that are readily dispersed, particularly in disturbed areas (Johnston 2011). Consequently, it is likely that climate change will tend to favor species such as *Bromus tectorum* L. (downy brome or cheatgrass) and *Bromus rubens* L. (red brome).

Spatial patterns of *Bromus* response to climate will depend in large part on how climate changes across the Intermountain West. Temperature rise, which is a direct consequence of rising greenhouse gases, is fairly consistently projected by climate models (and therefore has relatively low uncertainty, IPCC 2013). In contrast, precipitation changes are an indirect consequence of rising greenhouse gases, dependent on not only overall temperature changes but also spatial patterns of temperature change and prominent ocean circulation patterns such as the El Niño Southern Oscillation and the Pacific Decadal Oscillation. As a result, precipitation change is less consistently projected by climate models (and therefore has higher uncertainty, IPCC 2013). Thus, species response to temperature alone can be predicted more confidently than species response to temperature and precipitation. Because *Bromus* germination, growth, and reproduction are limited by temperature and water availability, the uncertainty in precipitation makes it difficult to make specific geographic predictions of how distribution and abundance are likely to change. As trends in precipitation change come into better focus, adaptive management practices (e.g., Pyke et al. 2015) will add important flexibility for implementing new control efforts in areas where *Bromus* species are likely to expand. In this chapter, we review projected climate changes for the Western USA and highlight studies of *Bromus* response to specific climate factors to identify how climate changes might influence *B. tectorum* or *B. rubens* expansion and/or contraction. We also evaluate *Bromus* interactions with native ecosystems and feedbacks with a dominant ecosystem driver, fire, under projected climate changes. We conclude with a discussion of management implications and research needs.

9.1.1 Climate Projections for the Western USA

The magnitude of climate change into the future is largely dependent on rates of anthropogenic greenhouse gas emissions, which are highly uncertain (IPCC 2013). The standard scenario presented by the IPCC Fifth Assessment Report is RCP4.5, or a "representative concentration pathway" associated with a scenario where greenhouse gas emissions stabilize at a radiative forcing of 4.5 W/m^2 by 2100. Radiative forcing relates to how much of the sun's radiation the Earth absorbs vs. emits back to space, with higher values associated with more absorption and greater warming. RCP4.5 is roughly equivalent to an atmospheric CO_2 concentration of 650 ppm by 2100 (van Vuuren et al. 2011). (Preindustrial CO_2 concentration was 280 ppm, and levels as of 2014 were roughly 400 ppm.) This may be an optimistic scenario, as emissions trajectories measured between 1990 and 2009 suggest a pathway closer to RCP6.0 or a CO_2 equivalent of roughly 850 ppm by 2100 (Le Quere et al. 2009). Higher emissions by 2100 increase the amount of projected temperature change.

IPCC temperature projections are modeled based on summer (Jun–Aug) vs. winter (Dec–Feb). Based on the RCP4.5 emissions scenario, summer temperatures (Jun–Aug) are projected to rise throughout the Western USA by 1–3 °C by 2050 and by 2–4 °C by 2100 (IPCC 2013). Winter temperatures (Dec–Feb) are projected to rise by 1–3 °C by 2050 and by 1.5–4 °C by 2100, with larger increases in winter temperature forecast for more northern climates (IPCC 2013). All projected future temperatures are outside the current range of variability of mean seasonal temperatures.

IPCC precipitation projections are modeled based on half-year summer (Apr–Sep) vs. winter (Oct–Mar) precipitation. Based on the RCP4.5 scenario, summer precipitation (Apr–Sep) is projected to decrease in the southwest by 0–20 % in both the 2050 and 2100 scenarios, while increasing from 0 to 10 % in the northwest. In both cases, the model ensemble mean is not significantly different from present-day variability of seasonal precipitation (i.e., both projections encompass zero change; IPCC 2013). However, water stress is expected to increase throughout the west because warming temperatures increase evaporation rates even if precipitation remains constant or increases slightly (Seager et al. 2007). Winter precipitation (Oct–Mar) is projected to remain the same in the southwest and increase by 0–10 % by 2050 and 0–20 % by 2100 in the northwest. However, rising temperatures will lead to more precipitation falling as rain vs. snow during the winter and will likely increase drought stress throughout areas historically dependent on winter snowmelt

(Barnett et al. 2005). Overall soil moisture, which is driven by the combination of precipitation and temperature, is projected to decline significantly throughout the west in all seasons (IPCC 2013).

9.1.2 Climate Effects on Native Ecosystems in the Western USA

The relative dominance of exotic annual *Bromus* species and their effects on native ecosystems will be significantly influenced by climate effects on native species. Projected current and future geographic distribution of biomes in North America created by Rehfeldt et al. (2012) generally indicate the following changes in climate suitable area for the Western USA: (1) expansion to higher latitudes and elevations for cold and warm desert biomes, California evergreen forest and woodland and coastal scrub, Great Basin montane scrub, and Rocky Mountain montane conifer forest; (2) contraction of California chaparral, evergreen forests and woodlands and valley grasslands, and Great Basin conifer woodland and shrub-grassland; and (3) relative stability for Great Plains grasslands. For example, the Mojave and the Great Basin Deserts, including the Colorado Plateau ecoregion, are expected to increase in area by 40 % and 45 % by 2060, respectively, and Great Basin montane scrub is expected to increase by 87 %, while Great Basin conifer woodland and shrub-grassland decline by 21 % and 33 %, respectively. But, even in cases where biomes are projected to expand, range shifts in climatically suitable areas are likely to lead to considerable biome turnover. By 2060, only 65 % of the current Mojave Desert and 67 % of the Great Basin Desert will still have climate conditions that exist within the biome extents today (Rehfeldt et al. 2012).

Specific ecosystems that are currently invaded by *Bromus* species are also likely to be increasingly stressed by climate change. Bioclimate envelope models for *Artemisia tridentata* Nutt. (big sagebrush) and other *Artemisia* species, which *B. tectorum* frequently invades, project large decreases in southern latitudes and lower elevations but relatively small increases in northern latitudes and higher elevations (Bradley 2010; Schlaepfer et al. 2012; Still and Richardson 2015; Fig. 9.1). For *A. tridentata* Nutt. ssp. *wyomingensis* Beetle & Young (Wyoming big sagebrush), which occupies the warmest and driest portions of the species range, a 39 % reduction (66 million hectares) in suitable climate is predicted by mid-century (Still and Richardson 2015). Regions predicted to be most vulnerable to loss of climate suitability include the trailing edge (i.e., the southern periphery of the species), and lower elevation areas of the Columbia and Great Basin. Regions that retain or gain climate suitability include higher elevations in the Great Basin, and the northern Great Plains (Bradley 2010; Schlaepfer et al. 2012).

In the Mojave Desert and Colorado Plateau, regions typically invaded by *B. rubens* and to a lesser degree *B. tectorum*, a dominant native shrub *Coleogyne ramosissima* Torr. (blackbrush) is expected to lose much of its current climatic habitat. While the potential range of *C. ramosissima* expands by an estimated 52 % by 2060, most of this

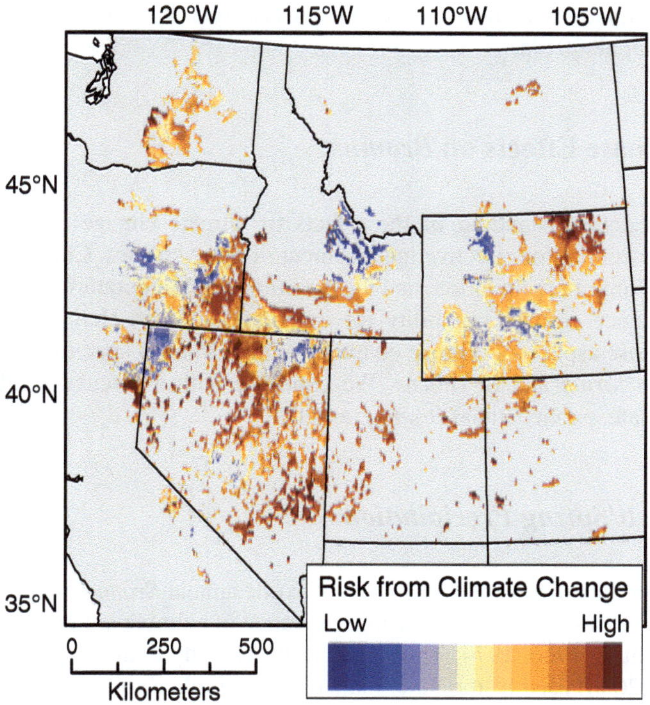

Fig. 9.1 Risk to existing sagebrush (*Artemisia* spp.) populations in the Great Basin associated with climate change. Cooler colors are lower risk, defined as higher likelihood of maintained climatic suitability under multiple general circulation model projections. Warmer colors have high risk from climate change, with few or zero general circulation models projecting future climatic suitability (Adapted from Bradley 2010)

is new range in the Great Basin Desert (Richardson et al. 2014). Climatic habitat loss within the Mojave Desert is nearly 100 % (Richardson et al. 2014). Cold air drainage in certain areas of the topographically diverse Great Basin may result in lower minimum temperatures than generally predicted from climate models and favor cold-adapted ecotypes of *C. ramosissima* (Richardson et al. 2014). Although more studies are needed for native species, case studies for *A. tridentata* and *C. ramosissima* suggest that native ecosystems could be strongly negatively affected by climate change, potentially reducing their resistance to *Bromus* invasion (Chambers et al. 2014a).

In addition to shifting mean climate conditions, greater climate variability likely will favor invasion of annual weeds (Bradley et al. 2010) and negatively affect native species persistence in areas that remain otherwise climatically suitable. Reduced soil moisture availability coupled with greater climate variability can result in reduced resilience or recovery potential of native ecosystems following disturbances such as wildfire (Brooks and Matchett 2003; Brooks and Chambers 2011; Chambers et al. 2014a). In turn, decreased resilience can lower the resistance of these ecosystems to invasive annual grasses like *B. tectorum* and *B. rubens* (Chambers et al. 2014a).

9.2 Climate Effects on *Bromus*

An increased understanding of the effects of climate change on exotic annual *Bromus* is essential for effective management of these species. Changes in the timing and amount of precipitation, and the magnitude of temperature increase, significantly influence climate suitability for individual species. Here we examine the effects of these aspects of climate on exotic annual *Bromus* species with an emphasis on *B. tectorum* and *B. rubens*. We conclude with a discussion of feedbacks among climate, exotic annual *Bromus*, and fire.

9.2.1 Fall/Spring Precipitation

Germination, establishment, and growth of exotic annual *Bromus* are predominantly limited by soil water availability within the range of suitable temperatures for these species (Beatley 1966, 1974; Thill et al. 1979; Roundy et al. 2007). Bradford and Lauenroth (2006) used historical climate data to simulate soil moisture availability over 1000 years within sagebrush steppe ecosystems. They estimated that over the 1000-modeled years, conditions were suitable for *B. tectorum* germination and establishment 72 % of the time in the spring and 25 % of the time in the fall. However, this general finding has large spatial variation. In the Great Basin, the timing of precipitation shifts from primarily winter/spring in the west toward relatively more spring/summer or summer in the east (Chambers et al. 2015). Most of the available information that we review on the life history and physiological traits of *B. tectorum* is from the western portion of the Great Basin, where exotic annual *Bromus* species tend to be more problematic. *B. tectorum* establishment depends on receiving adequate precipitation during periods that are favorable for germination and growth and can occur in fall following precipitation events, and/or in spring when soil moisture increases as a result of snowmelt or spring rains (Mack and Pyke 1984). Survival following fall germination is strongly influenced by short-term drought in the fall (Mack and Pyke 1984). In warmer and drier areas, *B. tectorum* can fail to recruit during low precipitation years, and population growth is constrained by longer-term drought (Meyer et al. 2001; Chambers et al. 2007). Hence, changes in either fall precipitation or winter and spring precipitation could affect germination and establishment of *B. tectorum* across various portions of its range.

High precipitation during the winter/spring in semiarid ecosystems increases the biomass of *Bromus* species (Meyer et al. 2001; Chambers et al. 2007). The accumulation of *B. tectorum* biomass following wet winter/spring conditions can be so dramatic that it can be seen from satellites (Bradley and Mustard 2005). In the Mojave and southern Great Basin, wet winters are associated with strong El Niño events (caused by warm surface waters in the Pacific). Higher *B. tectorum* biomass has been observed for 2 years following El Niño events in areas dominated by the invader. Even if second-year climate conditions are conducive to only average growth, the strong growth response during the second year might be due to prolific seed availability (Bradley and Mustard 2005) and/or higher nutrient availability as a result of more favorable conditions for decomposition during years with wet winter/spring conditions (Jones et al. 2014). El Niño years can greatly increase reproduction and spread of exotic invasive *Bromus* into marginal environments as has been observed for *B. rubens* in the Mojave Desert (Salo 2005) and *B. tectorum* in salt desert ecosystems of the cold desert (Meyer et al. 2001). Conversely, drought during winter/spring can inhibit annual invasive *Bromus* establishment, growth, and reproduction in a variety of ecosystems (Beatley 1966, 1974; Thill et al. 1979; Roundy et al. 2007; Zelikova et al. 2013). Ultimately, the effects of both El Niño years and drought on *Bromus* invasion are strongly moderated by the productivity and species composition of the native community (Chambers et al. 2014a, 2015). Other global change factors that generally promote *Bromus* growth, including rising CO_2 (Smith et al. 1987, 2000; Ziska et al. 2005), increasing N deposition (Hulbert 1955; Adair et al. 2008), and rising temperatures (Thill et al. 1979; Compagnoni and Adler 2014), are all highly dependent on soil water availability, and their effects will largely depend on the magnitude of change in both temperature and precipitation.

While IPCC precipitation projections focus on winter (Oct–Mar) vs. summer (Apr–Sep) precipitation, *Bromus* species are mainly influenced by fall precipitation and/or winter/spring precipitation for germination and growth. The lack of seasonal congruence between projected precipitation changes and the *Bromus* growing season increases uncertainty as to how precipitation change might affect *Bromus*. One consistency in the climate models is a likely increase in winter precipitation in the northern part of the Intermountain West (above ~37° latitude) and an associated increase in anomalously wet winters (Abatzoglou and Kolden 2011; IPCC 2013). This increase could result in expansion of *B. tectorum* into drier areas that are currently marginal habitat but are projected to become wetter at critical times for *Bromus* germination and growth (and, therefore, more suitable). In southern portions of the Intermountain West (below ~37° latitude), winter climate is likely to become drier (Abatzoglou and Kolden 2011; IPCC 2013), which for *B. tectorum* could lead to reduced establishment and growth, lower rates of invasion across much of the area, and potential range contraction where water is more limiting. *Bromus rubens* occurs in areas with slightly lower mean annual precipitation (Brooks et al. 2015) and may be able to expand into some of the areas that are no longer suitable for *B. tectorum* (see Sect. 9.2.5).

9.2.2 Summer Precipitation

The potential range of *Bromus* species depends not only on the life history and physiological traits of *Bromus* but also on the climatic response of potentially invaded ecosystems (Chambers et al. 2014a). *B. tectorum* is established in all 50 US states and almost all Canadian provinces (USDA-NRCS 2014) but typically is not considered invasive and rarely becomes abundant in areas that receive plentiful precipitation in summer (Bradford and Lauenroth 2006; Bradley 2009, 2013). Seasonality of precipitation has important consequences for soil water balance during the year and, thus, plant functional type dominance (Sala et al. 1997) and competitive interactions with *B. tectorum* (Bradford and Lauenroth 2006). The amount of precipitation that is received during the period when temperature, and thus potential evapotranspiration, is low influences the amount of water that is stored in deep soil layers and therefore the relative dominance of woody vs. herbaceous species (Sala et al. 1997). In the west, areas that receive more winter/spring precipitation typically have greater deep soil water storage and are dominated by woody species, which are more effective at using deep soil water. In contrast, areas that receive predominantly summer precipitation tend to be dominated by perennial grasses.

In a regional species distribution modeling study, Bradley (2009) showed that *B. tectorum's* current range within the Great Basin is best explained by summer precipitation. Areas with higher summer precipitation were less likely to be invaded. Increasing summer precipitation may result in less favorable conditions for establishment of *Bromus* and strong competition from native grass species that dominate under this precipitation regime (Bradford and Lauenroth 2006; Bradley et al. 2009). The strength of these competitive interactions likely increases as precipitation increases, and the native community becomes more productive (see Chambers et al. 2015). If average summer moisture availability declines as a result of climate change, Bradley (2009) projected that the land area susceptible to *B. tectorum* (based on climatic suitability) may increase by up to 45 %, particularly in sagebrush steppe in Montana and higher elevation areas of the Colorado Plateau (Box 9.1).

Box 9.1

Figure 9.2 illustrates current and future distributions of *B. tectorum* predicted by Bradley (2009). Here, we used Maxent (Phillips et al. 2006), an implementation of maximum entropy modeling that predicts climatic suitability based on geographic occurrences of a species, to model climatic suitability for *B. tectorum*. Climatic suitability for occurrence implies that the combination of precipitation and temperature conditions at a given location support the germination, growth, and establishment of the species. We used mean temperature in the hottest and coldest months and quarterly precipitation derived from PRISM (averaged for 1951–2006) as our climatic predictors (Daly et al. 2002). Occurrence data were plots surveyed by the Southwest and Northwest Gap Analysis Programs (GAP) (2003–2005) with greater than 25 % reported *B. tectorum* cover. We applied a mask in Maxent to focus only on the surveyed region within the geographic range encompassing all GAP analysis points. We chose a model threshold value that encompassed 90 % of the *B. tectorum* training data with high abundance. This threshold also resulted in the correct identification of 76 % of testing data with *B. tectorum* absence according to the GAP analysis dataset. Loss of average summer precipitation or prolonged summer droughts could enable *B. tectorum* invasion into sagebrush steppe that is currently resistant to invasion and resilient to fire disturbance (Fig. 9.2).

(continued)

Box 9.1 (continued)

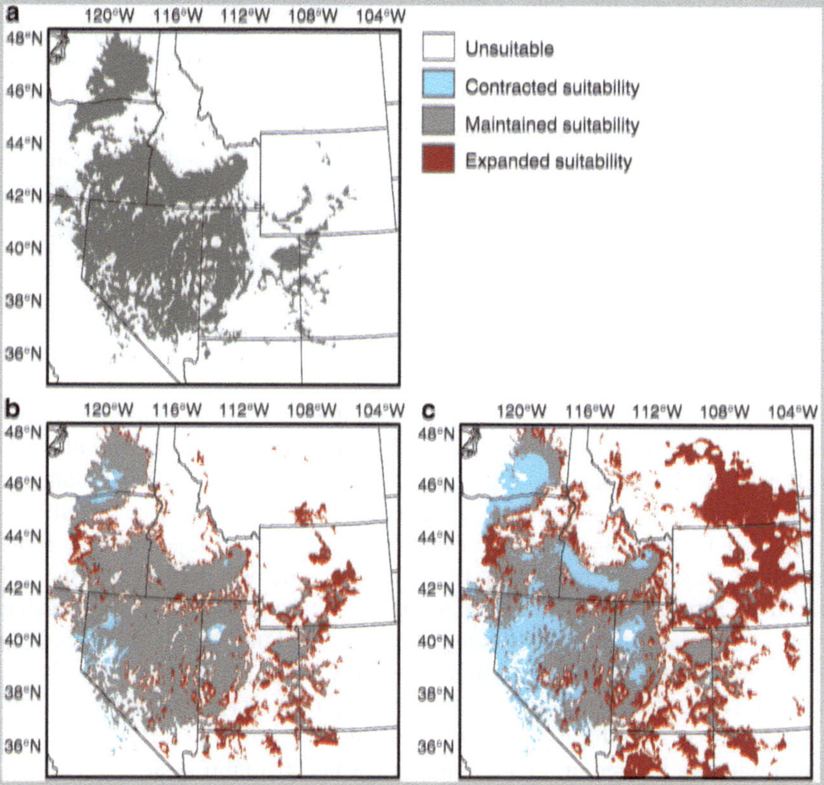

Fig. 9.2 Climatic suitability for high abundance (>25 % cover) of *B. tectorum* in the Intermountain West with varying summer precipitation, specifically (**a**) current, average climate conditions, (**b**) a 25 % loss of average summer precipitation, or (**c**) a 50 % loss of average summer precipitation. *White* remains unsuitable, *light blue* becomes unsuitable, *medium gray* remains suitable, and *dark red* gains suitability for *B. tectorum*

9.2.3 Winter Precipitation

Several studies have tested the response of *B. tectorum* germination and growth to winter snowfall (Griffith and Loik 2010; Concilio et al. 2013; Compagnoni and Adler 2014). At higher elevations, prolonged snowpack delayed *B. tectorum* emergence (Concilio et al. 2013) and reduced seedling survival for individuals that germinated in the fall (Griffith and Loik 2010; Compagnoni and Adler 2014), although these effects did not reduce overall biomass and seed production later in the spring

(Concilio et al. 2013). Colder temperatures in general and colder and wetter winters resulted in lower growth and seed production of *B. tectorum* across elevation gradients in Nevada and Utah (Chambers et al. 2007). Conversely, warmer temperatures increase *B. tectorum* germination rates, survival, growth, and reproduction (Thill et al. 1979; Gasch and Bingham 2006; Zelikova et al. 2013; Compagnoni and Adler 2014). But, the effect of warmer temperatures is only apparent with sufficient water availability (Thill et al. 1979; Zelikova et al. 2013; Compagnoni and Adler 2014). Projected winter warming (IPCC 2013) and loss of snowpack (Mote et al. 2005) at mid elevations are likely to favor *B. tectorum* under non-drought conditions.

9.2.4 Winter and Spring Temperature

Winter and spring temperatures are important predictors of rate and timing of spring germination. Colder temperatures generally decrease germination rates (Thill et al. 1979; Gasch and Bingham 2006; Roundy et al. 2007), plant establishment, and growth and reproduction of *B. tectorum* (Chambers et al. 2007). Spring soil temperature, based on accumulated degree days when soil water availability was sufficient for seed hydration, was the best predictor of the timing and rate of *B. tectorum* germination across elevation gradients in sagebrush ecosystems of Nevada and Utah (Roundy et al. 2007). Plant establishment and reproduction across the same elevation gradients were constrained by low soil temperatures at relatively high elevation (frigid to cryic soils), dependent on growing season conditions at mid elevation (frigid/mesic soils), and optimal under relatively moderate temperature and water availability at lower elevation (xeric/aridic to xeric/mesic soils) (Chambers et al. 2007). At lower elevations with warmer soils (warm-xeric), such as in salt desert ecosystems, constraints shift primarily to precipitation limitation (Meyer et al. 2001; Zelikova et al. 2013). Because *B. tectorum* has a fairly broad temperature tolerance range and high phenotypic plasticity, wide variation in timing of green-up can occur across its range that is dependent on local climates (Hulbert 1955).

In areas where *B. rubens* and *B. tectorum* germinate in the fall, cold winter temperatures could lead to mortality. Both species gain considerable cold temperature tolerance once hardened, so late fall cold snaps before hardening occurs are most likely to kill seedlings (O'Connor et al. 1991; Bykova and Sage 2012). *B. tectorum* has a freezing tolerance of between −30 °C (O'Connor et al. 1991) and −22 °C (Bykova and Sage 2012). *B. rubens* is relatively more susceptible to sudden cold extremes than *B. tectorum* (Bykova and Sage 2012), and cold winter temperatures likely limit *B. rubens*' northern range (Hulbert 1955) and elevational distribution (Salo 2005). Warmer winter temperatures will reduce the likelihood that either *B. tectorum* or *B. rubens* populations that germinate in the fall will experience mortality during the winter over much of their range. *Bromus tectorum* will likely spread upward in elevation, and *B. rubens* might expand northward and/or increase its abundance in the Great Basin Desert and Colorado Plateau (Curtis and Bradley 2015).

9.2.5 Future Range Shifts with Climate Change

Previous distribution modeling studies for *B.tectorum* have suggested that climate change could create large-scale restoration opportunities for invaded areas as climate conditions become unsuitable for the invasive annual grass (Bradley 2009; Bradley et al. 2009). Geographically, these opportunities are most likely at the hotter and drier edge of the species' range, although *B. tectorum* shows local adaptation to warmer range margins (Leger et al. 2009), which may enable persistence for longer than anticipated by climatic suitability models (e.g., Bradley 2009). However, *B. rubens* dominates under hotter and drier conditions characteristic of the Mojave and Sonoran Deserts and may well expand to fill any range that *B. tectorum* vacates. Indeed, while *B. tectorum* is currently more abundant across the southern Great Basin, *B. rubens* is already present at low levels of abundance at lower elevations across much of the Great Basin (Salo 2005) and may be able to spread if *B. tectorum* declines. Hence, rapid *B. rubens* invasion could be possible with altered climate conditions (Box 9.2).

Box 9.2

Figure 9.3 compares climatic suitability models for *B. tectorum* and *B. rubens* under current and future climate changes. Distribution data included over 1,900 locations with abundant (>25 %) *B. tectorum* cover recorded on GAP analysis plots and over 12,000 occurrences of *B. rubens* from a combination of surveys conducted during the 2000s (Southwest Exotic Plant Mapping Program, Nevada Natural Heritage Program, California Invasive Plant Council, Southwest Environmental Information Network, surveys by managers from the Mojave Network Parks). Distribution data for *B. tectorum* were locations with greater than 25 % cover, but points for *B. rubens* were all available occurrences because cover was not collected in any of the compiled survey data. We used Maxent (Phillips et al. 2006) to model current climatic suitability based on the temperature and precipitation variables from PRISM described above. For future conditions, we used an ensemble of average climate projections from 2040 to 2069 obtained from Climate Wizard (Girvetz et al. 2009) for the following climate models: CGCM 3.1-T47, CSIRO MK3.0, ECHAM/MPI-OM, GFDL CM2.1, IPSL CM4, and UKMO HadCM3. Model projections were based on the A2 emissions scenario from the IPCC Fourth

(continued)

Box 9.2 (continued)

Assessment Report, which is consistent with current emissions trajectories (Le Quere et al. 2009). We chose a threshold suitability value that encompassed 90 % of *B. tectorum* or *B. rubens* points and applied the same threshold to each future climate projection. Projections of future climatic suitability based on the climate model scenarios were combined to create a single model in which any pixel predicted to have suitable climate conditions by at least one climate model is considered potential habitat.

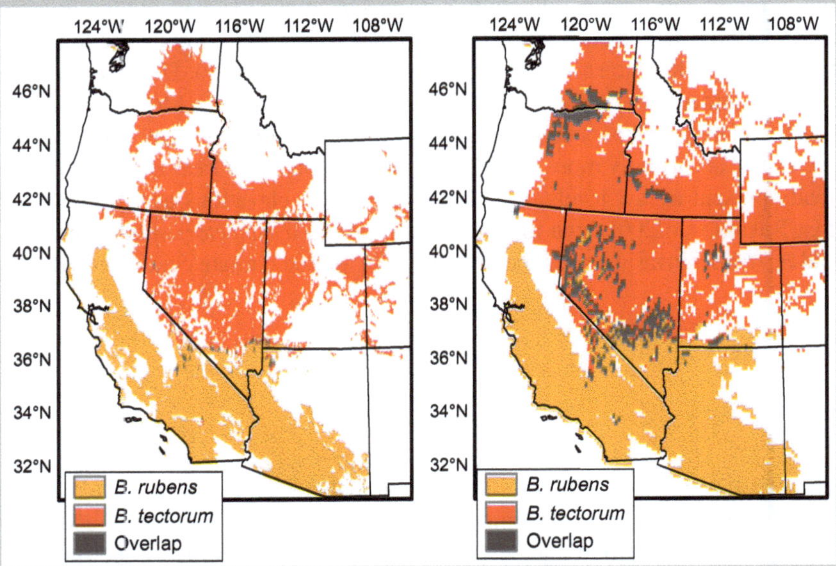

Fig. 9.3 Climatic suitability for *B. tectorum* and *B. rubens* under current (*left*) and future (*right*) climate conditions. *B. rubens* could expand northward into the Great Basin and Colorado Plateau with climate change, potentially negating any restoration opportunities in areas currently invaded by *B. tectorum*. Note: current climate conditions are interpolated to a 4 km spatial resolution, while future climate conditions are scaled to a 12 km spatial resolution

As illustrated by Box 2, climate change could result in a substantial expansion of *B. rubens* northward into the southern Great Basin Desert and Colorado Plateau. The potential area of overlap between the two species also expands considerably, from 2700 km^2 to 86,000 km^2, primarily in the drier portions of the Great Basin Desert. Based on this analysis, it appears unlikely that any restoration opportunities afforded by climate change will last for long. If *B. tectorum* retreats, *B. rubens* may occupy those vacated lands given sufficient available water.

9.2.6 Climate-Fire Feedbacks with Growing Season Precipitation

Invasion of *B. tectorum* and *B. rubens* in semiarid ecosystems greatly increases the amount and continuity of fine fuels and can result in more frequent and larger fires (Brooks 1999; Brooks et al. 2004, 2015). In uninvaded arid and semiarid shrublands where productivity is low, lack of continuous cover decreases the ability of fire to carry from plant to plant (Brooks 1999). The Mojave Desert and lower elevation shrubland in the cold desert historically experienced low fire frequencies. For example, charcoal sediments in low elevation *A. tridentata* shrubland suggest that on average fires burned every 1000 years (Mensing et al. 2006). Estimates from Balch et al. (2013) found similar fire frequencies in sagebrush ecosystems by using remotely sensed data to measure more recent fire frequency. They calculated fire return intervals (i.e., the length of time between repeated burning of the same location) for sagebrush across the Great Basin of between 100 and 300 years. Fires in *B. tectorum*-invaded ecosystems were larger and more frequent, with fire return intervals averaging between 50 and 80 years regionally (Balch et al. 2013) and even more frequently in highly invaded landscapes (Whisenant 1990).

Fires in *B. tectorum*- and *B. rubens*-invaded landscapes are strongly linked to climate and short-term weather conditions and, therefore, are likely to be influenced by climate change. In forested ecosystems, hot and dry weather during the fire season leads to drying of fuels and increased fire frequency (Westerling et al. 2006; Littell et al. 2009). But in low productivity shrubland and desert, fire activity is influenced more by the production of fine fuels, and significant buildup of annual grass and fine fuels occurs in years with wet winters and springs. As a result, the best predictor of fire size and frequency in low productivity shrubland may be wet winters in the year or two prior to the fire (Knapp 1998; Littell et al. 2009; Abatzoglou and Kolden 2013). Similarly, Balch et al. (2013) found that the frequency and size of fires on *B. tectorum*-dominated land cover were correlated with precipitation during the preceding winter. These findings suggest that wet winter climatic conditions at lower elevations where *B. tectorum* is most problematic promote *Bromus* biomass production, increase fine fuels, and will also promote fires. Fires, in turn, result in mortality of many of the dominant native shrub species and an increase in resource availability that can result in increased growth and reproduction of *Bromus* and further enable its expansion (Chambers et al. 2014a), especially in drier regions where shrub recovery is less likely to recover (Taylor et al. 2014).

Climate changes that bring wetter winters to highly invaded regions, as predicted for the Northern Great Basin, are therefore likely to strengthen the fire-*Bromus* feedback, especially under relatively warm conditions. Consistently higher *Bromus* biomass could lead to more frequent and larger fires, facilitating the spread of these exotic invasive grasses. More frequent *Bromus*-driven fires in the northern Intermountain West are also likely if wet winters are accompanied by warmer summers with reduced precipitation, as forecasted with climate change (Diffenbaugh et al. 2008; Abatzoglou and Kolden 2011; IPCC 2013). Conversely, drier winters, or less frequent anomalously wet years in the southwest, could weaken the fire-Bromus feedback.

9.3 Management Implications

The combination of observations, experimental, and modeling studies is increasing our scientific knowledge of *Bromus* species and native community response to climate. However, uncertainty in climate projections coupled with the heterogeneous landscapes of the Intermountain West makes location-specific forecasts a challenge. Management decisions must embrace multiple possible future climate pathways and rely on adaptive management to adjust responses appropriately. Part of *Bromus* response to climate change will depend on the species physiology and life history traits directly, and part will depend on the response of native competitors. Loss of native vegetation increases available resources for *Bromus* species' growth (Roundy et al. 2007; Prevéy et al. 2010) and enables faster dispersal (Johnston 2011). Hence, managing for native competitors and reducing stress and disturbance could limit *Bromus* dispersal and establishment.

Vegetation changes in response to climate are most likely to occur first at the margins of their distribution. For example, *Bromus* species may become more competitive, and therefore more abundant, at higher elevations and latitudes as temperatures warm, while native species may become less competitive and more susceptible to climate extremes at their lower elevation and latitudinal ranges. Monitoring mortality of native perennial species as well as cover of *Bromus* at range margins may give advanced warning of ongoing range shifts and enable adaptive management.

Resilient native ecosystems are important for increasing resistance to *Bromus* invasion (Chambers et al. 2014a). Management actions that promote native species diversity and abundance under changing climate conditions could reduce invasion rates. But historical rates of native plant migration have been estimated to be only 10–30 km per century (McLachlan et al. 2005; Yansa 2006), and many native species may not be able to expand into newly suitable climate within the short period of time in which climate changes are likely to occur (ca. 40–50 years). If native species are unable to colonize newly suitable areas, assisted migration is one possible solution (McLachlan et al. 2007; Richardson et al. 2009; Vitt et al. 2010). Assisted migration can be defined as the purposeful movement of individuals or propagules of a species to facilitate or mimic natural range expansion or long-distance gene flow within the current range, as a direct management response to climate change (Havens et al. 2015). Plant sources adapted to the new areas would need to be used to ensure successful assisted migration. Also, soil conditions and other environmental characteristics would need to be suitable to the new species (Richardson et al. 2014).

9.4 Research Needs

While a number of studies have focused on the physiological tolerance of *Bromus* species (primarily *B. tectorum*) to climate conditions, fewer studies have considered climate effects on interactions of *Bromus* with native species. Changes in native plants' germination, growth, and mortality rates due to climate will alter overall ecosystem resistance to invasion and resilience following drought events and/or fires. Experimental and modeling studies that test biotic interactions across existing environmental gradients (e.g., Brooks et al. 2015) might provide better insight about overall invasion risk with climate change than studies of *Bromus* independently.

Important information about likely species' response to climate change will come from long-term monitoring sites, experimental studies, and observations across environmental gradients. The high interannual variability of climate conditions across the west creates a challenge for forecasting likely species' response. Many short-to-intermediate-term studies report very different responses of *Bromus* species from 1 year to the next due to different precipitation conditions (e.g., West and Yorks 2002; Chambers et al. 2014b), which makes it difficult to infer longer-term patterns. Prioritizing longer-term, multi-year experiments will improve our ability to project future ecological changes. Further, the importance of collecting and analyzing vegetation response data from long-term monitoring sites at university experimental stations, national parks, US Forest Service monitoring sites, and other public lands cannot be overstated, as they will be critical for understanding vegetation response across a broader range of climatic conditions.

References

Abatzoglou JT, Kolden CA (2011) Climate change in western US deserts: potential for increased wildfire and invasive annual grasses. Rangel Ecol Manag 64:471–478

Abatzoglou JT, Kolden CA (2013) Relationships between climate and macroscale area burned in the western United States. Int J Wildland Fire 22:1003–1020

Adair EC, Burke IC, Lauenroth WK (2008) Contrasting effects of resource availability and plant mortality on plant community invasion by *Bromus tectorum* L. Plant Soil 304:103–115

Balch JK, Bradley BA, D'Antonio CM et al (2013) Introduced annual grass increases regional fire activity across the arid western USA (1980–2009). Glob Change Biol 19:173–183

Barnett TP, Adam JC, Lettenmaier DP (2005) Potential impacts of a warming climate on water availability in snow-dominated regions. Nature 438:303–309

Beatley JC (1966) Ecological status of introduced brome grasses (*Bromus* spp.) in desert vegetation of southern Nevada. Ecology 47:548–554

Beatley JC (1974) Phenological events and their environmental triggers in Mojave Desert ecosystems. Ecology 55:856–863

Bradford JB, Lauenroth WK (2006) Controls over invasion of *Bromus tectorum*: the importance of climate, soil, disturbance and seed availability. J Veg Sci 17:693–704

Bradley BA (2009) Regional analysis of the impacts of climate change on cheatgrass invasion shows potential risk and opportunity. Glob Change Biol 15:196–208

Bradley BA (2010) Assessing ecosystem threats from global and regional change: hierarchical modeling of risk to sagebrush ecosystems from climate change, land use and invasive species in Nevada, USA. Ecography 33:198–208

Bradley BA (2013) Distribution models of invasive plants over-estimate potential impact. Biol Invasions 15:1417–1429

Bradley BA, Blumenthal DM, Wilcove DS et al (2010) Predicting plant invasions in an era of global change. Trends Ecol Evol 25:310–318

Bradley BA, Mustard JF (2005) Identifying land cover variability distinct from land cover change: cheatgrass in the Great Basin. Remote Sens Environ 94:204–213

Bradley BA, Oppenheimer M, Wilcove DS (2009) Climate change and plant invasions: restoration opportunities ahead? Glob Change Biol 15:1511–1521

Brooks ML (1999) Alien annual grasses and fire in the Mojave Desert. Madroño 46:13–19

Brooks ML, Belnap J, Brown CS et al (2015) Exotic annual *Bromus* invasions – comparisons among species and ecoregions in the western United States. In: Germino MJ, Chambers JC, Brown CS (eds) Exotic brome-grasses in arid and semiarid ecosystems of the Western USA: causes, consequences, and management implications. Springer, New York, NY (Chapter 2)

Brooks ML, Chambers JC (2011) Resistance to invasion and resilience to fire in desert shrublands of North America. Rangel Ecol Manag 64:431–438

Brooks ML, D'Antonio CM, Richardson DM et al (2004) Effects of invasive alien plants on fire regimes. Bioscience 54:677–688

Brooks ML, Matchett JR (2003) Plant community patterns in burned and unburned blackbrush (*Coleogyne ramossisima* Torr.) shrublands in the Mojave Desert. Western NA Natur 63:283–298

Bykova O, Sage RF (2012) Winter cold tolerance and the geographic range separation of *Bromus tectorum* and *Bromus rubens*, two severe invasive species in North America. Glob Change Biol 18:3654–3663

Chambers JC, Bradley BA, Brown CS et al (2014a) Resilience to stress and disturbance, and resistance to *Bromus tectorum* L. invasion in cold desert shrublands of western North America. Ecosystems 17:360–375

Chambers JC, Miller RF, Board DI et al (2014b) Resilience and resistance of sagebrush ecosystems: implications for state and transition models and management treatments. Rangel Ecol Manag 67:440–454

Chambers JC, Roundy BA, Blank RR et al (2007) What makes Great Basin sagebrush ecosystems invasible by *Bromus tectorum*? Ecol Monogr 77:117–145

Chambers JC, Germino MJ, Belnap J et al (2015) Plant community resistance to invasion by *Bromus* species – the roles of community attributes, *Bromus* interactions with plant communities, and *Bromus* traits. In: Germino MJ, Chambers JC, Brown CS (eds) Exotic brome-grasses in arid and semiarid ecosystems of the Western USA: causes, consequences, and management implications. Springer, New York, NY (Chapter 10)

Chen IC, Hill JK, Ohlemueller R et al (2011) Rapid range shifts of species associated with high levels of climate warming. Science 333:1024–1026

Compagnoni A, Adler PB (2014) Warming, soil moisture, and loss of snow increase *Bromus tectorum*'s population growth rate. Elem Sci Anth 2:000020

Concilio AL, Loik ME, Belnap J (2013) Global change effects on *Bromus tectorum* L. (Poaceae) at its high-elevation range margin. Glob Change Biol 19:161–172

Curtis CA, Bradley BA (2015) Climate change may alter both establishment and high abundance of red brome (*Bromus rubens*) and African mustard (*Brassica tournefortii*) in the semiarid southwest United States. Invasive Plant Sci Manag 8:341–352

Daly C, Gibson WP, Taylor GH et al (2002) A knowledge-based approach to the statistical mapping of climate. Climate Res 22:99–113

Diffenbaugh NS, Giorgi F, Pal JS (2008) Climate change hotspots in the United States. Geophys Res Lett 35:63–87

Gasch C, Bingham R (2006) A study of *Bromus tectorum* L. seed germination in the Gunnison Basin, Colorado. Bios 77:7–12

Girvetz EH, Zganjar C, Raber GT et al (2009) Applied climate-change analysis: the climate wizard tool. PLoS One 4(12), e8320

Griffith AB, Loik ME (2010) Effects of climate and snow depth on *Bromus tectorum* population dynamics at high elevation. Oecologia 164:821–832

Havens K, Vitt P, Still S et al (2015) Seed sourcing for restoration in an era of climate change. Nat Area J 35:122–133

Hulbert LC (1955) Ecological studies of *Bromus tectorum* and other annual bromegrasses. Ecol Monogr 25:181–213

IPCC (2013) Climate Change 2013: The physical science basis: working group I contribution to the fifth assessment report of the intergovernmental panel on climate change. Cambridge University Press, Cambridge, UK and New York

James JJ, Drenovsky RE, Monaco TA et al (2011) Managing soil nitrogen to restore annual grass-infested plant communities: effective strategy or incomplete framework. Ecol Appl 21:490–502

Johnston DB (2011) Movement of weed seeds in reclamation areas. Restor Ecol 19:446–449

Jones R, Chambers JC, Johnson DW et al (2014) Effect of repeated burning on plant and soil carbon and nitrogen in cheatgrass (*Bromus tectorum*) dominated ecosystems. Plant Soil 386:47–64

Knapp PA (1998) Spatio-temporal patterns of large grassland fires in the Intermountain West, USA. Glob Ecol Biogeogr 7:259–272

Le Quere C, Raupach MR, Canadell JG et al (2009) Trends in the sources and sinks of carbon dioxide. Nat Geosci 2:831–836

Leger EA, Espeland EK, Merrill KR et al (2009) Genetic variation and local adaptation at a cheatgrass (*Bromus tectorum*) invasion edge in western Nevada. Mol Ecol 18:4366–4379

Littell JS, McKenzie D, Peterson DL et al (2009) Climate and wildfire area burned in western US ecoprovinces, 1916–2003. Ecol Appl 19:1003–1021

Mack RN, Pyke DA (1983) The demography of *Bromus tectorum*: variation in space and time. J Ecol 71:69–93

Mack RN, Pyke DA (1984) The demography of *Bromus tectorum*: the role of microclimate, grazing and disease. J Ecol 72:731–748

McLachlan JS, Clark JS, Manos PS (2005) Molecular indicators of tree migration capacity under rapid climate change. Ecology 86:2088–2098

McLachlan JS, Hellmann JJ, Schwartz MW (2007) A framework for debate of assisted migration in an era of climate change. Conserv Biol 21:297–302

Mensing S, Livingston S, Barker P (2006) Long-term fire history in Great Basin sagebrush reconstructed from macroscopic charcoal in spring sediments, Newark Valley, Nevada. West N Am Nat 66:64–77

Meyer SE, Garvin SC, Beckstead J (2001) Factors mediating cheatgrass invasion of intact salt desert shrubland. In: McArthur EDF, Daniel J (eds) Shrubland ecosystem genetics and biodiversity, 13–15 June 2000, Provo, UT. USDA, Forest Service, Rocky Mountain Research Station, Ogden, UT, pp 224–232

Monaco TA, Johnson DA, Norton JM et al (2003) Contrasting responses of Intermountain West grasses to soil nitrogen. J Range Manag 5:282–290

Morisette JT, Richardson AD, Knapp AK et al (2009) Tracking the rhythm of the seasons in the face of global change: phenological research in the 21st century. Front Ecol Environ 7:253–260

Mote PW, Hamlet AF, Clark MP et al (2005) Declining mountain snowpack in western North America. Bull Am Meteorol Soc 86:39–49

O'Connor BJ, Paquette SP, Gusta LV (1991) A comparison of the freezing tolerance of downy brome, Japanese brome and norstar winter-wheat. Can J Plant Sci 71:565–569

Parmesan C, Yohe G (2003) A globally coherent fingerprint of climate change impacts across natural systems. Nature 421:37–42

Phillips SJ, Anderson RP, Schapire RE (2006) Maximum entropy modeling of species geographic distributions. Ecol Model 190:231–259

Prevey JS, Germino MJ, Huntly NJ et al (2010) Exotic plants increase and native plants decrease with loss of foundation species in sagebrush steppe. Plant Ecol 207:39–51

Pyke DA, Beck JL, Brooks ML et al (2015) Land uses, fire, and invasion – exotic annual *Bromus* and human dimensions. In: Germino MJ, Chambers JC, Brown CS (eds) Exotic brome-grasses in arid and semiarid ecosystems of the Western USA: causes, consequences, and management implications. Springer, New York, NY (Chapter 11)

Rehfeldt GE, Crookston NL, Cuauhtemoc Saenz-Romero et al (2012) North American vegetation model for land use planning in a changing climate: a solution to large classification problems. Ecol Appl 22:119–141

Richardson BA, Kitchen SG, Pendleton RL et al (2014) Adaptive responses reveal contemporary and future ecotypes in a desert shrub. Ecol Appl 24:413–427

Richardson DM, Hellmann JJ, McLachlan JS et al (2009) Multidimensional evaluation of managed relocation. Proc Natl Acad Sci U S A 106:9721–9724

Root TL, Price JT, Hall KR et al (2003) Fingerprints of global warming on wild animals and plants. Nature 421:57–60

Roundy BA, Hardegree SP, Chambers JC et al (2007) Prediction of cheatgrass field germination potential using wet thermal accumulation. Rangel Ecol Manag 60:613–623

Salo LF (2005) Red brome (*Bromus rubens* subsp. *madritensis*) in North America: possible modes for early introductions, subsequent spread. Biol Invasions 7:165–180

Salo OE, Lauenroth WK, Galluscio RA (1997) Plant functional types in semi-arid regions. In: Smith TM, Shugart HH, Woodward FI (eds) Plant functional types: their relevance to ecosystem properties and global change. Cambridge University press, Cambridge, MA.

Schlaepfer DR, Lauenroth WK, Bradford JB (2012) Effects of ecohydrological variables on current and future ranges, local suitability patterns, and model accuracy in big sagebrush. Ecography 35:374–384

Seager R, Ting M, Held I et al (2007) Model projections of an imminent transition to a more arid climate in southwestern North America. Science 316:1181–1184

Smith SD, Huxman TE, Zitzer SF et al (2000) Elevated CO_2 increases productivity and invasive species success in an arid ecosystem. Nature 408:79–82

Smith SD, Strain BR, Sharkey TD (1987) Effects of CO_2 enrichment on four great basin grasses. Funct Ecol 1:139–143

Still SM, Richardson BA (2015) Projections of contemporary and future climate niche for Wyoming big sagebrush (*Artemisia tridentata* subsp. *wyomingensis*): a guide for restoration. Nat Area J 35:30–43

Taylor K, Brummer T, Rew LJ et al (2014) *Bromus tectorum* response to fire varies with climate conditions. Ecosystems 17:960–973

Thill D, Schirman R, Appleby A (1979) Influence of soil moisture, temperature, and compaction on the germination and emergence of downy brome (*Bromus tectorum*). Weed Sci 27:625–630

USDA, NRCS (2014) The PLANTS Database (http://plants.usda.gov). National Plant Data Team, Greensboro, NC 27401-4901 USA

van Vuuren DP, Edmonds J, Kainuma M et al (2011) The representative concentration pathways: an overview. Clim Change 109:5–31

Vitt P, Havens K, Kramer AT et al (2010) Assisted migration of plants: changes in latitudes, changes in attitudes. Biol Conserv 143:18–27

West NE, Yorks TP (2002) Vegetation responses following wildfire on grazed and ungrazed sagebrush semi-desert. J Range Manag 55:171–181

Westerling AL, Hidalgo HG, Cayan DR et al (2006) Warming and earlier spring increase western US forest wildfire activity. Science 313:940–943

Whisenant S (1990) Changing fire frequencies on Idaho's Snake River plains: ecological and management implications. In: McArthur ED, Romney, Smith SD et al (eds) Symposium on cheatgrass invasion, shrub die-off, and other aspects of shrub biology and management, 5–7 April 1989. Gen Tech Rep INT-276. USDA, Forest Service, Intermountain Research Station, Ogden, UT, pp 4–10

Willis CG, Ruhfel B, Primack RB et al (2008) Phylogenetic patterns of species loss in Thoreau's woods are driven by climate change. Proc Natl Acad Sci U S A 105:17029–17033

Yansa CH (2006) The timing and nature of late quaternary vegetation changes in the northern Great Plains, USA and Canada: a re-assessment of the spruce phase. Quat Sci Rev 25:263–281

Zelikova TJ, Hufbauer RA, Reed SC et al (2013) Eco-evolutionary responses of *Bromus tectorum* to climate change: implications for biological invasions. Ecol Evol 3:1374–1387

Ziska LH, Reeves JB, Blank B (2005) The impact of recent increases in atmospheric CO_2 on biomass production and vegetative retention of cheatgrass (*Bromus tectorum*): implications for fire disturbance. Glob Change Biol 11:1325–1332

Chapter 10
Plant Community Resistance to Invasion by *Bromus* Species: The Roles of Community Attributes, *Bromus* Interactions with Plant Communities, and *Bromus* Traits

Jeanne C. Chambers, Matthew J. Germino, Jayne Belnap, Cynthia S. Brown, Eugene W. Schupp, and Samuel B. St. Clair

Abstract The factors that determine plant community resistance to exotic annual *Bromus* species (*Bromus* hereafter) are diverse and context specific. They are influenced by the environmental characteristics and attributes of the community, the traits of *Bromus* species, and the direct and indirect interactions of *Bromus* with the plant community. Environmental factors, in particular ambient and soil temperatures, have significant effects on the ability of *Bromus* to establish and spread. Seasonality of precipitation relative to temperature influences plant community

J.C. Chambers (✉)
USDA Forest Service, Rocky Mountain Research Station,
Reno, NV 89512, USA
e-mail: jchambers@fs.fed.us

M.J. Germino
US Geological Survey, Forest and Rangeland Ecosystem Science Center,
Boise, ID 83706, USA
e-mail: mgermino@usgs.gov

J. Belnap
US Geological Survey, Southwest Biological Science Center, Moab, UT 84532, USA
e-mail: jayne_belnap@usgs.gov

C.S. Brown
Department of Bioagricultural Sciences and Pest Management, Colorado State University,
Fort Collins, CO 80523-1177, USA
e-mail: cynthia.s.brown@colostate.edu

E.W. Schupp
Department of Wildland Resources, Utah State University, Logan, UT 84322-5230, USA
e-mail: eugene.schupp@usu.edu

S.B. St. Clair
Plant and Wildlife Sciences, Brigham Young University, Provo, UT 84602, USA
e-mail: stclair@byu.edu

M.J. Germino et al. (eds.), *Exotic Brome-Grasses in Arid and Semiarid Ecosystems of the Western US*, Springer Series on Environmental Management,
DOI 10.1007/978-3-319-24930-8_10

resistance to *Bromus* through effects on soil water storage, timing of water and nutrient availability, and dominant plant life forms. Differences among plant communities in how well soil resource use by the plant community matches resource supply rates can influence the magnitude of resource fluctuations due to either climate or disturbance and thus the opportunities for invasion. The spatial and temporal patterns of resource availability and acquisition of growth resources by *Bromus* versus native species strongly influence resistance to invasion. Traits of *Bromus* that confer a "priority advantage" for resource use in many communities include early-season germination and high growth and reproductive rates. Resistance to *Bromus* can be overwhelmed by high propagule supply, low innate seed dormancy, and large, if short-lived, seed banks. Biological crusts can inhibit germination and establishment of invasive annual plants, including several annual *Bromus* species, but are effective only in the absence of disturbance. Herbivores can have negative direct effects on *Bromus*, but positive indirect effects through decreases in competitors. Management strategies can be improved through increased understanding of community resistance to exotic annual *Bromus* species.

Keywords Ecosystem resilience • Plant traits • Resource dynamics • Species interactions • Herbivory

10.1 Introduction

The resistance or conversely susceptibility of plant communities to exotic annual *Bromus* species (*Bromus* hereafter) and other invasive species is a function of environmental factors, community attributes and processes, life history and ecophysiological traits of the invader, and direct and indirect interactions of the invader (see reviews in Chesson 2000; Shea and Chesson 2002). Resistance to invasion varies both spatially and temporally (Chesson 2000; Seabloom et al. 2003) and is closely related to the resilience of a community (Chambers et al. 2014a). Resilience is the capacity of a community to *regain* its fundamental structure, processes, and functioning when altered by stressors like drought, and disturbances like inappropriate grazing and altered fire regimes (Holling 1973; Folke 2006); resistance is the capacity of a community to *retain* its fundamental structure, processes, and functioning despite stresses, disturbances, or invasive species (Folke et al. 2004). While resilience is a measure of the recovery potential of a community following stress or disturbance (Chambers et al. 2014a), resistance to invasion is a measure of the capacity of a community to limit the population growth of an invading species (D'Antonio and Thomsen 2004). The relationships of environmental factors, including climate, topography, and soils, to the abiotic and biotic attributes of communities, and to resilience to disturbance and resistance to *Bromus*, are illustrated in Fig. 1.1 (Germino et al. 2015).

In this chapter, we examine the interactions of four focal, invasive annual *Bromus* species—*B. diandrus* Roth (ripgut brome), *B. hordeaceus* L. (soft brome), *B. rubens*

L. (red brome), and *B. tectorum* L. (cheatgrass or downy brome)—with plant communities in the Western USA. We emphasize the factors that influence plant community resistance to invasion, specifically, environmental characteristics, community attributes and processes, life history and ecophysiological traits of the invader, and direct and indirect interactions of the invader with the community. We integrate the role of disturbance and stressors in influencing community interactions with *Bromus* into each section. We address the implications for management and research needs in the final sections.

10.2 Effects of Environmental Factors on Plant Community Resistance to *Bromus*

Resistance of plant communities to *Bromus* is strongly affected by environmental factors such as climate and soils. Environmental factors affect (1) the ability of *Bromus* to establish and persist, (2) the interactions of *Bromus* with the plant community, and (3) the attributes of plant communities such as life form dominance and productivity that determine their capacity to resist *Bromus* invasion. *Bromus* species differ in the climatic regimes where they can establish and persist, and these differences are well illustrated by soil temperature and moisture regimes (Brooks et al. 2015). For example, resistance to *B. tectorum* in the cold desert varies strongly over elevation gradients. *Bromus tectorum* germination, growth, and/or reproduction is limited in relatively hot and dry environments by low and sporadic precipitation (thermic to xeric/aridic soils), constrained in cold and moist environments by low soil temperatures (frigid to cryic soils), and optimal under relatively moderate temperature and water availability (xeric/aridic to xeric/mesic soils) (Meyer et al. 2001; Chambers et al. 2007; Leger et al. 2009). In contrast, *B. rubens* L. (red brome) is less cold tolerant than *B. tectorum*, and *B. rubens* establishment and persistence occurs primarily at lower latitudes and altitudes characterized by warmer environments (thermic to xeric soils) (Salo 2005; Bykova and Sage 2012).

At the scale of the western USA, differences in overlap between seasonality of precipitation and temperature, and onset of the dry season have important consequences for plant functional type dominance (Sala et al. 1997), and competitive interactions with *Bromus* (Bradford and Lauenroth 2006). Amount of precipitation that is received during the period when temperature, and thus potential evapotranspiration, is low influences the amount of water stored in deep soil layers and therefore the balance between woody and herbaceous species (Sala et al. 1997). Areas that receive more winter/spring precipitation typically have greater deep soil water storage and are dominated by woody species, which are more effective at using deep soil water (Fig. 10.1a, b). In contrast, areas that receive predominantly summer precipitation are dominated by grasses. Also, seasonality of wet soil during the period when temperatures are favorable for plant growth is an important control on the balance between C3 and C4 species with C3 species dominating in areas with cool, wet springs and C4 species dominating in areas with warm, wet summers

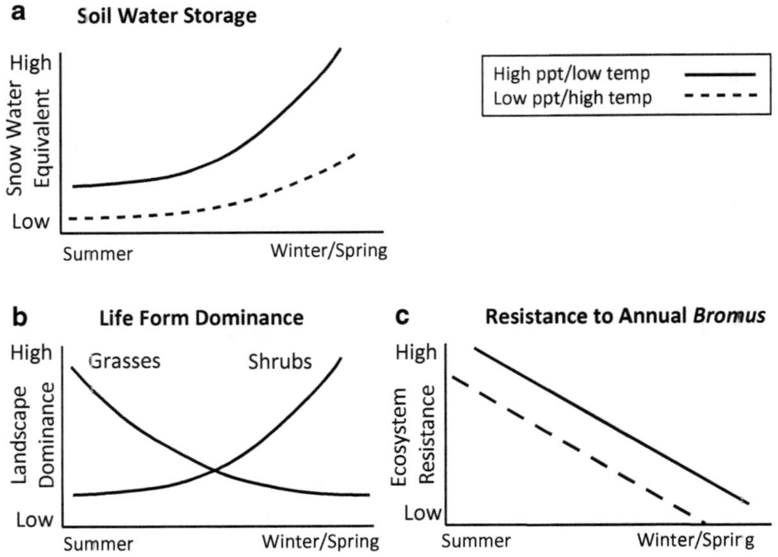

Fig. 10.1 Changes in soil water storage, life form dominance, and resistance to annual *Bromus* as seasonality of precipitation transitions from primarily summer to winter. (**a**) Soil water storage increases as winter/spring precipitation and snow water equivalent increase and these changes are relatively greater for areas with relatively high precipitation and low temperature. (**b**) Landscape dominance of grasses is highest with primarily summer precipitation; shrub dominance is greatest with primarily winter/spring precipitation. (**c**) Resistance to *Bromus* is higher in areas where soil water storage is low and grasses dominate largely due to strong resource competition. Decreases in effective precipitation can increase resource fluctuations and lower resistance to *Bromus* At more local scales, resistance also is influenced by nutrient availability and disturbance

(Paruelo and Lauenroth 1996; Sala et al. 1997). Resistance to *Bromus* generally increases with increasing summer precipitation (Fig. 10.1c) as a function of increasing grass dominance. This appears to be due to less favorable conditions for establishment of *Bromus* and strong competition from grass species that dominate under this precipitation regime (Bradford and Lauenroth 2006; Bradley 2009). *Bromus* also may be constrained in regions with predominant summer precipitation due to low soil phosphorus availability (Belnap et al. 2015).

At more local scales, resistance to *Bromus* varies over environmental gradients. Increases in effective precipitation are typically associated with greater resource availability and more favorable environmental conditions for plant growth and reproduction. Higher resource availability and plant productivity are associated with higher resilience to disturbance or recovery potential in the cold desert (Chambers et al. 2014a). These conditions translate into greater plant community resistance to *B. tectorum* on more mesic sites in the cold desert (Chambers et al. 2014b) and to *B. diandrus* and *B. hordeaceus* in California grasslands (Corbin et al. 2007). Resistance is likely decreased by low precipitation or high temperatures, which decrease soil water availability and plant productivity regardless of seasonality of precipitation. Low effective precipitation can result in increases in water and nutrient resource fluctuations and decreased resistance to invasion in arid and semiarid ecosystems (Davis et al. 2000). This has been observed following El Niño years for *B. rubens* in the Mojave Desert (Salo 2005) and *B. tectorum* in salt desert ecosystems of the cold desert (Meyer et al. 2001), as well as over elevation gradients in the central Basin and Range (Chambers et al. 2007).

10.3 Interactions of *Bromus* with Plant Communities

10.3.1 **Bromus** *Plant Traits and Resource Dynamics*

Community resistance to *Bromus* invasion is strongly influenced by spatial and temporal patterns of resource availability that result from differences in temperature and precipitation regimes, and by effects of these patterns on acquisition or use of growth resources by *Bromus* and natives. Like many invasive annuals, *Bromus* is generally more effective than native species at using limiting soil resources on short timescales largely due to its life history strategies and rapid growth rates. Soil resource uptake can vary between *Bromus* and competitors as a function of differences in root growth and surface area, rooting depth, spatial distribution or timing of activity, uptake capacity, or resource use efficiency (RUE) (see Smith et al. 1997; Leffler and Ryel 2012).

In communities dominated by winter/spring precipitation, which often exhibit the greatest impacts of *Bromus*, the spring growth period and shallow soil layers are the most significant factors for nutrient uptake and growth of all species, including *Bromus*. Nutrients are typically concentrated in shallow soils, and soil water availability that is sufficient for mineralization and transport of nutrients to plant roots occurs at these depths primarily in spring (Ryel et al. 2008). High growth rates, early-season root growth, and use of this shallow resource pool are important aspects of the "seasonal priority advantage" of *Bromus* compared to a wide range of native perennials, but competitive outcomes depend on life form, life stage, and season of growth. For example, although *B. tectorum* roots occur at depths of nearly 2 m in some situations, *B. tectorum* root densities are typically highest in the top 20–30 cm of soil (see Thill et al. 1984). Roots of *Bromus* tend to grow more rapidly and at lower temperatures than those of native species, as shown for *B. tectorum* compared to bunchgrasses (Aguirre and Johnson 1991). Consequently, *Bromus* has relatively higher water use from shallow soils (e.g., 0–30 cm vs. 30–200 cm depth) than many native perennials, which often have deeper roots and also extract soil water from greater depths (see Wilcox et al. 2012; Leffler and Ryel 2012). *Bromus tectorum* has been shown to compete effectively with woody plants such as *Chrysothamnus* Nutt. spp. (rabbitbrush) and *Artemisia tridentata* Nutt. (big sagebrush) for near-surface soil water and nutrients (e.g., nitrogen and phosphorus) during spring (Melgoza et al. 1990; Booth et al. 2003). However, uptake of soil water by deeper-rooted woody plants from greater depths during summer drought can facilitate their persistence (Leffler and Ryel 2012). This growth/maintenance resource pool helps explain coexistence of *B. tectorum* and woody species in areas with deep-water storage, but intensity of competitive interactions likely depends on temperature and precipitation regimes and shrub rooting depths. Use of shallow soil water by *B. tectorum* also helps explain its coexistence with deep-rooted exotic forbs in areas with deep-water storage as shown in *Bromus*-dominated areas using soil and plant water balance (Hill et al. 2006), in shrub steppe with winter rainout shelters (Prevey et al. 2010) and stable isotope tracers of soil water use (Kulmatiski et al. 2006), and in California grasslands with water balance studies (Enloe et al. 2004).

Mature, perennial grasses with relatively high densities of roots in shallow soils and with similar phenologies are typically strong competitors with *Bromus* for shallow resource pools as shown for *B. tectorum* (Booth et al. 2003; Blank and Morgan 2013). However, competitiveness of *Bromus* against these grasses appears to increase with soil nitrogen as illustrated in the central Basin and Range (Beckstead and Augspurger 2004), short-grass steppe (Lowe et al. 2003), and California grasslands (Corbin et al. 2007). Also, *B. tectorum* and other *Bromus* often preempt establishment of shallow-rooted native seedlings. A meta-analysis of studies that examined effects of nitrogen manipulation (fertilizer, straw, sawdust) on seedling growth of annual *Bromus* and native perennial species in a wide variety of ecological types showed that *Bromus* maintained higher growth rates, biomass, and tiller production than natives under both low and high nitrogen availability (James et al. 2011). Greenhouse studies with *B. tectorum* and *B. rubens* further showed that *Bromus* tends to have greater root length and root biomass (James et al. 2008) and nitrogen content (Defalco et al. 2003; Monaco et al. 2003) than seedlings of native species over a range of nitrogen availability if other nutrients are not limiting. Effects of phosphorus and potassium on *Bromus* interactions with native species are detailed in Belnap et al. (2015).

Aboveground canopy and litter effects also influence interactions of *Bromus* with its plant community. *Bromus* are high-light adapted and generally exhibit low acclimation and tolerance of shade under dense overhead canopies (Pierson et al. 1990). Overhead canopy and standing litter can have either negative or positive effects on *B. tectorum*, depending on factors such as weather and species (Newingham et al. 2007; Bansal et al. 2014). Where *B. diandrus* and *B. hordeaceus* dominate community cover, increased competition for light during spring when growth of annuals is most rapid suppresses growth and reproduction of perennial native species (Dyer and Rice 1999). Litter accumulation in *Bromus*-dominated areas appears to have a positive feedback effect on *B. tectorum* growth through decreases in soil water loss, changes in nutrient cycling, and effects on seed entrapment and germination (e.g., Bansal et al. 2014). *Bromus* litter can promote emergence of *B. tectorum* on smooth, hard surfaces of fine-textured soils (Young and Evans 1975). However, thick *Bromus* litter likely prevents germination and establishment of small-seeded native species (Chambers 2000).

10.3.2 Interactions of **Bromus** *with Perennial Grasses*

Competitiveness of established perennial grasses with *Bromus* differs among and within ecoregions as a function of environmental attributes and resource availability and is strongly affected by disturbance. Invasive *Bromus* cover is typically negatively related to the amount of cover of native species, such as grasses of the cold desert and western Great Plains. An inverse relationship of *B. tectorum* to native or naturalized bunchgrass cover (Reisner et al. 2013; Chambers et al. 2014b) and increases in *B. tectorum* biomass upon experimental reduction of native perennial herbaceous species have been observed across the elevation range with suitable climate for *B. tectorum* in the Great Basin (Chambers et al. 2007). At local scales, *B. tectorum* cover is positively related to size of bare soil gaps separating bunchgrasses, negatively correlated with biological soil crust (biocrust) cover, and increases with grazing pressure or soil water deficits (Reisner et al. 2013). At landscape scales, *B. tectorum* cover generally increases with fire. However, *B. tectorum* cover varies depending on topographic position and is positively associated with solar radiation and negatively associated with perennial herbaceous species (Condon et al. 2011).

In areas dominated by winter/spring precipitation, where *B. tectorum* is often most problematic, cool-season, early-seral bunchgrasses, such as *Elymus elymoides* Raf. (Swezey) (squirreltail) that have traits similar to *Bromus*, can quickly occupy disturbed sites and appear to have a better overall ability to compete against *B. tectorum* than other native grass species (Booth et al. 2003; McGlone et al. 2011). Greenhouse studies show that mature plants of long-lived, cool-season grasses that occur over a range of soil temperature and moisture regimes, including *Elymus wawawaiensis* J. Carlson & Barkworth (Snake River wheatgrass), *Achnatherum hymenoides* (Roem. & Schult.) Barkworth (Indian ricegrass), and *Leymus triticoides* Buckley Pilg. (creeping wild rye), strongly suppress growth of *B. tectorum*, likely due to co-opting biological soil space and reducing nitrogen availability (Blank and Morgan 2012). As with all plant types, seedling recruitment of native grasses is very low in dense, established *Bromus* stands, regardless of native seed abundance as seen for *B. tectorum* (Mazzola et al. 2011). Introduced bunchgrasses, such as *Agropyron cristatum* (L.) Gaertn. (crested wheatgrass), are often more competitive with *B. tectorum* in the seedling stage than native bunchgrasses (e.g., James et al. 2008). This has led to widespread efforts to seed introduced bunchgrasses such as *A. cristatum* following fires or other disturbances, even though these introduced species also outcompete native seedlings over a wide range of climate conditions in the cold desert (Knutson et al. 2014).

In Mediterranean California grasslands characterized by winter/spring precipitation and cool-season grasses, both above- and belowground factors help explain competitive interactions of *Bromus* with native bunchgrasses. Established perennial grasses preempt sunlight for exotic annual seedlings, including *B. diandrus* and *B. hordeaceus*, and inhibit carbon gain, growth, and rooting depth to the degree that plants cannot survive the onset of later spring and summer soil drying (Corbin and D'Antonio 2004; Seabloom et al. 2003). The same effect occurs for perennial seedlings in *Bromus*-dominated grassland (Dyer and Rice 1999). All life stages of *Nassella pulchra* Hitchc. (Barkworth) (purple needlegrass) respond negatively to *Bromus* and other invasive annuals in the relatively dry Central Valley, but *N. pulchra* and other perennial native grasses have greater resistance in areas with higher precipitation (Corbin et al. 2007). Competition also occurs among *Bromus* in annual grasslands of California. *B. diandrus* is able to exclude *B. hordeaceus* in fertile microsites (see Callaway 2007). Short-grass steppe is dominated by summer precipitation and warm-season grasses such as *Bouteloua* Lag. spp. (grama). In the high plains east of the Colorado Front Range, short-grass steppe is largely resistant to *B. tectorum* (Kotanen et al. 1998), but nitrogen addition increases the competitive ability of *B. tectorum* (Lowe et al. 2003) and disturbance can promote invasion (Bradford and Lauenroth 2006). In the northwestern Great Plains, which is also dominated by winter precipitation but has cooler soils and both warm- and cool-season grasses, species are active throughout the growing season and soil nitrogen fluctuations are minimal (McCulley et al. 2009). Despite this, *B. tectorum* and *Bromus japonicus* (syn. *Bromus arvensis* L. [field brome]) are common invaders in areas used for forage production (Ogle et al. 2003).

On the Colorado Plateau, which has a mixed precipitation regime, perennial grasslands dominated by another warm-season, rhizomatous grass, *Pleuraphis jamesii* Torr. (syn. *Hilaria jamesii*; James' galletta), showed a significant negative association with *B. tectorum* after fire (Getz and Baker 2008). However, in another location with sandy loam soils that receives about 35 % of annual precipitation in summer, grasslands dominated by cool-season *A. hymenoides* and *Hesperostipa comata* (Trin. & Rupr.) Barkworth (needle-and-thread grass) remained uninvaded while adjacent communities dominated by *P. jamesii* were highly invaded. Soils in cool-season grasslands that lacked *B. tectorum* were slightly coarser and less fertile and contained less soil biotic richness than invaded soils (Belnap and Phillips 2001).

10.3.3 Interactions of Bromus *with Woody Plants*

Environmental factors, particularly temperature, strongly influence resistance of plant communities dominated by woody plants to *Bromus*. In the cold desert, a seeding experiment demonstrated that *B .tectorum* growth and reproduction were consistently higher at low-elevation *Artemisia tridentata* Nutt. ssp. *wyomingensis* Beetle and Young (Wyoming big sagebrush) sites with warmer soils, were variable and dependent on growing season temperatures at mid-elevation *A. tridentata* Nutt. ssp. *vaseyana* (Rydb.) Beetle (mountain big sagebrush) sites, and were lowest at high-elevation *A. tridentata* ssp. *vaseyana* /mountain shrub sites (Chambers et al. 2007). These differences were apparent in a regional study of management treatments (prescribed fire, *Pinus monophylla* Torr. & Frém. [singleleaf pinyon] and *Juniperus* L. spp. [juniper] cutting, and *A. tridentata* mowing) where resistance to *B. tectorum* and thus capacity to recover was higher on sites with lower soil temperatures (Chambers et al. 2014b). Similarly, undisturbed conifer forests in the northern Rockies tended to exhibit progressively higher resistance to introduced *B. tectorum* as elevation increased (*Pinus ponderosa* Lawson and C. Lawson [ponderosa pine], *Pseudotsuga menziesii* Mirb. Franco [Douglas-fir], *Abies grandis* Douglas ex D. Don Lindl. [grand fir], *Thuja plicata* Donn ex D. Don [western redcedar]) due to limitations on its growth and reproduction at lower temperatures (Pierson and Mack 1990).

Within plant communities, a variety of factors influence whether *Bromus* has negative or positive associations with woody plants, including direct interactions between species, moderated microclimate, enhanced nutrients or higher litter under shrubs, and livestock grazing that alters species composition and spatial distribution. In the cold desert, *B. tectorum* frequently occurs under *A. tridentata* and appears to be facilitated by the shrub at the scale of the individual plant. In an experimental seeding study in the Basin and Range, Chambers et al. (2007) found that *B. tectorum* had higher biomass and seed production in under-shrub microsites than in interspace microsites on a per plant basis, likely due to higher resource availability. However, interspaces typically had higher emergence and plant densities and thus greater total biomass and seed production than under-shrub microsites likely due to experimental seed burial. In an observational study in the eastern Sierra Nevada, Griffith (2010) found that *B. tectorum* plant density and seed production were naturally greater under *A. tridentata* Nutt. ssp. *tridentata* (basin big sagebrush) and *Purshia tridentata* Pursh DC. (antelope bitterbrush) compared to bare interspaces, probably because conditions for seed burial were lacking in interspaces. Seedling survival of *B. tectorum* is generally high following emergence (Mack and Pyke 1983). However, seed production and root growth of *B. tectorum* can be suppressed by *A. tridentata* ssp. *wyomingensis*, as revealed by experimental exclusions (Reichenberger and Pyke 1990), and removal of *A. tridentata* can result in large increases in *B. tectorum* (Chambers et al. 2007; Prevey et al. 2010). Thus, establishment of *Bromus* in association with shrubs in the cold desert is determined by factors that affect seed burial and seedling emergence, such as location and depth of litter or occurrence of biocrusts, while *Bromus* biomass and seed production are strongly influenced by soil nutrients, which are typically more available under shrubs even in competitive environments.

Indirect influences of shrubs on *B. tectorum* include soil and vegetation legacy effects in which microsite conditions created by the shrub continue to affect the plant community beyond the life of the shrub (Sankey et al. 2012). In historically grazed areas of the Basin and Range, native bunchgrasses (e.g., *Pseudoroegneria spicata* (Pursh) Á. Löve [bluebunch wheatgrass] and *Achnatherum thurberianum* (Piper) Barkworth [Thurber's needlegrass]) often have greater association with shrub coppice mounds than *B. tectorum*, which is more evenly distributed among mounds and interspaces (Hoover and Germino 2012; Reisner et al. 2013). Bunchgrass mortality during fire is typically higher under shrubs, especially in dense stands with high levels of woody fuels (Miller et al. 2013). *Bromus tectorum* exhibits a greater growth response than bunchgrasses to fertile shrub mounds compared to interspace soils (Hoover and Germino 2012), especially after removal of native herbaceous species and burning (Chambers et al. 2007). Consequently, *B. tectorum* can rapidly dominate shrub islands after burns in the absence of adequate cover of perennial herbaceous competitors.

In the warmer and drier Mojave Desert, competitive and facilitative relationships with *B. rubens* vary among shrub species (Abella et al. 2011) and over environmental gradients (Brooks 2009). In a survey of seven Mojave Desert sites (710–1367 m), *B. rubens* was more common under shrubs than in interspaces but had relatively lower cover under *Encelia farinosa* A. Gray ex Torr. (brittlebush) and *Bebbia juncea* (Benth.) Greene (sweetbush) and higher cover under *Krameria erecta* Willd. ex Schult. (littleleaf ratany) (Abella et al. 2011). However, in another study, biomass of *B. rubens* beneath canopies of perennial plants generally increased with elevation likely due to more mesic conditions rather than any specific local conditions created by the perennial species (Brooks 2009). Positive effects of *Ambrosia dumosa* (A. Gray) Payne (burrobush) on *B. rubens*' survival and especially reproduction have been attributed mainly to canopy effects with little evidence for belowground competition (Holzapfel and Mahall 1999).

Trees and forest canopies also can exert either positive or negative effects on *Bromus*. *Juniperus* spp., *P. monophylla*, and *P. edulis* Engelm. (two-needle pinyon) on warmer and drier soils typically inhibit herbs, including *Bromus*, underneath their crowns and canopies, but invasion and prolific growth often occurs after the tree has died or burned (e.g., Kane et al. 2011). Similarly, *B. tectorum* increases rapidly after fire in *P. ponderosa* forests in the southern Sierra Nevada (Keeley and McGinnis 2007), on the Colorado Plateau (McGlone et al. 2011), and in the Northern Rockies (Gundale et al. 2008). In contrast, in the coastal mountains and inland foothill woodlands of California, *B. diandrus* and *B. hordeaceus* densities are enhanced by living oak crowns due to greater soil fertility (reviewed in Callaway 2007).

10.3.4 Interactions of Bromus *with Forbs*

Only a few studies have focused on interactions between native forbs and *Bromus*, and they indicate that forbs can affect various aspects of *Bromus* growth and reproduction through competition or facilitation. In the greenhouse, *B. rubens* used water more rapidly and had greater biomass and nitrogen content than two native Mojave Desert annuals, the grass *Vulpia octoflora* (Walter) Rydb. (6 weeks fescue) and forb *Descurainia pinnata* (Walter) Britton (western tansymustard), partly because of *B. rubens*' greater root-surface area and exploitation of deeper soils (DeFalco et al. 2003). In the Mojave Desert and central Basin and Range, native annual forbs, such as *Amsinckia tessellata* A. Gray (bristly fiddleneck), decreased biomass or seed density of *B. tectorum* when forbs were grown at similar densities with *B. rubens* in a greenhouse (Abella et al. 2011) and when grown at higher densities than *B. tectorum* in greenhouse and field settings (Leger et al. 2014). Also, relatively high transplant densities of *Sphaeralcea ambigua* A. Gray (desert globemallow, a short-lived perennial forb) in the Mojave Desert significantly decreased *B. rubens* biomass when compared to controls (Abella et al. 2012). Ability of these native annual or short-lived forbs to compete with *Bromus* has been examined to determine their potential as restoration species, but difficulty in routinely establishing these forbs makes their use in restoration problematic.

Interactions between perennial forbs and *Bromus* can be complex. A field experiment in the Basin and Range indicated that litter of adult *Lupinus argenteus* Pursh (silvery lupine) increased soil nitrogen and resulted in higher *B. tectorum* biomass and reproduction, despite overall negative effects of *L. argenteus* on emergence and survival (Goergen and Chambers 2012).

Several studies describe interactions between exotic forbs and *Bromus*. In sagebrush shrublands of the central California foothills, *Bromus* can form a competitive hierarchy with exotic forbs that leads to exclusion of native forbs (Cox and Allen 2011). In California grassland, annual grass and forb dominance varies over time and is weather dependent (Corbin et al. 2007). In the cold desert, *B. tectorum* can replace exotic mustards (e.g., *Sisymbrium* L. spp. [hedge mustard]) or *Salsola kali* L. [Russian thistle]) on disturbed sites over time, but in other cases dominance depends on weather, fire or other disturbance (Piemeisel 1951; Chambers et al. 2014b), or die-off events caused by pathogens (see Meyer et al. 2015). Many exotic, tap-rooted perennial forbs are particularly invasive in mid- to high-elevation cold desert communities that are disturbed and otherwise colonized by *B. tectorum* or other *Bromus*. These secondary, tap-rooted invaders include *Cirsium* L. spp. (thistles), *Centaurea* L. spp. (knapweeds), *Chondrilla* L. spp. (chondrilla), and nearly 50 other perennial or biennial species (Hill et al. 2006; Kulmatiski et al. 2006). Similarly, *Centaurea solstitialis* L. (yellow star-thistle) has invaded annual (including *Bromus*) grasslands in central California (Enloe et al. 2004). This secondary invasion by tap-rooted forbs or the metastable coexistence of biennial or annual forbs with *Bromus* appears related to soil water partitioning as described earlier.

10.4 *Bromus* Reproductive Traits, Propagule Supply, and Priority Advantage

10.4.1 Seed Production and Propagule Supply

Bromus exhibits highly plastic growth and reproduction, which enables it to produce sufficient seeds to maintain low, persistent populations under marginal conditions and to increase seed production under favorable conditions. Plant biomass and seed production differ among communities with different environmental conditions and among growing seasons within communities (Mack and Pyke 1983; Chambers et al. 2007). In a relatively dry *A. tridentata* ssp. *wyomingensis* community on the Columbia Plateau, most *B. tectorum* plants produced between 0 and 70 viable seeds, but one individual produced 555 viable seeds (Mack and Pyke 1983). During the subsequent growing season, the maximum number of seeds produced by any one plant was 32. In comparison with warmer and drier communities, higher elevation communities with cool to cold soil temperature regimes routinely produce the least seeds per plant (Mack and Pyke 1983; Chambers et al. 2007). Local adaptations may exist in populations from communities with environmental extremes as shown for an arid saline site dominated by *Sarcobatus vermiculatus* (Hook.) Torr. (greasewood) and for clearings within a cool, mesic *Tsuga heterophylla* (Raf.) Sarg. (western hemlock) forest habitat (Rice and Mack 1991).

High densities of *Bromus* seedlings can result in intraspecific competition and thinning, although this may vary among *Bromus* species. *Bromus rubens* exhibits less plasticity in its response to canopy crowding (i.e., shade intolerance) than *B. hordeaceus* leading to greater density-induced mortality of individuals albeit without loss of seed production at the population level (Wu and Jain 1979). Seedling survival of *B. tectorum* is seldom density dependent, particularly in populations from mesic habitats, but plant and seed mass decrease with greater plant density (Rice and Mack 1991). In high-density stands of *B. tectorum*, plants have only 1–2 culms and do not form basal rosettes, but in low-density stands, plants can resemble small bunchgrasses (Mack and Pyke 1983). This same effect on plant mass and, consequently, seed production is seen when *B. tectorum* is growing in intact *Artemisia* communities with high densities of native perennial herbaceous species (Chambers et al. 2007).

Seed production of *Bromus* can increase dramatically with increased resource availability. Nitrogen (N) fertilization increased the number of *B. tectorum* seeds from 350 seeds/dm^2 without added N to 1200 seeds/dm^2 with 7.2 g N/m^2 (Hulbert 1955). In warm and cold deserts, disturbances that increase available water and nutrients are wildfire and removal of native perennial grasses and forbs through inappropriate grazing by livestock. Experimental removal of native perennial grasses and forbs in big sagebrush communities can increase *B. tectorum* seed production 2- to 3-fold, burning 3- to 6-fold, and removal combined with burning 10- to 30-fold (Chambers et al. 2007). These studies underscore the importance of perennial grasses and forbs as competitors with *Bromus* that decrease not only growth but also seed production.

10.4.2 Seed Characteristics and Dispersal

All four focal *Bromus* species have lemma awns that can facilitate seed dispersal to new locations and subsequent seed burial. Lemma awns are wirelike appendages attached to the lemma, the membrane that surrounds the caryopsis on the dorsal side (Hitchcock 1971). Species in section Genea, which include *B. diandrus*, *B. rubens*, and *B. tectorum*, have "sharp calluses and barbed lemmas and awns" (Hulbert 1955, p 183). Lemmas of species in section Bromus, which includes *B. hordeaceus*, lack barbs on lemmas and awns and their calluses are not as sharp (Hulbert 1955).

Diaspores of *B. tectorum* may include one or more florets in what are essentially multiple-awned seed units. Awns can promote adhesion of seeds to animal fur and wool and human clothing (Mack 1981), but they often function primarily to anchor seeds to soil and facilitate germination (Sorensen 1986). Awns can assist seed burial due to hygroscopic characteristics that result in twisting with wetting and drying cycles (Peart 1979). On the Colorado Plateau, most *B. tectorum* diaspores had either a single floret with an awn and filled seed (simple diaspore) or one or more sterile florets with awns that were attached to the floret with the filled seed (complex diaspores) (Monty et al. 2013). Even though complex diaspores were slightly heavier, they traveled more than five times farther and their maximum dispersal distance was more than 15 times greater than simple diaspores. Increased surface area created by sterile florets likely resulted in greater lift forces to carry them longer distances. Similarly, Chambers (2000) found that seeds with greater surface area were transported longer distances despite having structures that appear designed to promote burial (e.g., hygroscopic awns) and retention (e.g., sharp tip of diaspore and directional hairs that encourage movement into materials such as fur and fabric as in many *Bromus* species).

Both physical and biotic factors influence *Bromus* dispersal. Wind and water are important dispersal agents of *Bromus* (Kelrick 1991), especially over short distances (Hulbert 1955). Wind usually carries *Bromus* diaspores across soil surfaces, moving seeds 1–2 m from parent plants (Kanarek and Kao 2011). At times, “dust whirls” may carry diaspores longer distances (Hulbert 1955). Johnston (2011) found *B. tectorum* diaspores traveled 2.4 m on average over bare, mostly level soil with depressions less than 10 cm deep. Only 5 % of seeds traveled farther than 7.6 m and maximum dispersal distance was 20.8 m.

Diaspores accumulate with litter in cracks and depressions in soil and near obstacles that slow wind speed like shrubs (Kelrick 1991). In *P. edulis* and *Juniperus osteosperma* (Torr.) Little (Utah juniper) plant communities, *B. tectorum* seeds traveled more than five times farther on average in burned areas with little vegetation to obstruct seed movement compared to unburned areas (Monty et al. 2013). Higher *B. tectorum* recruitment can occur under *A. tridentata* shrubs where more litter and seeds accumulate than in bare interspaces (Kelrick 1991). Favorable moisture and temperature conditions where seeds and litter are deposited may enhance germination (Reichman 1979; Chambers 2000). Burial in relatively bare soil may be facilitated by precipitation events. Johnston (2011) reported that *B. tectorum* diaspores were visible only by their awns protruding from the soil after a 3.2 mm rainfall event.

Animals are likely to play a role in dispersal of *Bromus* over distances farther than a few meters (Hulbert 1955; Klemmedson and Smith 1964). Although *Bromus* diaspores lack adaptations to facilitate adhesion such as barbs or hooks (Sorensen 1986), the sharp callus and backward directed hairs that some *Bromus* species possess can facilitate attachment to animals and humans (Mack 1981). Diaspores remain attached until they are discovered and removed by the animal or fall off passively (Sorensen 1986). When and where detachment from the animal occurs is important to understanding long-distance dispersal of *Bromus* and other plant species, but is difficult to study.

Seeds of *Bromus* species are dispersed by humans intentionally and unintentionally. Common gardens with ecotypes of *B. tectorum* from different regions of the world planted in Pullman, WA, and Lewiston, ID, USA, may have been one origin of future spread (Mack 1981). Although animals and humans disperse seeds of *Bromus* through adhesion, this mechanism is likely minor compared to the scale and extent of spread through contamination of hay and crop seed, in particular seed of forage and small grain crops such as wheat (Mack 1981).

Grass seed can survive digestive tracts of herbivores. Survival seems to be related to hard seededness (hard-seeded species have higher survival) (Gardener et al. 1993a, b), seed size (Pakeman et al. 2002), and transit time through the gut (the shorter the transit time, the greater the survival) (Janzen 1984). In general, *Bromus* do not have characteristics associated with high survival after bovine ingestion. Despite this, Wells and Lauenroth (2007) found that *B. tectorum* seedlings emerged from recent horse manure collected from a trail used for backcountry access in the Colorado Rocky Mountains.

Birds eat seeds of *Bromus* species and could serve as dispersal vectors should the seed survive, although information as to this possibility is lacking. In western Utah, *Alectoris chukar* (chukars), an exotic game bird species widely introduced in the western USA, had *B. tectorum* seed in 76.3 % of their crops with an average of 522 *B. tectorum* seeds per crop (Larsen et al. 2007). Despite this, no *B. tectorum* germinated from chukar fecal samples after being treated to break seed dormancy.

La Tourrette et al. (1971) studied the role of heteromyid rodents in seed dispersal in degraded *A. tridentata* communities. Cheek pouches of *Dipodomys* spp. (kangaroo rats) contained over 1500 *B. tectorum* seeds on average, more than any other species, although an introduced forage grass had higher seed mass. Cheek pouches of *Perognathus parvus* (pocket mice) also contained *B. tectorum* seeds. Rodents stored caryopses of *B tectorum* in soil depressions approximately 5 cm deep and 3 cm wide, and *B. tectorum* seeds in caches germinated earlier than non-cached seed. Likewise, Reichman (1979) found that nearly twice as many seedlings grew from seed caches than from nearby soils. Higher seedling emergence was attributed primarily to higher seed densities in caches compared to controls, but microsite conditions of seed caches also may have contributed to enhanced germination (Reichman 1979). Emergence of *B. tectorum* was 100 times higher for seeds placed in 9 mm depressions in soil compared to seeds placed on bare soil (Evans and Young 1987 in Chambers and MacMahon 1994).

10.4.3 Seed Banks

A seed bank is a reserve of mature viable seeds located in fruits (or cones) on the plant (aerial seed bank), on the soil surface, or buried in soil, duff, or litter (Roberts 1981). In annual *Bromus* the majority of the seed bank is stored on the soil surface or in litter. Seed burial studies show that most seeds of *B. tectorum* germinate in the first year after dispersal. Few *B. tectorum* seeds are viable after 2–3 years of burial, but some seeds survive as long as 5 years (Thill et al. 1984).

Seed bank information is limited for *B. diandrus* and *B. hordeaceus*, but in heavily invaded communities seed bank densities of all four *Bromus* species often reach numbers in the thousands following dispersal. More intact communities with minimal *Bromus* also can have significant *Bromus* in the seed bank. In southern California annual grasslands, *B. diandrus* seed density estimates range from 569 to 1473 m^{-2}, some of the lowest peak densities of the four species (Moyes et al. 2005; Cox and Allen 2008). *Bromus rubens* seed bank densities range from 2837 to 4085 m^{-2} in southern California annual grasslands (Cox and Allen 2008) and >3000 to >5800 m^{-2} in Mojave Desert communities (Abella et al. 2009; Esque et al. 2010; Jurand and Abella 2013). Estimates of *B. hordeaceus* seed bank densities span the greatest range, from >1900 seeds m^{-2} in southern California to >103,500 seeds m^{-2} in the northern California's Central Valley (Young and Evans 1989; Cox and Allen 2008). *Bromus tectorum* seed banks are highly variable in invaded ecosystems, with a low of 35 seeds m^{-2} in northern mixed prairie of Montana (Karl et al. 1999) and a high of >30,000 seeds m^{-2} in a shadscale ecosystem in the Great Basin (Meyer et al. 2007), with frequent counts from 2000 to 20,000 seeds m^{-2} (e.g., Mack and Pyke 1983; Humphrey and Schupp 2001; Meyer et al. 2007). Even in more intact communities, *Bromus* are in the seed bank in low (e.g., <50 seeds m^{-2} for *B. tectorum* in the Great Basin; Hassan and West 1986; Allen et al. 2008) to moderate (e.g., 806 seeds m^{-2} for *B. hordeaceus* in coastal sage scrub of southern California; Cox and Allen 2008) densities. Higher densities in relatively intact communities appear due to high propagule pressure from adjacent heavily invaded communities (Cox and Allen 2008).

Peak seed densities are short lived, and the seed bank is rapidly depleted, primarily by germination but also by seed predation (Borchert and Jain 1978; Beattie 1989) and pathogens (Crist and Friese 1993; Meyer et al. 2007, 2015). Consequently, seed bank carryover to the following year is frequently less than 10 % for *B. tectorum* in the Great Basin (Meyer et al. 2007; Smith et al. 2008), *B. rubens* in the Mojave Desert (Jurand et al. 2013), and *B. hordeaceus* in the northern Central Valley, California (Young and Evans 1989). However, the actual proportion surviving depends on the environment, and carryover of *B. tectorum* in the Great Basin can be greater in drier years and on drier sites due to reduced germination (Smith et al. 2008), or less on drier sites due to ungerminated seeds having longer exposure to pathogens (Meyer et al. 2007). Nonetheless, high levels of yearly loss from the seed bank mean that survival for more than 3 years is unlikely (Meyer et al. 2007). Although no studies have been conducted on carryover for *B. diandrus*, germination patterns suggest it also has little carryover (Harradine 1986).

Abiotic and biotic factors contribute to variation in seed bank densities, although relative strengths of different factors are unknown. *Bromus* must largely refill the seed bank annually, and ability to refill depends largely on precipitation—*B. tectorum* input to a seed bank in western Utah varied from 3567 seeds m^{-2} in a drier year to 13,942 seeds m^{-2} in a wetter year (Smith et al. 2008). Prescribed burns and wildfires frequently reduce *Bromus* seed bank densities by 94–99 % (Humphrey and

Schupp 2001; Allen et al. 2008; Cox and Allen 2008; Diamond et al. 2012), although reductions can be less depending on timing of fire relative to seed shatter (Hassan and West 1986; Rasmussen 1994). *Bromus tectorum* seed bank densities in the Great Basin can completely recover to pre-burn densities in two growing seasons (Humphrey and Schupp 2001; Allen et al. 2008). In contrast, recovery of *B. rubens* seed banks in the Mojave Desert appears to take 4 or more years (Brooks 2002; Abella et al. 2009). Livestock grazing also influences seed banks and targeted grazing by livestock can reduce *B. tectorum* inputs to the seed bank (Diamond et al. 2012), but may not be sufficient to influence *B. tectorum* populations (Hempy-Mayer and Pyke 2008). At cheatgrass-dominated sites in the Basin and Range, plants clipped short at the boot stage and again 2 weeks later had among the lowest mean seed densities, but still had 123–324 seeds m^{-2} at one site, and 759–2256 seeds m^{-2} at a second site (Hempy-Mayer and Pyke 2008).

In highly invaded areas, *Bromus* often dominate the seed bank. In a degraded *B. tectorum*-dominated community in western Utah, over 90 % of the seed bank was *B. tectorum*; in contrast to >12,800 *B. tectorum* seeds m^{-2}, there were <3 native perennial seeds m^{-2} (Humphrey and Schupp 2001). Similarly, in a Mojave Desert community, *B. rubens* comprised >90 % of the seed bank (Jurand and Abella 2013). In highly invaded areas, when *Bromus* do not dominate the seed bank, it is often because other invasive annuals, not natives, are more dominant (Cox and Allen 2008; Esque et al. 2010). Low densities of *B. tectorum* in Montana grasslands (35 m^{-2}) are likely at least partially a consequence of site domination by *B. japonicus*, with >10,000 seeds m^{-2} (Karl et al. 1999). Even more intact communities can have significant representation of *Bromus* in the seed bank. In an *A. tridentata/J. osteosperma* community in western Utah, *B. tectorum* seed densities were low but still comprised 47 % of the seed bank (Hassan and West 1986). In an intact Wyoming big sagebrush community in Utah, *B. tectorum* comprised only 7.5 % of the seed bank, but was the fourth most abundant species (out of 47) with higher density than all but one native (Pekas and Schupp 2013).

10.4.4 Seed Germination

Seeds of all four species are nondormant following a period of after-ripening and readily germinate when soil water and temperature are not limiting (Corbineau et al. 1992; Andersson et al. 2002; Beckstead et al. 1996; Del Monte and Dorado 2011). Differences in length of dormancy and germination responses to environmental conditions exist among annual *Bromus* species and influence competitive interactions (e.g., Andersson et al. 2002), but are not well quantified. Adaptive germination and growth responses in *B. tectorum* populations from contrasting habitats indicate both genetic and environmental controls on germination response (Beckstead et al. 1996; Dyer et al. 2012), thus partially explaining invasibility of this and likely other invasive annual *Bromus* species.

Timing of germination and seasonal activity (phenology) of *Bromus* allows these invaders to use different resource pools than species in many of the native communities where they are most problematic. In those ecoregions where most precipitation arrives in winter and spring, invasive *Bromus* species that germinate and grow early in the growing season may preempt resources and attain competitive dominance. This aspect of the "seasonal priority advantage" of *Bromus* has been shown for *B. diandrus* and *B. hordeaceus* in Mediterranean California annual grassland (Dyer and Rice 1999; Abraham et al. 2009; Wainwright et al. 2012), *B. tectorum* in the cold desert (Mack and Pyke 1983; Booth et al. 2003; Kulmatiski et al. 2006), and *B. rubens* in the Mojave Desert (DeFalco et al. 2007). Timing of germination depends on having adequate soil water availability when temperatures are high enough for physiological activity and consequently varies among ecoregions and over environmental gradients. In cold desert sagebrush steppe, germination of *B. tectorum* was predicted to occur more often in spring than in fall and much more often than in winter based on simulated soil moisture availability from historical climate records (Bradford and Lauenroth 2006) and soil temperature and water data measured over elevation gradients in the central Basin and Range (Roundy et al. 2007). However, on the Columbia Plateau germination was higher in late summer and fall than winter or spring (Mack and Pyke 1983). Earlier germination in fall or winter vs. spring may increase competitiveness of *Bromus* with natives, as shown for *B. rubens* and native annual species in the Mojave Desert (DeFalco et al. 2007). However, *B. diandrus* and *B. hordeaceus* in California coastal sage scrub had higher mortality when emergence resulted from a late-summer watering event than when emergence followed ambient winter rainfall due to factors such as increased herbivory (Wainwright et al. 2012).

10.5 Effects of Biological Crusts on Seed Dispersal and Germination

Roughness, texture, cracking, and other aspects of the soil surface influence the ultimate destination of plant seeds and most of these variables are heavily affected by biocrusts. There are two general morphologies of biocrusts: smooth biocrusts found in warm deserts or recently disturbed areas and pinnacled biocrusts found in cold deserts. Smooth biocrusts are found in regions where soils do not freeze; they are characterized by a heavy dominance of cyanobacteria and physical/chemical crusting. Because these biocrusts actually smooth the soil surface, most seeds are easily moved across plant interspaces by wind or water unless the seed has specialized mechanisms for adhesion to smooth soil surfaces (e.g., mucilaginous coats) (Gutterman 1994). *Bromus* seeds lack these adaptations and *Bromus* seeds and plants are typically found under or near obstacles such as rocks or plants rather than in plant interspaces.

In contrast, surfaces in cold deserts are characterized by lichen-moss biocrusts, which, combined with soils that freeze in winter, create a highly roughened and cracked soil surface. Many studies have shown that these rough surfaces trap seeds more effectively than smooth soil surfaces (Harper and St. Clair 1985; Eckert et al. 1986; Harper and Marble 1988; Mücher et al. 1988; Prasse 1999). In such settings, *Bromus* seeds can often be seen on the soil surface in both plant interspaces and under plant canopies. This surface effect was verified using a portable wind tunnel where seeds with large appendages, including *B. tectorum*, quickly left smooth interspace soils but were retained by roughened biocrusted surfaces (Belnap, unpubl. data).

Multiple studies in warm and cold deserts show that well-developed biocrusts can inhibit germination and establishment of invasive annual plants including several *Bromus* species in the USA and Israel (Evans and Young 1984; St. Clair et al. 1984; Eckert et al. 1986; Kaltenecker et al. 1999; Larsen 1995; Howell 1998; Prasse 1999). In contrast, crushed biocrusts can stimulate growth of invasive annual plants when left in place (Crisp 1975; Larsen 1995; Howell 1998). It has been speculated in the literature that *Bromus* and *Schismus* P. Beauv (Mediterranean grass) evolved with animal herds whose hooves break up biological and physical soil crusts facilitating seed burial and annual grass success (Mack and Thompson 1982).

The mechanism by which intact biocrusts inhibit *Bromus* and other annual plants is not known, but several factors are likely involved. Germination and emergence are facilitated in arid systems when seeds are buried due to a higher probability of sufficient soil moisture for germination. Also, buried seeds are hidden from consumers, potentially reducing predation. Small cracks found in biocrusts allow small seeds to enter the soil, but seeds with large appendages, such as many *Bromus*, are prevented from entering these cracks. Well-developed biocrusts also stabilize soils and thus prevent moving sediment from burying seeds (Belnap 2003).

Biocrusts cannot protect all areas from *Bromus* invasion. In a southeast Utah grassland covered by well-developed biocrusts, a combination of low seed predation and unusual climate conditions facilitated a large *B. tectorum* invasion (Belnap and Phillips 2001). Extremely low (<0.01 %) cover of *B. tectorum* was first documented in this grassland in 1964 (Kleiner and Harper 1977) and stayed at a low level until 1995. In late August-early September 1995, almost daily rain events resulted in *B. tectorum* establishment and invasion over a 50 ha area. This "instant" large-scale invasion implied that sufficient seed were in the seed bank to facilitate plant establishment and population growth under favorable weather conditions despite the presence of biocrusts.

10.6 Herbivory

Most of the available information on herbivory is from research on *B. tectorum*, and there is still much to learn about the other three focal species. Soil and plant community disturbance often precede and reinforce *Bromus* invasions. Herbivory often has less immediately apparent effects on *B. tectorum* establishment and expansion than fire, but its influence can vary depending on the intensity and timing of herbivory and climate conditions. Negative effects of herbivory on *B. tectorum* can be exceeded by indirect positive effects through competitive release when herbivores preferentially target native grasses and forbs over *B. tectorum*.

10.6.1 Grazing Impacts on B. tectorum

There has been interest in livestock grazing effects on *B. tectorum* because of its potential as a biocontrol. Livestock and wildlife forage on *B. tectorum* particularly during the winter and spring vegetative period (Murray 1971; Wikeem and Pitt 1992). *Bromus tectorum* has good nutritional value during winter/spring as indicated by feeding preference by wildlife (Austin et al. 1994) and livestock weight gain in *B. tectorum*-dominated pastures (Murray and Klemmedson 1968). However, preference for *B. tectorum* tends to decrease during dry years (Murray 1971) and in late spring/summer when senescence and seed production reduce its nutritional quality and palatability (Cook and Harris 1952; Wikeem and Pitt 1992).

Intensive grazing of *B. tectorum* by cattle (80–90 % removal) or experimental clipping reduces *B. tectorum* biomass and seed production, particularly when defoliation is repeated or occurs early in development (Hempy-Mayer and Pyke 2008; Schmelzer 2009; Diamond et al. 2012). However, clipping plants to a 2.5 cm height at the boot stage and again 2 weeks later still resulted in 123–324 seeds m^{-2} at one site and 769–2256 seeds m^{-2} at a second site, calling into question the potential for using livestock grazing as a biocontrol in *B. tectorum*-dominated areas (Hempy-Mayer and Pyke 2008). Grazing tolerance in *B. tectorum* may result from continued growth of its root system despite defoliation (Arredondo and Johnson 2009).

Available data indicate that the response to high-intensity grazing is generally negative, but that the magnitude of change and degree of recovery depends on climate and site conditions. In a high-elevation, semiarid grassland near Flagstaff, AZ, high-intensity grazing had strong directional effects that led to a decline in perennial forb cover and an increase in annual plants, particularly *B. tectorum* (Loeser et al. 2007). Following a severe drought in the sixth year of the study, plant cover of exotic species increased significantly and this increase was greatest in the high-impact grazing plots where native cover had been reduced by one-half (Loeser et al. 2007). A multivariate study in *A. tridentata* ssp. *wyomingensis* communities in the Basin and Range showed that cattle grazing reduced resistance to invasion by decreasing bunchgrass cover, increasing the size of gaps between perennial herbaceous plants, and reducing biological soil crusts (Reisner et al. 2013). Cheatgrass cover was positively associated with sandy soils and negatively associated with high heat loads due to negative effects on bunchgrass cover (Reisner et al. 2013). Analyses of long-term datasets from sagebrush steppe in the Northern Basin and Range clearly indicate that site conditions and climate influence *B. tectorum* abundance and plant community dynamics over time regardless of grazing history (West and Yorks 2002; Bagchi et al. 2013).

Grazing history can interact with herbivory to modulate its influence on *B. tectorum* success. A factorial study in an *A. tridentata* ssp. *vaseyana* site that examined effects of grazing and fire showed that long-term grazing exclusion followed by fire stimulated much higher levels of *B. tectorum* cover, density, and biomass than grazing exclusion without recent fire or grazing followed by fire due to higher fuel biomass and fire severity (Davies et al. 2009).

10.6.2 Native Herbivores and Community Resistance

Native herbivores consisting primarily of small mammals and insects can decrease invasive plant establishment and performance and thus increase community resistance to invasion (Levine et al. 2004). In a sagebrush community, folivory (leaf eating) by small mammals had little effect on seedling survival of *B. tectorum*, but significantly reduced biomass and seed production (Pyke 1986). In a follow-up study, small mammal folivory strongly increased *B. tectorum* seedling mortality, but tolerance to herbivory increased as seedlings matured (Pyke 1987). Grasshoppers consume *B. tectorum* (Rogers and Uresk 1974; Beckstead et al. 2008), but their preference for it decreases from late spring to summer (Fielding and Brusven 1992). Harvester ants can be very efficient at controlling *B. tectorum* establishment around their mounds by removing seedlings, but their influence outside of their nest area is not well studied (Clark and Comanor 1975). Better understanding of the relationships between *B. tectorum* and native folivores could result in management approaches that strengthen community resistance to *B. tectorum* establishment and expansion.

Herbivory can promote establishment and expansion on *B. tectorum* through competitive release if herbivores preferentially target other plant species (Strand et al. 2014). Studies of insect and ungulate herbivory demonstrate preference for native plant species over *B. tectorum* (Rogers and Uresk 1974; Loeser et al. 2007). Also, there is evidence of indirect competition between *B. tectorum* and native grasses in which high-density *B. tectorum* populations increase grasshopper populations that consume native grasses more intensively than *B. tectorum* (Beckstead et al. 2008).

10.6.3 Potential Effects on Disturbance Regimes and Community Resilience

Herbivory in *B. tectorum*-dominated communities has the potential to mitigate invasive grass-fire regimes by altering biomass composition, and reducing fine fuels and their continuity (Strand et al. 2014). High-impact cattle grazing in successive post-fire years reduced *B. tectorum* biomass and cover resulting in reduced flame lengths and fire spread in subsequent years (Diamond et al. 2010). However, given the size of recent wildfires in the western USA (e.g., 80,000 ha and larger; Murphy et al. 2013) relative to the size of areas where high-impact grazing can be successfully implemented, applicability is likely limited to fuel breaks or protecting property. Because of the central role that fire has in *B. tectorum* invasions, reducing fire severity and extent has the potential to increase resilience of the plant community to invasion (Chambers et al. 2014a). However, any potential gains resulting from fine fuel removal by livestock may be counterbalanced by decreased resistance to

B. tectorum due to herbivory of native plants that compete with *B. tectorum*, increased soil disturbance, and damage to biocrusts (Reisner et al. 2013).

10.7 Management Implications

This chapter illustrates that environmental factors, community attributes and processes, life history and ecophysiological traits of the invader, and direct and indirect interactions of the invader with the plant community are important determinants of community resistance to *Bromus*. Understanding the effects of climate and soils, especially soil temperature and moisture regimes, on interactions of *Bromus* and plant communities can be used to prioritize management activities across the landscape and determine appropriate activities at local scales (Miller et al. 2013; Chambers et al. 2014a, c). For example, in higher elevation communities with cooler soil temperatures and greater resistance to *B. tectorum*, it may be possible to successfully use treatments like prescribed fire that likely would result in *B. tectorum* dominance in communities with warmer soil temperatures and lower resistance. Also, recognition of the importance of community attributes and processes on *Bromus* invasion, such as the role of biocrusts in preventing *Bromus* seed entrapment and seedling establishment, can be used as management incentives to conserve communities where these attributes and processes are still intact and to restore those where they have been degraded.

Rethinking effects of resource availability on competitive interactions within plant communities may allow development of more effective management strategies. *Bromus* is generally more effective than native species at using limiting soil resources on short timescales, and manipulating soil nitrogen through carbon additions or repeated fire is largely ineffective in more arid ecosystems (Monaco et al. 2003; James et al. 2011; Jones et al. 2015a, b). However, perennial herbaceous species, especially those with similar phenologies, are highly effective competitors (Booth et al. 2003; Blank and Morgan 2012). This indicates that in more highly invaded areas and areas with high climate suitability to *Bromus*, restoration species need to have the capacity to both establish and compete, specifically similar phenology and use of shallow resource pools. In less highly invaded areas and areas with lower climate suitability to *Bromus*, both passive and active management activities may need to focus on increasing perennial herbaceous species, especially grasses.

Recognition of the importance of propagule supply as influenced by seed production and seed banking, seed dispersal, and priority effects also informs management. In highly invaded areas where restoration/rehabilitation or fire prevention are objectives, herbicides and high-impact livestock grazing can decrease seed production and seed banks. However, caution is advised in applying these management strategies because native species can also be negatively affected by preemergent herbicides (Pyke et al. 2014) and heavy livestock grazing (Reisner et al. 2013).

10.8 Research Needs

Understanding community resistance to invasion is complex because it is affected by many interacting variables about which we are still learning. Both manipulative studies and longer-term observational studies that examine a suite of variables are needed in order to better clarify the environmental characteristics and community attributes and processes that influence resistance to *Bromus*. Although most research to date has focused on *B. tectorum*, in a warming environment other invasive species adapted to warmer temperatures such as *B. rubens* may become more problematic (see Bradley et al. 2015). An experimental approach that examines broad-scale environmental gradients and multiple species is likely to provide the most useful information. Research is needed in several areas to improve our understanding of resistance to *Bromus*, and our ability to develop more effective management strategies for increasing resistance. These needs include:

- Better define the climate suitability (space) of *Bromus* species and of co-occurring native species to understand (1) effects of environmental factors on community resistance at both large landscape scales and local scales and (2) potential expansion in a warming environment.
- Increase our understanding of resource pools (sensu Leffler and Ryel 2012) and their influence on resistance to *Bromus*. How does resource availability vary across environmental gradients and in response to disturbance and management treatments? How do changes in resource availability across environmental gradients and in response to disturbance and management treatments influence interactions of *Bromus* with plant communities?
- Increase our understanding of how native species' life history and ecophysiological traits influence competitive interactions with *Bromus*.
- Increase our understanding of community processes, such as herbivory by native herbivores and livestock, on resistance to *Bromus*.

Acknowledgments Jayne Belnap was supported by the USGS Ecosystem Invasives Program. Any use of trade names is for descriptive purposes only and does not imply endorsement by the US Government.

References

Abella SR, Craig DJ, Chiquoine LP et al (2011) Relationships of native desert plants with red brome (*Bromus rubens*): toward identifying invasion-reducing species. Invasive Plant Sci Manag 4:115–124

Abella SR, Craig DJ, Smith SD et al (2012) Identifying native vegetation for reducing exotic species during the restoration of desert ecosystems. Restor Ecol 20:781–787

Abraham JK, Corbin JD, D'Antonio CM (2009) California native and exotic perennial grasses differ in their response to soil nitrogen, exotic annual grass density, and order of emergence. Plant Ecol 201:445–456

Allen EA, Chambers JC, Nowak RS (2008) Effects of a spring prescribed burn on the soil seed bank in sagebrush steppe exhibiting pinyon-juniper expansion. West N Am Nat 68:265–277

Andersson L, Milberg P, Schütz W et al (2002) Germination characteristics and emergence time of annual *Bromus* species of differing weediness in Sweden. Weed Res 42:135–147

Aguirre L, Johnson DA (1991) Influence of temperature and cheatgrass competition on seedling development of two bunchgrasses. J Range Manag 4:347–354

Arredondo J, Johnson DA (2009) Root responses to short-lived pulses of soil nutrients and shoot defoliation in seedlings of three rangeland grasses. Rangel Ecol Manag 62:470–479

Austin DD, Stevens R, Jorgensen KR et al (1994) Preferences of mule deer for 16 grasses found on Intermountain winter ranges. J Range Manag 47:308–311

Bagchi S, Briske DD, Bestelmeyer BT et al (2013) Assessing resilience and state-transition models with historical records of cheatgrass *Bromus tectorum* invasion in North American sagebrush-steppe. J Appl Ecol 50:1131–1141

Bansal S, Sheley RL, Blank B et al (2014) Plant litter effects on soil nutrient availability and vegetation dynamics: changes that occur when annual grasses invade shrub-steppe communities. Plant Ecol 215:367–378

Beattie AJ (1989) The effects of ants on grasslands. In: Huenneke LF, Mooney HA (eds) Grassland structure and function: California annual grassland. Springer, Netherlands, pp 105–116

Beckstead J, Augspurger CK (2004) An experimental test of resistance to cheatgrass invasion: limiting resources at different life stages. Biol Invasions 6:417–432

Beckstead J, Meyer SE, Allen PS (1996) *Bromus tectorum* seed germination: between-population and between-year variation. Can J Bot 74:875–882

Beckstead J, Meyer SE, Augsperger CK (2008) The indirect effects of cheatgrass invasion: grasshopper herbivory on native grasses determined by neighboring cheatgrass. In: Kitchen S, Pendleton R, Monaco T et al (eds) Shrublands under fire: disturbance and recovery in a changing world, 6–8 June 2006. Gen Tech Rep RMRS-P-52. USDA, Forest Service, Rocky Mountain Research Station, Cedar City, UT, pp 41–48

Belnap J (2003) Biological soil crusts and wind erosion. In: Belnap J, Lange OL (eds) Biological soil crusts: structure, function, and management. Springer, Berlin, pp 339–347

Belnap J, Phillips SL (2001) Soil biota in an ungrazed grassland: response to annual grass (*Bromus tectorum*) invasion. Ecol Appl 11:1261–1275

Belnap J, Stark JM, Rau BJ (2015) Soil moisture and biogeochemical factors influence the distribution of annual *Bromus* species. In: Germino MJ, Chambers JC, Brown CS (eds) Exotic brome-grasses in arid and semiarid ecosystems of the Western USA: causes, consequences, and management implications. Springer, New York, NY (Chapter 8)

Blank RR, Morgan T (2012) Suppression of *Bromus tectorum* L. by established perennial grasses: potential mechanisms-part one. Appl Environ Soil Science. Article ID 632172

Booth MS, Caldwell MM, Stark JM (2003) Overlapping resource use in three Great Basin species: implications for community invasibility and vegetation dynamics. J Ecol 91:36–48

Borchert MI, Jain S (1978) The effect of rodent seed predation on four species of California annual grasses. Oecologia 33:101–113

Bradford JB, Lauenroth WK (2006) Controls over invasion of *Bromus tectorum*: the importance of climate, soil, disturbance and seed availability. J Veg Sci 17:693–704

Bradley BA (2009) Regional analysis of the impacts of climate change on cheatgrass invasion shows potential risk and opportunity. Glob Change Biol 15(1):196–208

Bradley BA, Curtis CA, Chambers JC (2015) *Bromus* response to climate and projected changes with climate change. In: Germino MJ, Chambers JC, Brown CS (eds) Exotic brome-grasses in arid and semiarid ecosystems of the Western USA: causes, consequences, and management implications. Springer, New York, NY (Chapter 9)

Brooks ML (2002) Peak fire temperatures and effects on annual plants in the Mojave Desert. Ecol Appl 12:1088–1102

Brooks ML (2009) Spatial and temporal distribution of non-native plants in upland areas of the Mojave Desert. The Mojave Desert: ecosystem processes and sustainability. University of Nevada Press, Reno, NV

Brooks ML, Belnap J, Brown CS et al (2015) Exotic annual *Bromus* invasions - comparisons among species and ecoregions in the western United States. In: Germino MJ, Chambers JC, Brown CS (eds) Exotic brome-grasses in arid and semiarid ecosystems of the Western USA: causes, consequences, and management implications. Springer, New York, NY (Chapter 2)
Bykova O, Sage RF (2012) Winter cold tolerance and the geographic range separation of *Bromus tectorum* and *Bromus rubens*, two severe invasive species in North America. Glob Change Biol 18:3654–3663
Callaway RM (2007) Positive interactions and interdependence in plant communities. Springer, Netherlands
Chambers JC, MacMahon JA (1994) A day in the life of a seed: movements and fates of seeds and their implications for natural and managed systems. Ann Rev Ecol Syst 25:263–292
Chambers JC (2000) Seed movements and seedling fates in disturbed sagebrush steppe ecosystems: implications for restoration. Ecol Appl 10:1400–1413
Chambers JC, Bradley BA, Brown CS et al (2014a) Resilience to stress and disturbance, and resistance to *Bromus tectorum* L. invasion in cold desert shrublands of western North America. Ecosystems 17:360–375
Chambers JC, Miller RF, Board DI et al (2014b) Resilience and resistance of sagebrush ecosystems: implications for state and transition models and management treatments. Rangel Ecol Manag 67:440–454
Chambers JC, Pyke DA, Maestas JD et al (2014c) Using resistance and resilience concepts to reduce impacts of invasive annual grasses and altered fire regimes on the sagebrush ecosystem and greater sage-grouse: a strategic multi-scale approach. Gen Tech Rep RMRS-GTR-326. USDA, Forest Service, Rocky Mountain Research Station, Fort Collins, CO, p 73
Chambers JC, Roundy BA, Blank RR et al (2007) What makes Great Basin sagebrush ecosystems invasible by *Bromus tectorum*? Ecol Monogr 77:117–145
Chesson P (2000) General theory of competitive coexistence in spatially-varying environments. Theor Popul Biol 58:211–237
Clark WH, Comanor PL (1975) Removal of annual plants from the desert ecosystem by western harvester ants, *Pogonomyrmex occidentalis*. Environ Entomol 4:52–56
Condon L, Weisberg PJ, Chambers JC (2011) Abiotic and biotic influences on *Bromus tectorum* invasion and *Artemisia tridentata* recovery after fire. Int J Wildl Fire 20:597–604
Cook CW, Harris LE (1952) Nutritive value of cheatgrass and crested wheatgrass on spring ranges of Utah. J Range Manag 5:331–337
Corbin J, Dyer A, Seabloom E (2007) Competitive interactions. In: Stromberg MR, Corbin JD, D'Antonio CM (eds) California grasslands: ecology and management. UC Press, Berkeley, CA, pp 156–168
Corbin JD, D'Antonio CM (2004) Competition between native perennial and exotic annual grasses: implications for an historical invasion. Ecology 85:1273–1283
Corbineau F, Belaid D, Côme D (1992) Dormancy of *Bromus rubens* L. seeds in relation to temperature, light and oxygen effects. Weed Res 32:303–310
Cox RD, Allen EB (2008) Stability of exotic annual grasses following restoration efforts in southern California coastal sage scrub. J Appl Ecol 45:495–504
Cox RD, Allen EB (2011) The roles of exotic grasses and forbs when restoring native species to highly invaded southern California annual grassland. Plant Ecol 212:1699–1707
Crisp M (1975) Long term change in arid zone vegetation. Dissertation, University of Adelaide, Australia
Crist TO, Friese CF (1993) The impact of fungi on soil seeds: implications for plants and granivores in a semiarid shrub-steppe. Ecology 74:2231–2239
D'Antonio CM, Thomsen M (2004) Ecological resistance in theory and practice. Weed Technol 18:1572–1577
Davies KW, Svejcar TJ, Bates JD (2009) Interaction of historical and nonhistorical disturbances maintains native plant communities. Ecol Appl 19:1536–1545
Davis MA, Grime JP, Thompson K (2000) Fluctuating resources in plant communities: a general theory of invasibility. J Ecol 88:528–534

DeFalco LA, Bryla DR, Smith-Longozo V et al (2003) Are Mojave Desert annual species equal? Resource acquisition and allocation for the invasive grass *Bromus madritensis* subsp. *rubens* (Poaceae) and two native species. Am J Bot 90:1045–1053

DeFalco LA, Fernandez GC, Nowak RS (2007) Variation in the establishment of a non-native annual grass influences competitive interactions with Mojave Desert perennials. Biol Invasions 9:293–307

Del Monte JP, Dorado J (2011) Effects of light conditions and after-ripening time on seed dormancy loss of *Bromus diandrus* Roth. Weed Res 51:581–590

Diamond JM, Call CA, Devoe N (2010) Effects of targeted cattle grazing on fire behavior of cheatgrass-dominated rangeland in the northern Great Basin, USA. Int J Wildl Fire 18:944–950

Diamond JM, Call CA, Devoe N (2012) Effects of targeted grazing and prescribed burning on community and seed dynamics of a downy brome (*Bromus tectorum*)-dominated landscape. Invasive Plant Sci Manag 5:259–269

Dyer AR, Hardison JL, Rice KJ (2012) Phenology constrains opportunistic growth response in *Bromus tectorum* L. Plant Ecol 213:103–112

Dyer AR, Rice KJ (1999) Effects of competition on resource availability and growth of a California bunchgrass. Ecology 80:2697–2710

Eckert RE Jr, Peterson FF, Meurisse MS et al (1986) Effects of soil-surface morphology on emergence and survival of seedlings in big sagebrush communities. J Range Manag 39:414–420

Enloe SF, DiTomaso JM, Orloff SB et al (2004) Soil water dynamics differ among rangeland plant communities dominated by yellow starthistle (*Centaurea solstitialis*), annual grasses, or perennial grasses. Weed Biol Ecol 52:929–935

Esque TC, Young JA, Tracy CR (2010) Short-term effects of experimental fires on a Mojave Desert seed bank. J Arid Environ 74:1302–1308

Evans RA, Young JA (1984) Microsite requirements for downy brome (*Bromus tectorum*) infestation and control on sagebrush rangelands. Weed Sci 32:13–17

Fielding DJ, Brusven M (1992) Food and habitat preferences of *Melanoplus sanguinipes* and *Aulocara elliotti* (Orthoptera: Acrididae) on disturbed rangeland in southern Idaho. J Econ Entomol 85:783–788

Folke C (2006) Resilience: the emergence of a perspective for social-ecological systems analyses. Glob Environ Change 16:253–267

Folke C, Carpenter S, Walker B et al (2004) Regime shifts, resilience, and biodiversity in ecosystem management. Ann Rev Ecol Syst 35:557–581

Gardener CJ, McIvor JG, Jansen A (1993a) Passage of legume and grass seeds through the digestive tract of cattle and their survival in faeces. J Appl Ecol 30:63–74

Gardener CJ, McIvor JG, Jansen A (1993b) Survival of seeds of tropical grassland species subjected to bovine digestion. J Appl Ecol 30:75–85

Germino MJ, Chambers JC, Brown CS (2015) Introduction: exotic annual *Bromus* in the Western USA. In: Germino MJ, Chambers JC, Brown CS (eds) Exotic brome-grasses in arid and semiarid ecosystems of the Western USA: causes, consequences, and management implications. Springer New York, NY (Chapter 1)

Getz HL, Baker WL (2008) Initial invasion of cheatgrass (*Bromus tectorum*) into burned piñon-juniper woodlands in western Colorado. Am Midl Nat 159:489–497

Goergen E, Chambers JC (2012) Facilitation and interference of seedling establishment by a native legume before and after wildfire. Oecologia 168:199–211

Griffith AB (2010) Positive effects of native shrubs on *Bromus tectorum* demography. Ecology 91:141–154

Gundale MJ, Sutherland S, DeLuca TH (2008) Fire, native species, and soil resource interactions influence the spatio-temporal invasion pattern of *Bromus tectorum*. Ecography 31:201–210

Gutterman Y (1994) Strategies of seed dispersal and germination in plants inhabiting deserts. Bot Rev 60:373–425

Harper KT, Clair LLS (1985) Cryptogamic soil crusts on arid and semiarid rangelands in Utah: effects on seedling establishment and soil stability. Department of Botany and Range Science, Brigham Young University, Provo, UT

Harper KT, Marble JR (1988) A role for nonvascular plants in management of arid and semiarid rangelands. In: Tueller PT (ed) Vegetation science applications for rangeland analysis and management. Springer, pp 135–169
Harradine AR (1986) Seed longevity and seedling establishment of *Bromus diandrus* Roth. Weed Res 26:173–180
Hassan MA, West NE (1986) Dynamics of soil seeds pools in burned and unburned sagebrush semi-deserts. Ecology 67:269–272
Hempy-Mayer K, Pyke DA (2008) Defoliation effects on *Bromus tectorum* seed production: implications for grazing. Rangel Ecol Manag 61:116–123
Hill JP, Germino MJ, Wraith JM et al (2006) Advantages in water relations contribute to greater photosynthesis in *Centaurea maculosa* compared with established grasses. Int J Plant Sci 167:269–277
Hitchcock AS (1971) Manual of the grasses of the United States, vol 2. Courier Dover Publications, New York, NY
Holzapfel C, Mahall BE (1999) Bidirectional facilitation and interference between shrubs and annuals in the Mojave Desert. Ecology 80:1747–1761
Holling CS (1973) Resilience and stability of ecological systems. Ann Rev Ecol Syst 4:1–2
Hoover AN, Germino MJ (2012) A common-garden study of resource-island effects on a native and an exotic, annual grass after fire. Rangel Ecol Manag 65:160–170
Howell W (1998) Germination and establishment of *Bromus tectorum* L. in relation to cation exchange capacity, seedbed, litter, soil cover and water. MS thesis, Prescott College, Arizona
Hulbert LC (1955) Ecological studies of *Bromus tectorum* and other annual bromegrasses. Ecol Monogr 25:181–213
Humphrey LD, Schupp EW (2001) Seed banks of *Bromus tectorum*-dominated communities in the Great Basin. West N Am Nat 61:85–92
James JJ (2008) Effect of soil nitrogen stress on the relative growth rate of annual and perennial grasses in the Intermountain West. Plant Soil 310:201–210
James JJ, Drenovsky RE, Monaco TA et al (2011) Managing soil nitrogen to restore annual grass-infested plant communities: effective strategy or incomplete framework? Ecol Appl 21:490–502
Janzen DH (1984) Dispersal of small seeds by big herbivores: foliage is the fruit. Am Nat 123:338–353
Johnston DB (2011) Movement of weed seeds in reclamation areas. Restor Ecol 19:446–449
Jones R, Chambers, JC, Board DI et al (2015a) Understanding the role of resource limitation in restoration of sagebrush ecosystems dominated by cheatgrass – a mechanistic approach. Ecosphere. 6:107
Jones R, Chambers JC, Johnson DW et al (2015b) Effect of repeated burning on plant and soil carbon and nitrogen in cheatgrass (*Bromus tectorum*) dominated ecosystems. Plant Soil 386:1–18
Jurand BS, Abella SR, Suazo AA (2013) Soil seed bank longevity of the exotic annual grass *Bromus rubens* in the Mojave Desert, USA. J Arid Environ 94:68–75
Jurand BS, Abella SR (2013) Soil seed banks of the exotic annual grass *Bromus rubens* on a burned desert landscape. Rangel Ecol Manag 66:157–163
Kaltenecker JH, Wicklow-Howard M, Pellant M (1999) Biological soil crusts: natural barriers to *Bromus tectorum* L. establishment in the northern Great Basin, USA. In: Eldridge D, Freudenberger D (eds) People and rangelands building the future. Aitkenvale, Queensland, pp 109–111
Kanarek AR, Kao RH (2011) The relationships among plant cover, density, seed rain, and dispersal of *Bromus tectorum* in high-elevation populations. West N Am Nat 71:131–136
Kane JM, Meinhardt KA, Chang T et al (2011) Drought-induced mortality of a foundation species (*Juniperus monosperma*) promotes positive afterlife effects in understory vegetation. Plant Ecol 212:733–741
Karl MG, Heitschmidt R, Haferkamp MR (1999) Vegetation biomass dynamics and patterns of sexual reproduction in a northern mixed-grass prairie. Am Midl Nat 141:227–237

Kotanen PM, Bergelson J, Hazlett DL (1998) Habitats of native and exotic plants in Colorado shortgrass steppe: a comparative approach. Can J Bot 76:664–672

Keeley JE, McGinnis TW (2007) Impact of prescribed fire and other factors on cheatgrass persistence in a Sierra Nevada ponderosa pine forest. Int J Wildl Fire 16:96–106

Kelrick MI (1991) Factors affecting seeds in a sagebrush-steppe ecosystem and implications for the dispersion of an annual plant species, cheatgrass (*Bromus tectorum* L.). Dissertation, Utah State University, Logan, UT

Kleiner EF, Harper K (1977) Occurrence of four major perennial grasses in relation to edaphic factors in a pristine community. J Range Manag 30:286–289

Klemmedson JO, Smith JG (1964) Cheatgrass (*Bromus tectorum* L.). Bot Rev 30:226–262

Knutson KC, Pyke DA, Wirth TA et al (2014) Long-term effects of seeding after wildfire on vegetation in Great Basin shrubland ecosystems. J Appl Ecol 51:1414–1424

Kulmatiski A, Beard KH, Stark JM (2006) Exotic plant communities shift water-use timing in a shrub-steppe ecosystem. Plant Soil 288:271–284

La Tourrette JE, Young JA, Evans RA (1971) Seed dispersal in relation to rodent activities in seral big sagebrush communities. J Range Manag 24:118–120

Larsen KD (1995) Effects of microbiotic crusts on the germination and establishment of three range grasses. MS thesis, Boise State University, Boise, ID

Larsen RT, Flinders JT, Mitchell DL et al (2007) Conservation risks of exotic chukars (*Alectoris chukar*) and their associated management: implications for a widely introduced phasianid. Wildl Res 34:262–270

Leffler AJ, Ryel RJ (2012) Resource pool dynamics: conditions that regulate species interactions and dominance. Invasive plant ecology and management. Linking processes to practice. CABI International, Cambridge, MA

Leger EA, Espeland EK, Merrill KR et al (2009) Genetic variation and local adaptation at a cheatgrass (*Bromus tectorum*) invasion edge in western Nevada. Mol Ecol 18:4366–4379

Leger EA, Goergen EM, de Queiroz TF (2014) Can native annual forbs reduce *Bromus tectorum* biomass and indirectly facilitate establishment of a native perennial grass? J Arid Environ 102:9–16

Levine JM, Adler PB, Yelenik SG (2004) A meta-analysis of biotic resistance to exotic plant invasions. Ecol Lett 7:975–989

Loeser MR, Sisk TD, Crews TE (2007) Impact of grazing intensity during drought in an Arizona grassland. Conserv Biol 21:87–97

Lowe PN, Lauenroth WK, Burke IC (2003) Effects of nitrogen availability on competition between *Bromus tectorum* and *Bouteloua gracilis*. Plant Ecol 167:247–254

Mack RN (1981) Invasion of *Bromus tectorum* L. into Western North America: an ecological chronicle. Agro-Ecosystems 7:145–165

Mack RN, Pyke DA (1983) The demography of *Bromus tectorum*: variation in time and space. J Ecol 71:69–93

Mack RN, Thompson JN (1982) Evolution in steppe with few large, hooved mammals. Am Nat 119:757–773

Mazzola MB, Chambers JC, Blank RR et al (2011) Effects of resource availability and propagule supply on native species recruitment in sagebrush ecosystems invaded by *Bromus tectorum*. Biol Invasions 13:513–526

McCulley RL, Burke IC, Lauenroth WK (2009) Conservation of nitrogen increases with precipitation across a major grassland gradient in the Central Great Plains of North America. Oecologia 159:571–581

McGlone CM, Sieg CH, Kolb TE (2011) Invasion resistance and persistence: established plants win, even with disturbance and high propagule pressure. Biol Invasions 13:291–304

Melgoza G, Nowak RS, Tausch RJ (1990) Soil water exploitation after fire: competition between Bromus tectorum (cheatgrass) and two native species. Oecologia 83:7–13

Meyer SE, Beckstead J, Pearce JL (2015) Community ecology of fungal pathogens on *Bromus tectorum*. In: Germino MJ, Chambers JC, Brown CS (eds) Exotic brome-grasses in arid and

semiarid ecosystems of the Western USA: causes, consequences, and management implications. Springer, New York, NY (Chapter 7)
Meyer SE, Garvin SC, Beckstead J (2001) Factors mediating cheatgrass invasion of intact salt desert shrubland. In: McArthur D, Fairbanks DJ (eds) Shrubland ecosystem genetics and biodiversity. Gen Tech Rep RMRS-P-21. USDA, Forest Service, Rocky Mountain Research Station, Fort Collins, CO, pp 224–232
Meyer SE, Quinney D, Nelson DL, Weaver J (2007) Impact of the pathogen *Pyrenophora semeniperda* on *Bromus tectorum* seedbank dynamics in North American cold deserts. Weed Res 47:54–62
Miller R, Chambers J, Pyke D et al (2013) Fire effects on vegetation and soils in the Great Basin Region: response and site characteristics. Gen Tech Rep RMRS-GTR-308. USDA, Forest Service, Rocky Mountain Research Station, Fort Collins, CO, p 126
Monaco TA, Johnson DA, Norton JM et al (2003) Contrasting responses of Intermountain West grasses to soil nitrogen. J Range Manag 56:282–290
Monty A, Brown CS, Johnston DB (2013) Fire promotes downy brome (*Bromus tectorum* L.) seed dispersal. Biol Invasions 15:1113–1123
Moyes AB, Witter MS, Gamon JA (2005) Restoration of native perennials in a California annual grassland after prescribed spring burning and solarization. Restor Ecol 13:659–666
Mücher H, Chartres C, Tongway DJ et al (1988) Micromorphology and significance of the surface crusts of soils in rangelands near Cobar, Australia. Geoderma 42:227–244
Murray R, Klemmedson J (1968) Cheatgrass range in southern Idaho: seasonal cattle gains and grazing capacities. J Range Manag 42:308–313
Murray RB (1971) Grazing capacity, sheep gains: cheatgrass, bunchgrass ranges in southern Idaho. J Range Manag 21:407–410
Murphy T, Naugle DE, Eardley R et al (2013) Trial by fire: improving our ability to reduce wildfire impacts to sage-grouse and sagebrush ecosystems through accelerated partner collaboration. Rangelands 35:2–10
Newingham B, Vidiella P, Belnap J (2007) Do soil characteristics or microhabitat determine field emergence and success of *Bromus tectorum*? J Arid Environ 70:389–402
Ogle SM, Reiners WA, Gerow KG (2003) Impacts of exotic annual brome grasses (*Bromus* spp.) on ecosystem properties of northern mixed grass prairie. Am Midl Nat 149:46–58
Pakeman RJ, Digneffe G, Small JL (2002) Ecological correlates of endozoochory by herbivores. Funct Ecol 16:296–304
Paruelo JM, Lauenroth WK (1996) Relative abundance of plant functional types in grasslands and shrublands of North America. Ecol Appl 6:1212–1224
Peart MH (1979) Experiments on the biological significance of the morphology of seed-dispersal units in grasses. J Ecol 67:843–863
Pekas KM, Schupp EW (2013) Influence of aboveground vegetation on seed bank composition and distribution in a Great Basin Desert sagebrush community. J Arid Environ 88:113–120
Piemeisel RL (1951) Causes affecting change and rate of change in a vegetation of annuals in Idaho. Ecology 32:53–72
Pierson EA, Mack RN (1990) The population biology of *Bromus tectorum* in forests: distinguishing the opportunity for dispersal from environmental restriction. Oecologia 84: 519–525
Pierson EA, Mack RN, Black RA (1990) The effect of shading on photosynthesis, growth, and regrowth following defoliation for *Bromus tectorum*. Oecologia 84:534–543
Prasse R (1999) Experimentelle untersuchungen an gefäßpflanzenpopulationen auf verschiedenen geländeoberfl_chen in einem sandwüstengebiet (Experimental studies with populations of vascular plants on different soil surfaces in a sand desert area). Dissertation, Universitätsverlag Rasch, Osnabrück, Germany
Prevéy JS, Germino MJ, Huntly NJ (2010) Loss of foundation species increases population growth of exotic forbs in sagebrush steppe. Ecol Appl 20:1890–1902
Pyke DA (1986) Demographic responses of *Bromus tectorum* and seedlings of *Agropyron spicatum* to grazing by small mammals: occurrence and severity of grazing. J Ecol 75:739–754

Pyke DA (1987) Demographic responses of *Bromus tectorum* and seedlings of *Agropyron spicatum* to grazing by small mammals: the influence of grazing frequency and plant age. J Ecol 75:825–835

Pyke DA, Shaff SE, Lindgren A et al (2014) Region-wide ecological responses of arid Wyoming Big Sagebrush communities to fuel treatments. Rangel Ecol Manag 67:455–467

Rasmussen G (1994) Prescribed burning considerations in sagebrush annual grassland communities. In: Monsen S, Kitchen SG (eds) Ecology and management of annual rangelands. Gen Tech Rep INT-GTR-313. USDA, Forest Service Ogden, UT, pp 69–70

Reichenberger G, Pyke DA (1990) Impact of early root competition on fitness components of four semiarid species. Oecologia 85:159–166

Reichman OJ (1979) Desert granivore foraging and its impact on seed densities and distributions. Ecology 60:1086–1092

Reisner MD, Grace JB, Pyke DA et al (2013) Conditions favouring *Bromus tectorum* dominance of endangered sagebrush steppe ecosystems. J Appl Ecol 50:1039–1049

Rice KJ, Mack RN (1991) Ecological genetics of *Bromus tectorum*. Oecologia 88(1):91–101

Roberts HA (1981) Seed banks in soils. Adv Appl Biol 6:1–55

Rogers LE, Uresk DW (1974) Food plant selection by the migratory grasshopper (*Melanoplus sanguinipes*) within a cheatgrass community. Northwest Sci 48:230–234

Roundy BA, Hardegree SP, Chambers JC et al (2007) Prediction of cheatgrass field germination potential using wet thermal accumulation. Rangel Ecol Manag 60:613–623

Ryel RJ, Ivans CY, Peek MS et al (2008) Functional differences in soil water pools: a new perspective on plant water use in water-limited ecosystems. In: Lüttge U, Beyschlag W, Murata J (eds) Progress in botany. Springer, Berlin, pp 397–422

Sala OE, Lauenroth WK, Golluscio RA (1997) Plant functional types in temperate semi-arid regions. In: Smith TM, Shugart HH, Woodward FI (eds) Plant functional types. Cambridge University Press, Cambridge, MA, pp 217–233

Salo LF (2005) Red brome (*Bromus rubens* subsp. *madritensis*) in North America: possible modes for early introductions, subsequent spread. Biol Invasions 7:165–180

Sankey JB, Germino MJ, Sankey TT et al (2012) Fire effects on the spatial patterning of soil properties in sagebrush steppe, USA: a meta-analysis. Int J Wildl Fire 21:545–556

Schmelzer L (2009) Reducing fuel load of key cheatgrass (*Bromus tectorum* L.) dominated range sites by the use of fall cattle grazing. MS thesis, University of Nevada, Reno, NV

Seabloom EW, Harpole WS, Reichman OJ et al (2003) Invasion, competitive dominance, and resource use by exotic and native California grassland species. Proc Natl Acad Sci U S A 100:13384–13389

Shea K, Chesson P (2002) Community ecology theory as a framework for biological invasions. Trends Ecol Evol 17:170–176

Smith DC, Meyer SE, Anderson V (2008) Factors affecting *Bromus tectorum* seed bank carryover in western Utah. Rangel Ecol Manag 61:430–436

Smith S, Monson R, Anderson J (1997) Physiological ecology of North American desert plants. Springer, New York, NY

Sorensen AE (1986) Seed dispersal by adhesion. Ann Rev Ecol Syst 17:443–463

St Clair L, Webb BL, Johansen JR et al (1984) Cryptogamic soil crusts: enhancement of seedling establishment in disturbed and undisturbed areas. Reclam Reveg Res 3:129–136

Strand EK, Launchbaugh KL, Limb RF et al (2014) Livestock grazing effects on fuel loads for wildland fire in sagebrush dominated ecosystems. J Rangel Appl 1:35–57

Thill DC, Beck KG, Callihan RH (1984) The biology of downy brome (*Bromus tectorum*). Weed Sci 32:7–12

Wainwright CE, Wolkovich EM, Cleland EE (2012) Seasonal priority effects: implications for invasion and restoration in a semi-arid system. J Appl Ecol 49:234–241

Wells FH, Lauenroth WK (2007) The potential for horses to disperse alien plants along recreational trails. Rangel Ecol Manag 60:574–577

West NE, Yorks TP (2002) Vegetation responses following wildfire on grazed and ungrazed sagebrush semi-desert. J Range Manag 55:171–181

Wikeem B, Pitt M (1992) Diet of California bighorn sheep, *Ovis canadensis californiana*, in British Columbia: assessing optimal foraging habitat. Can Field-Nat 106:327–335
Wilcox BP, Turnbull L, Young MH et al (2012) Invasion of shrublands by exotic grasses: ecohydrological consequences in cold versus warm deserts. Ecohydrology 5:160–173
Wu K, Jain S (1979) Population regulation in *Bromus rubens* and *B. mollis*: life cycle components and competition. Oecologia 39:337–357
Young JA, Evans RA (1975) Germinability of seed reserves in a big sagebrush community. Weed Sci 23:358–364
Young JA, Evans RA, Palmquist DE (1989) Big sagebrush (*Artemisia tridentata*) seed production. Weed Sci 37:47–53

Part IV
Relating the Science to Human Uses and Restoration of Western Rangeland Landscapes

Chapter 11
Land Uses, Fire, and Invasion: Exotic Annual *Bromus* and Human Dimensions

David A. Pyke, Jeanne C. Chambers, Jeffrey L. Beck, Matthew L. Brooks, and Brian A. Mealor

Abstract Human land uses are the primary cause of the introduction and spread of exotic annual *Bromus* species. Initial introductions were likely linked to contaminated seeds used by homesteading farmers in the late 1880s and early 1900s. Transportation routes aided their spread. Unrestricted livestock grazing from the 1800s through the mid-1900s reduced native plant competitors leaving large areas vulnerable to *Bromus* dominance. Ecosystems with cooler and moister soils tend to have greater potential to recover from disturbances (resilience) and to be more resistant to *Bromus* invasion and dominance. Warmer and drier ecosystems are less resistant to *Bromus* and are threatened by altered fire regimes which can lead to *Bromus* dominance, impacts to wildlife, and alternative stable states. Native Americans used fire for manipulating plant communities and may have contributed to the early dominance of *Bromus* in portions of California. Fire as a tool is now limited to site

D.A. Pyke (✉)
US Geological Survey, Forest and Rangeland Ecosystem Science Center, Corvallis, OR 97331, USA
e-mail: david_a_pyke@usgs.gov

J.C. Chambers
USDA Forest Service, Rocky Mountain Research Station, Reno, NV 89512, USA
e-mail: jchambers@fs.fed.us

J.L. Beck
Department of Ecosystem Science and Management, University of Wyoming, Laramie, WY 82071, USA
e-mail: jlbeck@uwyo.edu

M.L. Brooks
US Geological Survey, Western Ecological Research Center, Oakhurst, CA 93644, USA
e-mail: matt_brooks@usgs.gov

B.A. Mealor
Sheridan Research and Extension Center, University of Wyoming, Sheridan, WY 82801, USA
e-mail: bamealor@uwyo.edu

M.J. Germino et al. (eds.), *Exotic Brome-Grasses in Arid and Semiarid Ecosystems of the Western US*, Springer Series on Environmental Management,
DOI 10.1007/978-3-319-24930-8_11

preparation for revegetation in most ecosystems where *Bromus* is a significant problem. Once *Bromus* dominates, breaking annual grass/fire cycles requires restoring fire-tolerant perennial grasses and forbs, which can compete with *Bromus* and resist its dominance. Current weed management policies often lack regulations to prevent further expansion of *Bromus*. Research is needed on how and where livestock grazing might help increase perennial grass and forb cover and density to create ecosystems that are more resistant to *Bromus*. Also, studies are needed to ascertain the role, if any, of oil and gas development in contributing to the spread of *Bromus*.

Keywords Energy development • Farming • Grazing • Management policies • Wildlife responses

11.1 Introduction

Although human occupation of the western USA began 12,000–30,000 years before present (Henige 1998), degradation of ecosystems by humans was likely minimal and localized (e.g., Anasazi culture; Janssen and Scheffer 2004). Not until Euro-American settlements did human land use begin to reshape ecosystems. Spanish exploration of New Mexico and California began in 1540, but settlements did not occur until 1598 and 1769, respectively. Additional Euro-American settlements associated with the fur industry began to appear in northern California, Oregon, and Washington in the early 1800s. Establishment of the Santa Fe and Oregon Trails in 1821 and 1842, followed by a series of travel routes into California, Utah, and the Northwest, resulted in migration to the West, and initiation of agrarian and mining communities throughout the region (Duffus 1972; Olson 2004). With these settlements came changes in land use from largely nomadic fur trading to ranching, farming, mining, and commerce. Rapid population growth in the twentieth century through present resulted in conversion of many native ecosystems to farmlands or urban and exurban population centers. To support these areas, development of water use, energy extraction, and vast transportation and utility networks became necessary (e.g., Chambers and Wisdom 2009; Smith et al. 2009). In less than a century, these lands have become fragmented and highly altered.

Land uses by humans modify ecosystem resilience to disturbance and resistance to invasion through modifications of abiotic attributes (e.g., nutrient deposition and hydrologic and geomorphic processes) along with biotic attributes (e.g., productivity, species composition, and species interactions) (see Fig. 1.1 in Germnio et al. 2015b; Brooks et al. 2015; Chambers et al. 2014a, 2015). Resilience of native ecosystems varies over environmental gradients and depends largely on soil temperature and moisture regimes and ecosystem productivity. Resistance to invasive species, specifically exotic annual *Bromus* species (*Bromus* hereafter), is a function of (1) suitable climate for establishment and persistence of *Bromus* and (2) interactions with the plant community such as competition for resources, herbivory, etc. (Chambers et al. 2015).

Over the last century and a half, *Bromus* has colonized, established, and become dominant over large expanses of the western USA (USDA NRCS 2014) largely as a function of human modification of lands (Bangert and Huntley 2010). Governmental regulations and policies may promote or constrain degradation. For example, the Mexican General Colonization Acts of 1824 in California and the various USA homestead acts of the late 1800s brought settlements, crops, and livestock into the region. Limited restrictions on use of public land coupled with policies to maximize livestock production during major wars contributed to widespread degradation. The recognition of the degradation, in turn, brought about conservation measures aimed at improving lands (Donahue 1999). Continued interpretations and changes in policy and land uses still shape these lands today.

In this chapter, we discuss the role of land uses and policies regarding land management and the sustainable use of natural resources that may play a role in *Bromus* invasion and potential dominance. We relate these uses and policies to ecosystems' resilience to disturbances and resistance to invasions by *Bromus* (Brooks et al. 2015; Chambers et al. 2015). Fire is a potential natural disturbance in nearly all ecosystems and can shift the resistance of communities to *Bromus* invasion leading to altered fire regimes in several western ecosystems (Germino et al. 2015a). We briefly discuss these changes and then discuss how fire remains a potential vegetation management tool that may assist in restoration and adjustments in fuels for wildfire management, but that carries inherent risks for management of other species in the ecosystem.

Since most of the available research on *Bromus* introductions and their shift in community dominance relative to land uses comes largely from studies on *Bromus tectorum* L. (cheatgrass or downy brome) in the Cold Deserts through the western Wyoming Basin, we focus the majority of our discussion on this species and these regions. The shift in dominance to *Bromus* in California was largely complete before ecologists understood the community associations and the impacts of land uses on vegetation. Therefore, we know less about resilience to disturbance and resistance to *Bromus* in those ecosystems (Bartolome et al. 2007). *Bromus hordeaceus* L. (soft brome or soft chess; syn. *B. mollis*) and *B. diandrus* Roth (ripgut brome; syn. *B. diandrus* ssp. *rigidis* or *B. rigidus*) occur in the moister northern portions of the Mediterranean California region, and *Bromus madritensis* L. (foxtail brome or compact brome; syn. *B. madritensis* ssp. *madritensis*) and *B. rubens* L. (red brome; syn. *B. madritensis* ssp. *rubens*) in arid southern portions of the Mediterranean California region and in the Mojave Basin and Range of the Warm Deserts. *Bromus* invasion and spread in the Wyoming Basin of the Cold Deserts and the Great Plains has been more recent than other regions with fewer studies. *Bromus arvensis* L. (field brome or Japanese brome; syn. *B. japonicus*) tends to replace or co-dominate with *B. tectorum* as the Wyoming Basin grades into the Northwestern Great Plains. We supplement the discussion of *B. tectorum* with information on other species and regions when available.

We conclude with management implications and research needs. In the management section we include some potential policy considerations that may enhance resilience of ecosystems and strengthen resistance to *Bromus* invasions. Research needs are focused on gaining a better understanding of early-warning indicators for effective adaptive management of land uses.

11.2 Fire Regimes and Uses Across Regions

Ecosystems with relatively high productivity and fuel continuity, and with climates conducive to seasons with dry fuel, had more frequent pre-settlement fires and typically evolved more fire-tolerant species (Pausas and Bradstock 2007). In contrast, ecosystems with limited fuel production and continuity tended to have smaller and less frequent fires. Species in these ecosystems evolved in the near absence of fire and were largely fire intolerant (Brooks and Minnich 2006). The relationship of fire frequency to fire-tolerant life forms can be observed along environmental/productivity gradients; resprouting species tend to increase while obligate seeders decrease with increasing productivity (Pausas and Bradstock 2007; Davies et al. 2012).

The annual life form tends to be fire tolerant due to the ability of seeds to escape high fire temperatures after they disperse on or below the soil surface. In both the Mediterranean California region and the Palouse Prairies of Washington and Oregon in the Cold Desert region, where productivity and plant cover was high, native annual plants may have coexisted with perennial grasses or shrubs contributing to a continuous fuel source (Bartolome et al. 2007; Keeley et al. 2012). Keeley et al. (2012) implied that Native American's use of fire in Mediterranean California may have tipped the balance from native perennials toward native annuals making some areas more prone to rapid expansion of exotic annual grasses (*B. hordeaceus*, *B. diandrus*, *B. madritensis*, and *B. rubens*) after their introduction (Keeley et al. 2012).

Before the introduction of exotic annual grasses into the Mojave Basin and Range of the Warm Deserts, the abundance of native annual plants varied yearly depending on winter and summer precipitation (Keeler-Wolf 2007), but did not likely contribute a continuous fuel source. Also, ignition sources were generally infrequent (Brooks 1999). Similarly, the Cold Deserts lacked fuel from annual plants, but perennials provided adequate amounts and continuity of fuel to carry fires in locations with higher annual precipitation (cool and moist communities). In contrast, warm and dry communities lacked adequate fuel amounts and continuity to carry fires (Germino et al. 2015a).

Plant invasions have the potential to alter fuelbed conditions, fire behavior, and fire regimes in a self-perpetuating process referred to as the invasive plant/fire regime cycle or annual grass/fire cycle for areas invaded by *Bromus* (D'Antonio and Vitousek 1992; Brooks et al. 2004). *Bromus tectorum* and *B. rubens* invaded arid and semiarid woody ecosystems of the Cold and Warm Deserts. They increased fine surface vegetation and horizontal fuel continuity and ignitability, and decreased coarse canopy fuels, leading to increased frequency, extent, and seasonal window of fires (Fig. 11.1, Brooks et al. 2006, 2015). The potential for these changes is driven by characteristics of the invaded ecoregion, specifically its resilience

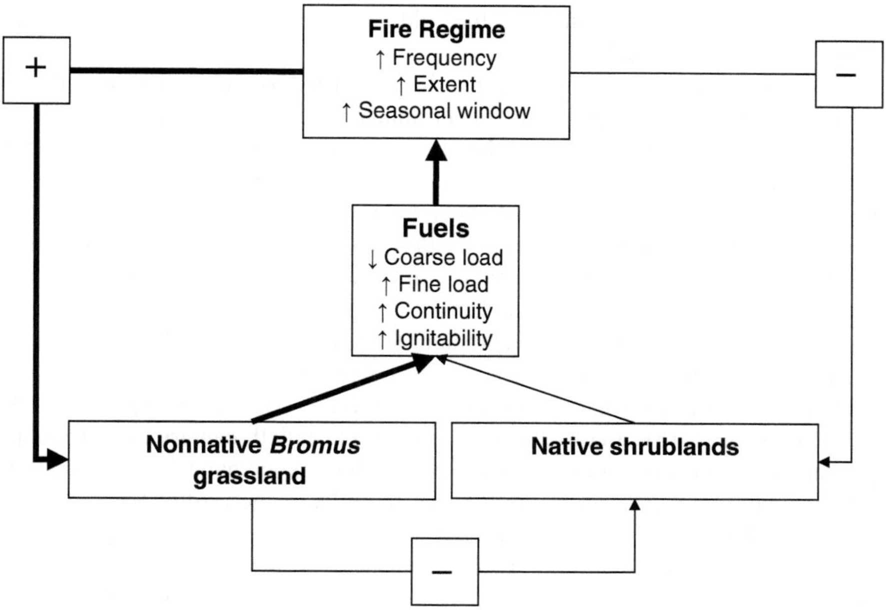

Fig. 11.1 Changes in fuelbed and fire regime properties caused by the invasion of exotic annual *Bromus* species into woody shrubland ecosystems. These new conditions are characterized by a positive feedback between *Bromus* grassland fuels and the altered fire regime, and a negative feedback between native shrubland fuels and the new regime. As successive, short-interval fires occur and fuels type-convert from native shrubland to *Bromus* grassland, the positive feedback becomes more prominent as highlighted by heavier weighting of that feedback loop in the figure (modified from Brooks 2006, Fig. 3.7)

to fire as influenced by the reference communities' fire regime without *Bromus* (Brooks et al. 2015).

Ecoregions with the greatest potential for an annual grass/fire cycle and vegetation type conversion to *Bromus* grasslands are those that experienced limited historical fire and possessed native species with low tolerance to fire, but are not so warm and dry (thermic to mesic and aridic) that resistance to *Bromus* invasion is high (Brooks et al. 2015). These include the more moist ecosystems of the Warm Desert and the more arid ecosystems of the Cold Desert and Mediterranean California ecoregions. Relatively hot and dry conditions in these ecosystems lead to low native plant productivity, growth rates, and fuel loads, and thus infrequent historical fire and low tolerance of native plants to fire.

Ecosystems with the least potential for an annual grass/fire cycle are those that have either high resistance to *Bromus* or high resilience to fire (Brooks et al. 2015). These include the hottest and driest ecosystems of the Warm Desert ecoregion where high temperatures and low precipitation levels exceed the tolerances of *Bromus* species (Fig. 2.2 in Brooks et al. 2015). They also include the cooler and moister ecosystem types of the Western Forest, Northwestern Great Plains, and Mediterranean California ecoregions where low temperature and high precipitation levels exceed *Bromus* tolerances on the cold and wet end of the climate spectrum (Fig. 2.2 in Brooks et al. 2015).

Land uses that reduce ecosystem resistance to *Bromus* and resilience to disturbance may increase the potential for an annual grass/fire cycle and altered fire regimes. A primary mechanism in this process is alteration of the historical reference disturbance regime. Fire regimes may be altered by increasing the amount of fire in space and time, as described above for the grass/fire cycle where land uses have reduced ecosystem resistance to *Bromus*. Alternatively, fire suppression, another form of human land use, reduces the amount of fire and can lead to fuel accumulation and high severity and intensity fires. This alternative is currently playing out in the Western Forest ecoregion (Hurteau et al. 2013). This situation has been greatly facilitated by forest clear-cutting and fire suppression that reduced forest resilience and ultimately led to conditions that favor *Bromus* species (e.g., Fig. 2.15 in Brooks et al. 2015). A consequence has been a twentieth-century loss of foothill pine and blue oak woodland and increased extent of exotic annual grasses (mostly *Bromus* and *Avena* species) in the western foothills of the central Sierra Nevada mountains during the 1900s (Thorne et al. 2008).

Fire, as a land use tool, may increase resistance to *Bromus* in relatively intact systems by enhancing the competitive ability of perennial herbaceous species (Chambers et al. 2014b) although temporary (3–5 years) increases in *Bromus* may occur (Miller et al. 2013). In systems where *Bromus* dominates, this reduction is only temporary, generally only lasting 3 years in the Mediterranean California (Bartolome et al. 2007) and in the Cold Deserts ecoregions (Miller et al. 2013). A fire-induced reduction of *Bromus* may be coupled with revegetation in an integrated weed management and perennial plant restoration project where fire is a site preparation tool for restoration (Monaco et al. 2015).

11.3 Land Uses

11.3.1 Native American Land Uses before Exotic Annual Bromus *Introduction*

It is difficult to say what native vegetation in the western USA was like before Euro-American colonization, especially before disease-induced population declines of Native Americans (pre-1500s; Lovell 1992). Paleo-botanical and archeological records provide insights into Native American land uses along with plant and animal dominance before Euro-American influences. The ratio of settled to nomadic lifestyle of each culture depended on climate within their homelands, sources of water, and the availability of seasonal foods and natural resources for subsistence. Wildlife were used by both nomadic and agrarian cultures as primary food sources, and wildlife were suspected to have fluctuated due to Native American hunting (Kay 1994; Hart 2001). Woody plants that were preferred by wildlife, and more sensitive to browsing, were likely reduced when wildlife populations were high and Native American populations were low resulting in plant community compositional changes (Laliberte and Ripple 2003).

The use of agriculture by the Ancient Pueblo (also termed Anasazi) cultures of the southwest began centuries before the current era and ranged from the Arizona/New Mexico Plateau and Colorado Plateau of the Cold Desert region to the Mojave Basin and Range of the Warm Deserts (Hard et al. 1996; Vlaisch 2005; Raish 2013). Areas were cleared of existing vegetation and a variety of methods were used to irrigate crops (Vlaisch 2005). Tree removal for fire wood and building was common and was evidenced by reductions in juniper pollen counts from some locations during increases in human populations (Hevly 1988). Extended and repeated droughts in the twelfth and thirteenth centuries likely contributed to crop failures and eventually to community abandonment by these cultures (Benson et al. 2007). However, many of these people continued to subsist off of these lands.

Many western tribes used fire, provided fuel was adequate, to reduce woody plant dominance and encourage desired food plants, wildlife, or access for farming, transportation, or hunting (Keeley 2002). Evidence for such uses comes primarily from Mediterranean California, northern Cold Deserts, and Great Plains ecoregions (Gruell 1985; Keeley 2002), and not the less productive lands of the southern Cold Deserts or Mojave Basin and Range of the Warm Deserts (Brooks et al. 2013).

This brief review indicates that native Americans likely had significant impacts on ecosystems prior to Euro-American colonization, and may have influenced longer-term ecosystem trajectories including the fire-induced reductions in ecosystem resistance to *Bromus* species mentioned above (Keeley et al. 2012). The difference between Native-American and Euro-American land use, especially at the time of *Bromus* invasion, is that Native-American use tended to be more localized, severe disturbances were smaller in extent, and post-disturbance successional processes, before Euro-Americans, took place in the absence of invasive species like *Bromus*.

11.3.2 Role of Land Uses and Regulatory Policies in Exotic Annual **Bromus** *Colonization and Spread*

A common theme between Native American and Euro-American settlements was their associations with lower elevation lands near perennial water sources that typically had deeper soils and warmer temperatures. This is a global theme of people residing in continental interior locations (Small and Cohen 2004) that continues to shape human land use and impacts in the western USA and that often has negative effects on ecosystem resilience and resistance. These lower elevation lands, especially those farther from perennial water sources, are often warmer and drier locales that are least resilient to disturbances and most vulnerable to *Bromus* invasion (Brooks et al. 2015; Chambers et al. 2015).

Land policies from the 1850s through 1916 yielded a variety of homestead acts that contributed to land degradation through their impacts on the least resilient lands. These acts gave federal lands for settlement and required farming or ranching on a portion of those lands, but restricted the amount of land to 259 ha (640 ac) or less (Donahue 1999). Initially, lands with readily available irrigation water were settled. Later, dryland farming opened new lands for settlement on even warmer and drier, less resilient lands. Within these ecoregions, 259 ha were recognized as largely insufficient to sustain a family since productivity was low (Donahue 1999). Some ranchers were able to get around this restriction by owning lands with water and grazing livestock on adjacent public lands to take advantage of the extended forage base. However, these practices led to degradation of these common lands and set the stage for the introduction and expansion of *Bromus*.

Introductions of *Bromus* into the western states were first recorded during the homestead era (Mack 1981; Salo 2005). *Bromus tectorum* and *B.rubens* were first noted in the northern Cold Deserts and Mediterranean California, respectively, at transfer points along transportation corridors such as sea ports, river depots, and train stations (Mack 1981; Salo 2005). Both species, along with *B. madritensis*, are suspected to have been unintentional contaminants of seed grains for crops (Mack 1981, 1986, 1991; Salo 2005). Less is known about introductions of other *Bromus*, but since they increase with disturbance like their congeners, they may have been crop seed contaminants as well. Seeds of *Bromus* that were contaminants with purchased grain seeds also grew along with the crop and were later harvested and planted on the same or other farms (Mack 2000). As lives became more settled, people began ordering plants through seed catalogs for landscaping, ornamental uses, and erosion control. Several *Bromus* were available from these sources (Mack 1991).

By the mid-1900s, *B. tectorum* and *B. rubens* colonized most of their current ranges (Mack 1981; Salo 2005). The human footprint (impact of human presence and their action on ecosystems; Sanderson et al. 2002) across the western USA expanded during this period. By 2000, a minimum estimate of 13 % of the land area in the western USA was impacted by humans (Leu et al. 2008). Nearly 10 % was due to crop production, about 2 % resulted from populated areas, and slightly over 1 % was due to improved roads (excluding unmaintained single or two-track roads). Infrastructures to sustain human life (power lines, irrigation canals, railroads, and energy development in decreasing magnitude of importance) represented 0.01–0.05 % of the land (Leu et al. 2008; Bangert and Huntley 2010). Disturbances related to each land use increase the likelihood of colonization and spread of invasive species, especially in areas of low inherent resilience to disturbance and resistance to *Bromus*.

11.3.2.1 Cropland

Crop growth was encouraged by homestead laws, but failures of dryland farming in the early 1900s resulted in millions of hectares of abandoned land in the Intermountain West (Stewart 1938), where *B. tectorum* was likely already present (Mack 1981). Morris et al. (2011) found differences in shrub and forb cover on previously cultivated and never cultivated lands, but the direction of the difference may depend on the ecological site. Surprisingly, they found little *B. tectorum* on both never cultivated and previously cultivated areas, but noted their result was an anomaly relative to similar studies in the Cold Desert region where *B. tectorum* has dominated previously cultivated sites for up to 50 years.

Technological improvements to current seed-harvesting equipment now prevent most contamination of crop seed with weeds (however, see weed policies below) and thus prevent direct seeding of weeds with crops. Yet, activities associated with croplands, such as livestock grazing, fertilization, or herbicide drift, still increase the likelihood of weeds in lands surrounding crops. Within several kilometers of cropland, *B. tectorum* is more likely to exceed cover levels found in areas not associated with cropland (Bradley and Mustard 2006; Bradley 2010). Bradley and Mustard (2006) found that *B. tectorum* appeared to become more abundant near cultivation between 1973 and 2001, indicating that associated activities in adjacent lands (listed above) might have reduced resistance near these fields. Farmers are required to treat noxious weeds on their property, but those weeds not classified as noxious, including *Bromus*, may remain untreated while nontarget native species might be unintentionally harmed. A common practice to control weeds on roadsides, ditches, and field margins is by using fire. However, this may favor *Bromus* since it responds well to the post-fire environment and to the lack of competitors (Miller et al. 2013).

11.3.2.2 Weed Policies and Exotic Annual *Bromus*

Noxious weeds are plants declared by a government to be injurious to public health, agriculture, recreation, wildlife, or property. Property owners must control noxious weeds or face potential penalties. Generally, these are weeds that have been rarely seen in the locale for which agencies are attempting early detection and rapid response for weed eradication or control. Once weeds become common in an area, they may be excluded from the noxious category because the cost of control may exceed the potential economic threat to an individual or county (county exceptions do occur in Wyoming; Mealor et al. 2013). However, when *Bromus* dominance changes fire regimes, the resulting costs of fire management may exceed a billion dollars annually in the western USA (based on 1/3 of the 2009 appropriation to federal agencies; USGAO 2009). These costs are paid by Federal and State funds and are never considered regarding noxious weed status. A 2015 Department of the Interior Secretarial Order (3336) attempts to shift priorities and begins to examine *Bromus* control and restoration as a means of reducing these costs and combating this threat. Ecoregions where *Bromus* species have the potential to change fire regimes may be treated with fuel management funds and receive higher priority for fire suppression in the future.

Other regulatory programs encourage reductions in the spread of weeds from croplands to new locations. Weed-free forage programs are similar to noxious weed lists because states that participate have a goal of controlling noxious weed dispersal associated with sales of livestock forage. Public land management agencies (US Forest Service and Bureau of Land Management [BLM]) require use of weed-free forage on many of their lands. However, *Bromus* species are not represented as part of the minimum standards for weed-free forage (North American Invasive Species Management Association [NAISM]; http://www.naisma.org/images/WFFStandards.pdfAccessed 06/09/2015). The inclusion of *Bromus* may incur additional costs of inspection and treatment, thus making regulatory programs prohibitive to implement.

11.3.2.3 Land Management Policies Addressing Public Land Uses

Degradation of native ecosystems has resulted in regulatory and legal changes geared at conserving and restoring resilient and resistant native ecosystems. Land uses are regulated to some degree to halt the tragedy of the commons (Hardin 1968) by largely declaring that natural ecosystems are not available for common uses, but also fall under certain rules and regulations to sustain goods and services. The overuse of lands in the West by livestock during the early 1900s was a clear example of the tragedy of the commons (US Department of Agriculture, Secretary of Agriculture 1936).

After the advent of homesteading, those lands not claimed by individuals remained in the public trust, but the USA lacked management policies in the late 1800s and early 1900s to prevent degradation of lands and thus the loss of perennial plants, soils, or water and ultimately resilience. The degradation was recognized at the turn of the twentieth century and some livestock grazing restrictions were implemented on forest reserves in 1902, but non-forested lands largely were unregulated until the passage of the Taylor Grazing Act in 1934. This act was intended to halt the progression of overuse and allow the orderly use and improvement of public rangelands in the west.

Degradation was severe by the time this act was passed and *Bromus* had colonized most of the west, thus filling the void left by the loss of perennial grasses due to overgrazing by livestock (US Department of Agriculture, Secretary of Agriculture 1936; Leopold 1941; Mack 1981; Salo 2005). However, it took until the late 1960s and 1970s for passage of the Federal Land Policy and Management Act of 1976 and the Public Rangeland Improvement Act of 1978. These Acts recognized that (1) public lands should be managed for multiple uses, (2) permits for livestock grazing should be connected with the condition of the land, and (3) land condition should be monitored and reported.

Some legislation during that period led to increases of animal use on rangelands because of the Wild and Free-roaming Horses and Burros Act of 1971. This law provided protection for feral horses and burros and allowed their use on lands year-around. Season of use is regulated for domestic livestock, but not for feral horses and burros which use rangeland at will, regardless of the resilience of the plant community to defoliation or resistance to invasive plants (see grazing tolerance below). Feral horse and burro numbers are regulated through annual roundups of animals, but restricting them to specific areas and maintaining their populations at desired levels are difficult.

More recent legislation has attempted to expand the assessment of land condition beyond the key plants for livestock. Rangeland Reform of 1994 implemented new standards for rangelands that followed recommendations of a National Academy panel of experts (National Research Council 1994) that lands be evaluated relative to the lands' potential to support plant production and composition, and that soil, water, and biological components be included. This recognized that lands dominated by invasive plants, such as *Bromus*, may no longer have the ability to recover even if disturbances are removed since they have essentially lost their resilience and resistance. The dominance of *Bromus*, even if it could be used as livestock forage, was recognized as a degradation of the status of the land.

In 2015, Secretarial Order 3336 was released by the Department of the Interior to reduce sizes of fires and the spread of invasive plants like *B. tectorum* in an attempt to sustain sagebrush-grassland ecosystems and *Centrocercus urophasianus* Bonaparte (greater sage-grouse) in the western USA. This is the first Federal policy, to our knowledge, that has directly addressed *Bromus* management.

11.3.2.4 Livestock Grazing

The most ubiquitous land use in the western USA is livestock grazing. This use cannot be ruled out as contributing to the initial range expansion of *Bromus*. In the California Mediterranean ecoregion, The General Colonization Laws of Mexico granted large tracts of land for ranches in California in 1824 beginning overuse of western lands by livestock. For example, governmental reports on the status of rangelands made specific mentions of dominance of *Bromus* and depletion of native perennials as a result of uncontrolled livestock in the Mediterranean California and Cold Desert ecoregions (called California Foothills and Pacific Bunchgrass in the report; US Department of Agriculture, Secretary of Agriculture 1936).

Millions of hectares of rangelands in the Western USA are grazed by wildlife, livestock, and feral horses or burros each year. Grazing responses differ widely among individual plant species and communities largely due to inherent levels of resistance and resilience of plant communities to grazer-induced stresses. Differences in grazing (including defoliation and hoof impacts) by various herbivores, and timing and intensity of grazing, undoubtedly play a large role in the susceptibility of rangelands to invasion by *Bromus* species.

Grazing Tolerance Controlling *Bromus* on lands grazed by livestock requires maintaining perennial plants in sufficient densities and distributions (Chambers et al. 2007, 2014a; Reisner et al. 2013). Grazing tolerance, or a plant's ability to survive and compete for resources with other plants while being defoliated, is formed by adaptations often selected through evolution with grazing animals (Strauss and Agrawal 1999). For native plants, this relates to adaptive traits developed through grazing selection at the location, but for exotic species it depends on the evolutionary adaptations to grazing where they initially evolved. The interactions among the grazing responses of exotic and native species will determine the community's response to current grazing (Fig. 11.2).

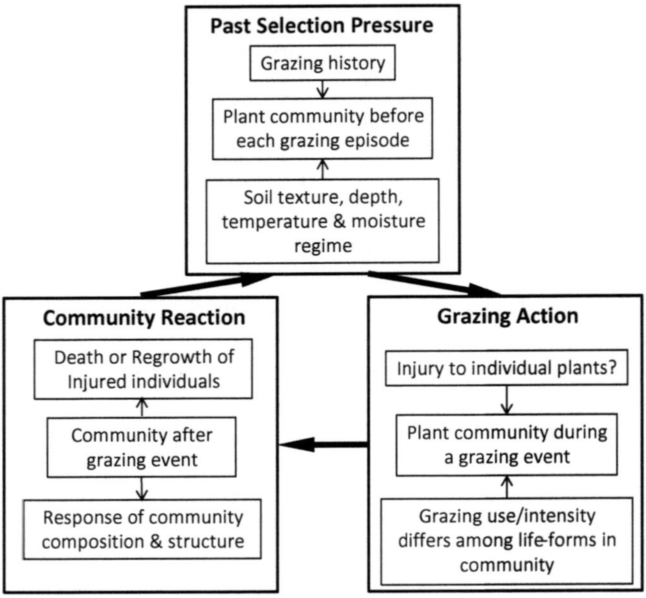

Past Selection Pressure

	Annual *Bromus*		Native Herbaceous		Community Reaction
	Environment	Graze	Environment	Graze	
Great Plains	-	+	+	+	NH => AB
Cold Desert	+	+	+	--	AB >> NH
Warm Desert	+	+	+	--	AB >> NH
Med. California	+	+	+	--	AB >> NH

Fig. 11.2 Plant community structure (above) as shaped by grazing during grazing events, by the community's response to the grazing, and by the past selection pressures including the environment and grazing among the grazed plants at the site of evolution (modified from Milchunas et al. 1988). In the table (below), negative (–) and positive (+) signs indicate degree of adaptation for annual *Bromus* species (AB) and native herbaceous species (NH) to environments of four ecoregions and to grazing during the active growing season. Under community reaction, equal than and greater than signs (>) reflect dominance strength of AB and NH based on these adaptations

The geographic center of evolution for the *Bromus* species that are invasive in the Western USA is between the Middle East and eastern Europe (Sales 1994; Atkinson and Brown 2015), the region of origins of plant and animal domestication (about 10,000 years BP; Flannery 1969). *Bromus* seeds were associated with humans from at least 5000 years BP and likely earlier (Sales 1994; Marinova 2003; White et al. 2014). Thus, they had ample time to evolve with both crops and domestic livestock grazing. The annual life cycle is one adaptation that imparts grazing tolerance to these species. As populations of *B. tectorum* decrease in one generation, and in the absence of competition from native perennial plants, the remaining individuals tend to produce more seeds for the next generation compensating for temporary population reductions (Mack and Pyke 1983; Hempy-Mayer and Pyke 2008). Seed banks in soil may not be impacted directly by grazing intensities (Clements et al. 2007); therefore, once *Bromus* becomes abundant within plant communities, their seed densities tend to dominate seed banks (Chambers et al. 2015).

In addition, a morphological adaptation of *B. tectorum* is the ability to develop decumbent tillers by activating new buds after multiple grazing occurrences. This is an avoidance mechanism that results in placement of inflorescences and seeds below the lowest grazing level of many livestock (Hempy-Mayer and Pyke 2008). This combination of grazing avoidance and tolerance makes it difficult to use livestock to eliminate *B. tectorum* without perennial plant competitors to access available resources.

Native grasses in the western USA vary in their evolved traits for climate which may have determined the likelihood of frequent grazing by animals and influenced the evolution of adaptations to tolerate grazing (Mack and Thompson 1982; Milchunas et al. 1988, Fig. 11.2). A gradient in precipitation seasonality exists from east to west and from south to north across the western USA. The Northwestern Great Plains consistently receives higher amounts of summer precipitation than other western ecoregions. In addition, these prairies evolved with large herds of bison that persisted in this area due to the abundance of nutritional forage throughout most of the year. These plants adapted to grazing because they had predictable resources for regrowth and had grazing tolerance mechanisms such as clonal growth. In contrast, most precipitation arrives in winter in the Mojave Basin and Range portion of the Warm Deserts, in the Cold Deserts of the Intermountain West, and in the Mediterranean California ecoregions. In these winter-wet regions, native grasses do not have consistent growing season moisture. This kept plant production and nutrition low especially in the summer which in turn kept populations of grazing animals low and not distributed widely throughout these regions (Mack and Thompson 1982; Milchunas et al. 1988).

Plant communities within all western ecoregions except the Northwestern Great Plains are less grazing resistant and those with suitable climate regimes are subject to *Bromus* dominance once these communities lose deep-rooted perennial grasses. Inappropriate grazing that leads to degradation often entails excessive defoliation repeated over time resulting from overstocking the land with animals (i.e., overgrazing; Briske et al. 2011) or repeated defoliation of perennial grasses between the middle and late growing season weakening the grasses' ability to regrow in the following year (i.e., season of use; Briske and Richards 1995)—both of which are likely related to patchy grazing distribution across the landscape. This may eventually lead to reductions in perennial plants, increases in *Bromus* dominance, and ultimately in a type conversion to annual grasslands, where perennial plants have low grazing tolerance. Some speculate this happened early in the Mediterranean California region due to the early Spanish/Mexican ranches with year-round grazing that was continued when this area became part of the USA. The same levels or timing of grazing may play less of a role in *Bromus* dominance in regions with higher grazing tolerance, such as Great Plains grasslands, but may still contribute to *Bromus* presence within this ecoregion. The full potential of *Bromus* invasion in the Northwestern Great Plains is unknown. *Bromus arvensis* is a persistent component of large areas of the cooler and wetter Northwestern Great Plains (Haferkamp et al. 1992). *Bromus tectorum* and *B. arvensis* are significant components at the intersection of the warm and dry to cool and dry Cold Deserts and Northwestern Great Plains in Wyoming (Mealor, unpublished data). They may both continue to spread and become more dominant as land uses expand in this region (see Energy Development below).

Trampling Impacts Trampling may also affect *Bromus* dynamics. Perhaps one of the most important indirect impacts of livestock trampling that appears to benefit *B. tectorum* is the reduction and fragmentation of biological soil crusts in many Cold Desert plant communities (Warren and Eldridge 2001; Reisner et al. 2013). Biological soil crusts aid in soil stabilization and may reduce susceptibility to invasion by *Bromus* and other weedy species in arid ecosystems (Chambers et al. 2015). Physical damage to crusts, caused by hooves of grazing animals (and recreationalist's shoes or OHVs), increases the number of safe sites (sensu Harper 1977) where annual grasses can emerge and establish (Chambers et al. 2015). This may produce gradients of biological soil crust cover that are negatively correlated with livestock grazing intensity and *B. tectorum* cover as shown in eastern Oregon (Reisner et al. 2013). The degree that this relationship can be extended to other *Bromus* has not been tested, but it appears to be a reasonable outcome since many of the *Bromus* have similar traits (Atkinson and Brown 2015).

Grazing and Exotic Annual *Bromus* Species Invasion and Dominance Mismanagement of livestock grazing in the late 1800s and early to mid-1900s is considered a primary driver in the widespread introduction and dominance of *B. tectorum* into the Cold Deserts (Mack 1981; Young and Clements 2007) and potentially *B. rubens* in California Mediterranean and the Mojave Basin and Range of the Warm Deserts (Salo 2005). However, it is sometimes difficult to disentangle impacts of historic livestock grazing and current grazing practices. Surprisingly, there is a dearth of long-term studies relating livestock stocking rates or grazing seasons to plant compositional changes, especially in the Cold Deserts and the Mojave Basin and Range in the Warm Deserts where *Bromus* continues to be problematic. Briske et al. (2011) only cite three studies from these ecoregions with native plants and none reported data on annual grasses (all lumped them into other grasses). Thus, we lack good studies examining how stocking rates or grazing seasons may influence *Bromus* relative to native perennials.

Increased grazing intensity through livestock trampling and defoliation has been studied once in the Cold Desert (*B. tectorum*) and once in the Mojave Basin and Range of the Warm Deserts ecoregion. Both studies used proximity to watering sites to create grazing gradients that revealed similar effects at the community level, but opposite effects on individual *Bromus* species (Brooks et al. 2006; Reisner et al. 2013). Absolute and relative cover of exotic annual plants increased with proximity to water, whereas cover and species richness of native plants decreased. In the Cold

Deserts, cover of *B. tectorum* and the upright annual forb, *Lepidium perfoliatum* L. (clasping pepperweed), increased with nearness to water (Reisner et al. 2013). In the Mojave Basin and Range, cover of the *B. madritensis* ssp. *rubens* decreased with nearness to water and cover of the exotic rosette-forming forb *Erodium cicutarium* (L.) L'Hér. ex Aiton (redstem stork's bill) and a short-statured, nonnative annual grass *Schismus* spp. P. Beauv. (Mediterranean grass) increased (Brooks et al. 2006). The differences between these annual species may reflect the greater grazing avoidance of *E. cicutarium* and *Schismus* relative to *B. rubens* (Brooks et al. 2006). In the Cold Deserts, *B. tectorum* is rarely competing with plants with traits that avoid livestock grazing. A similarity between both studies was reduction in perennial plant cover nearer to water (Brooks et al. 2006; Reisner et al. 2013). Reisner et al. (2013) found that with this lower cover of perennial plants, distances between perennial plants increased. This distance was the strongest factor relating to *B. tectorum* cover, while biological soil crust cover was the second strongest. These results indicate that although livestock use may not generally promote *Bromus* invasions and dominance, repeated livestock use at high intensity can tip the balance toward *Bromus* species unless other species with better grazing avoidance mechanisms exist to replace *Bromus* if it is grazed by livestock.

Rangeland managers have promoted water developments to better distribute livestock across lands; however, each water location, especially when they are permanently located, creates another gradient such as these noted in the studies above. Heady and Child (1999) recommended placing water no farther than 1.3 km apart. Given the gradient of annual plants, including *Bromus* found near the water, each new water source may enhance *Bromus* spread and dominance. Strategic placement to provide livestock water but minimize the dominance of *Bromus* could use investigations.

Reductions in the biomass or seed production of *Bromus* species with livestock grazing are commonly used as evidence that livestock can control and potentially tip the balance in favor of desired perennial species (Mosley and Roselle 2006). The strongest evidence supports the potential for using livestock to reduce fuel levels and fire behavior of *Bromus*-dominated communities (Strand et al. 2014). However, this use must be repeated annually, since *Bromus* always has plants that produce high numbers of viable seeds (Hempy-Mayer and Pyke 2008). Diamond et al. (2012) used combinations of cattle grazing and fire to reduce *B. tectorum* seed densities, but their results were still two to three times higher than the 330 seeds m^{-2} estimated by Hempy-Mayer and Pyke (2008) that would be necessary to keep *B. tectorum* from competing with seedlings of native perennials. More details on livestock as a control for *Bromus* are provided in Chambers et al. (2015).

11.3.2.5 Urban, Suburban, and Exurban Development

The western USA has increased in population by 60.5 % between 1980 and 2006 with 8 of the 11 fastest growing states found in regions where *Bromus* can dominate lands (Nevada, Arizona, Utah, Idaho, Washington, California; Albrecht 2008). Five of the ten fastest-growing cities between 2000 and 2010 were St. George, Utah (53 % increase), Las Vegas, Nevada (42 %), Orem-Provo, Utah (40 %), Greeley,

Colorado (40 %) and Bend, Oregon (37 %) (Mackun and Wilson 2011). A population rise results in increases in use of fossil fuels and changes in atmospheric gases that may potentially impact plant communities and benefit growth of *Bromus*. The annual rate of increase in atmospheric carbon dioxide (CO_2) has doubled since 1975 (data from Mauna Loa Observatory, Hawaii, http://www.esrl.noaa.gov/gmd/ccgg/trends/Accessed 23 March 2015). Elevated CO_2 increases water use efficiency of plants, and may result in slower rates of water depletion, improved plant water relations, and greater biomass production (Polley et al. 2011). These biomass increases may be the greatest in invasive *Bromus* (Nowak et al. 2004). However, advantages to growth and reproduction of *Bromus* due to elevated CO_2 depend on the availability of other key resources, like water and N, and are possibly offset by droughts (Nowak et al. 2004; Smith et al. 2014).

In contrast to increased atmospheric CO_2, atmospheric nitrogen deposition is more localized and tends to increase closer to or downwind from population centers. The greatest source of depositional nitrogen is air pollutants from burning fossil fuels in vehicles or power plants (Galloway et al. 2003). Deposition near cities can be 3.5–10 times higher than away from these centers and this can contribute to high biomass of *Bromus* in Mediterranean California and Warm and Cold Desert ecoregions (Allen and Geiser 2011; Fenn et al. 2011). The higher biomass could lead to higher fuel loads contributing to more or larger fires in these areas of deposition.

Population spread from the city and its suburban communities has led to exurban development that is defined as low-density residential housing outside of city and suburban limits, with one house for each 4–16 ha (Theobald 2003), and that typically results from sale of ranch or farm lands for subdivision development. In northwestern Colorado, *B. tectorum* cover was higher on exurban and wildlife reserves than on existing ranches with similar soils (Maestas et al. 2003). These reserves were highly visited recreational areas with trails and roads that may have contributed to the increase in *B. tectorum*. In addition, 8 of 23 exotic plant species were unique to exurban land uses (Maestas et al. 2003).

Education of urban and exurban populations is important for detection and control of invasive plants. Mealor et al. (2011) found that greater than half of the exurban landowners surveyed in Wyoming correctly identified photographs of *B. tectorum*. In addition, these landowners ranked weeds as a concern on their property and information about those weeds and their control as a desired need (Mealor et al. 2011). However, regulations or incentives for control of *Bromus* generally do not exist except in local areas since regulations are restricted to noxious weed laws discussed above.

11.3.2.6 Transportation Corridors and Vehicles

Transportation corridors for trains and automobiles are potential conduits for invasive species dispersal. Improved roads and railroads cover 1.22 % and 0.02 %, respectively, of the land area in the western USA (Leu et al. 2008). Current construction or maintenance of these routes may also be a source of *Bromus* species dispersal through the gravel used to construct road bases and verges (strips between the road shoulder and the vegetation). Some state and federal agencies are requiring weed-free gravel, yet, similar to weed-free forage, the minimum standard does not preclude *Bromus* species (NAISM; http://www.naisma.org/images/Gravelpit_inspect_stdrs.pdf. Accessed 06/09/2015).

Transportation routes have broader ecological effects on *Bromus* than the direct land area of the route itself. Road density was the greatest human-related predictor of wildfire probability after fuel load and climate across the western USA (Parisien et al. 2012). Sources (e.g., vehicles, equipment, or smoking) of ignition may vary, but roads were a common location associated with the ignition (Syphard and Keeley 2015). Any increase in road density may contribute to wildfires, and provided *Bromus* exists in the plant community, fires may contribute to changes in fire regimes.

Transportation corridors may provide safe sites for establishment and spread of plants. Seeds dispersed along roadsides or railroads, if capable of establishing and reproducing, are likely to colonize surrounding vegetation (Gelbard and Belnap 2003; Gelbard and Harrison 2005). Improved roads increase the probability of *B. tectorum* being found within 500 m of roads in Nevada (Bradley 2010). Improved roads tend to have wider verges with deeper soils and different soil texture and chemistry than surrounding native soils because of road development and maintenance (Brooks and Lair 2009). Runoff from roads carries nutrients that become available for plant growth within and adjacent to verges. Reductions in perennial plants along these verges through maintenance (e.g., grading or herbicides) may also lead to annual grass establishment and growth. A negative correlation between cover of perennial plants and *Centaurea solstitialis* L. (yellow star-thistle), an annual forb, occurs as distance from roads increases (Gelbard and Harrison 2005).

Most state and county highway and transportation departments attempt to manage vegetation along roads for human safety and the control of designated noxious weeds, water pollutants and runoff, and fires (Transportation Research Board 2005). Management includes herbicides or buried root barriers to maintain a plant-free zone between 2 and 10 ft from pavement. Mechanical cutting of vegetation may be applied beyond the vegetation-free zone to reduce fuel loads while maintaining sufficient plant cover for erosion and runoff control. Herbicide spraying for noxious weeds in roadside right-of-ways is conducted as needed in accordance with vegetation management plans. However, *Bromus* are rarely on the list of species for control since they are not considered noxious. Timing of road maintenance is not geared to reducing seed production of *Bromus*. Often this timing is related to maximizing noxious weed controls or lowering vegetation height for fuel reductions both of which typically occur after *Bromus* reproduction. In addition, herbicidal control using dicot-specific herbicides may reduce target weeds, but may allow *Bromus* to fill the void.

Vehicles are common dispersal agents of seeds. The average automobile is estimated to be carrying 2–4 seeds per car at any given time. A systematic review of seed transport found that 96 % of the 626 species on cars were considered weedy, and of those, grasses were the most common life form (Ansong and Pickering 2013). Seeds mixed with soil are common on unimproved roads and driving conditions may influence dispersal distances of seeds. High percentages of seeds stuck to vehicles when they were mixed with soil. If the soil dried on the vehicle and the vehicle was driven under dry road conditions, seeds were retained on vehicles (86–99 %) for a distance of at least 256 km regardless of road surface (Taylor et al. 2012). If driven during wet road conditions, fewer seeds were retained on vehicles when driven on paved relative to unpaved roads, since water on the road tended to wash soil and seeds from the cars (Zwaenepoel et al. 2006; Taylor et al. 2012). Cleaning vehicles moving between areas may be a potential method for reducing spread of weeds (Fleming 2005).

11.3.2.7 Mineral Extraction, Energy Development, and Their Infrastructures

Extracted natural resources (minerals, gas, and oil) and renewable energy development (wind, solar, and geothermal) are subject to a wide variety of regulations. Each type of resource has an immediate footprint that may impact *Bromus* species' colonization and spread. Hard rock or mineral mining is often more localized with roads directly to the mines and then accessing major transportation corridors. Surface mining, where the soil and rock above the mineral is removed to extract the mineral, tends to have a broader footprint, but the post-mining reclamation is regulated (Surface Mining Control and Reclamation Act of 1977). A Montana Greater Sage-grouse Habitat Conservation Advisory Council (2014) recently recommended that control of *B. tectorum* and *B. arvensis* should be added to the reclamation requirements when mines occur in habitat for the *C. urophasianus*. Implementation of these recommendations by Montana still remains.

Other energy developments, however, have the direct footprint associated with well pads, wind power towers, and solar panel arrays, but also include either buried pipelines and cables or electrical transmission towers and corridor rights-of-way. For example, oil and gas wells in the western USA represent 0.01 % of the human footprint, while power lines represent a 3- to 5-fold greater impact (Leu et al. 2008; Knick et al. 2011). The land area impacted by oil and gas wells, relative to their associated pipelines and including their buffers, was nearly 4:1, but this ratio increases to 10:1 if roads to connect wells are included and the potential buffer around these roads is counted in the calculation (Knick et al. 2011). These roads and buffers are likely conduits for the spread of *Bromus* and for fire ignitions (see Transportation above).

Well pads alone may contribute to increased distances of *B. tectorum* seed dispersal. The maximum dispersal distances for *B. tectorum* reported occurred on simulated well pads with dispersal reaching nearly 20 m, nearly 50-fold higher than previously recorded maximum distances in intact sagebrush ecosystems (Johnston 2011). However, Bergquist et al. (2007) found no difference in *B. tectorum* or *B. arvensis* occurrences or cover between coal bed methane sites and similar control sites. They attributed the similarity among treatments of *Bromus* cover to previous disturbances such as livestock grazing as *Bromus* already dominated the sites before coal bed methane development. They did note a lack of perennial plants on well pads and warned that these disturbances would be prone to potential future invasions of *Bromus*.

Power line rights-of-ways and their associated roads tend to have a greater likelihood of *B. tectorum* presence than distant locations (Bradley and Mustard 2006). Vegetation clearing is one of many disturbances found in these corridors; however, it may be difficult to separate the impact of power line corridor maintenance from other impacts associated with corridors such as service roads, livestock grazing, and OHVs. In the Hells Canyon Complex of the Idaho Power Company, a botanical survey found *B. tectorum* was common along with several noxious weeds (Dumas et al. 2003) and listed vehicle use as a likely contributing factor to occurrence of these weeds.

Power lines may directly contribute to fire ignitions when they are located in areas with high winds that may directly damage power lines or cause materials to come in contact with the power line and create an electrical arc. This was recently noted in Mediterranean California (Syphard and Keeley 2015) and was the source of a major Australian fire that began with a downed power line on grazed grassland (Cruz et al. 2012). In Australia, this resulted in a shift to buried power lines to prevent future fires in similar areas.

11.3.3 Interrelated Impacts of Anthropogenic Infrastructure and Exotic Annual **Bromus** *Species on Wildlife*

The expansion of anthropogenic infrastructure is one of the largest impacts to wildlife in the Intermountain West (Copeland et al. 2011). Not only do these infrastructures encourage establishment and spread of invasive species including *Bromus* (Gelbard and Belnap 2003; Bradley and Mustard 2006), they also cause habitat fragmentation and loss (Kirol et al. 2015). In addition, changes to habitat may result in behavioral avoidance of once preferentially used habitat (Buchanan et al. 2014) and in concomitantly reduced vital rates (Dinkins et al. 2014; LeBeau et al. 2014). A related and unintended consequence of increasing infrastructure in native ecosystems is increases in predators. *Corvus corax* L. (common ravens), for example, are provisioned with nesting structure and food resources in formerly unsuitable habitat. They have increased predation on a variety of animals including juvenile *Gopherus agassizii* Cooper (desert tortoise; Boarman 1993) and *C. urophasianus* (Coates and Delehanty 2010) which may also become more visible to predators because of habitat shifts to *Bromus*.

Invasion of annual grasses such as *B. rubens* in the Mojave Basin and Range of the Warm Deserts has led to increased wildfire (Salo 2005), which has direct effects on mortality and loss of cover for *G. agassizii* and other small vertebrates, and hypothetical indirect effects on diet composition and greater exposure to extreme temperatures (Esque et al. 2003). Recreational activities such as target shooting and OHV activity in the Mojave Desert suggest these disturbances may also lead to greater invasion by *B. rubens* and other annuals, along with concomitant impacts to wildlife and habitats in this arid region (Vollmer et al. 1976).

11.3.3.1 Impacts of Land Uses, Exotic Annual *Bromus*, and Fire on Sagebrush-Dependent Wildlife

The sagebrush biome is one of the largest in the USA, but it is threatened by the combination of *Bromus* and wildfires (Noss et al. 1995). Sagebrush communities provide habitat for a diverse assemblage of wildlife (Paige and Ritter 1999; Welch 2005), and it is likely that hundreds of vertebrate species use sagebrush habitats in many landscapes at some point during a year. For example, Wyoming's sagebrush lands provide habitat to nearly 450 avian, mammalian, herptile, and fish species (Wyoming Game and Fish Department 2010: III-9-5), with approximately 6 % (25) of these sagebrush-associated species identified as species of greatest conservation need (Wyoming Game and Fish Department 2010: IV-i-1).

Currently, large contiguous fires pose a significant risk to habitats on which sagebrush-obligate species depend, because they tend to burn uniformly over an area. The dominant sagebrush species do not survive fires (Pechanic et al. 1965) and little suitable sagebrush habitat remains for sagebrush-obligate species (Bukowski and Baker 2013). Nine wildlife species in the western USA are commonly considered sagebrush obligates that depend on sagebrush communities for resources required during critical life stages. These include *Spizella breweri* Cassin (Brewer's sparrow), *C. urophasianus* Bonaparte and *C. minimus* Young, Braun, Oyler-McCance, Hupp & Quinn (Gunnison sage-grouse), *Antilocapra americana* Ord (pronghorn), *Brachylagus idahoensis* Merriam (pygmy rabbit), *Oreoscoptes montanus* Townsend (sage thrasher), *Sceloporus graciosus* Baird and Girard (sagebrush lizard), *Artemisiospiza nevadensis* Ridgway (sagebrush sparrow), and *Lemmiscus curtatus* Cope (sagebrush vole) (Braun et al. 1976; Paige and Ritter 1999; Welch 2005). Sagebrush-associated wildlife tend to decrease while grassland-associated species tend to increase with sagebrush conversion to grassland (Olson et al. 2003; Reinkensmeyer et al. 2007; Larrucea and Brussard 2008; Holmes 2010). Two of these sagebrush-dependent species (*A. americana* and *C. urophasianus*) along with two facultative sagebrush habitat species, *Cervus elaphus* L. (elk) and *Odocoileus hemionus* Rafinesque (mule deer), are of sporting and economic interest in most western states. Further loss of habitat through *Bromus*-fueled fires may directly impact recreation in those states.

Woody plant reductions are often management actions for enhancing wildlife habitat and reducing woody fuels (Heady and Child 1999). Research indicates that prescribed burning as a management tool to achieve either goal should be avoided in *A. tridentata* ssp. *wyomingensis* Beetle & Young (Wyoming big sagebrush) habitats because it encourages spread and increase in *Bromus*, slows recovery of shrubs, and often does not enhance insects and forbs which are often management objectives of prescribed fires (Beck et al. 2009, 2012; Rhodes et al. 2010; Davies et al. 2011; Hess and Beck 2012a, 2014). Hess and Beck (2012a) reported that sagebrush cover and height needed to meet guidelines for nesting and brood-rearing *C. urophasianus* (Connelly et al. 2000) had not recovered as much as 19 years post-burning. Mechanical forms of tree removal may be more beneficial for fuel reduction goals while also improving both shrub and herbaceous components and decreasing the likelihood of *Bromus* becoming dominant (Chambers et al. 2014b). Mowing may be a more promising treatment for enhancing perennial herbaceous plants while reducing but not completely removing sagebrush cover (Hess and Beck 2012a); however, insects and forbs in north-central Wyoming were not more abundant, nor was forb nutritional quality enhanced following mowing in *A. tridentata* ssp. *wyomingensis* (Hess and Beck 2014). Furthermore, mowing *A. tridentata* ssp. *wyomingensis* communities with intact herbaceous understories in several locations had mixed results on perennial herbaceous species cover, density, or biomass. Mowing increased cover, density, and production of annual forbs and grasses compared to untreated sites in some comparisons (Davies et al. 2011; Pyke et al. 2014), but not others (Chambers et al. 2014b). Sagebrush reduction treatments appear to be more promising in *Artemisia tridentata* ssp. *vaseyana* (Rydb.) Beetle (mountain big sagebrush), which are more resilient and resistant and respond to disturbance better than *A. tridentata* ssp. *wyomingensis* (Dahlgren et al. 2006; Beck et al. 2012).

Some recent studies provide insights as to how *C. urophasianus* populations may respond at landscape and micro-habitat scales to invasion of sagebrush habitats by exotic annual grasses. At the micro-habitat scale, *C. urophasianus* nest site selection in south-central Wyoming was negatively correlated with the presence of *B. tectorum*, but positively correlated with *A. tridentata* ssp. *vaseyana* and *A. tridentata* ssp. *wyomingensis* canopy cover (Kirol et al. 2012). At the landscape scale, exotic or introduced grasslands (*B. tectorum* or *Agropyron cristatum* (L.) Gaertn. [crested wheatgrass]) in central Nevada had a negative influence on per-capita recruitment of male *C. urophasianus* to leks where recruitment was consistently low in areas with substantial exotic or introduced grasslands even following years with favorable rainfall (Blomberg et al. 2012). These authors also reported males breeding at leks with substantial exotic grass species had lower annual survival compared to males at leks surrounded by native sagebrush (Blomberg et al. 2012). Abandonment of *C. urophasianus* leks in north-central Wyoming was related to numbers of oil and gas well pads within a 1-km radius of leks, the percent area of wildfire within 1 km, and the variability in shrub height within 1 km of leks (Hess and Beck 2012b). Additive factors stemming from increasing disturbance in this area due to energy development and wildfire were believed to be related to a reduction in habitat quality, at least partially due to establishment and spread of *Bromus*, leading to loss of some leks (Hess and Beck 2012b).

11.4 Management Implications

Human land uses across the western USA have reshaped the structure and function of many ecosystems, especially in the warmer drier ecosystems of the Mediterranean California, the Mojave Basin and Range of the Warm Deserts, and the Cold Deserts. This has led to changes in fire regimes that contribute to feedbacks ensuring the maintenance of annual grass dominance once it occurs. Changes in land use planning, management, and policy may be necessary to impede further loss of native ecosystems to *Bromus* and degradation of wildlife habitats, and to restore native plant communities that are resilient to stresses and disturbances such as fire, land uses, and global changes and that are resistant to increases in *Bromus*. Such changes may require eliminating or mitigating land uses that are incompatible with this goal, while retaining land uses that are compatible or could become compatible with adjustments in land management.

Understanding which lands are more resistant to *Bromus* and resilient to land disturbances provides an opportunity to prioritize land uses relative to the threat of *Bromus* invasion and dominance. The detailed research to determine which ecosystems are more prone to *B. tectorum* dominance in the Cold Deserts has led to ranking of resilience and resistance based on soil temperature and moisture regimes (Chambers et al. 2014c). Similar rankings for other species and regions would improve land use planning and policy development.

A 2015 Secretarial Order from the Department of the Interior (SO 3336 on Rangeland Fire Prevention, Management, and Restoration of 5 January 2015) specifically lists *B. tectorum* as contributing to the "increased threat of rangeland fires." These fires pose "a significant threat to ranchers, livestock managers, sportsman, and outdoor recreation enthusiasts who use sagebrush-steppe ecosystems, and puts at risk their associated economic contributions across this landscape that support and maintain the American way of life in the West." This Secretarial Order is causing reexamination of land use planning and policies relating to management of fire, invasive plants, and restoration of functional, resilient ecosystems in Cold Deserts. This action was a direct response to conserving habitat for *C. urophasianus* and other wildlife species currently threatened by the increasing dominance of *Bromus* throughout the region. Future changes may occur relating to policy and management of these lands with the goal of reducing future fires and a concomitant reduction of *Bromus*.

Planning and policies in other areas of land use and management will likely require similar revamping to stem this tide of *Bromus* increases in response to land uses. For example, weed management policy and planning from national to county levels vary with each successive and finer level of resolution (Ielmini et al. 2015), indicating a need for more consistency among different levels of government, especially related to weed species such as *Bromus* that contribute fuel for fires that may harm wildlife habitats and human safety.

An emphasis on increased communication and education regarding impacts of land uses on *Bromus* could shift the public from the view that "nothing can be done, so why bother," to the view that "our actions, individually and corporately, can focus on activities that do not increase *Bromus*." Education and information regarding potential land use actions that can promote or restrict weeds, including *Bromus*, may aid in changing policies or providing incentives for land managers.

Road and railroad maintenance often uses gravel. Encouraging the use of weed-free gravel will aid in reducing this as a source of weed colonization. But weed-free hay and gravel would need to exclude *Bromus* as well as designated noxious weeds and these may need testing and certification programs. In addition, vegetation treatments along road verges could consider the timing of *Bromus* reproduction, to potentially reduce *Bromus* populations along with other vegetation.

Consideration could be given to closing some roads or to minimizing use to essential vehicles while new road construction is kept to a minimum to reduce new sources of *Bromus* spread. Education of all users of public lands about the benefits of cleaning vehicles as they move from one location to another is important to control the transport of *Bromus*. Courtesy inspections for weed seeds may aid this education similar to courtesy boat inspections done currently in some regions are attempting to reduce the transport of invasive aquatic species. Where practical, consideration could be given to using weed washing technology (e.g., Fleming 2005) for vehicles that must go off-roads to perform their land uses. Some Australian states and territories provide online education on how to wash vehicles and standards for these cleaning facilities to prevent weed spread (Queensland Government 2008).

Livestock grazing management that focuses on increasing resilience of communities by encouraging increased density and cover of deep-rooted perennial grasses while reducing distances among them will likely benefit ecosystem recovery in most ecoregions, except the Mediterranean California system which is now managed as an annual grassland. Grazing systems that encourage light stocking levels and grazing seasons that allow seed and tiller production of perennial species are likely to be complementary with increasing resilience, especially for grass species that did not evolve with repeated grazing.

Maintaining sustainable feral horse and burro populations without degrading habitat is difficult without being able to manage animal stocking rates for both livestock and feral equids. Recommendations for moving forward with an adaptive management study of newly developed contraceptives (Garrott and Oli 2013) may aid in future herd management and halt further contributions to land degradation from equids.

11.5 Research Needs

Resistance to invasion and dominance of *Bromus* coupled with resilience to disturbances such as fire are dependent on the cover, density, and spatial relationships among perennial plants, especially perennial grasses. Understanding the values for these measures for differing ecological sites or plant communities is necessary. These values are most likely related to soil temperature and moisture regimes (Chambers et al. 2015) and, thus, are specific to ecoregions and ecological sites within ecoregions. Progress is being made in using such factors to evaluate the recovery potential of Great Basin ecosystems to management actions and wildfire (Miller et al. 2014). However, our current understanding of these values is inadequate to predict outcomes (e.g., *Bromus* or native plant dominance) of disturbances like wildfire with assurance. In addition, an understanding of these factors may provide managers with early-warning indicators of the potential loss in plant community resilience in areas impacted by *Bromus* or provide quantitative goals for restoration objectives.

BLM and USDA Natural Resources Conservation Service are currently collecting data on these plant measures across the western USA using a standardized protocol with replicated sites in ecoregions (Toevs et al. 2011). These should provide a baseline for relating *Bromus* cover to these factors across a range of sites differing in resilience and resistance.

The rapid expansion of oil and gas wells and renewable energy in the western USA is cause for more study into impacts these types of land uses and their infrastructures have on *Bromus*. We are only aware of one study in one locale in Wyoming that addressed the impacts of these wells on the immediate plant community (Bergquist et al. 2007), but the control sites already had *Bromus*. Is this a consistent finding or an anomaly? Further studies in other energy development sites are needed to confirm or refute this finding.

The surprising lack of research relating livestock grazing intensity and season of use to *Bromus* and perennial grass dominance makes these areas ripe for research. Since long-term grazing studies are expensive and hard to replicate, we suggest using additional studies examining plant compositional changes along grazing intensity gradients. It would be necessary to use a covariate such as ecological sites to begin to tease apart the livestock grazing component from the plant community potential component. Since a complex of factors can impact the results for both *Bromus* and perennial grasses, the use of multivariate statistical procedure will likely be necessary (Reisner et al. 2013, 2015). Examining livestock water placement and *Bromus* cover across the landscape may aid in placing and using water strategically to control *Bromus* spread.

The use of livestock as a tool for reducing fire risk and for controlling *Bromus* has been proposed. The utility of using targeted grazing, especially in locations with varied levels of resistance and resilience and in areas where perennial grasses already exist, warrants further investigation. Studies demonstrating the achievement of goals to reduce *Bromus* while maintaining or increasing deep-rooted perennial grasses are currently lacking. There is no evidence that this goal can be achieved without controlled and replicated studies.

There is a great need to better understand the impacts of *Bromus* to wildlife habitats in many ecosystems in the western USA. Specific information such as predicted changes in cover and dietary resources (e.g., insect availability and forage nutritional quality), as well as population responses for sagebrush obligates such as *C. urophasianus* and other species of concern (e.g., *G. agassizii*), will be useful to assist in developing practical and successful restoration protocols for burned or *Bromus*-dominated habitats. Research is also needed to evaluate vital rates and habitat selection for wildlife during critical life stages (e.g., juveniles) and seasons (e.g., breeding and winter) when wildlife populations are most vulnerable to large-scale habitat changes and restoration practices. These evaluations will be useful to ascertain the effectiveness of restorative actions to improve conditions for wildlife populations in areas impacted by *Bromus*.

References

Albrecht DE (2008) Population brief: the changing west – a regional overview. Western Rural Development Center, Utah State University, Logan, UT, p 4

Allen EB, Geiser LH (2011) North American deserts. In: Pardo LH, Robin-Abbott MJ, Driscoll CT (eds) Assessment of effects of N deposition and empirical critical loads for nitrogen for ecoregions of the United States. Gen Tech Rep NRS-80. USDA, Forest Service, Northern Research Station, Newtown Square, pp 133–142

Ansong M, Pickering C (2013) Are weeds hitchhiking a ride on your car? A systematic review of seed dispersal on cars. PLoS One 8(11), e80275

Atkinson SY, Brown CS (2015) Attributes that confer invasiveness and impacts across the large genus *Bromus* – lessons from the *Bromus* REEnet database. In: Germino MJ, Chambers JC, Brown CS (eds) Exotic brome-grasses in arid and semiarid ecosystems of the Western USA: causes, consequences, and management implications. Springer, New York, NY (Chapter 6)

Bangert R, Huntley N (2010) Distribution of native and exotic plants in a naturally fragmented landscape. Biol Invasions 12:1627–1640

Bartolome JW, Barry WJ, Griggs T et al (2007) Valley grassland. In: Barbour MG, Keeler-Wolf T, Schoenherr AA (eds) Terrestrial vegetation of California. UC Press, Los Angeles, CA, pp 635–710

Beck JL, Connelly JW, Wambolt CL (2012) Consequences of treating Wyoming big sagebrush to enhance wildlife habitats. Rangel Ecol Manag 65:444–455

Beck JL, Connelly JW, Reese KP (2009) Recovery of Greater Sage-Grouse habitat features in Wyoming big sagebrush following prescribed fire. Restor Ecol 17:393–403

Benson L, Petersen K, Stein J (2007) Anasazi (Pre-Columbian Native American) migrations during the middle-12th and late 13th centuries – Were they drought induced? Climate Change 83:187–213

Bergquist E, Evangelista P, Stohlgren TJ et al (2007) Invasive species and coal bed methane development in the Powder River Basin, Wyoming. Environ Monit Assess 128:381–394

Blomberg EJ, Sedinger JS, Atamian MT et al (2012) Characteristics of climate and landscape disturbance influence the dynamics of greater sage-grouse populations. Ecosphere 3(6):55

Boarman WI (1993) When a native predator becomes a pest: a case study. In: Majumdar SK, Miller EW, Miller DE et al (eds) Conservation and resource management. Pennsylvania Academy of Sciences, Philadelphia, PA

Bradley BA (2010) Assessing ecosystem threats from global and regional change: hierarchical modeling of risk to sagebrush ecosystems from climate change, land use, and invasive species in Nevada, USA. Ecography 33:198–208

Bradley BA, Mustard JF (2006) Characterizing the landscape dynamics of an invasive plant and risk of invasion using remote sensing. Ecol Appl 16:1132–1147

Braun CE, Baker MF, Eng RL et al (1976) Conservation committee report on effects of alteration of sagebrush communities on the associated avifauna. Wilson Bull 88:165–171

Briske DD, Richards JH (1995) Plant responses to defoliation: a physiological, morphological, and demographic evaluation of individual plants to grazing: current status and ecological significance. In: Bedunah DJ, Sosebee RE (eds) Wildland plants: physiological ecology and developmental morphology. Society for Range Management, Denver, CO, pp 635–710

Briske DD, Sayre NF, Huntsinger L et al (2011) Origin, persistence, and resolution of the rotational grazing debate: integrating human dimensions into rangeland research. Rangel Ecol Manag 64:325–334

Brooks ML (1999) Alien annual grasses and fire in the Mojave Desert. Madroño 46:13–19

Brooks ML, Belnap J, Brown CS et al (2015) Exotic annual *Bromus* invasions – comparisons among species and ecoregions in the western United States. In: Germino MJ, Chambers JC, Brown CS (eds) Exotic brome-grasses in arid and semiarid ecosystems of the Western USA: causes, consequences, and management implications. Springer, New York, NY (Chapter 2)

Brooks ML, Chambers JC, McKinley RA (2013) Fire history, effects, and management in southern Nevada. In: Chambers JC, Brooks ML, Pendleton BK et al (eds) The Southern Nevada agency partnership science and research synthesis: science to support land management in Southern Nevada. Gen Tech Rep RMRS-GTR-303. USDA, Forest Service, Rocky Mountain Research Station, Fort Collins, CO, pp 75–96

Brooks ML, D'Antonio CM, Richardson DM et al (2004) Effects of invasive alien plants on fire regimes. BioScience 54:677–688

Brooks ML, Lair BM (2009) Ecological effects of vehicular routes in a desert ecosystem. In: Webb RH, Fenstermaker LF, Heaton JS et al (eds) The Mojave Desert: ecosystem processes and sustainability. University of Nevada, Las Vegas, NV, pp 168–195

Brooks ML, Matchett JR, Berry KH (2006) Effects of livestock watering sites on alien and native plants in the Mojave Desert, USA. J Arid Environ 67:125–147

Brooks ML, Minnich RA (2006) Southeastern deserts bioregion. In: Sugihara NG, van Wagtendonk JW, Shaffer KE et al (eds) Fire in California's ecosystems. UC Press, Berkeley, CA, pp 391–414

Buchanan CB, Beck JL, Bills TE et al (2014) Seasonal resource selection and distributional response by elk to development of a natural gas field. Rangel Ecol Manag 67:369–379

Bukowski BE, Baker WL (2013) Historical fire regimes, reconstructed from land-survey data, led to complexity and fluctuation in sagebrush landscapes. Ecol Appl 23:546–564

Chambers JC, Bradley BA, Brown CS et al (2014a) Resilience to stress and disturbance, resistance to *Bromus tectorum* L. invasion in cold desert shrublands of western North America. Ecosystems 17:360–375

Chambers JC, Miller RF, Board DI et al (2014b) Resilience and resistance of sagebrush ecosystems: implications for state and transition models and management treatments. Rangel Ecol Manag 67:440–454

Chambers JC, Pyke DA, Maestas JD et al (2014c) Using resistance and resilience concepts to reduce impacts of invasive annual grasses and altered fire regimes on the sagebrush ecosystem and greater sage-grouse: a strategic multi-scale approach. Gen Tech Rep RMRS-GTR-326. USDA, Forest Service, Rocky Mountain Research Station, Fort Collins

Chambers JC, Roundy BA, Blank RR et al (2007) What makes Great Basin sagebrush ecosystems invasible by *Bromus tectorum*? Ecol Monogr 77:117–145

Chambers JC, Wisdom MJ (2009) Priority research and management issues for the imperiled Great Basin of the western United States. Restor Ecol 17:707–714

Chambers JC, Germino MJ, Belnap J et al (2015) Plant community resistance to invasion by *Bromus* species – the roles of community attributes, *Bromus* interactions with plant communities, and *Bromus* traits. In: Germino MJ, Chambers JC, Brown CS (eds) Exotic brome-grasses in arid and semiarid ecosystems of the Western USA: causes, consequences and management implications. Springer, New York, NY (Chapter 10)

Clements DR, Krannitz PG, Gillespie SM (2007) Seed bank responses to grazing history by invasive and native plant species in a semi-desert shrub-steppe environment. Northwest Sci 81:37–49

Coates PS, Delehanty DJ (2010) Nest predation of greater sage-grouse in relation to microhabitat factors and predators. J Wildl Manag 74:240–248

Connelly JW, Schroeder MA, Sands AR et al (2000) Guidelines to manage sage-grouse populations and their habitats. Wildl Soc B 28:967–985

Copeland HE, Pocewicz A, Kiesecker JM (2011) Geography of energy development in western North America: potential impacts on terrestrial ecosystems. In: Naugle DE (ed) Energy development and wildlife conservation in western North America. Island Press, Washington, DC, pp 7–22

Cruz MG, Sullivan AL, Gould JS et al (2012) Anatomy of a catastrophic wildfire: the Black Saturday Kilmore East fire in Victoria, Australia. Forest Ecol Manag 284:269–285

D'Antonio CM, Vitousek PM (1992) Biological invasions by exotic grasses, the grass/fire cycle, and global change. Ann Rev Ecol Syst 23:63–87

Dahlgren DK, Chi R, Messmer T (2006) Greater sage-grouse response to sagebrush management in Utah. Wildl Soc B 34:975–985
Davies GM, Bakker JD, Dettweiler-Robinson E et al (2012) Trajectories of change in sagebrush-steppe vegetation communities in relation to multiple wildfires. Ecol Appl 22:1562–1577
Davies KW, Boyd CS, Beck JL et al (2011) Saving the sagebrush sea: an ecosystem conservation plan for big sagebrush plant communities. Biol Conserv 144:2573–2584
Diamond JM, Call CA, Devoe N (2012) Effects of targeted grazing and prescribed burning on community and seed dynamics of downy brome (*Bromus tectorum*)-dominated landscape. Invasive Plant Sci Manag 5:259–269
Dinkins JB, Conover MR, Kirol CP et al (2014) Greater sage-grouse hen survival: effects of raptors, anthropogenic and landscape features, and hen behavior. Can J Zool 92:319–330
Donahue DL (1999) The western range revisited: removing livestock from public lands to conserve native biodiversity. University of Oklahoma Press, Norman
Duffus RL (1972) The Sante Fe Trail. First University of New Mexico Press, Albuquerque, NM
Dumas BC, Holmstead GL, Kerr MJJ et al (2003) Effects of road and transmission-line rights-of-way on botanical resources. Hells Canyon Complex federal energy regulatory commission No. 1971, Gen Tech Rep Appendix E.3.3-4 (Revised). Idaho Power Company, Boise, ID
Esque TC, Schwalbe CR, DeFalco LA et al (2003) Effects of desert wildfires on desert tortoise (*Gopherus agassizii*) and other small vertebrates. Southwest Nat 48:103–111
Fenn ME, Allen EB, Geiser LH (2011) Mediterranean California. In: Pardo LH, Robin-Abbott MJ, Driscoll CT (eds) Assessment of effects of N deposition and empirical critical loads for nitrogen for ecoregions of the United States. USDA, Forest Service, Northern Research Station, Gen Tech Rep NRS-80, Newtown Square, pp 143–170
Flannery KV (1969) Origins of ecological effects of early domestication in Iran and the near East. In: Ucko PJ, Dimbleby GW (eds) The domestication and exploitation of plants and animals. Aldine Publishing, Chicago, IL, pp 73–100
Fleming J (2005) Vehicle cleaning technology for controlling the spread of noxious weeds and invasive species. USDA, Forest Service, San Dimas Technology and Development Center, San Dimas, CA
Galloway JN, Aber JD, Erisman JW et al (2003) The nitrogen cascade. BioScience 53:341–356
Garrott RA, Oli MK (2013) A critical crossroad for BLM's wild horse program. Science 341:847–848
Gelbard JL, Belnap J (2003) Roads as conduits for exotic plant invasions in a semiarid landscape. Conserv Biol 17:420–432
Gelbard JL, Harrison S (2005) Roadless habitats as refuges for native grassland diversity: interactions with soil type, aspect, and grazing. Ecol Appl 15:1570–1580
Germino MJ, Belnap J, Stark JM et al (2015) Ecosystem impacts of exotic annual invaders in the genus *Bromus*. In: Germino MJ, Chambers JC, Brown CS (eds) Exotic brome-grasses in arid and semiarid ecosystems of the Western USA: causes, consequences, and management implications. Springer, New York, NY (Chapter 3)
Germino MJ, Chambers JC, Brown CS (2015) Introduction: exotic annual *Bromus* in the western USA. In: Germino MJ, Chambers JC, Brown CS (eds) Exotic brome-grasses in arid and semiarid ecosystems of the Western USA: causes, consequences, and management implications. Springer, New York, NY (Chapter 1)
Gruell GE (1985) Indian fires in the interior West: a widespread influence. In: Lotan JE, Kilgore BM, Fischer WC et al (eds) 1983 wilderness fire symposium. Gen Tech Rep 182. USDA, Forest Service, Intermountain Research Station, Ogden, UT, pp 68–74
Haferkamp, MR, Grings EE, Karl MG et al (1992) Japanese brome in the northern Great Plains. In: Monsen SB, Kitchen SG (comps) Proceedings of the symposium on ecology and management of annual rangelands. Gen Tech Rep INT-GTR-313. USDA, Forest Service, Intermountain Research Station, Ogden, UT
Hard RJ, Mauldin RP, Raymond GR (1996) Mano size, stable carbon isotope ratios, and macrobotanical remains as multiple lines of evidence of maize dependence in the American southwest. J Archaeol Method Theory 3:253–318

Hardin G (1968) Tragedy of the commons. Science 162:1243–1248
Harper JL (1977) Population biology of plants. Academic, London
Hart R (2001) Where the buffalo roamed – or did they? Great Plains Res 11:83–102
Heady HF, Child RD (1999) Rangeland ecology and management. Westview Press, Boulder, CO
Hempy-Mayer K, Pyke DA (2008) Defoliation effects on *Bromus tectorum* seed production: implications for grazing. Rangel Ecol Manag 61:116–123
Henige DP (1998) Numbers from nowhere: the American Indian contact population debate. University of Oklahoma Press, Norman, OK
Hess JE, Beck JL (2012a) Burning and mowing Wyoming big sagebrush: do treated sites meet minimum guidelines for greater sage-grouse breeding habitats? Wildl Soc B 36:85–93
Hess JE, Beck JL (2012b) Disturbance factors influencing greater sage-grouse lek abandonment in north-central Wyoming. J Wildl Manag 76:1625–1634
Hess JE, Beck JL (2014) Forb, insect, and soil response to burning and mowing Wyoming big sagebrush in greater sage-grouse breeding habitat. Environ Manag 53:813–822
Hevly RH (1988) Prehistoric vegetation and paleoclimates on the Colorado Plateaus. In: Gumerman GJ (ed) The Anasazi in a changing environment. Cambridge University Press, New York, NY, pp 92–118
Holmes AL (2010) Small mammal and bird abundance in relation to post-fire habitat succession in mountain big sagebrush (*Artemisia tridentata* ssp. *vaseyana*) communities. PhD thesis, Oregon State University, Corvallis, OR
Hurteau MD, Bradford JB, Fule PZ et al (2013) Climate change, fire management, and ecological services in the southwestern US. Forest Ecol Manag 327:280–289
Ielmini M, Hopkins T, Mayer KE et al (2015) Invasive plant management and greater sage-grouse conservation: a review and status report with strategic recommendations for improvement. Western Association of Fish and Wildlife Agencies, Unpublished Report, Cheyenne, Wyoming, p 47
Janssen MA, Scheffer M (2004) Overexploitation of renewable resources by ancient societies and the role of sunk-cost effects. Ecol Soc 9:6
Johnston DB (2011) Movement of weed seeds in reclamation areas. Restor Ecol 19:446–449
Kay CE (1994) Aboriginal overkill: the role of Native Americans in structuring western ecosystems. Hum Nat 5:359–398
Keeler-Wolf T (2007) Mojave desert scrub vegetation. In: Barbour MG, Keeler-Wolf T, Schoenherr AA (eds) Terrestrial vegetation of California. UC Press, Los Angeles, CA, pp 609–656
Keeley JE (2002) Native American impacts on fire regimes of the California coastal ranges. J Biogeogr 29:303–320
Keeley JE, Bond WJ, Bradstock RA et al (2012) Fire in Mediterranean ecosystems: ecology, evolution and management. Cambridge University Press, New York, NY
Kirol CP, Beck JL, Dinkins JB et al (2012) Microhabitat selection for nesting and brood rearing by the greater sage-grouse in xeric big sagebrush. The Condor 114:75–89
Kirol CP, Beck JL, Huzurbazar SV et al (2015) Identifying greater sage-grouse source and sink habitats for conservation planning in an energy development landscape. Ecol Appl 25:968–990
Knick ST, Hanser SE, Miller SF et al (2011) Ecological influence and pathways of land use in sagebrush. In: Knick ST, Connelly JW (eds) Greater sage-grouse: ecology and conservation of a landscape species and its habitats, vol 38, Studies in Avian Biology. UC Press, Berkeley, CA, pp 203–251
Laliberte AS, Ripple WJ (2003) Wildlife encounters by Lewis and Clark: a spatial analysis of interactions between Native Americans and wildlife. BioScience 53:994–1003
Larrucea ES, Brussard PF (2008) Habitat selection and current distribution of the pygmy rabbit in Nevada and California, USA. J Mammal 89:691–699
LeBeau CW, Beck JL, Johnson GD et al (2014) Short-term impacts of wind energy development on greater sage-grouse fitness. J Wildl Manag 78:522–530
Leopold A (1941) Cheat takes over. The Land 1:310–313

Leu M, Hanser SE, Knick ST (2008) The human footprint in the west: a large-scale analysis of anthropogenic impacts. Ecol Appl 18:1119–1139

Lovell WG (1992) Heavy shadows and black night: disease and depopulation in colonial Spanish America. Ann Assoc Am Geogr 82:426–443

Mack RN (1981) Invasion of *Bromus tectorum* L. into western North America: an ecological chronicle. Agro-Ecoystems 7:145–165

Mack RN (1986) Alien plant invasion into the Intermountain West: a case history. In: Mooney HA, Drake JA (eds) Ecology of biological invasions of North America and Hawaii. Springer, New York, NY, pp 191–213

Mack RN (1991) The commercial seed trade: an early disperser of weeds in the United States. Econ Bot 45:257–73

Mack RN (2000) Cultivation fosters plant naturalization by reducing environmental stochasticity. Biol Invasions 2:111–122

Mack RN, Pyke DA (1983) Demography of *Bromus tectorum*: variation in time and space. J Ecol 71:69–93

Mack RN, Thompson JN (1982) Evolution in steppe with few large, hooved mammals. Am Nat 119(6):757–773

Mackun P, Wilson S (2011) Population distribution and change: 2000–2010. 2010 Census Briefs, March 2011, C2010BR-01. US Department of Commerce, Economic and Statistics Administration, US Census Bureau. http://www.census.gov/prod/cen2010/briefs/c2010br-01.pdf. Accessed on 10 Mar 2015

Maestas JD, Knight RL, Gilgert WC (2003) Biodiversity across a rural land-use gradient. Conserv Biol 17:1425–1434

Marinova E (2003) Paleoethnobotanical study of Early Bronze II in the Upper Stryama Valley (Dubene – Sarovka IIB). BAR International Series No 1139(2):499–504

Mealor B, Mealor RD, Kelley WK et al (2013) Cheatgrass management handbook: managing an invasive grass in the Rocky Mountain region. University of Wyoming, Laramie, WY

Mealor RD, Meiman PJ, Hild AL et al (2011) New rangeland residents in Wyoming? A survey of exurban landowners. Rangel Ecol Manag 64:479–487

Milchunas DG, Sala OE, Lauenroth WK (1988) A generalized model of the effects of large herbivores on grassland community structure. Am Nat 132:87–106

Miller RF, Chambers JC, Pellant M (2014) A field guide to selecting the most appropriate treatments in sagebrush and pinyon-juniper ecosystems in the Great Basin: evaluating resilience to disturbance and resistance to invasive annual grasses and predicting vegetation response. Gen Tech Rep RMRS-GTR-322. USDA, Forest Service, Rocky Mountain Research Station, Fort Collins, CO

Miller RF, Chambers JC, Pyke DA et al (2013) A review of fire effects on vegetation and soils in the Great Basin region: response and ecological site characteristics. Gen Tech Rep RMRS-GTR-308. USDA, Forest Service, Rocky Mountain Research Station, Fort Collins, CO, p 126

Monaco TA, Hardegree SP, Pellant M et al (2015) Assessing restoration and management needs for ecosystems invaded by exotic annual *Bromus* species. In: Germino MJ, Chambers JC, Brown CS (eds) Exotic brome-grasses in arid and semiarid ecosystems of the Western USA: causes, consequences, and management implications. Springer, New York, NY (Chapter 12)

Montana's greater sage-grouse habitat conservation advisory council (2014) Greater Sage-grouse habitat conservation strategy. Office of the Governor, State of Montana. http://governor.mt.gov/Portals/16/docs/GRSG%20strategy%2029Jan_final.pdf. Accessed 14 Jul 2015

Morris LR, Monaco TA, Sheley RL (2011) Land-use legacies and vegetation recovery 90 years after cultivation in Great Basin sagebrush ecosystems. Rangel Ecol Manag 64:488–497

Mosley JC, Roselle L (2006) Targeted livestock grazing to suppress invasive annual grasses. In: Launchbaugh K, Walker J (eds) Targeted grazing: a natural approach to vegetation management and landscape enhancement. American Sheep Industry Association, Englewood, CO, pp 67–76

National Research Council (1994) Rangeland health: new methods to classify, inventory, and monitor rangelands. National Academy Press, Washington, DC

Noss RF, LaRoe ET III, Scott JM (1995) Endangered ecosystems of the United States: a preliminary assessment of loss and degradation. National Biological Service Biological Report 28. Washington DC, p 58

Nowak RS, Ellsworth DS, Smith SD (2004) Functional responses of plants to elevated atmospheric CO_2 – do photosynthetic and productivity data from FACE experiments support early predictions? New Phytol 162:253–280

Olson RA, Perryman BL, Petersburg S et al (2003) Fire effects on small mammal communities in Dinosaur National Monument. Western N Am Nat 63:50–55

Olson SP (2004) The Oregon Trail: a primary source history of the route to the American West. Rosen Publishing, New York, NY

Paige C, Ritter SA (1999) Birds in a sagebrush sea: managing sagebrush habitats for bird communities. Partners in Flight Western Working Group, Boise, ID

Parisien M, Snetsigner S, Greenberg JA et al (2012) Spatial variability in wildfire probability across the western United States. Int J Wildl Fire 21:313–327

Pausas JG, Bradstock RA (2007) Fire persistence traits of plants along a productivity and disturbance gradient in Mediterranean shrublands of south-east Australia. Glob Ecol Biogeogr 16:330–40

Pechanic JF, Plummer AP, Robertson JH et al (1965) Sagebrush control on rangelands. USDA Handbook No. 277, Washington, DC

Polley HW, Morgan JA, Fay PA (2011) Application of a conceptual framework to interpret variability in rangeland responses to atmospheric CO_2 enrichment. J Agric Sci 149:1–14

Pyke DA, Shaff SE, Lindgren AI et al (2014) Region-wide ecological responses of arid Wyoming big sagebrush communities to fuel treatments. Rangel Ecol Manag 67:455–467

Queensland Government (2008) Queensland weed spread prevention strategy. State of Queensland, Department of Primary Industries and Fisheries, Brisbane, Australia

Raish CB (2013) Human interactions with the environment through time in southern Nevada. In: Chambers JC, Brooks ML, Pendleton BK et al (eds) The southern Nevada agency partnership science and research synthesis: science to support land management in southern Nevada. Gen Tech Rep RMRS-GTR-303. USDA, Forest Service, Rocky Mountain Research Station, Fort Collins, CO, pp 155–174

Reinkensmeyer DP, Miller R, Anthony RG et al (2007) Avian community structure along a mountain big sagebrush successional gradient. J Wildl Manag 71:1057–1066

Reisner MD, Doescher PS, Pyke DA (2015) Stress gradient hypothesis explains susceptibility to *Bromus tectorum* invasion and community stability in North America's semi-arid *Artemisia tridentata wyomingensis* ecosystems. J Veg Sci. doi:10.1111/jvs.12327

Reisner MD, Grace JB, Pyke DA et al (2013) Conditions favouring *Bromus tectorum* dominance of endangered sagebrush steppe ecosystems. J Appl Ecol 50:1039–1049

Rhodes EC, Bates JD, Sharp RN et al (2010) Fire effects on cover and dietary resources of sage-grouse habitats. J Wildl Manag 74:755–764

Salo LF (2005) Red brome (*Bromus rubens* subsp. *madritensis*) in North America: possible modes for early introductions, subsequent spread. Biol Invasions 7:165–180

Sales F (1994) Evolutionary tendencies in some annual species of *Bromus* (*Bromus* L. sect. *Genea* Dum. (Poaceae)). Bot J Linn Soc 115:197–210

Sanderson EW, Jaiteh M, Levy MA et al (2002) The human footprint and the last wild. BioScience 52:891–904

Small C, Cohen JE (2004) Continental physiography, climate, and the global distribution of human populations. Curr Anthropol 45(2):1–11

Smith SD, Charlet TN, Fenstermaker LF et al (2009) Effects of global change on Mojave Desert ecosystems. In: Webb RH, Fenstermaker LF, Heaton JS et al (eds) The Mojave Desert. University of Nevada Press, Reno, NV, pp 31–56

Smith SD, Charlet TN, Zitzer SF et al (2014) Long-term response of a Mojave Desert winter annual plant community to a whole-ecosystem atmospheric CO_2 manipulation (FACE). Glob Change Biol 20:879–892

Strand EK, Launchbaugh KL, Limb R et al (2014) Livestock grazing effects on fuel loads for wildland fire in sagebrush dominated ecosystems. J Rangel Appl 1:35–57
Strauss SY, Agrawal AA (1999) The ecology and evolution of plant tolerance to herbivory. Trends Ecol Evol 14:179–185
Stewart G (1938) Revegetating man-made deserts. J Forestry 36:853–855
Syphard AD, Keeley JE (2015) Location, timing and extent of wildfire vary by cause of ignition. Int J Wildl Fire 24:37–47
Taylor K, Brummer T, Taper ML et al (2012) Human-mediated long-distance dispersal: an empirical evaluation of seed dispersal by vehicles. Divers Distrib 18:942–951
Theobald DM (2003) Defining and mapping rural sprawl: examples from the Northwest US. Fort Collins, CO, USA: Growth Management Leadership Alliance white paper, p 6
Thorne JH, Morgan BJ, Kennedy JA (2008) Vegetation change over sixty years in the Central Sierra Nevada, California, USA. Madronno 55:223–237
Toevs GR, Karl JW, Taylor JJ et al (2011) Consistent indicators and methods and a scalable sample design to meet assessment, inventory, and monitoring information needs across scales. Rangelands 33:14–20
Transportation Research Board (2005) Integrated roadside vegetation management: a synthesis of highway practice. National Cooperative Highway Research Program Synthesis 341, Washington, DC
USDA NRCS (US Department of Agriculture Natural Resources Conservation Service) (2014) National resources inventory rangeland resource assessment – June 2014. Natural Resources Conservation Service. Washington, DC. http://www.nrcs.usda.gov/wps/portal/nrcs/detail/national/technical/nra/nri/?cid=stelprdb1253602
US Department of Agriculture, Secretary of Agriculture (1936) The western range: a great but neglected natural resource. US Senate, 74th Congress, Document No. 199. US Government Printing Office, Washington, DC
US Government Accountability Office (USGAO) (2009) Wildland fire management: federal agencies have taken important steps forward, but additional, strategic action is needed to capitalize on those steps. US Government Accountability Office, GAO-09-877, Washington, DC
Vlaisch JA (2005) Pueblo Indian agriculture. University of New Mexico Press, Albuquerque, NM
Vollmer AT, Maza BG, Medica PA et al (1976) The impact of off-road vehicles on a desert ecosystem. Environ Manag 1:115–129
Warren SD, Eldridge DJ (2001) Biological soil crusts and livestock in arid ecosystems: are they compatible? In: Belnap J, Lange OL (eds) Biological soil crusts: structure, function, and management. Springer, Berlin, pp 403–415
Welch BL (2005) Big sagebrush: a sea fragmented into lakes, ponds and puddles. USDA Forest Service, Rocky Mountain Research Station. Gen Tech Rep RMRS-GTR-144, Fort Collins, CO
White CE, Chesson MS, Schaub T (2014) A recipe for disaster: emerging urbanism and unsustainable plant economies at Early Bronze Age Ras an-Numayra, Jordan. Antiquity 88:363–377
Wyoming Game and Fish Department (2010) Wyoming state wildlife action plan. Wyoming Game and Fish Department, Cheyenne, WY
Young JA, Clements CD (2007) Cheatgrass and grazing rangelands. Rangelands 29:15–20
Zwaenepoel A, Roovers P, Hermy M (2006) Motor vehicles as vectors of plant species from road verges in a suburban environment. Basic Appl Ecol 7:83–93

Chapter 12
Assessing Restoration and Management Needs for Ecosystems Invaded by Exotic Annual *Bromus* Species

Thomas A. Monaco, Stuart P. Hardegree, Mike Pellant, and Cynthia S. Brown

Abstract Invasive annual grass research and management in arid and semiarid ecosystems of the Western United States (USA) have historically focused on reducing weed abundance as opposed to ecosystem restoration, which addresses the underlying processes responsible for their persistence. Given the current impact of invasive annual grasses and their continued spread in this region, we identified common characteristics responsible for persistence of the most problematic exotic annual *Bromus*. For heavily invaded areas, these include transient, yet typically large seed banks, altered soil resource availability and litter production, displacement of native species, and frequent disturbance from fire. To better address these common characteristics for future management, we illustrate how an adaptive management framework can reduce existing uncertainty associated with the restoration of arid and semiarid ecosystems. We also assess how research and management influence these characteristics with a case study of treatment outcomes for *Bromus tectorum* L. (cheatgrass or downy brome) control and a case study of current approaches used by

T.A. Monaco (✉)
US Department of Agriculture, Agricultural Research Service, Forage and Range Research Laboratory, Utah State University, Logan, UT 84322-6300, USA
e-mail: Tom.Monaco@ars.usda.gov

S.P. Hardegree
US Department of Agriculture, Agricultural Research Service, Northwest Watershed Research Center, Boise, ID 83712, USA
e-mail: stuart.hardegree@ars.usda.gov

M. Pellant
US Department of Interior, Bureau of Land Management, Boise, ID 83709, USA
e-mail: mpellant@blm.gov

C.S. Brown
Department of Bioagricultural Sciences and Pest Management, Colorado State University, Fort Collins, CO 80523-1177, USA
e-mail: cynthia.s.brown@colostate.edu

M.J. Germino et al. (eds.), *Exotic Brome-Grasses in Arid and Semiarid Ecosystems of the Western US*, Springer Series on Environmental Management,
DOI 10.1007/978-3-319-24930-8_12

the US Department of Interior's Bureau of Land Management to restore landscapes impacted by *B. tectorum* and improve greater sage-grouse (*Centrocercus urophasianus*) habitat in the Great Basin region of the Western USA. We conclude that adapting for the future will require continued integration between research and management, greater inclusion of ecological processes, and long-term experiments to inform adaptive management decisions.

Keywords *Bromus* characteristics • Adaptive management framework • Treatment outcomes • Bureau of Land Management strategies

12.1 Introduction

The evolution and proliferation of annual weeds are closely linked to human-mediated disturbances that accompany land use. Perhaps one of the clearest examples of coevolution in agroecosystems is illustrated by the adaptation of annual plants to frequent soil disturbance associated with land cultivation, sowing, and harvesting activities (Palumbi 2001; Guglielmini et al. 2007). These disturbances provided the necessary selective pressures to favor life history traits for dispersal, colonization, and reproduction within a highly modified, ruderal environment (Lytle 2001). Consequently, invasive annual grasses that infest Western North America are generally preadapted to thrive in arid and semiarid ecosystems that have undergone recurrent disturbance from dryland agriculture (Mack 1981; Morris et al. 2014), historical overgrazing (Daubenmire 1940), wildfire, energy development, urbanization, recreational activities, and drought (Brooks and Chambers 2011).

While plant invasions are considered a consequence of both species traits and ecosystem characteristics (Vilà et al. 2011), the impacts of exotic plant invasions on ecological processes and functioning are context dependent (Vilà et al. 2011; Pyšek et al. 2012). For example, many exotic species do not create ecological problems, while others radically transform ecosystems or proliferate within areas disturbed by human land use (Catford et al. 2012). For heavily invaded ecosystems, there is tremendous need to broaden our understanding of how systems function through greater integration of innovative research and holistic adaptive management frameworks that allow iterative learning, reduce uncertainty, and support the development of effective strategies (Herrick et al. 2012).

Based on a recent global assessment, ecological impacts associated with changes in species abundance, diversity, richness of resident biota, and soil attributes were most significant if the invasive species was an annual grass with many examples drawn from the genus *Bromus* (Pyšek et al. 2012). In accordance with this assessment, exotic annual *Bromus* (*Bromus* hereafter) invasion is impacting many arid and semiarid ecosystems in the Western USA (Bradley et al. 2006; Chambers et al. 2014a; Brooks et al. 2015; Germino et al. 2015). While the degree of invasion varies depending on site conditions (e.g., elevation and temperature/moisture regimes), in extreme cases, these invasions create vegetation states that are perpetuated by high litter production, frequent fire occurrence, and fundamentally altered nutrient cycling (Parker and Schimel 2010; Davies and Nafus 2013). Without intervention, these ecosystems may remain in a degraded vegetation state where grazing livestock, soil disturbances, and recurring fires will favor annual relative to perennial species (Hobbs et al. 2007). Furthermore, net primary production and annual precipitation are highly variable in arid and semiarid ecosystems (Clinton et al. 2010), which amplifies their vulnerability to plant community disruption associated with climate change (Bradley 2009).

Given the prevalence of *Bromus* invasions in the Western USA, the objectives of this chapter are to (1) identify common characteristics responsible for the persistence of problematic *Bromus* in this region; (2) illustrate how future research and management can better address underlying processes responsible for annual grass persistence through the adoption of adaptive management; (3) assess past research efforts to control the problematic species, *B. tectorum*, with a case study of treatment outcomes; and (4) assess management efforts to remediate landscapes impacted by *B. tectorum* and improve greater sage-grouse (*Centrocercus urophasianus* Bonaparte; sage-grouse hereafter) habitat with a case study from the Great Basin region of the Western USA. We conclude that adapting for the future will need continued integration between research and management, greater understanding of ecological processes, and long-term experiments to inform adaptive management decisions. Although we emphasize heavily invaded ecosystems that require long-term, continual management efforts, our assessment is also relevant to conservation and preventative restoration projects.

12.2 Exotic Annual *Bromus* Grasses in Western USA Ecosystems

Many *Bromus* impact semiarid and arid ecosystems of the Western USA (Brooks et al. 2015; Germino et al. 2015; Atkinson and Brown 2015). While species have unique distributions and climate suitability as described in previous chapters, here, we review the five most notable species in order to identify common characteristics that must be addressed by restoration practitioners.

12.2.1 Bromus hordeaceus

Bromus hordeaceus L. (soft brome or chess) is a common component of California's annual grasslands. Compared to other species in annual grasslands, *B. hordeaceus* has relatively shallow roots (Brown and Rice 2010), seed longevity is very short irrespective of depth and duration of burial in seed banks, and seedling survival is generally higher for seeds germinating at the soil surface (Jensen 2009). Like many annual grasses, *B. hordeaceus* has short-lived seeds and transient seed banks that experience high rates of mortality (Marshall and Jain 1967), yet seed banks remain large because of prolific seed production (Wainwright et al. 2012). *Bromus hordeaceus* also has invaded serpentine grasslands in Northern California (Bartolome et al. 1980) where herbicide applications have been relatively effective at reducing its percentage cover (Aigner and Woerly 2011).

12.2.2 Bromus diandrus

Seeds of *Bromus diandrus* Roth (ripgut brome) have a short-lived innate dormancy (Gill and Blacklow 1985), and similar to *B. hordeaceus*, seed banks are essentially transient (Cheam 1986). In California coastal prairie ecosystems, growth of *B. diandrus* is favored in nitrogen-rich patches (Maron and Connors 1996). Compared to *B. hordeaceus*, *B. diandrus* also responds favorably to increased nutrient availability and is likely less adapted to lower soil nutrient availability in oak woodlands of California (Rice and Nagy 2000). However, experimentally reducing soil nitrogen in California coastal prairie did not reduce the competitive ability of *B. diandrus* relative to the native grass, *Bromus carinatus* L. (California brome), suggesting that restoration will require more than manipulating soil resources (Kolb and Alpert 2003). Additional control methods include carefully timed grazing and prescribed fire to reduce *B. diandrus* biomass and spikelet production (Skaer et al. 2013). For example, *B. diandrus* is considered more susceptible to burning than other California annual grassland species (DiTomaso et al. 2006), and seeds have higher mortality when exposed to direct flames during burn experiments (Sweet et al. 2008). Consequently, in the absence of fire, *B. diandrus* reaches high abundance (Kyser and DiTomaso 2002), allowing it to deplete soil moisture early in the growing season and reduce the abundance of the native bunchgrasses in California annual grasslands (Moyes et al. 2005).

12.2.3 Bromus arvensis

The winter annual grass *Bromus arvensis* L. (field or Japanese brome; syn. *B. japonicus* Thunb.) occurs throughout the Western USA, but is most problematic in mixed-grass prairies of the Northern Great Plains where it lowers forage production and reduces native ungulate use (Haferkamp et al. 2001). In this region, *B. arvensis* also alters several ecosystem properties, including net annual productivity, aboveground litter decomposition, and soil moisture content (Ogle et al. 2003). The contribution of annual grasses to annual biomass production in this region can range from 6 % to 96 % depending on precipitation, and *B. arvensis* invasion is enhanced by wet autumn conditions and the presence of dense ground litter (Haferkamp et al. 1993). *Bromus arvensis* tends to become more abundant as soil moisture increases and temperatures become cooler (Hulbert 1955), and similar to *B. diandrus*, fire exclusion leads to greater plant density (Whisenant 1990). Reducing litter and stubble height with grazing, clipping, or burning negatively impacts *B. arvensis* density and biomass relative to native perennial grasses in the first year or two after fire (Harmoney 2007). Fire also reduces seedling emergence of *B. arvensis* by consuming seeds deposited on the soil surface (Vermeire and Rinella 2009). Similar to other annual grasses, *B. arvensis* control is difficult because of its capacity to reestablish from seed banks (Andersen et al. 1992). For example, herbicide applications must be repeated 2 or more years in order to significantly reduce populations (Hewlett et al. 1981). Given the difficulty in preventing establishment from seed banks, much research has focused on identifying herbicide types and application prescriptions that specifically reduce seed production as opposed to reducing biomass production (Rinella et al. 2013).

12.2.4 Bromus rubens

Bromus rubens L. (red brome; syn. *B. madritensis* L. ssp. *rubens* [L.] Durvin) has invaded portions of the Great Basin and Mojave Desert where it produces fine, flammable fuels that increase fire frequency and reduce the abundance of native species (Salo 2005). Similar to other annual grasses, *B. rubens* is highly responsive to experimental additions of soil nitrogen (Brooks 2003). *Bromus rubens* also outperforms native annual species in the Mojave Desert by growing more rapidly, exploiting deeper soils, and producing larger seeds with greater germination percentages than native desert annual species (DeFalco et al. 2003). Similar to other annual *Bromus* species, *B. rubens* seeds do not persist in soil seed banks, and its invasion in the Southwestern USA has been attributed to earlier germination and growth and high seedling density and biomass production (Jurand et al. 2013). In addition, abundant litter production by *B. rubens* increases the occurrence of wildfire in desert ecosystems where fire was once rare (Brooks and Matchett 2006). Litter accumulation also increases soil water content during drought and facilitates early growth of *B. rubens* (Violle et al. 2006). Although wildfire in the Mojave Desert can be lethal to *B. rubens* seeds in shrub understory microhabitats and cause reductions in annual plant production for up to 4 years, it continues to dominate after wildfire in years that receive above-average winter precipitation (Jurand et al. 2013). While few experiments have directly evaluated control methods for *B. rubens* (Munson et al. 2015), existing studies suggest that seed densities are reduced after wildfire relative to unburned areas and revegetation should be pursued to competitively suppress annual grass recolonization (Jurand and Abella 2013).

12.2.5 Bromus tectorum

Of all *Bromus* in the Western USA, *B. tectorum* L. (downy brome or cheatgrass) is the most widespread and extensively studied (Duncan et al. 2004). Its dominance varies widely across this region and is generally most abundant at lower elevations, where growing conditions are warmer and drier (Chambers et al. 2014a). Despite over 60 years of research, however, surprisingly little headway has been made in the development of effective restoration tools and strategies for areas dominated by *B. tectorum* (Mack 2011). Consequently, *B. tectorum* continues to expand its range and gain site dominance by promoting recurrent wildfires and altering soil moisture and nutrient availability (Balch et al. 2013; Blank and Morgan 2013). Given its propensity to invade under certain conditions (Chambers et al. 2007; Reisner et al. 2013), research has focused on reducing plant abundance (e.g., cover, density, and biomass) and seed bank density in the short term to prevent the carryover of viable caryopses (seed) from one year to the next (Young et al. 1969; Jones et al. 2015). In contrast, relatively less research has emphasized how environmental conditions (Young et al. 1969) and active control methods impact soil surface characteristics (Young et al. 1972a, b, 1976).

12.2.6 Common Characteristics of Exotic Annual **Bromus** *Species*

Based on our brief review, we suggest that population persistence from transient seed banks, altered soil resources and litter production, displacement of native species, and changed disturbance regimes are common characteristics associated with ecosystems heavily invaded by *Bromus*. Due to the annual life form, persistence within arid and semiarid ecosystems in the Western USA is primarily attributed to sufficient seed production and regeneration from seed banks to perpetuate populations. Interestingly, populations persist even when environmental conditions, pathogens, and active control treatments greatly reduce seed production and seedling emergence (Smith et al. 2008; Baughman and Meyer 2013; Jurand et al. 2013). Although this mechanism of persistence indicates that successful control strategies will require multiple, consecutive-year efforts to diminish seed banks (Davies and Johnson 2011), additional common characteristics should also be targeted. For example, heavily invaded ecosystems may experience frequent disturbances that increase resource availability and reduce the abundance of resident native species. Consequently, annual grass invasion may leave legacies such as altered disturbance regimes, soil microbial communities, and nutrient cycling (Belnap et al. 2005, 2015). When such legacies are present, simply removing *Bromus* may not result in ecosystem recovery, especially for heavily invaded ecosystems where native species have been displaced (Bauer 2012; Chambers et al. 2014a). If positive feedbacks among annual grass dominance, high litter production, and recurring wildfires exist, the frequency of disturbance must be lengthened in order for slower-growing native perennial grass and shrub species to recover and establish from restoration seedings (Knapp 1996; Gaertner et al. 2014). In addition, because increased litter abundance generally favors germination and emergence of annual grasses, which is highly variable—depending on site characteristics and interannual productivity and grazing regime (Bansal et al. 2014)—restoration efforts may need to address the roles of litter in creating short-term legacy effects (Grman and Suding 2010) and annual grass-fire cycles (Balch et al. 2013). Finally, in addition to decreasing seed banks and litter production, heavily invaded ecosystems have low resilience to disturbance, and native perennial species are incapable of competing with *Bromus* and resisting recolonization where they have been displaced (Chambers et al. 2007, 2014a). Accordingly, restoration must create conditions to augment native species recovery and/or establishment from seedings in order to increase competitive interactions with *Bromus* (Corbin and D'Antonio 2010; James et al. 2013a). Because common *Bromus* characteristics interact with a number of complex environmental and ecological processes, it becomes clear that ecosystem restoration, as well as conservation projects, must adopt a framework that can be tailored to prevailing site-specific conditions.

12.3 Adopting an Adaptive Management Framework for *Bromus*

Tremendous uncertainty exists when developing realistic management goals and restoration strategies for ecosystems dominated by annual grasses (Wisdom and Chambers 2009). While short-term restoration efforts may initiate favorable site recovery responses (Hirsch-Schantz et al. 2014), multiple interventions are typically needed before ecosystems regain resilience to prevailing disturbance regimes (Suding et al. 2004). Consequently, the development of effective management and restoration goals will require long-term, systematic, stepwise improvements that utilize ecological principles, processes, and practices (Sheley et al. 2010; Herrick et al. 2012). At its core, adaptive management provides the planning, implementation, and evaluation steps that are applicable to nearly any invasive species or habitat (Fig. 12.1) (Williams et al. 2009; Williams 2011; Rist et al. 2013). Importantly, it provides a mechanism for iterative learning through a feedback loop between research and management (Reever-Morghan et al. 2006; Boyd and Svejcar 2009). This feedback loop is most effective when uncertainty is reduced over time as information from research and monitoring are continually considered for management alternatives (Williams and Brown 2012, 2014).

The adaptive management framework entails two distinct phases: the setup phase, in which essential elements are put into place, and the iterative phase, in which the components are linked in a repeated decision-making cycle (Fig. 12.1) (Williams and Brown 2012). Technical learning about the biophysical system and its response to management takes place in the setup phase, and learning about the social and institutional aspects of the process takes place by periodically reviewing and adjusting elements. While adaptive management can be divided into additional sub-steps when applied to specific restoration sites (Reever-Morghan et al. 2006) or used to advance new conceptualizations of the restoration process (Allen et al. 2011), all versions of the adaptive management model have similar core elements. Specifically, successful implementation requires involving all stakeholders who will

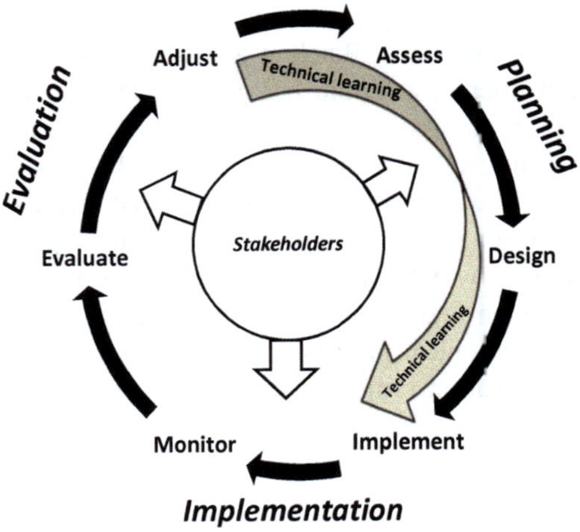

Fig. 12.1 The adaptive management cycle with three primary steps: planning, implementation, and evaluation. Stakeholder involvement has a central role and integrates elements of the cycle. Both technical learning (*gray arrow*) and institutional learning occur during the adaptive management cycle to improve decision-making

be affected by management actions and including local and scientific knowledge in all phases of the process (Fig. 12.1). Below, we discuss how the generalized steps of adaptive management—planning, implementation, and evaluation—can be applied to *Bromus* management in the Western USA.

12.3.1 Planning

The planning step involves assessing conditions of the restoration site to define management problems (Rist et al. 2013) and prioritize management questions (Reever-Morghan et al. 2006). In this step, both quantitative and qualitative data are evaluated to establish current knowledge about the system and identify uncertainties about its function and response to management activities (Rist et al. 2013). For example, acquiring a basic understanding of climate, soils vegetation, and plant community characteristics and indicators of ecological processes helps determine the site potential and limitations and the appropriate assessment and monitoring variables for an annual grass control project (Herrick et al. 2006; Miller et al. 2014). This information can be compared to site-specific standards to clarify the ecological processes that must be addressed at the project site (Sheley et al. 2011). Quantitative data from research conducted on similar ecological sites should also be obtained to forecast how potential land treatments might impact vegetation and soils (Karl et al. 2012).

Forecasting how restoration interventions may impact specific ecological processes and remediate the common characteristics identified for *Bromus* (see Sects. 12.1–12.5) is essential to planning (Herrick et al. 2012). However, due to the dynamic nature of plant community succession and extreme variability in weather from year to year (Hardegree et al. 2012), it may be difficult to identify which ecological processes to monitor and how they influence desired successional trajectories. To address this uncertainty, restoration practitioners have developed successional management concepts to understand how ecological processes operate over a range of spatial and temporal scales (Krueger-Mangold et al. 2006). These concepts also create a linkage between ecological processes that are not functioning as desired and potential tools and strategies that can be employed to reduce annual grass persistence, reestablish structural and functional diversity, and ultimately restore greater resilience to invaded ecosystems (Sheley et al. 2010; Chambers et al. 2014a). For example, the Ecologically-Based Invasive Plant Management framework (www.ebipm.org) has been adopted by numerous land managers and private landowners throughout the Western USA. The framework uses an iterative five-step program that incorporates (1) site assessment, (2) identification of the underlying causes of ecological processes that are not functioning as desired, (3) application of ecological principles to make decisions, (4) selection of appropriate tools and strategies based on principles, and (5) design and execution of a restoration plan using adaptive management principles (Sheley et al. 2010; Leffler and Sheley 2012). While these general principles help practitioners apply sequential interventions to reach a desired ecosystem state, unique management strategies will likely be needed for different *Bromus* species based on ecoregional and site-specific characteristics (Herrick et al. 2006; Suding and Hobbs 2008).

Planning also involves identifying realistic objectives and criteria for success of the project (Williams and Brown 2014). These elements are often overlooked or implied, which can make success elusive (Suding 2011). One challenge in identifying project objectives and criteria for success is that multiple alternative plant community pathways and outcomes can result from management options (Grman et al. 2013). Consequently, it is important to explicitly consider what these pathways and outcomes might be and recognize that nonequilibrium dynamics may best describe how ecosystems respond to management inputs (von Wehrden et al. 2012). Ecological Site Descriptions (ESDs), which are published reports that provide information regarding the suitability of land for specific uses and their capability to respond to management activities and disturbance (https://esis.sc.egov.usda.gov/Welcome/pgESDWelcome.aspx), can help address this uncertainty. Although ESDs do not currently exist for all areas invaded by *Bromus* in the Western USA, the US Department of Agriculture's Natural Resources Conservation Service houses draft descriptions for many rangeland and forestland ecosystems. These descriptions can also contain valuable information on successional dynamics of plant communities and management alternatives, which are depicted with state-and-transition models (STMs) (Provencher et al. 2015; Bestelmeyer et al. 2009). For example, based on site-specific monitoring data, STMs show restoration pathways and indicate differences in resilience to disturbance and resistance to *Bromus* (Bagchi et al. 2013; Chambers et al. 2014a; Miller et al. 2014). With an understanding of potential site dynamics, land managers, scientists, and other stakeholders can consider the costs, benefits, and opportunities for implementing alternative actions.

Clearly defining how changes in the ecosystem will be assessed is also a crucial component of adaptive management planning. Ideally, the ecosystem features chosen to monitor are closely linked to project objectives and are measured at appropriate spatial and temporal scales. Monitoring efforts should capture whether desired site conditions and standards are met and how management actions influence the common characteristics associated with *Bromus* persistence (see Sect. 12.2.6) and thus annual weed dominance (Sheley et al. 2010). Plans should also assess how restoration activities influence disturbance regimes, invasion resistance, and site resilience to stress and disturbance (Abella 2014; Chambers et al. 2014a).

12.3.2 Implementation

The next step in the adaptive management cycle is project implementation (Fig. 12.1). Multifaceted management actions are carried out as agreed upon by the managers and stakeholders followed by execution of the monitoring activities identified in the planning step. Monitoring data are summarized and analyzed as the cycle continues into the evaluation step.

12.3.3 Evaluation

After management intervention, data gathered through monitoring is used to evaluate the state of the system, as described in STMs (Williams and Brown 2012). Because criteria for success are defined in the planning process, the monitoring results are compared to these criteria to evaluate whether the trajectory of the system is consistent with the management objectives. If it is not, the management approach can be adjusted. These learning-based adjustments provide the necessary feedbacks for making iterative decisions regarding the choice of management actions (Svejcar and Boyd 2012; Williams and Brown 2012). For example, not every restoration decision will lead to success, but through application of monitoring, evaluating, and adjusting, management actions can be modified to correct site trajectories toward restoration goals. This feedback allows project leaders to learn from both successes and failures as they collectively learn and incorporate information about the fundamental successional processes responsible for changes within a restoration site (Walker et al. 2007).

Bromus control will also benefit from evaluating the influence of weather, which is a major contributor to the variability of restoration outcomes in arid and semiarid systems. Although weather is relatively unpredictable from year to year, it has historical characteristics that can be captured to remove significant uncertainty from management actions (Hardegree et al. 2011, 2012). While a number of models have been proposed to incorporate the accretion of knowledge through adaptive management (Allen et al. 2011), most comprehensive models do not specifically address the issue of annual and seasonal variability in weather (Hardegree et al. 2012). However, weather records, simulations of weather variability, and seasonal weather forecasting information could be incorporated into both the planning and evaluation steps of adaptive management. This incorporation reduces uncertainty by clarifying the role of weather variability within a climatic regime (Hardegree et al. 2012; Abatzoglou

2013) and distinguishing between treatment-dominated and weather-dominated effects (Hardegree et al. 2011, 2012).

Once climate regime and weather variability are considered, the evaluation step should determine how site aspect, slope, and soil characteristics, including texture and depth, modify restoration outcomes (Molinar et al. 2002). At a finer scale, microsite variability also affects specific demographic processes that ultimately explain the relative treatment success or failure in a given year (Hardegree et al. 2011; James et al. 2011). Thus, in addition to evaluating end-of-season biomass or plant density metrics (Hardegree et al. 2012; Svejcar et al. 2014), efforts should be made to integrate demographic data, including spatial distribution of plants (Reisner et al. 2013) and site-specific microsite standards when describing restoration outcomes (see Sect. 12.3.1, Pyke et al. 2002; Herrick et al. 2008).

The adaptive management model presented by Allen et al. (2011) can be modified to include a mechanism for selecting preferred treatment alternatives that accommodate or adjust for weather/management interactions (Fig. 12.2). Successful, partially successful, and unsuccessful management outcomes can be ranked relative to the climate variables and historical weather conditions at a given site (Hardegree et al. 2012; Enright et al. 2014). Unsuccessful management options are only abandoned if multiple treatment iterations result in failure in years with above-average

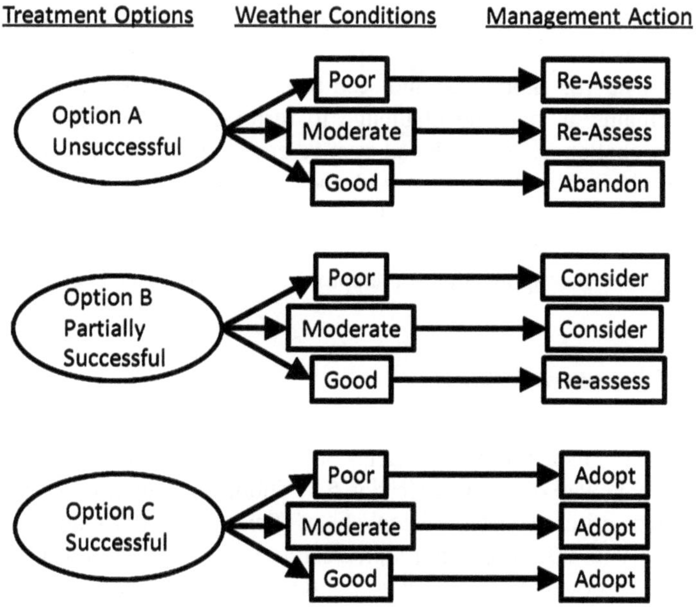

Fig. 12.2 Model for weather informed learning through adaptive management. Management actions should be interpreted within the context of weather conditions during a given treatment year. Unsuccessful treatment options in poor weather years should perhaps be reassessed in future years and possibly abandoned. Partially successful options may be more successful under better weather conditions and considered for future applications. Successful options that emerge become apparent under all weather conditions and should be adopted. Model adapted from the model for knowledge accretion through adaptive management (Allen et al. 2011)

weather for plant establishment. Partially successful management outcomes in an intermediate weather year may be more successful in an above-average year, and successful management outcomes may be artificially elevated in an above-average weather year. All decisions about management effectiveness, therefore, need to be gauged relative to weather favorability, and treatments should be evaluated over a range of weather conditions for a given climatic regime before general inferences are assigned (Sheley et al. 2010; Hardegree et al. 2011, 2012).

In the next two sections, we present two case studies to assess the possibilities of integrating past research results and current land management resources within an adaptive management framework for *Bromus* restoration projects in the Western USA. In the first case study, we explore how various experimental treatments impact the abundance of *B. tectorum*. In the second case study, we review how the US Department of the Interior's Bureau of Land Management (BLM) is applying adaptive management for *B. tectorum* reduction and conservation of sage-grouse habitat on Western USA arid and semiarid ecosystems.

12.4 Assessing *B. tectorum* Research: A Case Study of Treatment Outcomes

One of the most notable characteristics shared by *Bromus* is their capacity to reestablish from seed banks following the application of treatments. Although seeds are generally considered short-lived and seed banks are considered transient, reestablishment remains an ongoing challenge to restoration efforts (Mack 2011). To address this challenge, we analyzed past research efforts to reduce *B. tectorum* abundance within heavily invaded arid and semiarid ecosystems of the Western USA. From this analysis, we identify the relative effectiveness of control treatments, briefly discuss their strengths and weaknesses, and suggest future research needs to improve their application for adaptive management (see Sect. 12.7).

12.4.1 Literature Review

We identified published research articles on *B. tectorum* control treatments using a web-based tool (Web of Knowledge^{SM} v. 5.9; Thomson Reuters). In our search, we included two primary databases: Web of Science® (1975–2012) and CAB Abstracts® (1910–2012). Using a *topic* search, we identified all articles that contained the terms downy brome, cheatgrass, *Bromus tectorum*, and downy chess. We omitted articles that were not published in peer-reviewed journals; that did not report original data from arid and semiarid woodlands, shrublands, shrubland steppe, and grassland ecosystems; or that did not report mean values for at least one contrast between a *B. tectorum* reduction treatment and an untreated control. A total of 45 research articles, published between 1944 and 2012 (68 years), met our criteria (Table 12.1). Among these articles, we retained those that reported percentage

Table 12.1 Studies included in summary analysis of *Bromus tectorum* treatment methods

Study	Control method[a]	Response variables[b]	Timeframe[c]	Location (USA state)
Andersen et al. (1992)	R	B	E, L	MT
Banks and Baker (2011)	B	C	L	CO
Bates and Svejcar (2009)	B	C	E, L	OR
Beckstead et al. (2011)	B	B	E,	UT
Bruce et al. (2007)	D	C	E, L	NV
Chambers et al. (2007)	B, H	B	E, L	NV, UT
Collins (1987)	B, D	C	E, L	OK
Currie et al. (1987)	H	B	E, L	MT
Davies et al. (2008)	B	C	L	OR
Davies et al. (2009)	B, D	B, C	L	OR
Davison and Smith (2007)	H, R	B	E, L	NV
Diamond et al. (2009)	D	B, C	E, L	NV
Diamond et al. (2012)	B, D	B	E, L	NV
Eckert and Evans (1967)	H	B	E, L	CA, NV
Evangelista et al. (2004)	B	C	L	UT
Evans and Eckert (1965)	H	B	E	NV
Evans and Young (1977)	H	B	E, L	NV
Fahnestock et al. (2003)	D	C	E, L	SD
Gardner et al. (2009)	B	C	L,	UT
Gucker and Bunting (2011)	B	C	E, L	ID
Gundale et al. (2008)	B	B	L	MT
Hilty et al. (2004)	B	C	L	ID
Hoelzle et al. (2012)	R	B	L	CO
Hunter (1991)	B	B	L	NV
Jessop and Anderson (2007)	B	C	E, L	UT
Keeley and McGinnis (2007)	B	C	E, L	CA
Loeser et al. (2007)	D	C	E,	AZ
Mata-Gonzalez et al. (2008)	R	B	E, L	WA
McGlone et al. (2009)	B	C	E, L	AZ
McGlone et al. (2011)	D	C	E, L	AZ
Melgoza et al. (1990)	B	B	L	NV
Milchunas et al. (2011)	D	C	E	CO
Morrow et al. (1977)	H	B	E	NE
Owen et al. (2011)	H, R	B	E, L	AZ
Poreda and Wullstein (1994)	B	C	E	UT
Pyke (1986)	D	B	E	WA
Ralphs et al. (2007)	D	B	E	UT
Rose et al. (2001)	R	B	L	WY
Sorensen and McGlone (2010)	D	C	E, L	AZ
Thacker et al. (2009)	R	B	E, L	UT

(continued)

Table 12.1 (continued)

Study	Control method[a]	Response variables[b]	Timeframe[c]	Location (USA state)
Turner (1971)	D	C	L	CO
West and Yorks (2002)	B, D	C	L	UT
Wicks et al. (1965)	H	B	E, L	NE
Young et al. (1972a)	B	B, C	E	CA

[a]B, burn; D, defoliate; H, herbicide; R, revegetation
[b]C, *B. tectorum* cover; B, *B. tectorum* shoot biomass production
[c]Timeframe posttreatment application; E, early (≤1 year); L, late (≥2 year)

cover and annual biomass production data for *B. tectorum*. We also limited our inquiry to the following four *B. tectorum* treatment methods: burning (i.e., wildfire and/or prescribed fire), defoliation (i.e., grazing and/or mowing), herbicide application, and revegetation (i.e., seeding).

For the 45 articles that met our criteria, we extracted data from tables or estimated values from digital scans of figures using DataThief III Version 1.5 computer software (Myers and Harms 2009). In most cases, individual articles offered numerous control-treatment contrasts when they included more than one treatment and if experiments were conducted at different, independent study locations. When data were collected at the same location over time, we acquired both the first (1 year posttreatment application) and final census (>2 years posttreatment) to determine potential differences due to timeframe after treatment.

12.4.2 Analysis of B. tectorum *Treatment Outcomes*

To assess the effect of various treatment methods on *B. tectorum*, we used effect size estimates, an approach for synthesizing responses measured across a set of contrasting studies (Osenberg et al. 1999). Effect sizes were calculated with the following equation: $\ln RR = \ln(X_E/X_C)$, where lnRR is the natural log response ratio, X_E = treatment mean, and X_C = control mean (Goldberg et al. 1999). We report mean lnRR ± 95 % confidence intervals without considering within-study variance estimates (Hedges et al. 1999). Mean lnRR's were considered significant if 95 % confidence intervals did not overlap zero. While we recognize that the impact of *Bromus* control treatments are site specific and outcomes vary across precipitation and temperature regimes (Chambers et al. 2014a), our goal was to assess general, treatment outcomes, identify research gaps, and inform adaptive management planning (Fig. 12.1).

12.4.3 *Analysis Results*

12.4.3.1 Burning

Our analysis revealed that burning reduced percentage cover of *B. tectorum* in the early phases of studies, yet after a few years, this effect was not significant (Fig. 12.3). In contrast, burning clearly increased biomass production of *B. tectorum* on a per area basis in both the short and long term. The ability of *B. tectorum* to recover after being significantly reduced by wildfire has been attributed to survival of seeds on the soil surface that are not consumed by fire (Diamond et al. 2012). In addition, *B. tectorum* seedlings that emerge after burning experience lower intraspecific competition and produce plants with greater seed production (Evans and Young 1978), which explains why biomass production increased more than cover over time. Consequently, even though burning can reduce litter, increase bare ground, and temporarily reduce *B. tectorum* competition with seeded desirable perennial grasses (Young et al. 1969; Evans and Young 1970), these conditions appear to be transitory. In fact, recent research illustrates that a very short time period exists after wildfire when conditions are favorable for the establishment of perennial species (Pyke et al. 2013). Applying this knowledge to adaptive management would entail prioritizing seeding events as closely as possible after burning and incorporating an understanding of how burning interacts with site-specific differences in litter production and temperature/precipitation regimes (Bansal et al. 2014; Chambers et al. 2014a; Jones et al. 2015).

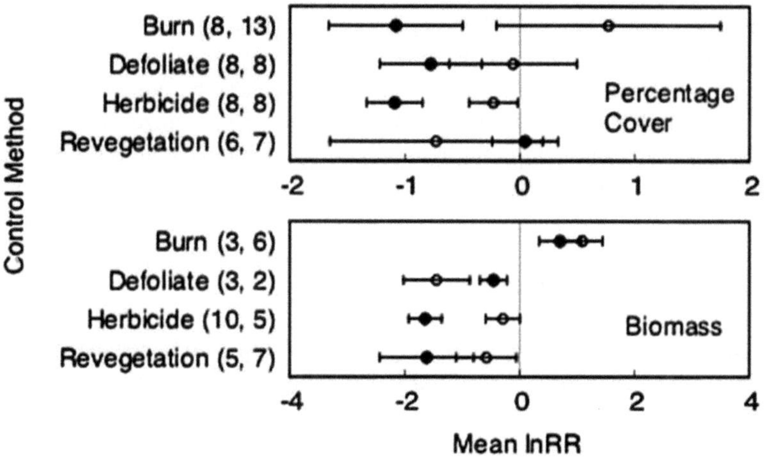

Fig. 12.3 Results of analysis comparing different *Bromus tectorum* control methods. Symbols represent mean lnRR effect sizes (±95 % confidence intervals) for percentage cover and biomass production for *B. tectorum* 1 year after treatment (*closed symbol*) and 2 or more years after treatment application (*open symbol*). Mean effects are considered significant when confidence intervals do not overlap the vertical zero line. Values in parentheses indicate number of articles (e.g., Table 12.1) used to calculate effect sizes for early and late responses, respectively

12.4.3.2 Defoliation

Our analysis showed that defoliation by grazing or mowing reduces *B. tectorum* biomass for up to 2 years and percentage cover for only 1 year (Fig. 12.3). This analysis agrees with previous results indicating that defoliation treatments need to be applied for more than 2 years because *B. tectorum* can regrow following defoliation under favorable environmental conditions (Hempy-Mayer and Pyke 2008; Diamond et al. 2009). To maximize the effect of defoliation, the intensity and timing of defoliation should be a primary consideration to reduce *B. tectorum* seed production and minimize negative effects on perennial grass establishment (Frost and Launchbaugh 2003; Young and Clements 2007). Consequently, when designing an adaptive management plan, the prudent use of defoliation and/or grazing in mixed herbaceous stands should include the application of time-controlled, short-duration, or high-intensity strategies to selectively graze annual grasses when they are most palatable and when perennial grasses are dormant and grazing tolerant (Frost and Launchbaugh 2003; Smith et al. 2012). For example, because cattle prefer to graze *B. tectorum* second to bunchgrasses (except in early spring), targeted grazing to reduce annual grass abundance should be applied cautiously to avoid undesirable impacts on perennial grasses (Ralphs et al. 2007). Heavy grazing can also decrease perennial forb cover and negatively impact perennial grasses in the short term, which may lead to an increase in *B. tectorum* cover (Loeser et al. 2007; Davies et al. 2009). Achieving longer-term defoliation impacts on *B. tectorum* abundance should also consider timing defoliation events when development of the annual grass is most sensitive to defoliation stress. For example, defoliation when the seed head is enclosed within the sheath of the flag leaf (i.e., during the boot stage) has greater impact on seed production than defoliation at later phenological stages (Hempy-Mayer and Pyke 2008; Diamond et al. 2009). However, because *B. tectorum*'s annual productivity is highly correlated with precipitation (Loeser et al. 2007), it may be more effective to plan and implement defoliation treatments in drier years to maximize the effect on seed production while minimizing regrowth potential—particularly in areas where defoliation will not be detrimental to residual native species (Hirsch-Schantz et al. 2014).

12.4.3.3 Herbicide Application

Of the four treatments we evaluated, only herbicide application consistently reduced both short- and longer-term *B. tectorum* percentage cover and biomass (Fig. 12.3). The efficacy of herbicide application as a treatment method may stem from a relatively longer research history relative to the other treatments. For example, the development of preemergent imidazolinone and sulfonylurea herbicides that target acetolactate synthase in plants provides high selectivity for invasive annual grasses at low application rates (Stidham 1991; Monaco and Creech 2004). However, recent studies indicate that seeded perennial grass density and cover can be reduced for up to 3 years after application with the soil-active herbicide imazapic (Pyke et al. 2014; Munson et al. 2015). For postemergent *B. tectorum* applications on semiarid ecosystems, the nonselective herbicide, glyphosate, also shows promise because it can be applied early in the spring during a period when exotic annual species are actively growing (yet slower-growing native species may be at risk) (Blackshaw 1991). Given the limitations of imazapic (Morris et al. 2009) and glyphosate (Wilson et al. 2010; Kyser et al. 2012), adaptive management of restoration sites should incorporate mechanisms to learn and adjust application rates and timings to maximize impact on *B. tectorum* persistence characteristics while minimizing nontarget effects on seeded and resident native plants.

12.4.3.4 Revegetation

Because revegetation entails seeding desirable perennial plant species after wildfires, prescribed burning, or burning in concert with herbicide treatments (Table 12.1), our analysis should be viewed in the context of how seeding provides additional benefits when applied in an integrated manner. Accordingly, our analysis illustrates that seeding reduced *B. tectorum* biomass production, but its effect on *B. tectorum* cover was variable, especially in the long term (Fig. 12.3). Reduced biomass may be a consequence of establishing perennial grasses competing for limiting soil resources and subsequent decreases in the growth of exotic annual grasses (Leonard et al. 2008; Hirsch-Schantz et al. 2014). Furthermore, the ability of seeding to reduce invasive species depends on initial seedling establishment, and this effect is more pronounced with greater time after seeding (Eiswerth et al. 2009; Pyke et al. 2013) and under cooler and moister environmental conditions (Knutson et al. 2014). For example, when successfully established, perennial grasses can stifle reinvasion of *B. tectorum* (Reisner et al. 2013; Munson et al. 2015), increase native vegetation structural diversity for wildlife habitat (Arkle et al. 2014), and increase resilience to wildfire disturbances (Chambers et al. 2014a). This interpretation of revegetation highlights the importance of adequate monitoring to identify the factors responsible for success and failure and anticipate how establishment will influence plant community dynamics on restoration projects. Effectively integrating monitoring results within an adaptive management framework may require numerous iterations combined with parallel research studies to identify effective restoration species (Jones and Monaco 2007; Leger and Baughman 2014) and make adjustments for the presence of environmental vagaries and edaphic limitations of the restoration site (Hardegree et al. 2012; Karl et al. 2012).

12.5 Assessing *B. tectorum* Management: A Case Study of the BLM

The BLM manages the majority of public rangelands in the USA outside of Alaska, the majority of these lands occur in the Great Basin region of the Western USA, where 30 million ha are managed by the BLM for multiple uses (livestock grazing, recreation, energy development, mining, etc.). These landscapes have highly variable topography, soils, and weather and support a wide variety of wildlife and native plant communities (West 1979). The climate is semiarid with the majority of the precipitation occurring in the winter and spring—over half of the Great Basin receives less than 305 mm of annual precipitation (West 1983).

The BLM is moving to an adaptive management approach to address the issues of nonnative species, habitat conservation and restoration, wildfires, and climate change at large landscape scales. This is a significant change from the historic "command and control" approach to a more flexible, collaborative, risk-tolerant, and landscape-level approach. Past resource management strategies have been characterized as a combination of (1) ad hoc management (e.g., "seat-of-the-pants" management with little foresight), (2) "wait-and-see" management (e.g., try something and wait and see what happens), (3) and steady-state management (e.g., maintain a static condition in the face of nonlinear disturbances) as described in Williams (1997). Adaptive management is now emphasized by the Department of Interior agencies (Williams and Brown 2012). In this section, we illustrate how the BLM is implementing the adaptive management framework (Fig. 12.1).

12.5.1 BLM Background

The BLM approach to addressing *B. tectorum* has historically included changing livestock grazing management practices, strategically reducing *B. tectorum* with the use of strips of less flammable vegetation to reduce the spread of wildfires (i.e., greenstrips), and rehabilitating burned areas to reduce postfire *B. tectorum* dominance. Extensive wildfires in 1999 in which nearly 0.7 million ha of public land were burned provided the stimulus for the BLM to launch a regional restoration strategy through the Great Basin Restoration Initiative to proactively address *B. tectorum* and wildfires (Pellant et al. 2004). More recently, the issues of sagebrush (*Artemisia* spp.) loss, increased wildfires, and *B. tectorum* expansion have coalesced into a concerted effort to conserve sagebrush habitat for sage-grouse in the Great Basin. Sagebrush ecosystems are some of the largest and most imperiled in North America (Noss et al. 1995), and sage-grouse is a sagebrush obligate bird species that has been considered for listing under the Federal Endangered Species Act in 2008, 2010, and 2015. The sections that follow provide a case study for the adaptive management approach that is currently being used by the BLM to manage, restore, and conserve sage-grouse habitat in the Great Basin.

12.5.2 Applying Adaptive Management to Reduce the Spread of B. tectorum *and Conserve Sage-Grouse Habitat*

12.5.2.1 Assess

Rapid Ecoregional Assessments (REAs) are used by the BLM to improve understanding of the existing condition of these landscapes and how these conditions may be altered by ongoing environmental changes and land use demands (http://www.

blm.gov/wo/st/en/prog/more/Landscape_Approach/reas.html). These assessments are conducted at the Level III Ecoregion scale (epa.gov/wed/pages/ecoregions/level_iii_iv.htm) and utilize existing information to describe and spatially portray conservation elements such as habitat for species of concern like sage-grouse (e.g., areas with high landscape cover of sagebrush). REAs then gauge the potential of these habitats to be affected by four *change agents:* climate, wildfires, invasive species, and development (both energy development and urban growth). In addition, REAs establish baseline ecological data to evaluate the effectiveness of future management actions and to serve as a benchmark for pre-project adaptive management assessments.

As an example, the Central Basin and Range REA encompasses large areas of Nevada and Utah and extends into California and Idaho for a total area of 359,869 km^2. Sagebrush ecosystems and sage-grouse are of major concern in this ecoregion. Several examples of REA products germane to restoration of degraded sagebrush habitat include:

1. Status of invasive annual grasses in occupied habitat
2. Current distribution of sagebrush habitat and sage-grouse strongholds
3. Potential areas to restore habitat connectivity based on current locations of change agents
4. Projections of climate change effects on current sagebrush habitat (contraction, overlap, or expansion)

These REA products provide managers a snapshot of information at a landscape scale, ranging from 4 to 39 million ha, which can be used to design and implement sound habitat restoration strategies. These products, in combination with accompanying geospatial data, maps, and predictive models, are openly available through the REA Data Portal (http://www.blm.gov/wo/st/en/prog/more/Landscape_Approach/reas/datadisclaimer.html). While REAs are primarily designed for land managers within an ecoregion to share science-based information, this information will need to be scaled down and integrated with concurrent project site monitoring data prior to making decisions at management scales.

12.5.2.2 Design

The BLM has a long history of designing land treatments and projects to improve or restore land health on public lands. However, management responses to resource issues have traditionally been at a local management unit level rather than at a regional scale. Designing restoration treatments positioned on the landscape in ways that maximize conservation effectiveness is especially important when addressing a broadly distributed and wide-ranging species like sage-grouse (Murphy et al. 2013). The BLM is working with other partners to develop a process that integrates both landscape-scale prioritization and management-scale decision tools to support habitat conservation (Chambers et al. 2014a, b; Miller et al. 2014).

The goal of this cooperative effort is to enhance the ability of land managers to maintain or improve the resilience (or recovery potential) of sagebrush habitat to disturbances and to improve resistance to exotic annual grasses that are partially responsible for altered fire regimes in the Great Basin (Chambers et al. 2014a). Resilience of sagebrush ecosystems to disturbance and resistance to *B. tectorum* invasion varies along climatic and topographic gradients, and ecosystems at higher elevation with cooler and moisture soil temperature/moisture regimes are typically more resilient and resistant than those at lower elevations with warmer and drier regimes. Soil temperature and moisture regimes are components of National Resources Conservation Service (NRCS) soil surveys that can be used to spatially portray resistance and resilience across large regions (Fig. 12.4). Landscape cover of sagebrush is the primary predictor of the probability of sage-grouse persistence (Knick et al. 2013) and, when combined with soil temperature and moisture regimes, can be used to identify areas on the landscape with the potential to support sage-grouse populations and their relative resilience to disturbance and resistance to invasive annual grasses (Chambers et al. 2014b). Portraying this information with respect to sage-grouse strongholds, termed Priority Areas for Conservation (USFWS 2013), provides managers with an effective decision tool to prioritize management and restoration activities across this large landscape (Miller et al. 2013; Chambers et al. 2014a). Specific management strategies include appropriate applications of fire operations (preparedness, prevention, and suppression activities), fuels management, postfire rehabilitation, and habitat recovery/restoration. For example, fire suppression and fuels management would be a high priority in sage-grouse habitat where landscape sagebrush cover is greater than 25 % and resistance to *B. tectorum* and resilience after disturbance are low because these areas can support persistent populations of sage-grouse but are at high risk of *Bromus* dominance. Fuels management such as conifer removal would be appropriate where sagebrush cover is over 25 % and resistance and resilience are moderate to high (Chambers et al. 2014a).

12.5.2.3 Implement

The assessment and design steps of adaptive management provide the context to implement appropriate restoration projects. Project implementation can be greatly improved by the use of ESDs, STMs (see Sect. 12.3.1), and the soil surveys that inform their development (Miller et al. 2013, 2014; Provencher et al. 2015). ESDs contain the biotic and abiotic characteristics that define the site potential (i.e., climate, physiographic, soil characteristics, and plant communities Caudle et al. 2013) and, in combination with current STMs, are useful in determining the target plant community and restoration pathways (Monaco et al. 2012; Chambers et al. 2014b).

Implementation includes project planning and application of *B. tectorum* control methods described above (see Sect. 12.4.3) as well as the seed and equipment to successfully carry out the restoration project. The Great Basin Native Plant Selection and Increase project (http://www.fs.fed.us/rm/boise/research/shrub/greatbasin.shtml) and the SageSTEP Project (www.sagestep.org) are two examples of regional research efforts assisting land managers in restoring native vegetation in the Great Basin.

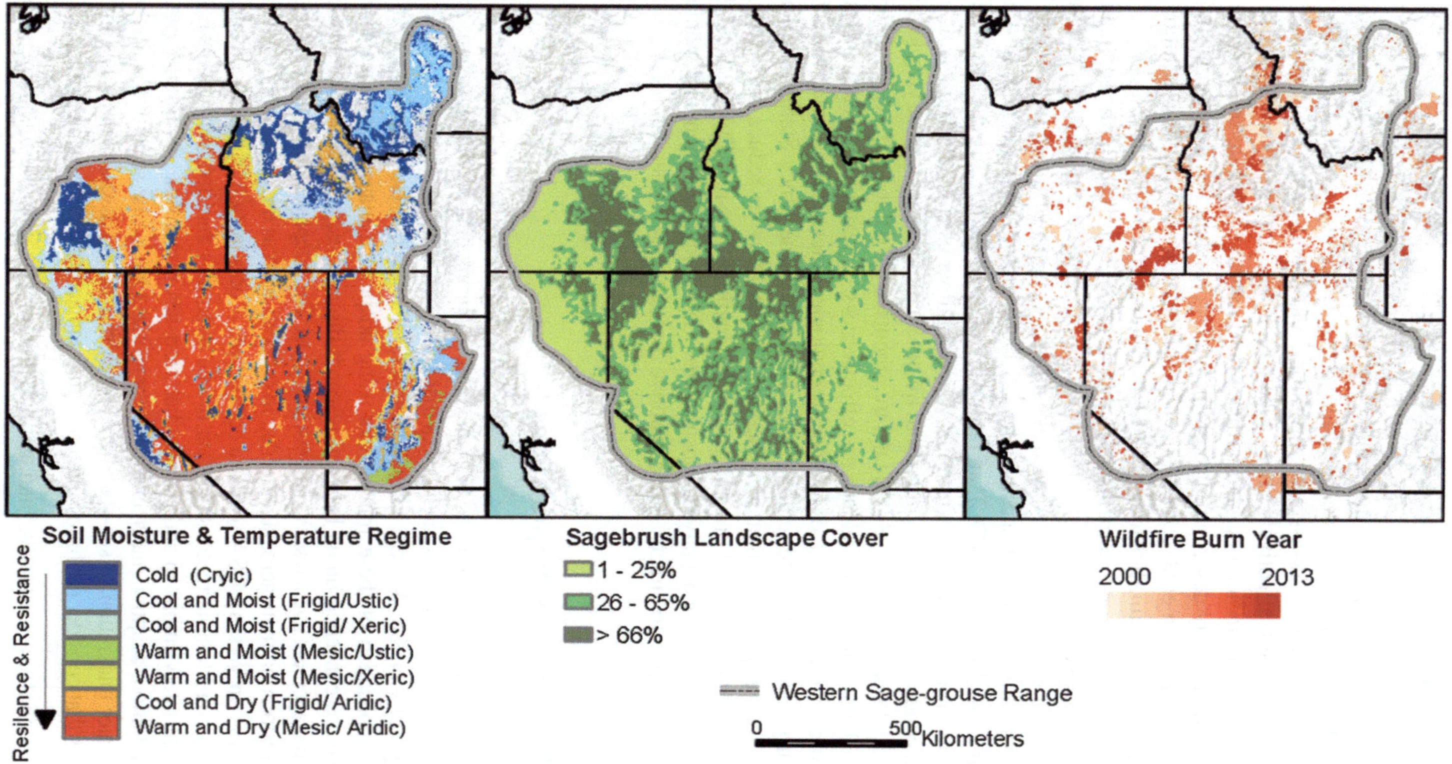

Fig. 12.4 The soil moisture temperature regimes, sagebrush landscape cover, and occurrence of wildfire within sage-grouse Management Zones III, IV, and V (Management Zones adapted from Stiver et al. 2006). See Chambers et al. (2014b) for additional explanation

12.5.2.4 Monitor, Evaluate, and Adjust

Monitoring is essential to determine the effectiveness of restoration strategies and projects and is the cornerstone of the adaptive management process. The BLM has recently implemented an Assessment, Inventory, and Monitoring (AIM) strategy to more efficiently and effectively meet local, regional, and national information needs. A set of core indicators, standardized field methods, remote sensing, and a statistically valid study design are used to provide consistent information at multiple scales (MacKinnon et al. 2001). The availability of monitoring data at multiple scales improves opportunities to apply adaptive management principles.

Monitoring restoration treatments at the project scale contributes to the application of adaptive management principles by providing the data necessary to evaluate the effectiveness of ongoing projects and make adjusts to improve future treatments. To assist in evaluating treatment success, the US Geological Survey created the Land Treatment Digital Library (LTDL), an online resource to catalog historic BLM land treatment information in the Western USA (www.ltdl.wr.usgs.gov/). The LTDL can be used to conduct queries, including about treatment effectiveness, at varying temporal and spatial scales for over 23,000 land treatments. This database provides managers with a large dataset to evaluate the effectiveness of restoration treatments and adjust future restoration strategies for specific ecological sites. For example, the LTDL was recently used to evaluate regional effectiveness of Emergency Stabilization and Rehabilitation project sites that were seeded after wildfire to reduce soil erosion, improve wildlife habitat, and prevent further *B. tectorum* invasion (Pyke et al. 2013; Knutson et al. 2014). Such analyses illustrate how data from assessment and monitoring steps can be used to directly guide adaptive management decisions.

12.6 Management Implications

Much research and discussion has been devoted to chronicling *Bromus* invasion, their effects on ecosystem processes, and factors that favor their dominance on arid and semiarid ecosystems in the Western USA (Chambers et al. 2007; Huttanus et al. 2011). However, application of these frameworks using adaptive management concepts will require a more concerted union between research discoveries, restoration practitioners, and land management agencies. We suggest that proactive, adaptive, and systematic application of restoration/management is needed in order to address the complexity of annual grass invasion on Western USA semiarid ecosystems. The adoption of adaptive management will require long-term consideration of site dynamics and iterative process-based management interventions to address the characteristics responsible for *Bromus* persistence. In addition, adaptive management that prioritizes landscape-scale restoration will require greater integration of site-specific research results and assessment tools. Finally, frameworks that consider the following topics may enhance adaptive management:

- Weather confounds adaptive management planning by introducing a large source of environmental variability that is difficult to capture in a scientific management framework. Recent advances in seasonal forecasting technology may reduce uncertainty in the prediction of restoration activities and increase the efficiency of programs that may have low probability of success in average or below-average precipitation years.

- Long-term restoration success in arid and semiarid ecosystems is challenging given environmental constraints and uncertainties regarding future system dynamics, which can present a big challenge for adaptive management (Beever et al. 2014). Consequently, restoration strategies must be applied strategically and effectively to maximize conservation values. For example, integrating adaptive management into planning and implementing restoration projects that meet sage-grouse habitat needs and promote resistance to annual grass invasion and resilience to disturbance is essential. Relatively low success rates may require redefining goals and accepting less ambitious outcomes.
- Furthermore, variability in arid and semiarid ecosystems and low probability of favorable weather in any given year may require acknowledging multiple acceptable restoration outcomes, relatively long-term management objectives, and an expectation of multiyear management treatments (Hardegree et al. 2012; Herrick et al. 2012). Maintenance of a positive trajectory toward a set of alternative desirable states based on biodiversity, functional, or sustainability goals will undoubtedly be a more feasible objective than the traditional goal of achieving a static and largely hypothetical, late-successional, historical plant community (Whisenant 1999; Hardegree et al. 2011).

12.7 Research Needs

Our general review of the exotic annual *Bromus* species impacting arid and semiarid ecosystems combined with our summary analysis for *B. tectorum* suggests a number of promising research directions. In particular, based on our assessment of common characteristics responsible for annual grass persistence (e.g., transient, yet typically large, seed banks, altered soil resources and litter production, displacement of native species, and frequent disturbance from fire), we suggest a few salient research needs to improve our understanding of specific ecological processes (Sheley et al. 2010; James et al. 2013b):

- Large seed banks are a highly effective reserve for *Bromus* that can perpetuate populations and compromise control efforts (Young et al. 1969; Diamond et al. 2012). This aspect of annual grass management suggests that much more research is needed to improve our understanding of how control methods impact mortality, germination and establishment, and plant reproduction (Davies and Johnson 2011). Consequently, integrated, and long-term control efforts are needed in conjunction with revegetation of desirable perennial species to further explore how restoration outcomes can be enhanced when seedling emergence, seed production, and seed dispersal of *Bromus* are minimized (Chambers et al. 2007; Brisbin et al. 2013).
- Soil surface characteristics such as plant litter, bare ground, and biological soil crusts strongly influence seed germination of annual grasses (Boudell et al. 2002;

Stavi et al. 2008). Further research is needed to specifically address how various control methods alter soil surface attributes and microtopography. Specifically, there is need to further define spatial and temporal patterns of germination, establishment, and growth of *Bromus*, resident native species, and seeded species under heterogeneous soil surface and vegetation conditions and over environmental gradients (Chambers and MacMahon 1994).

- Soil disturbances associated with mechanical control methods, revegetation, and seedbed preparation can have tremendous impacts on seed banks, resource availability, and competition between *B. tectorum* and perennial grasses (Smith 1970; Beckstead and Augspurger 2004). Timing, frequency, and duration of disturbances should be carefully studied for restoration sites to better define restoration opportunities. In particular, future research should emphasize how the short period of time following soil disturbance can be utilized more opportunistically for the establishment of desired species (Young et al. 1969; Giunta et al. 1975).
- The importance of soil resource availability in the invasion process is generally recognized (Davis et al. 2000), yet more mechanistic research is needed to develop a broader understanding of how invasive species management influences resource availability and recovery of native species (Jones et al. 2015). Although much research has emphasized experimental manipulation of soil nitrogen availability to reduce invasive species abundance (Perry et al. 2010), future research will also need to clarify how combined treatment applications influence soil nutrient concentrations and resident plant species (Chambers et al. 2007; Hirsch-Schantz et al. 2014; Pyke et al. 2014).

As research continually innovates by exploring mechanistic aspects of reducing *Bromus* populations and restoring ecological structure and function, adapting for the future will require bridging the gap between research and management. This may best be accomplished by in situ adaptive research and management experiments of how annual grasses, desirable resident plant species, environmental variables, and edaphic factors respond to management treatments and activities over environmental gradients and at multiple locations (James et al. 2013b).

References

Abatzoglou JT (2013) Development of gridded surface meteorological data for ecological applications and modelling. Int J Climatol 33:121–131

Abella SR (2014) Effectiveness of exotic plant treatments on National Park Service lands in the United States. Invasive Plant Sci Manag 7:147–163

Aigner P, Woerly R (2011) Herbicides and mowing to control barb goatgrass (*Aegilops triuncialis*) and restore native plants in serpentine grasslands. Invasive Plant Sci Manag 4:448–457

Allen CR, Fontaine JJ, Pope KL et al (2011) Adaptive management for a turbulent future. J Environ Manag 92:1339–1345

Andersen MR, Depuit EJ, Abernathy RH et al (1992) Value of mountain rye for suppression of annual bromegrasses on semiarid mined lands. J Range Manag 45:345–351

Arkle RS, Pilliod DS, Hanser SE et al (2014) Quantifying restoration effectiveness using multi-scale habitat models: implications for sage-grouse in the Great Basin. Ecosphere 5:31

Atkinson SY, Brown CS (2015) Attributes that confer invasiveness and impacts across the large genus *Bromus* – lessons from the *Bromus* REEnet database. In: Germino MJ, Chambers JC,

Brown CS (eds) Exotic brome-grasses in arid and semiarid ecosystems of the Western USA: causes, consequences, and management implications. Springer, New York, NY (Chapter 6)

Bagchi S, Briske DD, Bestelmeyer BT et al (2013) Assessing resilience and state-transition models with historical records of cheatgrass *Bromus tectorum* invasion in North American sagebrush-steppe. J Appl Ecol 50:1131–1141

Balch JK, Bradley BA, D'Antonio CM et al (2013) Introduced annual grass increases regional fire activity across the arid western USA (1980–2009). Glob Change Biol 19:173–183

Banks ER, Baker WL (2011) Scale and pattern of cheatgrass (*Bromus tectorum*) invasion in Rocky Mountain National Park. Nat Area J 31:377–390

Bansal S, Sheley RL, Blank B et al (2014) Plant litter effects on soil nutrient availability and vegetation dynamics: changes that occur when annual grasses invade shrub-steppe communities. Plant Ecol 215:367–378

Bartolome JW, Stroud MC, Heady HF (1980) Influence of natural mulch on forage production on differing California annual range sites. J Range Manag 33:4–8

Bates JD, Svejcar TJ (2009) Herbaceous succession after burning of cut western Juniper trees. West N Am Nat 69:9–25

Bauer JT (2012) Invasive species: "back-seat drivers" of ecosystem change? Biol Invasions 14:1295–1304

Baughman OW, Meyer SE (2013) Is *Pyrenophora semenipera* the cause of downy brome (*Bromus tectorum*) die-offs? Invasive Plant Sci and Manag 6:105–111

Beckstead J, Street LE, Meyer SE et al (2011) Fire effects on the cheatgrass seed bank pathogen *Pyrenophora semeniperda*. Rangel Ecol Manag 64:148–157

Beever EA, Mattsson BJ, Germino MJ et al (2014) Successes and challenges from formation to implementation of eleven broad-extent conservation programs. Conserv Biol 28:302–314

Belnap J, Phillips SL, Sherrod SK et al (2005) Soil biota can change after exotic plant invasion: does this affect ecosystem processes? Ecology 86:3007–3017

Belnap J, Stark JM, Rau BJ (2015) Soil moisture and biogeochemical factors influence the distribution of annual *Bromus* species. In: Germino MJ, Chambers JC, Brown CS (eds) Exotic brome-grasses in arid and semiarid ecosystems of the Western USA: causes, consequences, and management implications. Springer, New York, NY (Chapter 8)

Bestelmeyer BT, Tugel AJ, Peacock GL et al (2009) State and transition models for heterogeneous landscapes: a strategy for development and application. Rangel Ecol Manag 62:1–15

Blackshaw RE (1991) Control of downy brome (*Bromus tectorum*) in conservation fallow systems. Weed Technol 5:557–562

Blank RR, Morgan T (2013) Soil engineering facilitates downy brome (*Bromus tectorum*) growth – a case study. Invasive Plant Sci Manag 6:391–400

Boudell J, Link S, Johansen J (2002) Effect of soil microtopography on seed bank distribution in the shrub-steppe. West N Am Nat 62:14–24

Boyd CS, Svejcar TJ (2009) Managing complex problems in rangeland ecosystems. Rangel Ecol Manag 62:491–499

Bradley BA (2009) Regional analysis of the impacts of climate change on cheatgrass invasion shows potential risk and opportunity. Glob Change Biol 15:196–208

Bradley BA, Houghton RA, Mustard JF et al (2006) Invasive grass reduces aboveground carbon stocks in shrublands of the Western US. Glob Change Biol 12:1815–1822

Brisbin H, Thode A, Brooks M et al (2013) Soil Seed bank responses to postfire herbicide and native seeding treatments designed to control *Bromus tectorum* in a Pinyon-Juniper woodland at Zion National Park, USA. Invasive Plant Sci Manag 6:118–129

Brooks ML (2003) Effects of increased soil nitrogen on the dominance of alien annual plants in the Mojave Desert. J Ecol 40:344–353

Brooks ML, Chambers JC (2011) Resistance to invasion and resilience to fire in desert shrublands of North America. Rangel Ecol Manag 64:431–438

Brooks M, Matchett J (2006) Spatial and temporal patterns of wildfires in the Mojave Desert, 1980–2004. J Arid Environ 67:148–164

Brooks ML, Belnap J, Brown CS et al (2015) Exotic annual *Bromus* invasions – comparisons among species and ecoregions in the Western United States. In: Germino MJ, Chambers JC,

Brown CS (eds) Exotic brome-grasses in arid and semiarid ecosystems of the Western USA: causes, consequences, and management implications. Springer, New York, NY (Chapter 2)
Bruce LB, Perryman B, Conley K et al (2007) Case study: grazing management on seeded and unseeded post-fire public rangelands. Prof Anim Sci 23:285–290
Catford JA, Vesk PA, Richardson DM et al (2012) Quantifying levels of biological invasion: towards the objective classification of invaded and invasible ecosystems. Glob Change Biol 18:44–62
Caudle D, DiBenedetto J, Karl M et al (2013) Interagency ecological site handbook for rangelands. USDA, Natural Resource Conservation Service, p 109
Chambers JC, MacMahon JA (1994) A day in the life of a seed: movements and fates of seeds and their implications for natural land managed systems. Annu Rev Ecol Syst 25:263–292
Chambers JC, Roundy BA, Blank RR et al (2007) What makes Great Basin sagebrush ecosystems invasible by *Bromus tectorum*? Ecol Monogr 77:117–145
Chambers JC, Bradley BA, Brown CS et al (2014a) Resilience to stress and disturbance, and resistance to *Bromus tectorum* L. invasion in cold desert shrublands of western North America. Ecosystems 17:360–375
Chambers JC, Pyke DA, Maestas JD et al (2014b) Using resistance and resilience concepts to reduce impacts of invasive annual grasses and altered fire regimes on the sagebrush ecosystem and greater sage-grouse. A strategic multi-scale approach. Gen Tech Rep RMRS-GTR-326. USDA, Forest Service, Rocky Mountain Research Station, Fort Collins, CO, p 73
Cheam AH (1986) Patterns of change in seed dormancy and persistence of *Bromus diandrus* Roth (great brome) in the field. Aust J Agric Res 37:471–481
Clinton NE, Potter C, Crabtree B et al (2010) Remote sensing-based time-series analysis of cheatgrass (*Bromus tectorum* L.) phenology. J Environ Qual 39:955–963
Collins SL (1987) Interaction of disturbances in tallgrass prairie – a field experiment. Ecology 68:1243–1250
Corbin JD, D'Antonio CM (2010) Not novel, just better: competition between native and non-native plants in California grasslands that share species traits. Plant Ecol 209:71–81
Currie PO, Volesky JD, Hilken TO et al (1987) Selective control of annual bromes in perennial grass stands. J Range Manag 40:547–550
Daubenmire RF (1940) Plant succession due to overgrazing in the *Agropyron* bunchgrass prairie of southeastern Washington. Ecology 21:55–64
Davies KW, Johnson DD (2011) Are we "missing the boat" on preventing the spread of invasive plants in rangelands? Invasive Plant Sci Manag 4:166–172
Davies KW, Nafus AM (2013) Exotic annual grass invasion alters fuel amounts, continuity and moisture content. Int J Wildl Fire 22:353–358
Davies KW, Sheley RL, Bates JD (2008) Does fall prescribed burning *Artemisia tridentata* steppe promote invasion or resistance to invasion after a recovery period? J Arid Environ 72:1073–1082
Davies KW, Svejcar TJ, Bates JD (2009) Interaction of historical and nonhistorical disturbances maintains native plant communities. Ecol Appl 19:1536–1545
Davison JC, Smith EG (2007) Imazapic provides 2-year control of weedy annuals in a seeded Great Basin fuelbreak. Native Plants J 8:91–95
DeFalco LA, Bryla DR, Smith-Longozo V et al (2003) Are Mojave Desert annual species equal? Resource acquisition and allocation for the invasive grass *Bromus madritensis* subsp. *rubens* (Poaceae) and two native species. Am J Bot 90:1045–1053
Diamond JM, Call CA, Devoe N (2009) Effects of targeted grazing on fire behavior of cheatgrass-dominated rangeland in the northern Great Basin, USA. Int J Wildland Fire 18: 944–950
Diamond JM, Call CA, Devoe N (2012) Effects of targeted grazing and prescribed burning on community and seed dynamics of a downy brome (*Bromus tectorum*)-dominated landscape. Invasive Plant Sci Manag 5:259–269
DiTomaso JM, Kyser GB, Miller JR et al (2006) Integrating prescribed burning and clopyralid for the management of yellow starthistle (*Centaurea solstitialis*). Weed Sci 54:757–767

Duncan CA, Jachetta JJ, Brown ML et al (2004) Assessing the economic, environmental, and societal losses from invasive plants on rangeland and wildlands. Weed Technol 18:1411–1416

Eckert RE, Evans RA (1967) A chemical-fallow technique for control of brome and establishment of perennial grasses on rangeland. J Range Manag 20:35–41

Eiswerth ME, Krauter K, Swanson SR et al (2009) Post-fire seeding on Wyoming big sagebrush ecological sites: regression analyses of seeded nonnative and native species densities. J Environ Manag 90:1320–1325

Enright NJ, Fontaine JB, Lamont BB et al (2014) Resistance and resilience to changing climate and fire regime depend on plant functional traits. J Ecol 102:1572–1581

Evangelista P, Stohlgren TJ, Guenther D et al (2004) Vegetation response to fire and postburn seeding treatments in Juniper woodlands of the Grand Staircase-Escalante National Monument, Utah. West N Am Nat 64:293–305

Evans RA, Eckert RE (1965) Paraquat surfactant combinations for control of downy brome. Weeds 13:150–151

Evans RA, Young JA (1970) Plant litter establishment of alien annual weed species in rangeland communities. Weed Sci 18:697–703

Evans RA, Young JA (1977) Weed control-revegetation systems for big sagebrush-downy brome rangelands. J Range Manag 30:331–336

Evans RA, Young JA (1978) Effectiveness of rehabilitation practices following wildfire in a degraded big sagebrush downy brome community. J Range Manag 31:185–188

Fahnestock JT, Larson DL, Plumb GE et al (2003) Effects of ungulates and prairie dogs on seed banks and vegetation in a North American mixed-grass prairie. Plant Ecol 167:255–268

Frost RA, Launchbaugh KL (2003) Prescription grazing for rangeland weed management. Rangelands 25:43–47

Gaertner M, Biggs R, Te Beest M et al (2014) Invasive plants as drivers of regime shifts: identifying high-priority invaders that alter feedback relationships. Divers Distrib 20:733–744

Gardner ET, Anderson VJ, Johnson RL (2009) Arthropod and plant communities as indicators of land rehabilitation effectiveness in a semiarid shrubsteppe. West N Am Nat 69:521–536

Germino MJ, Belnap J, Stark JM et al (2015) Ecosystem impacts of exotic annual invaders in the genus *Bromus*. In: Germino MJ, Chambers JC, Brown CS (eds) Exotic brome-grasses in arid and semiarid ecosystems of the Western USA: causes, consequences, and management implications. Springer, New York, NY (Chapter 3)

Gill GS, Blacklow WM (1985) Variations in seed dormancy and rates of development of great brome, *Bromus diandrus* Roth, as adaptations to the climates of southern Australia and implications for weed control. Aust J Agric Res 36:295–304

Giunta BC, Christensen DR, Monsen SB (1975) Inter-seeding shrubs in cheatgrass with a browse seeder-scalper. J Range Manag 28:398–402

Goldberg DE, Rajaniemi T, Gurevitch J et al (1999) Empirical approaches to quantifying interaction intensity: competition and facilitation along productivity gradients. Ecology 80:1118–1131

Grman E, Suding KN (2010) Within-year soil legacies contribute to strong priority effects of exotics on native California grassland communities. Restor Ecol 18:664–670

Grman E, Bassett T, Brudvig LA (2013) Confronting contingency in restoration: management and site history determine outcomes of assembling prairies, but site characteristics and landscape context have little effect. J Appl Ecol 50:1234–1243

Gucker CL, Bunting SC (2011) Canyon grassland vegetation changes following fire in northern Idaho. West N Am Nat 71:97–105

Guglielmini A, Ghersa CM, Satorre EH (2007) Co-evolution of domesticated crops and associated weeds. Ecol Aust 17:167–178

Gundale MJ, Sutherland S, DeLuca TH (2008) Fire, native species, and soil resource interactions influence the spatio-temporal invasion pattern of *Bromus tectorum*. Ecography 31:201–210

Haferkamp MR, Volesky JD, Borman MM et al (1993) Effects of mechanical treatments and climatic factors on the productivity of northern Great Plains rangelands. J Range Manag 46:346–350

Haferkamp MR, Heitschmidt RK, Grings EE et al (2001) Suppression of annual bromes impacts rangeland: vegetation responses. J Range Manag 54:656–662

Hardegree SP, Jones TA, Roundy BA et al (2011) Assessment of range planting as a conservation practice. In: Briske DD (ed) Conservation benefits of rangeland practices: assessment, recommendations, and knowledge gaps. Allen Press, Lawrence, KS, pp 171–212

Hardegree SP, Schneider JM, Moffet CA (2012) Weather variability and adaptive management for rangeland restoration. Rangelands 34:53–56

Harmoney KR (2007) Grazing and burning Japanese brome (*Bromus japonicus*) on mixed grass rangelands. Rangel Ecol Manag 60:479–486

Hedges LV, Gurevitch J, Curtis PS (1999) The meta-analysis of response ratios in experimental ecology. Ecology 80:1150–1156

Hempy-Mayer K, Pyke DA (2008) Defoliation effects on *Bromus tectorum* seed production: implications for grazing. Rangel Ecol Manag 61:116–123

Herrick JE, Schuman GE, Rango A (2006) Monitoring ecological processes for restoration projects. J Nat Conserv 14:161–171

Herrick JE, Bestelmeyer BT, Crossland K (2008) Simplifying ecological site verification, rangeland health assessments, and monitoring. Rangelands 30:24–26

Herrick JE, Duniway MC, Pyke DA et al (2012) A holistic strategy for adaptive land management. J Soil Water Conserv 67:105A–113A

Hewlett DB, Johnson JR, Butterfield RI et al (1981) Japanese brome response to atrazine in combination with nitrogen fertilizer in the mixed prairie. J Range Manag 34:22–25

Hilty JH, Eldridge DJ, Rosentreter R et al (2004) Recovery of biological soil crusts following wildfire in Idaho. J Range Manag 57:89–96

Hirsch-Schantz MC, Monaco TA, Call CA et al (2014) Large-scale downy brome treatments alter plant-soil relationships and promote perennial grasses in salt desert shrublands. Rangel Ecol Manag 67:255–265

Hobbs R, Yates S, Mooney H (2007) Long-term data reveal complex dynamics in grassland in relation to climate and disturbance. Ecol Monogr 77:545–568

Hoelzle TB, Jonas JL, Paschke MW (2012) Twenty-five years of sagebrush steppe plant community development following seed addition. J Appl Ecol 49:911–918

Hulbert LC (1955) Ecological studies of *Bromus tectorum* and other annual brome grasses. Ecol Monogr 25:181–213

Hunter R (1991) *Bromus* invasions on the Nevada test site: present status of *B. rubens* and *B. tectorum* with notes on their relationships to disturbance and altitude. Great Basin Nat 51:176–182

James JJ, Svejcar TJ, Rinella MJ (2011) Demographic processes limiting seedling recruitment in arid grassland restoration. J Appl Ecol 48:961–969

James JJ, Boyd CS, Svejcar T (2013a) Seed and seedling ecology research to enhance restoration outcomes. Rangel Ecol Manag 66:115–116

James JJ, Sheley RL, Erickson T et al (2013b) A systems approach to restoring degraded drylands. J Appl Ecol 50:730–739

Jensen P (2009) Longevity of seeds of four annual grass and two dicotyledon weed species as related to placement in the soil and straw disposal technique. Weed Res 49:592–601

Jessop BD, Anderson VJ (2007) Cheatgrass invasion in salt desert shrublands: benefits of postfire reclamation. Rangel Ecol Manag 60:235–243

Jones R, Chambers JC, Johnson DW et al (2015) Effect of repeated burning on plant and soil carbon and nitrogen in cheatgrass (*Bromus tectorum*) dominated ecosystems. Plant Soil 386:47–64

Jones TA, Monaco TA (2007) A restoration practitioner's guide to the restoration gene pool concept. Ecol Restor 25:12–19

Jurand BS, Abella SR (2013) Soil seed banks of the exotic annual grass *Bromus rubens* on a burned desert landscape. Rangel Ecol Manag 66:157–163

Jurand BS, Abella SR, Suazo AA (2013) Soil seed bank longevity of the exotic annual grass *Bromus rubens* in the Mojave Desert, USA. J Arid Environ 94:68–75

Karl JW, Herrick JE, Browning DM (2012) A strategy for rangeland management based on best available knowledge and information. Rangel Ecol Manag 65:638–646

Keeley JE, McGinnis TW (2007) Impact of prescribed fire and other factors on cheatgrass persistence in a Sierra Nevada ponderosa pine forest. Int J Wildl Fire 16:96–106

Knapp PA (1996) Cheatgrass (*Bromus tectorum* L) dominance in the Great Basin Desert. Glob Environ Change 6:37–57

Knick ST, Hanser SE, Preston KL (2013) Modeling ecological minimum requirements for distribution of greater sage-grouse leks: implications for population connectivity across their western range, USA. Ecol Evol 3:1539–1551

Knutson KC, Pyke DA, Wirth TA et al (2014) Long-term effects of seeding after wildfire on vegetation in Great Basin shrubland ecosystems. J Appl Ecol 51:1414–1424

Kolb A, Alpert P (2003) Effects of nitrogen and salinity on growth and competition between a native grass and an invasive congener. Biol Invasions 5:229–238

Krueger-Mangold JM, Sheley RL, Svejcar TJ (2006) Toward ecologically-based invasive plant management on rangeland. Weed Sci 54:597–605

Kyser GB, DiTomaso JM (2002) Instability in a grassland community after the control of yellow starthistle (*Centaurea solstitialis*) with prescribed burning. Weed Sci 50:648–657

Kyser GB, Creech JE, Zhang J et al (2012) Selective control of medusahead (*Taeniatherum caput-medusae*) in California sagebrush scrub using low rates of glyphosate. Invasive Plant Sci Manag 5:1–8

Leffler AJ, Sheley RL (2012) Adaptive management in EBIPM: a key to success in invasive plant management. Rangelands 34:44–47

Leger EA, Baughman OW (2014) What seeds to plant in the Great Basin? Comparing traits prioritized in native plant cultivars and releases with those that promote survival in the field. Nat Area J 35:54–68

Leonard ED, Monaco TA, Stark JM et al (2008) Invasive forb, annual grass, and exotic shrub competition with three sagebrush-steppe growth forms: acquisition of a spring ^{15}N tracer. Invasive Plant Sci Manag 1:168–177

Lytle DA (2001) Disturbance regimes and life-history evolution. Am Nat 157:525–536

Mack RN (1981) Invasion of *Bromus tectorum* L. into western North America: an ecological chronicle. Agro-Ecosystems 7:145–165

Mack RN (2011) Fifty years of 'waging war on cheatgrass': research advances, while meaningful control languishes. In: Richardson DM (ed) Fifty years of invasion ecology: the legacy of Charles Elton. Wiley-Blackwell, Oxford, pp 253–265

MacKinnon WC, Karl JW, Toevs GR et al (2001) BLM core terrestrial indicators and methods. Tech Note 440. USDI, Bureau of Land Management, National Operations Center, Denver, CO

Maron JL, Connors PG (1996) A native nitrogen-fixing shrub facilitates weed invasion. Oecologia 105:302–312

Marshall DR, Jain SK (1967) Cohabitation and relative abundance of two species of wild oats. Ecology 48:656–659

Mata-Gonzalez R, Hunter RG, Coldren CL et al (2008) A comparison of modeled and measured impacts of resource manipulations for control of *Bromus tectorum* in sagebrush steppe. J Arid Environ 72:836–846

McGlone CM, Springer JD, Covington WW (2009) Cheatgrass encroachment on a ponderosa pine forest ecological restoration project in northern Arizona. Ecol Res 27:37–46

McGlone CM, Sieg CH, Kolb TE (2011) Invasion resistance and persistence: established plants win, even with disturbance and high propagule pressure. Biol Invasions 13:291–304

Melgoza G, Nowak RS, Tausch RJ (1990) Soil water exploitation after fire: competition between *Bromus tectorum* (cheatgrass) and two native species. Oecologia 83:7–13

Milchunas DG, Vandever MW, Ball LO et al (2011) Allelopathic cover crop prior to seeding is more important than subsequent grazing/mowing in grassland establishment. Rangel Ecol Manag 64:291–300

Miller RF, Chambers JC, Pyke DA et al (2013) A review of fire effects on vegetation and soils in the Great Basin Region: response and ecological site characteristics. Gen Tech Rep RMRS-GTR-308. USDA, Forest Service, Rocky Mountain Research Station, Fort Collins, CO, p 126

Miller RF, Chambers JC, Pellant M (2014) A field guide for selecting the most appropriate treatment in sagebrush and Pinyon-Juniper ecosystems in the Great Basin. Gen Tech Rep RMRS-GTR-322-rev. USDA, Forest Service, Rocky Mountain Research Station, Fort Collins, CO, p 72

Molinar F, Holechek J, Galt D et al (2002) Soil depth effects on Chihuahuan Desert vegetation. West N Am Nat 62:300–306

Monaco TA, Creech JE (2004) Sulfosulfuron effects on growth and photosynthesis of 15 range grasses. J Range Manag 57:490–496

Monaco TA, Jones TA, Thurow TL (2012) Identifying rangeland restoration targets: an appraisal of challenges and opportunities. Rangel Ecol Manag 65:599–605

Morris C, Monaco TA, Rigby CW (2009) Variable impacts of imazapic rate on downy brome (*Bromus tectorum*) and seeded species in two rangeland communities. Invasive Plant Sci Manag 2:110–119

Morris LR, Monaco TA, Sheley RL (2014) Impact of cultivation legacies on rehabilitation seedings and native species re-establishment in Great Basin shrublands. Rangel Ecol Manag 67:285–291

Morrow LA, Fenster CR, McCarty MK (1977) Control of downy brome on Nebraska rangeland. J Range Manag 30:293–296

Moyes A, Witter M, Gamon J (2005) Restoration of native perennials in a California annual grassland after prescribed spring burning and solarization. Restor Ecol 13:659–666

Munson SM, Long AL, Decker C et al (2015) Repeated landscape-scale treatments following fire suppress a non-native annual grass and promote recovery of native perennial vegetation. Biol Invasions 17:1915–1926

Murphy T, Naugle DE, Eardley R et al (2013) Trial by fire. Rangelands 35:2–10

Myers JA, Harms KE (2009) Seed arrival, ecological filters, and plant species richness: a meta-analysis. Ecol Lett 12:1250–1260

Noss RF, Peters RL (1995) Endangered ecosystems. Defenders of Wildlife, Washington, DC, p 133

Ogle SM, Reiners WA, Gerow KG (2003) Impacts of exotic annual brome grasses (*Bromus* spp.) on ecosystem properties of northern mixed grass prairie. Am Midl Nat 149:46–58

Osenberg CW, Sarnelle O, Cooper SD et al (1999) Resolving ecological questions through meta-analysis: goals, metrics, and models. Ecology 80:1105–1117

Owen SM, Sieg CH, Gehring CA (2011) Rehabilitating downy brome (*Bromus tectorum*)-invaded shrublands using imazapic and seeding with native shrubs. Invasive Plant Sci Manag 4:223–233

Palumbi SR (2001) Evolution – humans as the world's greatest evolutionary force. Science 293:1786–1790

Parker SS, Schimel JP (2010) Invasive grasses increase nitrogen availability in California grassland soils. Invasive Plant Sci Manag 3:40–47

Pellant M, Abbey B, Karl S (2004) Restoring the Great Basin Desert, USA: integrating science, management, and people. Environ Monit Assess 99:169–179

Perry L, Blumenthal D, Monaco T et al (2010) Immobilizing nitrogen to control plant invasion. Oecologia 163:13–24

Poreda SF, Wullstein LH (1994) Vegetation recovery following fire in an oakbrush vegetation mosaic. Great Basin Nat 54:380–383

Provencher L, Frid L, Czembor C et al (2015) State-and-transition models: conceptual vs. simulation perspectives, usefulness and breadth of use, and land management applications. In: Germino MJ, Chambers JC, Brown CS (eds) Exotic brome-grasses in arid and semiarid ecosystems of the western USA: causes, consequences, and management implications. Springer, New York, NY (Chapter 13)

Pyke DA (1986) Demographic responses of *Bromus tectorum* and seedlings of *Agropyron spicatum* to grazing by small mammals: occurrence and severity of grazing. J Ecol 74:739–754

Pyke DA, Herrick JE, Shaver P et al (2002) Rangeland health attributes and indicators for qualitative assessment. J Range Manag 55:584–597

Pyke DA, Wirth TA, Beyers JL (2013) Does seeding after wildfires in rangelands reduce erosion or invasive species? Restoration Ecol 21:415–421

Pyke DA, Shaff SE, Lindgren AI et al (2014) Region-wide ecological responses of arid Wyoming big sagebrush communities to fuel treatments. Rangel Ecol Manag 67:455–467

Pyšek P, Jarošík V, Hulme PE et al (2012) A global assessment of invasive plant impacts on resident species, communities and ecosystems: the interaction of impact measures, invading species' traits and environment. Glob Change Biol 18:1725–1737

Ralphs MH, Wiedmeier RD, Banks JE (2007) Decreasing forage allowance can force cattle to graze broom snakeweed (*Gutierrezia sarothrae*) as a potential biological control. Rangel Ecol Manag 60:487–497

Reever-Morghan KJ, Sheley RL, Svejcar TJ (2006) Successful adaptive management – the integration of research and management. Rangel Ecol Manag 59:216–219

Reisner MD, Grace JB, Pyke et al (2013) Conditions favouring *Bromus tectorum* dominance of endangered sagebrush steppe ecosystems. J Appl Ecol 50:1039–1049

Rice KJ, Nagy ES (2000) Oak canopy effects on the distribution patterns of two annual grasses: the role of competition and soil nutrients. Am J Bot 87:1699–1706

Rinella MJ, Masters RA, Bellows SE (2013) Effects of growth regulator herbicide on downy brome (*Bromus tectorum*) seed production. Invasive Plant Sci Manag 6:60–64

Rist L, Felton A, Samuelsson L et al (2013) A new paradigm for adaptive management. Ecol Soc 18:63

Rose KK, Hild AL, Whitson TD et al (2001) Competitive effects of cool-season grasses on re-establishment of three weed species. Weed Technol 15:885–891

Salo LF (2005) Red brome (*Bromus rubens* subsp *madritensis*) in North America: possible modes for early introductions, subsequent spread. Biol Invasions 7:165–180

Sheley R, James J, Smith B et al (2010) Applying ecologically based invasive-plant management. Rangel Ecol Manag 63:605–613

Sheley RL, James JJ, Vasquez EA et al (2011) Using rangeland health assessment to inform successional management. Invasive Plant Sci Manag 4:356–367

Skaer MJ, Graydon DJ, Cushman JH (2013) Community-level consequences of cattle grazing for an invaded grassland: variable responses of native and exotic vegetation. J Veg Sci 24:332–343

Smith AT (1970) Effects of disturbance on seed germination in some annual plants. Ecology 51:1106–1108

Smith DC, Meyer SE, Anderson VJ (2008) Factors affecting *Bromus tectorum* seed bank carryover in western Utah. Rangel Ecol Manag 61:430–436

Smith B, Sheley R, Svejcar T (2012) Grazing invasive annual grasses: the green and brown guide. USDA, Agricultural Research Service, USA, p 36

Sorensen CD, McGlone CM (2010) Ponderosa pine understory response to short-term grazing exclusion (Arizona). Ecol Res 28:124–126

Stavi I, Ungar EU, Lavee H et al (2008) Surface microtopography and soil penetration resistance associated with shrub patches in a semiarid rangeland. Geomorphology 94:69–78

Stidham MA (1991) Herbicides that inhibit acetohydroxyacid synthase. Weed Sci 39:428–434

Stiver SJ, Apa AD, Bohne JR et al (2006) Greater sage-grouse comprehensive conservation strategy. Western Association of Fish and Wildlife Agencies, Cheyenne, WY, p 28

Suding KN (2011) Toward an era of restoration in ecology: successes, failures, and opportunities ahead. Ann Rev Ecol Evol Syst 42:465–487

Suding KN, Hobbs RJ (2008) Threshold models in restoration and conservation: a developing framework. Trends Ecol Evol 24:271–279

Suding KN, Gross KL, Houseman GR (2004) Alternative states and positive feedbacks in restoration ecology. Trends Ecol Evol 19:46–53

Svejcar T, Boyd C (2012) The value of decision models: using Ecologically Based Invasive Plant Management as an example. Rangelands 34:2–5

Svejcar T, James J, Hardegree S et al (2014) Incorporating plant mortality and recruitment into rangeland management and assessment. Rangel Ecol Manag 67:603–613

Sweet SB, Kyser GB, DiTomaso JM (2008) Susceptibility of exotic annual grass seeds to fire. Invasive Plant Sci Manag 1:158–167

Thacker E, Ralphs MH, Monaco TA (2009) Seeding cool-season grasses to suppress broom snakeweed (*Gutierrezia sarothrae*), downy brome (*Bromus tectorum*), and weedy forbs. Invasive Plant Sci Manag 2:237–246

Turner GT (1971) Soil and grazing influences on a salt-desert shrub range in western Colorado. J Range Manag 24:31–37

Vermeire LT, Rinella MJ (2009) Fire alters emergence of invasive plant species from soil surface-deposited seeds. Weed Sci 57:304–310

Vilà M, Espinar JL, Hejda M et al (2011) Ecological impacts of invasive alien plants: a meta-analysis of their effects on species, communities and ecosystems. Ecol Lett 14:702–708

Violle C, Richarte J, Navas M (2006) Effects of litter and standing biomass on growth and reproduction of two annual species in a Mediterranean old-field. J Ecol 94:196–205

von Wehrden H, Hanspach J, Kaczensky P et al (2012) Global assessment of the non-equilibrium concept in rangelands. Ecol Appl 22:393–399

Wainwright CE, Wolkovich EM, Cleland EE (2012) Seasonal priority effects: implications for invasion and restoration in a semi-arid system. J Appl Ecol 49:234–241

Walker LR, Walker J, Moral RD (2007) Forging a new alliance between succession and restoration. In: Walker LR, Walker J, Hobbs RJ (eds) Linking restoration and ecological succession. Springer, New York, NY, pp 1–18

West NE (1979) Basic synecological relationships of sagebrush-dominated lands in the Great Basin and the Colorado Plateau. In: The sagebrush ecosystem: a symposium, Logan, Utah. Utah State University, pp 33–41

West NE (1983) Great Basin-Colorado plateau sagebrush semi-desert. In: West NE (ed) Temperate deserts and semi-deserts. Elsevier, Amsterdam, pp 331–349

West NE, Yorks TP (2002) Vegetation responses following wildfire on grazed and ungrazed sagebrush semi-desert. J Range Manag 55:171–181

Whisenant SG (1990) Postfire population dynamics of *Bromus japonicas*. Am Midl Nat 123:301–308

Whisenant SG (1999) Repairing damaged wildlands: a process-orientated, landscape-scale approach. Cambridge University Press, Cambridge, p 328

Wicks GA, Fenster CR, Burnside OC (1965) Chemical control of downy brome in grasslands of western Nebraska. Weeds 13:202–204

US Fish and Wildlife Service (2013) Greater sage-grouse (*Centrocercus urophasianus*) conservation objectives: Final Report. US Fish and Wildlife Service, p 92

Williams BK (1997) Approaches to the management of waterfowl under uncertainty. Wildl Soc Bull 25:714–720

Williams BK, Szaro RC, Shapiro RC (2009) Adaptive management: the US Department of the Interior Technical guide. Adaptive Management Working Group, USDI, Washington, DC, p 72

Williams BK (2011) Adaptive management of natural resources-framework and issues. J Environ Manag 92:1346–1353

Williams BK, Brown ED (2012) Adaptive management: the US Department of the Interior Applications guide. Adaptive Management Working Group, USDI, Washington, DC, p 120

Williams BK, Brown ED (2014) Adaptive management: from more talk to real action. Environ Manag 53:465–479

Wilson RG, Orloff SB, Lancaster DL et al (2010) Integrating herbicide use and perennial grass revegetation to suppress weeds in noncrop areas. Invasive Plant Sci Manag 3:81–92

Wisdom MJ, Chambers JC (2009) A landscape approach for ecologically based management of Great Basin shrublands. Restor Ecol 17:740–749

Young JA, Clements CD (2007) Cheatgrass and grazing rangelands. Rangelands 29:15–20

Young JA, Evans RA, Eckert RE (1969) Population dynamics of downy brome. Weed Sci 17:20–22

Young JA, Evans RA, Robison J (1972a) Influence of repeated annual burning on a medusahead community. J Range Manag 25:372–375

Young JA, Evans RA, Major J (1972b) Alien plants in the Great Basin. J Range Manag 25:194–201

Young JA, Evans RA, Weaver RA (1976) Estimating potential downy brome competition after wildfires. J Range Manag 29:322–325

Chapter 13
State-and-Transition Models: Conceptual Versus Simulation Perspectives, Usefulness and Breadth of Use, and Land Management Applications

Louis Provencher, Leonardo Frid, Christina Czembor, and Jeffrey T. Morisette

Abstract State-and-Transition Simulation Modeling (STSM) is a quantitative analysis method that can consolidate a wide array of resource management issues under a "what-if" scenario exercise. STSM can be seen as an ensemble of models, such as climate models, ecological models, and economic models that incorporate human dimensions and management options. This chapter presents STSM as a tool to help synthesize information on social–ecological systems and to investigate some of the management issues associated with exotic annual *Bromus* species, which have been described elsewhere in this book. Definitions, terminology, and perspectives on conceptual and computer-simulated stochastic state-and-transition models are given first, followed by a brief review of past STSM studies relevant to the management of *Bromus* species. A detailed case study illustrates the usefulness of STSM for land management. As a whole, this chapter is intended to demonstrate how STSM can help both managers and scientists: (a) determine efficient resource allocation for monitoring nonnative grasses; (b) evaluate sources of uncertainty in model simulation results involving expert opinion, and their consequences for

L. Provencher (✉)
The Nature Conservancy, Reno, NV 89501, USA
e-mail: lprovencher@tnc.org

L. Frid
Apex Resource Management Solutions Ltd., Bowen Island, BC, Canada
e-mail: leonardo.frid@apexrms.com

C. Czembor
Knight Piésold Ltd., Vancouver, BC, Canada
e-mail: c.czembor@gmail.com

J.T. Morisette
US Geological Survey, North Central Climate Science Center, Fort Collins, CO 80526, USA
e-mail: morisettej@usgs.gov

M.J. Germino et al. (eds.), *Exotic Brome-Grasses in Arid and Semiarid Ecosystems of the Western US*, Springer Series on Environmental Management,
DOI 10.1007/978-3-319-24930-8_13

management decisions; and (c) provide insight into the consequences of predicted local climate change effects on ecological systems invaded by exotic annual *Bromus* species.

Keywords State-and-transition • Conceptual model • Simulation • Uncertainty • Climate change

13.1 Introduction

Land managers dealing with impacts of exotic brome invasions are challenged with understanding how local, site-level actions across heterogeneous landscapes and time periods will ultimately scale up to the region where outcomes are measured against desired conditions. State-and-Transition Simulation Modeling (STSM) provides a quantitative framework to combine these various facets at both regional and local scales. Spatial or nonspatial analyses are used to investigate "what-if" scenarios that incorporate management options and evaluate sensitivity of systems to specific parameterizations or assumptions.

This chapter is written for those who want to understand how STSMs can integrate ecological and economic understanding into a simulation environment to provide insight into invasive species and vegetation management at landscape and regional scales. The focus is on using STSMs for management of vegetation, and particularly invasion by exotic annual *Bromus* (*Bromus* hereafter). We begin by offering background definitions, terminology, and perspectives. We then highlight the utility of STSMs through two published examples where STSMs have been used in other systems to determine efficient resource allocation for monitoring exotic grasses and to evaluate sources of uncertainty in STSMs involving expert opinion. We conclude with a new case study application of STSMs to provide insight into the consequences of predicted local climate change effects on ecological systems invaded by *Bromus*. For further reading, several publications offer excellent descriptions or reviews of STSM concepts (Czembor and Vesk 2009; Rumpff et al. 2011; Knapp et al. 2011a, b; Daniel and Frid 2012).

13.2 Definition of State-and-Transition Models

Conceptual state-and-transition models are presented as discrete, box-and-arrow representations of the continuous variation in vegetation composition and structure of an ecological system (Westoby et al. 1989; NRCS 2003; Stringham et al. 2003; Bestelmeyer et al. 2004). The classification of an ecological system is important for framing each state-and-transition model. One example of a classification is the

USDA Natural Resources Conservation Service's (NRCS) "ecological site description" (ESDs) system. Each ecological site represents "a distinctive kind of land with specific physical characteristics that differs from other kinds of land in its ability to produce a distinctive kind and amount of vegetation," and can have a unique conceptual state-and-transition model (STM) associated with it (NRCS 1998). STMs represent a method to organize and communicate complex information about the relationships among vegetation, soil, animals, hydrology, disturbances, and management actions on an ecological site (Caudle et al. 2013). They are comprised of states and transitions.

13.2.1 States

Within STMs, boxes represent the possible vegetation conditions of a parcel of land within an ecological system and include different (a) plant community *states* or (b) *phases* within a state (Fig. 13.1). A *state* is formally defined in the rangeland literature as a "recognizable, relatively resistant and resilient complex with attributes that include a characteristic climate, the soil resource including the soil biota, and the associated aboveground plant communities" (NRCS 2003; Caudle et al. 2013). The associated plant communities are phases of the same state that can be represented in a diagram with two or more boxes. Relatively reversible changes caused by disturbance or stress (e.g., fire, flooding, drought, insect outbreaks, herbivory, and others) and succession operate on phases within a state. *Phases* are most often recognizable steps of succession, which is a naturally continuous process. Phases can also occur among uncharacteristic vegetation classes as a result of succession. Different states are separated by at least one threshold. A transition across a threshold is often caused by an anthropogenic disturbance or species invasion. *Thresholds* are defined by conditions sufficient to modify ecosystem structure and function beyond the limits of ecological resilience, resulting in the formation of alternative states (Briske et al. 2008). Crossing of thresholds usually indicates that substantial management effort is required to restore ecosystem structure and function to another state. The reference state represents the dynamic vegetation phases resulting from a natural disturbance regime, including disturbances caused by indigenous populations, where vegetation returns to the pre-disturbance conditions via succession. A threshold often implies the creation of uncharacteristic vegetation classes, which often exist because of European post-settlement disturbance regimes, changes in climate, or species invasions. Moreover, thresholds can occur between different uncharacteristic states, usually signaling increasing degradation of the ecological system. A monoculture of *Bromus tectorum* L. (cheatgrass or downy brome) in a sagebrush shrubland is an example of an uncharacteristic vegetation class, which could be a phase or a state depending on model structure. Uncharacteristic vegetation classes can be formed of entirely native species (native uncharacteristic) or contain nonnative plant species (exotic uncharacteristic), such as *Bromus* (Rollins 2009).

a

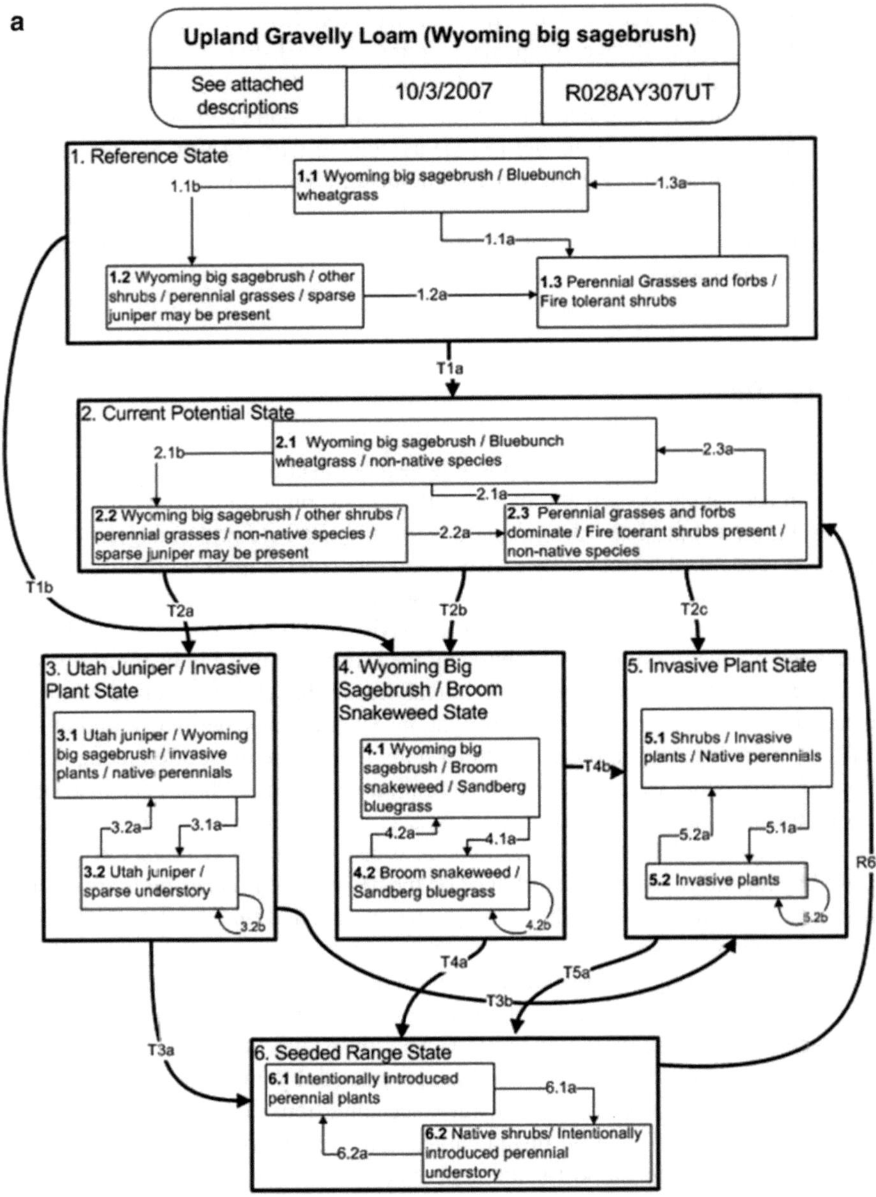

Fig. 13.1 (**a**) Conceptual state-and-transition model for NRCS ecological site R028AY307UT *A. tridentata* spp. *wyomingensis* growing on upland gravelly loam soils in Utah (Thacker et al. 2008). The *larger bold boxes* represent vegetation states with the *top* (#1) being the reference state and immediately below it (#2) the current potential state, which is similar to the reference with a non-dominant presence of nonnative plants. States #3–5 represent undesirable states that are invaded with *J. osteosperma*, *Gutierrezia sarothrae* (Pursh) Britton & Rusby (broom snakeweed), and invasive annual grasses or forbs, respectively. The final (#6) state represents a nonnative but more desirable vegetation community such as *Agropyron cristatum* (L.) Gaertn. (crested wheatgrass).

b

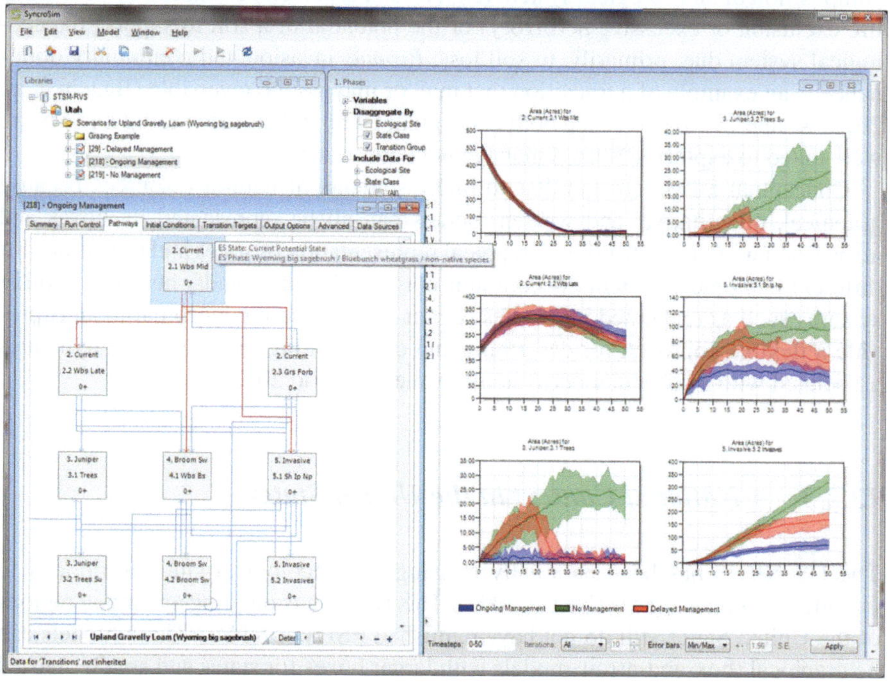

Fig. 13.1 (continued) Boxes within each state represent vegetation phases. *Arrows* represent either transitions between phases or states (i.e., 1.2a is a replacement event from late successional shrubs to grass forbs, while T2c represents invasion by annual grasses or invasive forbs). This conceptual state-and-transition diagram is accompanied by a detailed description of each state, phase, and transition in the NRCS source documentation. (**b**) Screen capture of STSM software showing a model developed for the same ecological site. The *box*-and-*arrow* "pathway diagram" shows phases for the current potential state (#2), and the three undesirable states (#3–5). *Arrows* in the diagram represent probabilistic or deterministic transitions driven by different process such as fire, grazing, or succession. The graphs on the *right* show model outputs for three different example management scenarios representing the amount of land in modeled states or phases. *Blue* represents ongoing management, *green* is no management, and *red* is delayed management. *Lines* represent the mean across multiple iterations and shaded areas represent model uncertainty

13.2.2 Transitions

The other fundamental component of a conceptual STM are *transitions* representing either succession between phases or disturbances that alter the structure or composition of phases and, eventually, states. Transitions can be natural (e.g., fire, flooding) or managed (e.g., prescribed burning). Furthermore, natural disturbances can represent pre-settlement (e.g., surface fire) and European post-settlement (e.g., *Bromus* invasion) events. Most transitions are reversible given succession, natural disturbances, or management actions; however, some transitions can result in crossing of biotic or abiotic thresholds that irreversibly change either the diagnostic species composition of an ecological system (e.g., loss of aspen clones caused by prolonged fire exclusion or excessive herbivory) or the potential of a soil to support the ecological system due, primarily, to soil loss. *Bromus* invasion and community dominance is an example of a transition that may or may not be reversible (Bagchi et al. 2013; Knapp 1996) and hence both conceptual STM and corresponding STSMs are well suited to exploring related management questions.

Conceptual STMs are familiar to many students of natural resources because graphical, quantitative, and written models can all be represented by boxes and arrows or a written description. Graphical representation of states and transitions for different ecological systems is common not only in rangelands, but also in other systems such as reclaimed mine sites (Grant 2006). These conceptual models provide a flexible approach for describing and documenting the vegetation dynamics associated with a particular ecosystem (Daniel and Frid 2012).

13.2.3 A National Context in the United States

The US NRCS has been nationally revising their ecological site descriptions to include conceptual STMs (NRCS 2003). This revision is ongoing and many regions of the United States still do not have published ESDs or STMs. These models can be graphical (box-and-arrow models with larger boxes for states and smaller nested boxes for phases), written descriptions of reference and uncharacteristic states, plus disturbances causing transitions between thresholds or a combination of both. The initial state depicted in NRCS models is the historic plant community (i.e., reference state [Rumpff et al. 2011]) from which all other states are derived through natural and managed transitions. The reference state is based on the natural range of conditions associated with natural disturbance regimes and often includes several plant communities (phases) that differ in dominant plant species relative to type and time since disturbance (Caudle et al. 2013).

NRCS ecological site descriptions are frequently used by US Department of Interior and Department of Agriculture staff for restoration project prescriptions (e.g., native seed mix) and US National Environmental Protection Act documentation. Conceptual STMs generate non-quantitative, general predictions about desirable and undesirable processes causing transitions between states at a site-specific level. A recent criticism of purely conceptual STMs developed for ecological sites is that they lack the ability to project state transitions that will be important in the future and to link these to levels of conservation funding for management and restoration actions (Twidwell et al. 2013). Consequently, there are currently efforts under way to digitize conceptual STMs from ESDs and convert them into STSMs that can be used to generate testable hypotheses. For example, quantitative models developed by Evers et al. (2013) explicitly consider how warmer, drier sites function differently from cooler, moister sites.

13.2.4 State-and-Transition Simulation Models

STSMs begin with conceptual models, such as the ones described above. Before the models are applied, the landscape being simulated is subdivided into simulation cells, which can be nonspatial or spatially represented using a map. These models can be quantified with the following additional information: (1) an inventory, either spatial or nonspatial, of the vegetation conditions of the landscape at the start of the simulation, which describes the ecological system, and state class (state and phase) of each simulation cell in the landscape and (2) a rate associated with each possible transition between state classes. Then, these transition rates can be further quantified using three general approaches: (2.1) probabilistic, with a specified probability at any point in time; (2.2) deterministic, occurring after a specified period of time in a state class has elapsed; or (2.3) with target areas assigned to occur on the landscape over time. The first two approaches are typically used to emulate natural processes such as disturbances and succession, whereas the last is typically applied for management actions such as herbicide application. Computer software then uses the inventory of starting vegetation conditions and rates associated with each transition to project future vegetation conditions of the landscape (Fig. 13.1b), as well as occurrence of transitions over time. The overall approach to applying STSM is described in detail in Daniel and Frid (2012).

In recent years there has been a proliferation of quantitative STSM applications to a diverse set of natural resource management problems (see Daniel and Frid 2012 for examples). This development has been driven in part by the model development training and awareness created by the Landscape Fire and Resource Management Planning Tools Project (LANDFIRE) in the United States (Rollins 2009; Blankenship et al. 2012) and the need for new management decision support tools. The popularity of this approach has been facilitated by the availability of flexible software tools, beginning with the Vegetation Dynamics Development Tool (VDDT) in the early 1990s for the Interior Columbia Basin Ecosystem Management Project (Barrett 2001; Hann and Bunnell 2001). The most recent of these tools, ST-Sim (www.syncrosim.com), has both nonspatial and spatially explicit capabilities. Note that while there are other modeling approaches and software packages for simulating landscape change, some of these are specifically tailored to forests (i.e., Landis II, Scheller et al. 2007) and many others are not documented or supported to the level available with ST-Sim (Keane et al. 2004). Prior to the availability of software, quantitative STMs have been either analytical (Horn 1975) or simulated with project-specific computer programs (Hardesty et al. 2000). Analytical STMs are rare because even the simplest models incorporate nonlinear step functions (i.e., age and time since past transitions) that render analysis difficult to intractable.

Many of the initial STSMs were created by US Forest Service ecologists and contractors (Merzenich et al. 1999; Barrett 2001; Hann and Bunnell 2001; Hemstrom et al. 2004) and ecologists of The Nature Conservancy (TNC; Hardesty et al. 2000; Forbis et al. 2006; Provencher et al. 2007) who were just starting to incorporate the conceptual developments and terminology proposed by rangeland ecologists

Table 13.1 Comparison of state-and-transition modeling terminology differentially used by rangeland ecologists, quantitative modelers, and simulation software

Rangeland ecologists	Quantitative modelers	Software
Ecological site	Ecological site(s) (TNC) Ecological system (NatureServe[a] and TNC) Potential natural vegetation type (US Forest Service) Biophysical setting (LANDFIRE and TNC)	Cover type (VDDT) Stratum (PATH[b], ST-Sim)
State	≥1 vegetation class	Cover type × Structural stage (VDDT & Path) State Class (ST-Sim)
Phase	Vegetation class (reference or uncharacteristic) Existing vegetation class (US Forest Service)	Cover type × Structural stage (VDDT & Path) State Class (ST-Sim)
Reversible transition	Transition (specified as natural, uncharacteristic, or managed)	Transition (specified as probabilistic or deterministic)
Irreversible transition	Transition (uncharacteristic or managed)	Transition (probabilistic or deterministic)
Threshold	Transition (uncharacteristic or managed)	Transition (probabilistic or deterministic)

[a]http://www.natureserve.org/library/usEcologicalsystems.pdf
[b]Path is the landscape simulation freeware platform that replaced VDDT: www.pathmodel.com. Currently, the ST-Sim simulator in the SyncroSim state-and-transition platform (www.syncrosim.com) is the latest nonspatial and spatial generation of software development started with VDDT

(Westoby et al. 1989) and NRCS (2003). Despite these nearly independent paths, both groups approached state-and-transition modeling concepts in a remarkably similar manner with differences mostly in jargon (Table 13.1). Fortunately, specific jargon and definitions matter little for actual STSMs because simulation software is flexible and can accommodate different terminology.

13.2.5 Nonspatial and Spatial Model Approaches

Many STSMs are nonspatial with the fate of each simulation cell being independent of the fate of any other cell, because they are simpler and faster to create and run, and require less data and fewer assumptions than spatially explicit models. Spatially explicit STSMs require at a minimum a polygon- or raster-based vegetation layer(s), size frequency distributions for each transition process (the frequency of very small to very large events, such as fire), and spatial constraint layers defining management zones (e.g., ownership polygons) or priority areas (e.g., no fire tolerated to large fire size allowed; Kurz et al. 2000; Provencher et al. 2007). These data can be difficult and expensive to obtain. Given the additional data and computational requirements of spatially explicit modeling, a compelling question or objective justifying the need to do spatially explicit modeling should be a prerequisite to developing a spatial STSM. That said, good cases for spatially explicit STSMs can usually be made for wildlife management, nonnative plant species control and monitoring (Frid and Wilmshurst 2009; Frid et al. 2013a, b), and wildland fire (Miller 2007). In these cases, the position and size of natural and managed transitions can critically affect project-specific metrics that track the condition of ecological systems studied. Extreme weather events and climate projection do vary across the landscape. Thus, studies looking to consider these issues ought to include a spatial component. There are a number of nonspatial (Evers et al. 2011, 2013; Provencher et al. 2013; Low et al. 2010; Forbis et al. 2006; Creutzburg et al. 2012) and spatially explicit (Provencher et al. 2007) STSM applications with a *Bromus* component.

13.2.6 Uses and Benefits of State-and-Transition Simulation Modeling

While conceptual STMs are useful tools for describing vegetation dynamics and identifying possible management prescriptions at the site-specific scale, they can fall short of providing prognostic information for landscape scale vegetation management efforts. Land managers are frequently faced with limited resources and competing objectives and an interest in knowing how their actions will play out on the landscape. For example, should restoration resources be applied toward areas affected by woody species encroachment or toward areas affected by annual species invasions? In addition to competing objectives such as these, land managers are often faced with uncertainties and incomplete information of the vegetation dynamics for the landscape of concern. Despite these challenges, land managers must frequently make decisions about the allocation of limited vegetation management resources. At landscape scales, STSMs are valuable tools for identifying robust management strategies, important trade-offs, and critical uncertainties for decision making.

STMs are popular among range and forest managers because they are easy to communicate and typically require less data to parameterize than more complex process-based models under different applications. However, STSMs can include information from process-based models and analysis (Halofsky et al. 2013). These models are management oriented and simulations can be useful to solve complex management questions. STSMs have the major benefits of being flexible and fostering stakeholder engagement and buy-in (Price et al. 2012; Nixon et al. 2014). Simulations have the ability to predict changes in vegetation under different scenarios considering alternative management actions and hypotheses about the response of natural systems to them. The variety of scenarios that can be explored—from climate change to single management actions (e.g., prescribed burning)—is very large and feasible. The social benefit of model building is that it allows land managers and scientists to explicitly document their understanding and assumptions about ecological processes, management actions, and the interactions between the two. Such a framework can help build support for proposed actions or provide useful direction for future experiments (Low et al. 2010; Price et al. 2012; Nixon et al. 2014).

13.2.7 *Examples of STSM Applications from Non-***Bromus** *Systems*

Here we present two examples of STSM applications to non-*Bromus* systems that could similarly be applied to *Bromus*. Many challenges in conservation science can be (1) traced to altered disturbance regimes, (2) caused by legacies of vegetation classes or phases that differ from reference conditions, and (3) effectively resolved with alternative management scenarios. Because the issues of exotic annual *Bromus* control involve a wide array of ecological processes as well as economics (Chambers 2008; Knapp 1996) and land management options (Monsen et al. 2004), it is useful to have a framework to integrate our current understanding, assumptions, and potential scenarios in a tractable and repeatable form. In the different applications of STSM presented in these examples, we hope to illustrate how STSM can integrate a wide breadth of ecological processes and management options. Specifically, these examples were chosen to illustrate two different management-oriented uses of STSM: (1) identification of the most cost-effective approach to control nonnative grasses using spatial STSMs designed to consider trade-offs between monitoring and treatment; and (2) quantification of model uncertainty when expert knowledge is used to parameterize an STSM.

13.2.7.1 State-and-Transition Simulation Models as a Tool for Guiding Invasive Plant Monitoring and Treatments

Despite the high economic impacts of exotic plant invasions (Pimentel et al. 2005), including those of *Bromus* (Knapp 1996), control and restoration resources available for *Bromus* management on any one landscape are often limited and intermittently available. These limited resources must be allocated to multiple activities including treatment of detected infestations, monitoring treatment success, and monitoring to discover previously undetected infestations. Cost-efficiency of control measures is greatly increased by discovering and treating new infestations ("nascent foci"), before propagule banks are established at or beyond the invasion site (Moody and Mack 1988; Maxwell et al. 2009; Frid et al. 2013a). However, (1) land management programs are often evaluated based on numbers of acres treated, which creates an incentive to focus all resources toward already detected infestations; (2) allocating resources toward monitoring and failing to detect any new infestations is often viewed as a waste of resources; and (3) most land managers lack tools to detect nascent foci or determine how much of their budget should be allocated to monitoring versus treatment.

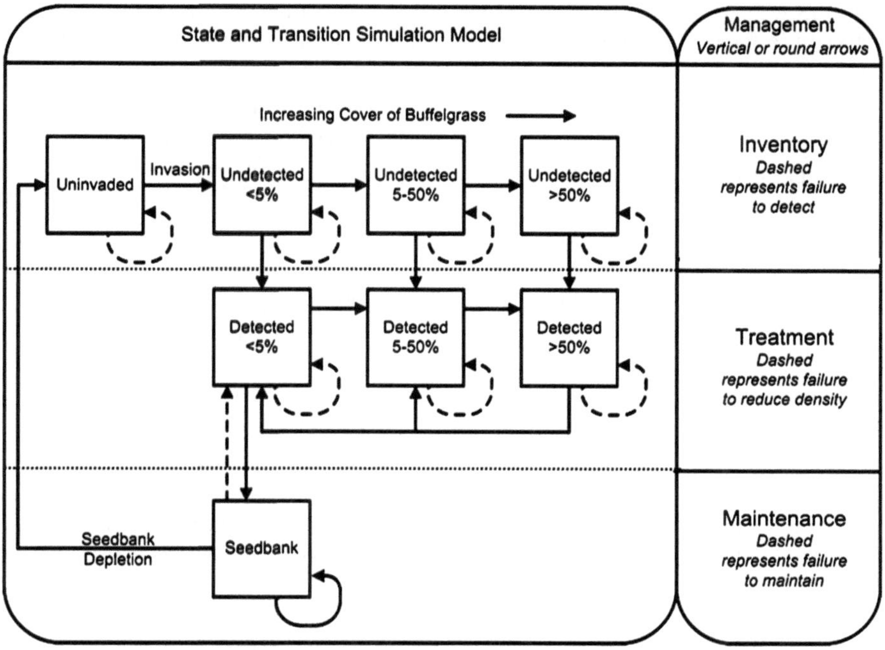

Fig. 13.2 State-and-transition model used by Frid et al. (2013b) to simulate alternative management strategies for *P. ciliare* in southern Arizona. The STSM categorizes each spatially explicit polygon dynamically over time based on both the presence and abundance of *P. ciliare* and on its detection status. The STSM has a total of eight possible states represented by five possible cover classes (absent, seedbank only, cover <5 %, cover 5–50 %, cover >50 %) and by whether the presence of live plants has been detected by managers

STSMs can help land managers determine appropriate strategies for resource allocation, including selection of management activities and locations. For this example, we draw upon a recent study of a non-*Bromus* invasive grass, *Pennisetum ciliare* (L.) Link (buffelgrass; synonymous with *Cenchrus ciliaris*), by Frid et al. (2013b) in the Sonoran Basin and Range, which demonstrates an application that could be applied in future work to *Bromus*. In this model, the phases distinguish between *P. ciliare* abundances (<5 %, 5–50 % and >50 % cover) and whether they are detected or not (Fig. 13.2).

This model was a spatially explicit STSM, developed using TELSA (Kurz et al. 2000), which included input and output maps of the *P. ciliare* over time, and distinguished between detected (and treatable) and undetected patches that would require resources to be detected. The model simulated dispersal of short- (i.e., neighbor-to-neighbor) and long-distance dispersal of *P. ciliare* to other locations on the landscape based on time series of spread determined from aerial photography (Olsson et al. 2012). A map of the current known distribution of *P. ciliare* was derived from the same aerial photographs. A habitat suitability model based on slope, aspect, and elevation was used to estimate where *P. ciliare* could grow, and at what densities.

Table 13.2 Simulated area invaded and cumulative area undergoing inventory treatment and maintenance for five scenarios showing area (ha) invaded by *P. ciliare* and cumulative area undergoing inventory, treatment, and maintenance at years 2030 and 2060

Scenario	Year	Hectares invaded	Cumulative area (ha)		
			Inventory	Treatment	Maintenance
Initial conditions	2010	82	0	0	0
No management	2030	1795	0	0	0
	2060	6263	0	0	0
Intermediate management: worst case	2030	997	82,281	682	9494
	2060	4952	236,142	3157	29,059
Intermediate management: best case	2030	603	104,520	971	18,244
	2060	3081	258,212	3364	46,843
Unlimited management: worst case	2030	864	81,897	837	10,060
	2060	637	358,643	11,543	99,986
Unlimited management: best case	2030	159	146,054	1460	27,851
	2060	54	392,803	3752	75,380

"Worst case" represents the low range and "best case" the high range of management efficiency as described in results of a survey conducted with *P. ciliare* managers. The total size of the landscape simulated was 46,000 ha

Treatment and monitoring effectiveness rates were derived from a survey of *P. ciliare* management practitioners.

Five alternative management options that varied in both the budget allocated toward monitoring and treatment and their effectiveness were evaluated in a binary fashion (high/low; Table 13.2). Budgets for treatments and monitoring were either unlimited, as a benchmark of what could be accomplished, or limited to half of what the unlimited scenario utilized. The model predicts that in the absence of management, the area infested with *P. ciliare* would grow exponentially until the ecological limit is reached within a 50-year period (Fig. 13.3).

Some of the key conclusions from these simulations were:

1. In the long term (2010–2060), unlimited budget scenarios used similar amounts of resources to those with a limited budget, yet resulted in less than an order of magnitude the amount of area infested per area treated because management was able to act when desirable plant community transitions required a lower energy input. In the short term (2010–2030), unlimited budgets used more resources than scenarios with a limited budget.
2. Monitoring for new infestations is a key component of the management strategy. When the monitoring budget or effectiveness was reduced, initial treatment costs were reduced because fewer detected patches of *P. ciliare* were treated. However, long-term treatment costs were much higher because plant community states that are difficult to transition from had been reached, requiring substantially greater investment. Reducing either the resources allocated to or the effectiveness of monitoring efforts results in larger, more distributed patches of *P. ciliare* on the landscape.

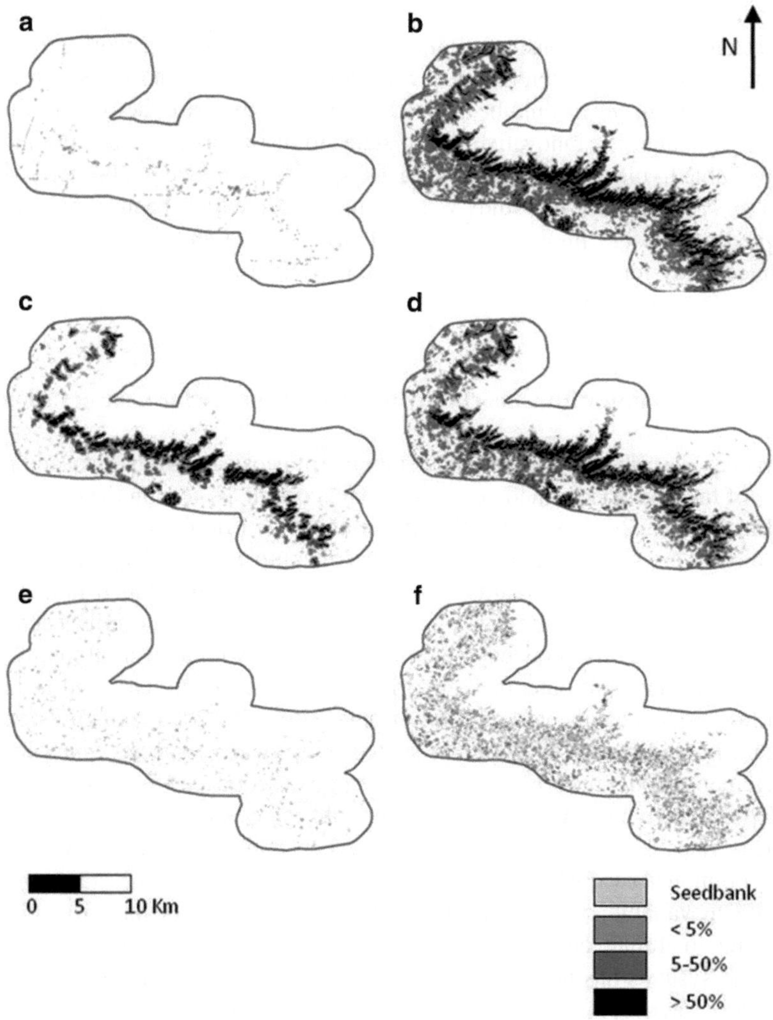

Fig. 13.3 Maps of the Santa Catalina Mountain Study area showing (**a**) mapped *P. ciliare* in 2010 and simulated *P. ciliare* invasion at year 2060 for five simulation scenarios: (**b**) no management, (**c**) intermediate management best case, (**d**) intermediate management worst case, (**e**) unlimited management best case, and (**f**) unlimited management worst case (From Frid et al. 2013b)

To our knowledge, the approach used by Frid et al. (2013b) using an STSM to simulate resource allocation to detection as well as treatment has not yet been applied to *Bromus*. Future work on *Bromus* species using this modeling approach could consider questions around resource allocation in space. For example, the model could consider trade-offs between monitoring and prioritizing treatment of more remote (and thus costly) areas versus accessible areas that require fewer resources to treat. In addition, the trade-off between monitoring and treatment could be explored further on a landscape and species-specific basis to determine which management strategy might be most effective. While other non-STSM modeling approaches could also consider similar questions, this existing work provides an example STSM that in combination with conceptual state-and-transition models and STSMs for *Bromus* could be adapted to consider similar trade-offs for this species. In particular, the innovative use of states to distinguish between the information available to managers about an invasive species' presence or absence is ideally suited to STSMs. Since this study was conducted, there have been significant technological and design improvements to STSM software and conducting a similar expanded study for *Bromus* should be more feasible using the latest available tools (ST-Sim, see www.syncrosim.com).

13.2.7.2 Modeling with Uncertainty and Consequence for Conservation

STSMs can describe the current understanding of ecosystem dynamics and predict effects of invasive species and vegetation management, as seen in the example provided in Sect. 13.2.7.1 (see also Rumpff et al. 2011; Frid et al. 2013a). However, since models are simplified characterizations of complex natural systems, their predictions will deviate from reality; this deviation is hereafter called model uncertainty. It is important to estimate how well, or poorly, a model describes ecosystem dynamics because this knowledge provides managers with a level of confidence in predicted management outcomes. Ignoring model uncertainty can lead to ineffective or wasted management (e.g., Johnson and Gillingham 2004) and, given the high cost and limited resources often associated with invasive species and vegetation management, model uncertainty can have large consequences for management.

Model uncertainty in STSM can arise from many sources, such as estimated effects or rates of transitions, or the use of expert opinion (see Regan et al. 2002). Expert opinion is used when empirical data of reference conditions, states, or transitions are unavailable; or when transition rates are expected to deviate from historical values due to climate change (e.g., Sect. 13.3). A lack of data on transition rates in rangeland systems has necessitated a high reliance on expert opinion (e.g., Forbis et al. 2006; Vavra et al. 2007; Evers et al. 2011, 2013). While there are examples of STSM for *Bromus* management that investigate certain sources of uncertainty (e.g., Evers et al. 2013; Creutzburg et al. 2014), there are no examples of characterization of uncertainty due to expert opinion in an STSM for *Bromus* or other invasive annual grass. Below, we demonstrate how to characterize this uncertainty in STSM, drawing from a published example from *Eucalyptus* forests in Australia (Czembor and Vesk 2009; Czembor et al. 2011). This example describes consequences for management that have direct applications to *Bromus* STSM and is summarized here as a model approach that should be integrated into STSM for *Bromus* or other invaders in the semiarid western United States.

The example considered three sources of uncertainty: variation among experts, imperfect knowledge, and system stochasticity. To incorporate variation among experts, experts were provided with an STM and they specified how transitions would affect state change and the rate of each transition occurrence. In this way, experts' own understanding of ecosystem dynamics was quantified and used to identify variation among experts.

Uncertainty due to imperfect knowledge occurs when an expert knows the approximate range for a model parameter, but is unsure of the exact, true value. Experts have provided single-value point estimates of transition rates in previous STSMs (e.g., Speirs-Bridge et al. 2010; McCarthy 2007) because, until recently, there was no option in STSM software to incorporate these bounds into models. However, using point estimates elicited from experts ignores imperfect knowledge and can lead to overconfidence in models and their predictions. Imperfect expert knowledge was addressed in the example by asking experts to estimate transition rates as a range of probable values, rather than as single-value point estimates. The ranges for each transition rate were converted to distributions; single points were sampled from distributions and compiled to create a set of replicate STSMs for each expert (Czembor et al. 2011).

Uncertainty due to system stochasticity arises because natural processes and disturbances occur randomly in space and time. It is independent of the uncertainty caused by using expert opinion and reflects the inherent variation in natural systems. The example incorporated system stochasticity using the VDDT software, which relies on Monte-Carlo random sampling methods where the occurrence of a transition to any one cell at a specific timestep is probabilistic and varies over multiple simulations (ESSA Technologies Ltd. 2007).

Once STSMs were constructed and simulations were complete, the example quantified which of the three sources of uncertainty contributed most to variance in model results (Quinn and Keough 2002). The authors conducted variance components analysis in R software using linear mixed-effects models in a maximum likelihood framework (Faraway 2006; Gelman and Hill 2007) to determine the contribution of each source of uncertainty to the variation in the proportion of cells in the desired vegetation state at the end of model simulations. Additional details regarding model parameterization and modeling can be found in Czembor et al. (2011).

All VDDT model results averaged together (i.e., with no consideration of uncertainty) indicated a slight increase in the desired vegetation state (low-density mature) over time, increasing from 6 % of the landscape to 15.6 % (Fig. 13.4). However, the model results for each expert separately are quite different from each other, with roughly 7.5–11.9 % of the landscape in the desired state for Experts 3–5, but up to 33.5 % predicted to occur in the desired state (Expert 2). Due to the similarity in results for Experts 1 and 2, it is interesting to note that these experts identified themselves as having expertise primarily in ecology, while Experts 3, 4, and 5 identified themselves as having expertise in natural resource and forest management. Variance due to imperfect knowledge (inner bars) was relatively constant over model simulations, while variance due to system stochasticity (outer bars) differed among experts' models and increased over time, particularly for Experts 1 and 5.

The results of the variance components analysis indicated that total variance in model results increased over time and reached equilibrium near the end of simulations (Fig. 13.5). The majority of total variance was due to the differences among

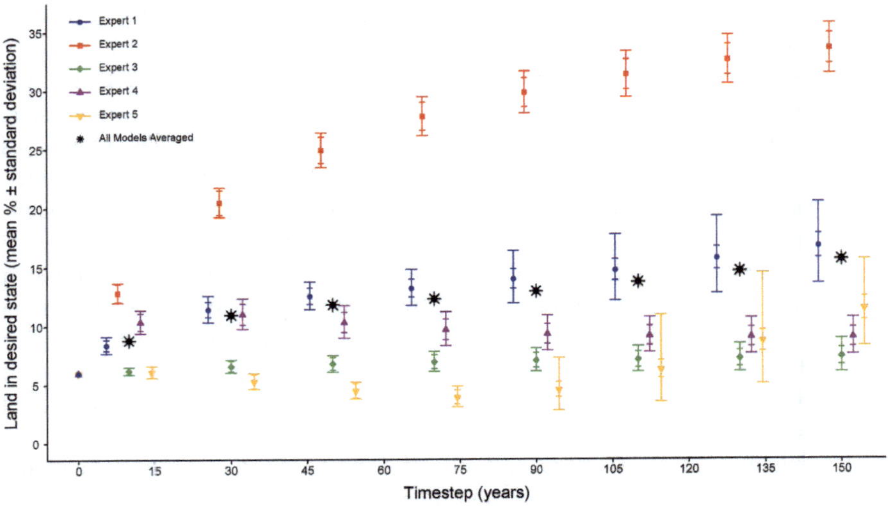

Fig. 13.4 The average percent of cells in the desired vegetation state (low-density mature) over time, for all models combined and for each expert. Error bars are stacked and represent the standard deviations for the variance due to imperfect knowledge (90 % confidence; inner bars) and system stochasticity (from end of inner bars to outer bars). Results adapted from Czembor et al. (2011)

experts (0.73), followed by system stochasticity (0.12), and then the imperfect knowledge component (0.01).

These results suggest that when a single or small group of similar experts provide the input to models, the results could deviate greatly from the true outcomes of vegetation management actions if the ecosystem dynamics are not well understood. However, even though the model results showed high variance, it is important to note that they are useful because they represent the current understanding of ecosystem dynamics and illustrate hypothetical depictions of landscape change. The large among-expert variance also makes selecting among management actions challenging because there is low confidence in model results. This variance could be reduced (i.e., masked) by subsampling similar individuals from available experts or aggregating opinions through consensus, both of which are very commonly done in rangeland management steering committees, at the risk of biasing results and to the detriment of management decisions. If experts are forced to form consensus, this uncertainty cannot be identified and management actions may be susceptible to unexpected results. Ascertaining which parameters are most variable among experts and collecting monitoring data to determine true rates and effects of transitions is therefore needed, and may be necessary for STSM applications to *Bromus* invaders.

There is no evidence provided in the example for the minimum number of experts needed to ensure adequate confidence in model predictions. Minimum expert sample sizes as high as 50 have been recommended, though typically less

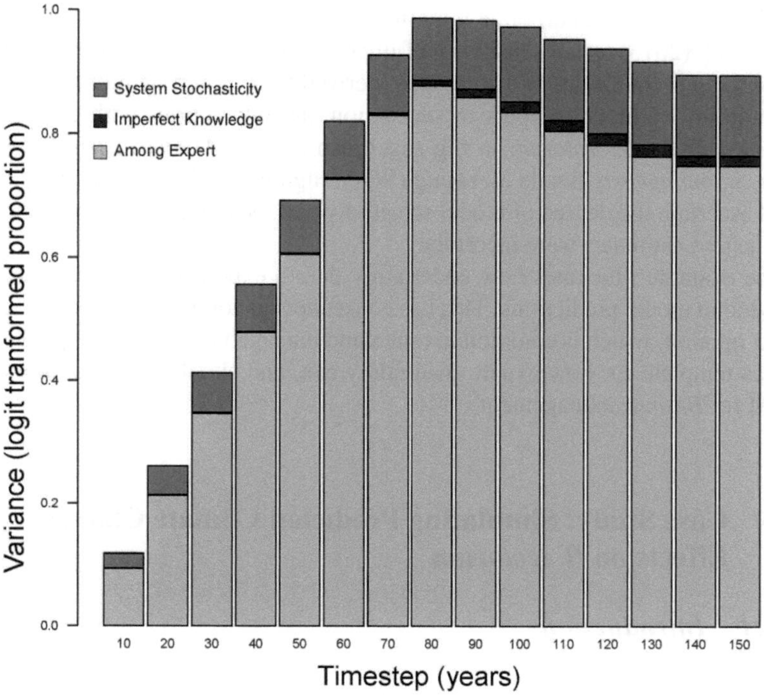

Fig. 13.5 Bar plots showing the variance due to among-expert uncertainty, imperfect knowledge, and system stochasticity every 10 years for 150 years. A 95 % Confidence Interval that corresponds to a logit transformed variance of 0.73 (the among-expert variance at timestep 150) would span 3–46 % of modeled cells in the desired vegetation state. Results adapted from Czembor et al. (2011)

than ten are sampled in practice because few appropriate experts are available (Czembor et al. 2011).

System stochasticity was a moderate source of variance in the model results for the *Eucalyptus* example. In VDDT, this variance reflects the stochastic sequence of disturbances over repeated simulations (i.e., transitions occur at random times within simulations, but with the same average probability through time across simulations). While stochasticity in the rate of disturbances over time was not evaluated in the *Eucalyptus* example, this source of uncertainty can be incorporated into "temporal multipliers" (see Sect. 13.3). Temporal transition multipliers were used to parameterize uncertainty in a study of the effects of the exotic annual grasses, *B. tectorum*, *Taeniatherum caput-medusae* L (medusahead), and *Ventenata dubia* (Leers) Coss. (North Africa grass), on native plant species in sagebrush steppe ecosystems across years differing in wildfire occurrence (Creutzburg et al. 2014).

The variance in model results due to imperfect knowledge was the least important source of uncertainty in this case study. The example incorporated imperfect knowledge using replicate models that used ranges of values for transition rates, but it is also possible to incorporate imperfect knowledge through sensitivity analyses. Evers et al. (2013) tested sensitivity of imperfect knowledge by adding, removing, or changing probabilities of disturbance (derived from expert opinion) to determine vulnerability of model outputs to succession and interactions with exotic annual grasses such as *B. tectorum* in big sagebrush steppe (*Artemisia tridentata* Nutt. subsp. *wyomingensis* Beetle & Young (Wyoming big sagebrush). In doing so, they could ascertain the degree of model sensitivity and the potential magnitude of error if the expert estimates were incorrect.

The example illustrates how uncertainty due to expert opinion can cause large variation in model predictions. This has consequences for models constructed using expert opinion, which is common in rangeland management. The example also provides a template for how expert uncertainty can, and should, be incorporated into STSM for *Bromus* management.

13.3 Case Study: Simulating Predicted Climate Change Effects on *B. tectorum*

13.3.1 Introduction

The US National Environmental Policy Act (NEPA 1969) requires federal agencies to integrate environmental values into their decision-making processes by considering the environmental impacts of their proposed actions and reasonable alternatives to those actions. Environmental values include climate change, although it was only recently added to lists of environmental concerns for the Bureau of Land Management (Manual 7300: Air Resource Management Program), US Forest Service (USFS, FSH 1909.12–Land Management Planning Handbook), and National Park Service (National Park Service: Climate Change Response Strategy). Rarely do federal planners have local data on climate change to make predictions; therefore, planners use more generic concepts to comment in NEPA documents. Climate change effects imply changing local trends and temporal variability for temperature, precipitation, evapotranspiration, and carbon dioxide. All of these variables can impact the growth and spread of invasive annual grasses, such as *B. tectorum* (Smith et al. 2000; Brown et al. 2004; Bradley 2009b), and native species, such as trees (Tausch and Nowak 1999) in the US arid Intermountain West.

The purpose of this case study is to propose a new methodology of incorporating readily available data on future CO_2 levels, precipitation, and temperature into the ecological processes of STSMs. To illustrate climate change effects on ecological systems and *B. tectorum*, we modeled *Artemisia tridentata* Nutt. spp. *vaseyana* (Rybd.) Beetle (mountain big sagebrush, 30–36 cm of average annual precipitation) sites within the boundary of Great Basin National Park in Nevada and adjacent *A. tridentata* spp. *wyomingensis* (25–30.4 cm of average annual precipitation) just outside the boundary of the Park (Provencher et al. 2013). The *A. tridentata* spp. *vaseyana* system occurs at higher elevations on cooler and moister soils than the *A. tridentata* spp. *wyomingensis* system; therefore, this study assumed that *A. tridentata* spp. *wyomingensis* would respectively replace the lowest and middle elevations of *A. tridentata* spp. *vaseyana* in the Park with climate change, especially warming. Both ecological systems are readily invaded by *B. tectorum*, especially on relatively warmer and drier soils, as well as by trees such as *Pinus monophylla* Torr. & Frém. (singleleaf pinyon) and *Juniperus osteosperma* (Torr.) Little (Utah juniper). As a result, both *A. tridentata* spp. *vaseyana* and *A. tridentata* spp. *wyomingensis* systems needed to be modeled together to predict climate change effects on vegetation class abundances using the Path Landscape Model (Path) software (see footnote of Table 13.1 about Path). Using climate change data to affect ecological processes in STSMs is rather new (Halofsky et al. 2013); therefore, both the novelty of our approach and forecasted values of precipitation and temperature from global circulation models introduced uncertainty in our simulation results. Another source of uncertainty was expert opinion, which was required as the scientific literature for Great Basin rangelands is data poor for model parameterization. We addressed uncertainty by using Monte-Carlo replicates that deliberately introduce strong variability into ecological processes. Therefore, this case study focuses more on demonstrating a new methodology and less on the accuracy of ecological results.

13.3.2 Methods

13.3.2.1 State-and-Transition Models

The *A. tridentata* spp. *vaseyana* STM presented here is part of a group of 21 such models previously developed with Path for cost-effective management of Great Basin National Park's ecological communities (Provencher et al. 2013). The *A. tridentata* spp. *wyomingensis* STM was obtained from concurrent STSM efforts in nearby Hamlin Valley and Pine Valley Mountains of southwest Utah's Great Basin, respectively, on lands managed by the US Department of Interior Bureau of Land Management and US Forest Service. Both STSM models originated during 2005–2007 from LANDFIRE's STSM development effort for the Great Basin mapping zone (Rollins 2009). These models were subsequently modified by (1) improving representation of fire disturbances, (2) adding uncharacteristic vegetation classes representing states or phases (e.g., invasive annual grassland), (3) incorporation of new disturbances observed in the field, including adding uncharacteristic ones such as invasive species expansion, and their rates, (4) adding management actions and budgets to the list of disturbances that affect states and transitions, and (5) introducing external sources of temporal variability that would modify disturbance rates over time (e.g., replacement fire). Models and results were reviewed by federal and state agency specialists, contractors, and academics during the Park's project workshops and in workshops of previous projects (Low et al. 2010). Informal but extensive sensitivity analyses were part of the review process.

13.3.2.2 Range Shifts

The replacement of "cooler or wetter" ecological systems and their indicator species by "warmer or drier" systems and their indicator species during climate change is called range shift. Theoretical bioclimatic envelope modeling (Rehfeldt et al. 2006; Bradley 2009a) and one field study (Kelly and Goulden 2008) provided widely conflicting conclusions on the speed of range shifts. Thus, in this modeling example, the percentage of the area shifting from *A. tridentata* spp. *vaseyana* to *A. tridentata* spp. *wyomingensis* systems over 100 years was first set at 10 %. We iteratively determined that a rate (probability per year) of 0.0604 year^{-1} (i.e., 604 virtual pixels shifted per 10,000 pixels per year) matched the 10 % range shift over 100 years. We further set values of 87 % and 13 % for the total area of *A. tridentata* spp. *vaseyana* that would be replaced by *A. tridentata* spp. *wyomingensis* and *A. nova* A. Nelson (black sagebrush), respectively, based on current sagebrush community proportions as found by Provencher et al. (2013). For simplicity, we only tracked the range shift between *A. tridentata* spp. *vaseyana* and *A. tridentata* spp. *wyomingensis* for this chapter. Moreover, we gradually introduced the range shift by setting the initial value of range shift to zero under the assumption of no climate change at year zero, and then linearly increased temporal multipliers for range shifts to a value of two by year 75 of the simulation. Therefore, the average rate of range shift of 0.0604 year^{-1} had a value of one over the 75 years of the time series to maintain rate integrity. In comparison, a control simulation without climate change would have a range shift temporal multiplier series equal to zero for all time intervals.

13.3.2.3 Climate Variability Effects on Ecological Processes

Temporal multipliers act as forcing factors of ecological processes in the STSM and also reflect hypotheses about the effects of climate variability on ecological processes. One temporal multiplier is a non-dimensional number ≥ 0 in a yearly time series that multiplies a base disturbance rate in the STSM. For example, for a given year, a temporal multiplier of one implies no change in a disturbance rate, whereas a multiplier of zero is a complete suppression of the disturbance rate, and a multiplier of three triples the disturbance rate. A temporal multiplier can be obtained from time series data or theoretically derived. In the current case, multipliers vary for scenarios with or without climate change. Temporal multipliers are determined by dividing each yearly value of the time series (for example, area burned) by the temporal average of the time series, thus creating a non-dimensional time series with an average of one. Division by the time series' average would remove the hypothesis of altered levels of the ecological process being modeled under climate change scenarios; thus each raw value of the new time series (e.g., future area burned) with climate change must be divided by the average of the time series not experiencing climate change.

The dominant ecological processes of big sagebrush models, fire, drought, invasive annual grass expansion, and tree expansion, all required temporal multipliers without and with climate change forcing. Forcing factors were based on future trends in atmospheric CO_2, local temperature, and local precipitation and included:

1. Increased expansion of invasive species (annual grasses, forbs, and trees) into uninvaded areas caused by CO_2 fertilization effects during wetter than average years (Smith et al. 2000; Brown et al. 2004; Bradley 2009b);
2. Decreased expansion of invasive species (annual grasses, forbs, and trees) into uninvaded areas during drier than average years regardless of CO_2 concentrations (Smith et al. 2000; Brown et al. 2004; Bradley 2009b);
3. Longer fire return intervals in shrubland systems due to increased drought frequency preventing fine fuel buildup (Westerling and Bryant 2008, Westerling 2009; Abatzoglou and Kolden 2011; Littell et al. 2009); and
4. Increased expansion of *P. monophylla* and *J. osteosperma* trees in shrublands caused by CO_2 fertilization during wetter-than-average years (Tausch and Nowak 1999).

The temporal multiplier for elevated CO_2 was calculated from time series for future CO_2 levels using the A2 emission scenario from IPCC's (2013) report, simply as change in CO_2 from time=0 to the end of the simulation period (i.e., division of each yearly CO_2 level by the level of the first year of simulation).

All simulations of temperature and precipitation effects were based on five replicate Global Circulation Models (GCM) forecasts available from the Downscaled Climate Projections Archive (of 37 GCMs available; http://gdo-dcp.ucllnl.org/downscaled_cmip_projections/dcpInterface.html, version 1.2, 06-August-2011) using the mean values for the Park and surrounding area. Normally, average values of an "ensemble" of many randomly selected GCM model outputs are used for simulations such as ours, given computing and cost limitations. The five model outputs were selected based on their marked differences for projected precipitation which varied much more that temperature among the models (listed in Fig. 13.6's caption). One GCM selected forecasted increasing precipitation, albeit from initial low levels (first replicate), three forecasted no change in precipitation but had different average precipitation levels (second, third, and fourth replicates), and one forecasted less precipitation over a century (fifth replicate). All data were displayed by year and month and our time series were 75 years into the future.

Five future time series replicates without climate change were created by using observed historic temperature and precipitation data obtained from the same library and for the same area using the same spatial averaging methods. We assumed that past climate reflected future climate without climate change and that recent warming of the past decades had not significantly affected the slow growing Great Basin vegetation as shown by Kelly and Goulden (2008) in a Mojave Desert elevation gradient. However, there was only one observed time series from 1950 to 1999, but five future replicates without climate change were needed. To create five replicates

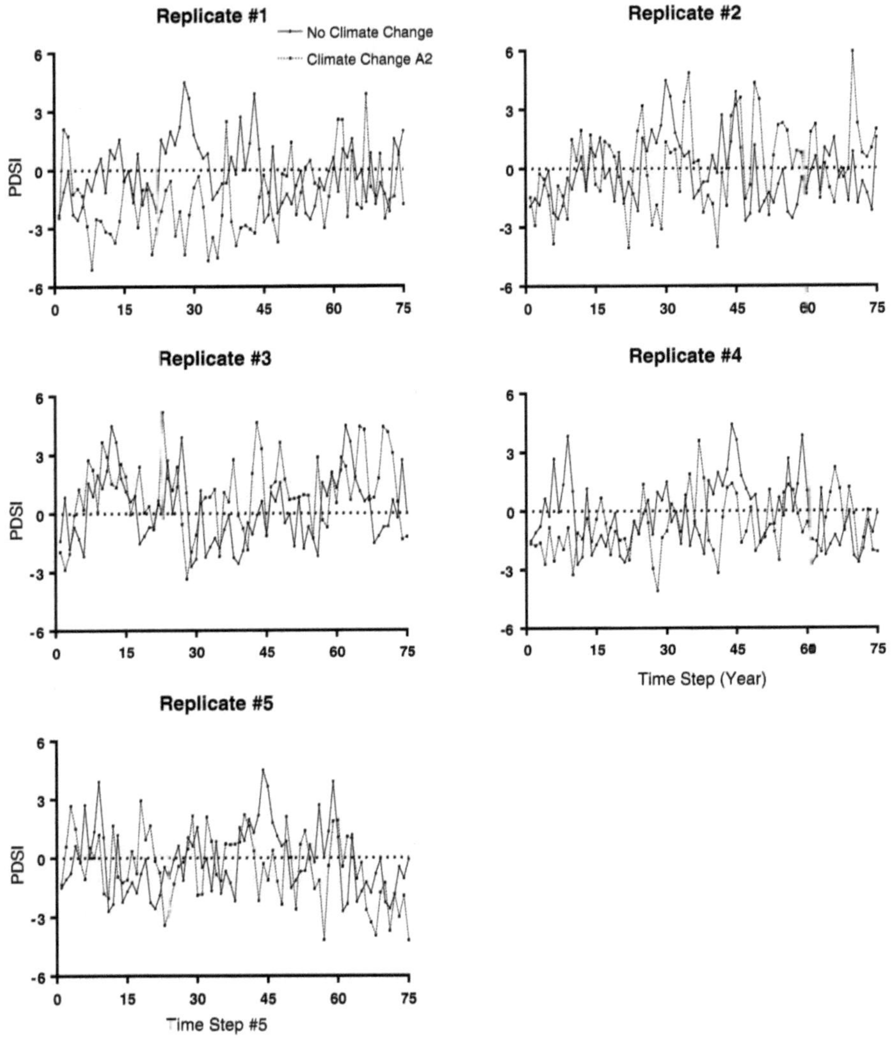

Fig. 13.6 Replicated and calculated PDSI time series using Eqs. (13.1) and (13.2) assuming no climate change (resampled historic precipitation and temperature time series) and assuming the A2 scenario from five Global Circulation Models as five 75-year replicates. Climate change legend: replicate #1 = mri_cgcm2_3_2a.2.sresa2; replicate #2 = ukmo_hadcm3.1.sresa2; replicate #3 = ncar_pcm1.1.sresa2; replicate #4 = ncar_ccsm3_0.1.sresa2; and replicate #5 = ncar_ccsm3_0.2.sresa2

and preserve potential seasonal and multiyear climate patterns in the data, the 1950 to 1999 time series was wrapped around as a time loop (i.e., year 1950 followed year 1999) and was resampled by randomly selecting five start years (replicate #1 = 1956; replicate #2 = 1954; replicate #3 = 1972; replicate #4 = 1960; replicate #5 = 1990) until 75 years of data were accumulated. The next step was to use the historic and projected CO_2, temperature, and precipitation to create the final temporal multipliers.

All projections of temperature and precipitation were integrated into the Palmer Drought Severity Index (PDSI; Palmer 1965; Heddinghaus and Sabol 1991) from which temporal multipliers for replacement fire, drought, annual grass expansion, and tree expansion were obtained when combined with future projections of CO_2. Because the literature offered no guidance on this subject, heuristic relationships were created to translate the variability of the PDSI into the local variability of drought, replacement fire, invasive annual grass expansion, and tree expansion. Hopefully our heuristic approach will spur research to improve upon our effort. Calculations of future values of the PDSI are found in the Appendix (Eqs. 13.1 and 13.2).

Drought was assumed to kill woody species (for trees; Pennisi 2010), sometimes mediated by triggering insect and disease attacks on trees, whereas wetter conditions suppressed this disturbance. In the STSMs for *A. tridentata* spp. *vaseyana* and *A. tridentata* spp. *wyomingensis*, the drought disturbance operated both by partial thinning of the dominant upper-layer lifeform (i.e., shrubs or trees that characterize the vegetation class) within a vegetation class without causing a transition to another state or phase (about 90 % of probabilistic outcomes) and by killing most woody individuals of the dominant upper-layer lifeform and thus causing a transition to a younger succession class (10 % of probabilistic outcomes). As drier (PDSI$<$0) or wetter (PDSI$>$0) conditions, respectively, were observed in the GCM time series, the base rate for the drought disturbance in the STSMs was increased ($>$1) or decreased ($<$1) by the yearly value of the temporal multiplier (Eq. 13.3).

Invasive annual grass expansion and tree expansion into uninvaded areas did not include infilling by invasive annual grasses and native trees, although that could be done in a more complicated STM. Rates of invasive plant advance in the STSMs could vary by vegetation classes and ecological systems based on the natural resistance of established vegetation (Chambers et al. 2014). We used a single temperature multiplier to relate moisture (precipitation) to greater dispersal (more seeds) and, thus, invasion (Eq. 13.4). Fertilization with elevated CO_2 was predicted to enhance the effect of a wetter condition but was a weaker effect overall (Nowak et al. 2004). We assumed that tree expansion was a much slower process than invasive annual grass expansion and also less responsive to PDSI (Eq. 13.5).

Fire frequency and total area burned have a complicated relationship to the PDSI in shrublands—they are more likely to burn if they first experience consecutive wetter-than-average years leading to accumulation of fine fuels that will more likely burn in a dry year immediately following the wet year sequence (Westerling and Bryant 2008, Littell et al. 2009; Westerling 2009). Area burned was first estimated by applying equations using PDSI and by assuming that the maximum fire size achieved under any scenario represents 10 % of the area sum of all shrubland–woodland ecological systems for the shrubland–woodland temporal multiplier. We chose 10 % of the area because managers considered fires exceeding 10 % of Great Basin National Park's area were very large and unusual according to the federal record. However, different managers may choose different percentages for different landscapes. The shrubland–woodland fire temporal multipliers considered the roles of 3 prior years of PDSI, more specifically that fine fuels will more likely burn in the current dry year immediately following 2 previous and consecutive wetter-than-average years during which fine fuels accumulated (Eq. 13.6).

13.3.3 Results

Three uncharacteristic states of the *A. tridentata* spp. *vaseyana* model (coded as MSu to represent Mountain Sagebrush upland) were invaded by *B. tectorum*: shrubland with mixed annual grass and perennial grass (MSu-SAP), tree-dominated shrubland with annual grass (MSu-TEA), and annual grassland (MSu-AG). Using remote sensing data to populate the initial conditions of the simulations (year = 0; Provencher et al. 2013), the MSu-TEA state was the most abundant (~2700 ha), followed by MSu-SAP (~1000 ha), whereas MSu-AG was nearly absent (~10 ha; Fig. 13.7). As a result of the simulated ecological processes with and without climate change, the area occupied by the MSu-SAP state gradually decreased over time as a result of fire, drought mortality, and tree encroachment. In the model's transitions, the first two disturbances caused the increase in area of the MSu-AG state, which closely matched the decrease in area of the MSu-SAP state (Fig. 13.7). The small area of the MSu-TEA state primarily loss to fire after year 40 also contributed to the increase in area of the MSu-AG state. The area of the MSu-TEA state was relatively stable compared to the other states because the area that burned was offset by the new area of the MSu-SAP state that became encroached by trees.

The simulated effect of climate change was nearly undetectable for *A. tridentata* spp. *vaseyana* (Fig. 13.7). Because of the strong variability in drought cycles in the Great Basin, the trends in ecological processes caused by climate change indicated here are far smaller than their natural variability; therefore, the effects of climate change in STSMs must become strong to be detected, and this takes several decadal iterations. Although climate change differences between simulations were not clearly observable for *A. tridentata* spp. *vaseyana*, they still incrementally occurred because states from *A. tridentata* spp. *wyomingensis* replaced those of *A. tridentata* spp. *vaseyana* starting on the fifth year of simulations (Fig. 13.8). Only range shifts caused this replacement in our models. Furthermore, as a result of model design, these range shifts will first be observed in all early-succession phases and classes and will occur more rapidly in phases or states with shorter fire intervals because we assumed stand-replacing events remove the biomass of original indicator species and allow the new indicator species to colonize in the same phase or state (see also Halofsky et al. 2013; Creutzburg et al. 2014). Range shifts do not usually change the phase or state, they only change the potential for certain dominant indicator species (e.g., from MSu-AG to the annual grassland state of *A. tridentata* spp. *wyomingensis*).

Starting with no area of *A. tridentata* spp. *wyomingensis* within the Park, simulated range shifts replacing *A. tridentata* spp. *vaseyana* with *A. tridentata* spp. *wyomingensis* (coded as WS) first caused new increases in area of two classes of vegetation: early-succession (WS-A) and annual grassland (WS-AG; Fig. 13.8). The cumulative area converted to *A. tridentata* spp. *wyomingensis* represented a large fraction of the area initially in *A. tridentata* spp. *vaseyana* (about 10 %, as built into the STSM—see *Range Shifts*). Following STM transitions, the other three phases and state emerged, albeit with low areas, as the product of succession (from

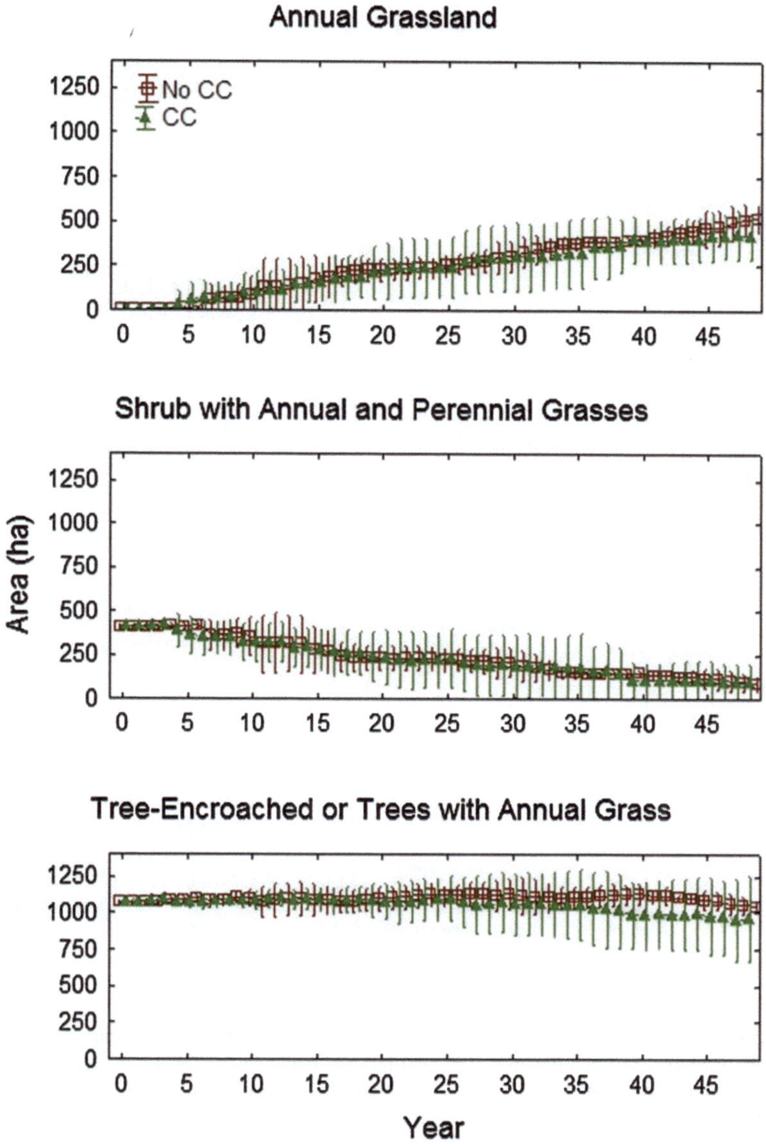

Fig. 13.7 The area of vegetation classes invaded by *B. tectorum* (ha) in the *A. tridentata* spp. *vaseyana* ecological system assuming no climate change (NoCC; based on resampled historic data) and assuming climate change for the A2 scenario (CC) from five Global Circulation Models (GCM). Legend: MSu-AG = annual grassland; MSu-SAP = shrubland with annual and perennial grasses; and MSu-TEA = tree encroached or wooded shrubland invaded by annual grass. Climate change GCMs replicates are mri_cgcm2_3_2a.2.sresa2, ukmo_hadcm3.1.sresa2, ncar_pcm1.1.sresa2, ncar_ccsm3_0.1.sresa2, and ncar_ccsm3_0.2.sresa2. Error bar represents the 95 % confidence interval for five replications

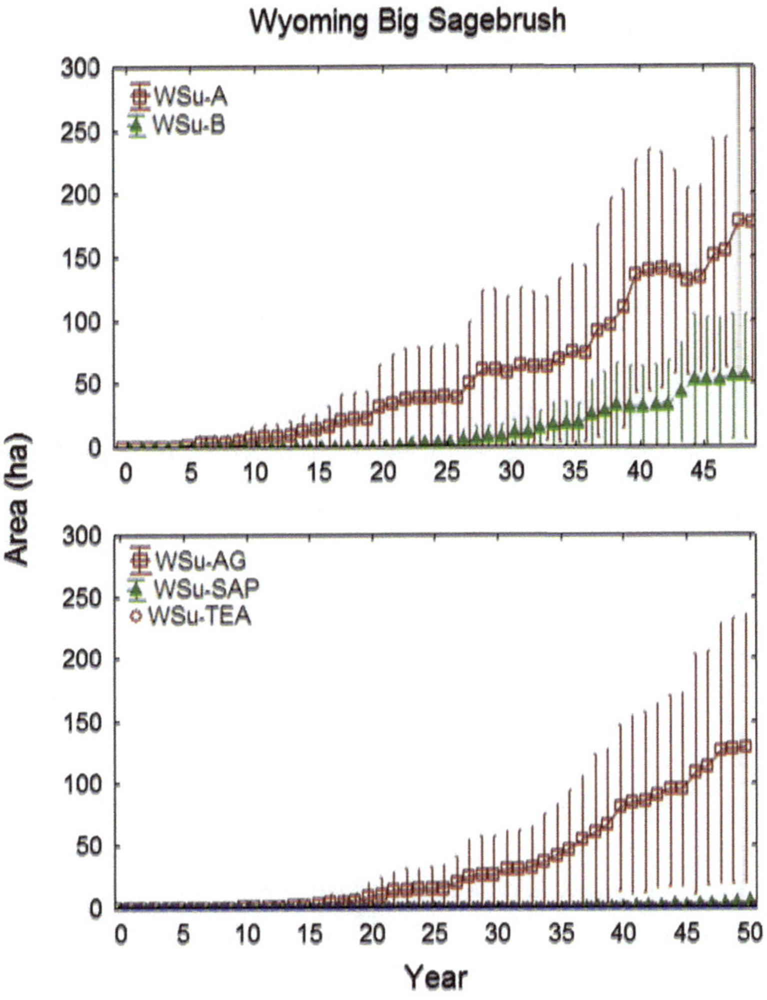

Fig. 13.8 The area of reference and *B. tectorum*-invaded vegetation classes (ha) in the *A. tridentata* spp. *wyomingensis* ecological system assuming climate change for the A2 scenario from five Global Circulation Models. Legend: WS-A = early-succession; WS-B = mid-succession; WS-AG = annual grassland; WS-SAP = shrubland with annual and perennial grasses; and WS-TEA = tree encroached or wooded shrubland invaded by annual grass. Climate change GCMs replicates are mri_cgcm2_3_2a.2.sresa2, ukmo_hadcm3.1.sresa2, ncar_pcm1.1.sresa2, ncar_ccsm3_0.1.sresa2, and ncar_ccsm3_0.2.sresa2. Error bar represents the 95 % confidence interval for five replications

WS-A to WS-B and from WS-SAP to WS-TEA) and invasion of uninvaded class by *B. tectorum* (from WS-B to WS-SAP; Fig. 13.8).

Three lessons learned from simulating *A. tridentata* spp. *vaseyana* and *A. tridentata* spp. *wyomingensis* community transitions were that (1) it may take decades to detect climate change effects on the distribution of vegetation classes within an ecological system because in the models natural "background" drought variability appears stronger than the variability caused by climate change in the Great Basin region of the United States, (2) range shifts between ecological systems, but not between phases or states, were accelerated by the short fire return interval of the annual grassland state, and (3) climate change did not cause more *B. tectorum* expansion 50 years into the future landscapes (e.g., MSu-AG in Fig. 13.7) because the trends in temperature and precipitation from the GCMs decreased soil moisture (i.e., increased drought intensity) and, as a result, decreased *B. tectorum* expansion (and tree expansion) regardless of the level of CO_2 fertilization. Range shifts, therefore, are predicted to occur more readily in the areas having an annual grassland state, because we hypothesized that range shifts occur through stand-replacing events in long-lived and drought-resistant shrublands and woodlands. The presence or dominance of *B. tectorum* shortens fire return intervals in landscapes, which in turn increase the likelihood of stand-replacing events. We have not, however, simulated the process of invasion by a new invasive annual grass species adapted to warmer conditions, such as *Bromus rubens* L. (red brome), although that would be feasible with additional data (e.g., Bradley et al. 2015).

This case study integrated STSM to predict distribution of *A. tridentata* spp. *vaseyana* and *A. tridentata* spp. *wyomingensis* plant community phases and states over time with relationships between GCM outputs and their effects on the number of ecological disturbance occurrences per year in STMs. Finding ways to incorporate climate change variability into STSM processes was the most difficult and time-consuming part of the case study, and furthermore this step introduced uncertainty. An alternative approach linking STSM to climate change effects was pioneered by the Integrated Landscape Assessment Project (ILAP; Halofsky et al. 2013; Creutzburg et al. 2014), which linked vegetation change and wildfire trend data from the GCMs and the MC1 dynamic vegetation model with STMs to inform watershed-level prioritization of fuel treatments in Arizona, New Mexico, Oregon, and Washington. Whereas we used a bottom-up approach based on precipitation, temperature, and CO_2 concentrations output from GCMs affecting disturbances and range shifts, ILAP was a top-down process where GCMs and MC1 determined range shifts and the variability of fire. ILAP's process required downscaling GCMs and MC1 subcontinental coarse resolution predictions of climate change to the project areas, and meshing processes from widely different spatial scales. This is major source of uncertainty because MC1 predicted (1) changes in general lifeforms groups (shrublands, grasslands, and forest), but not different ecological systems within a group, and (2) changes in general fire activity regardless of how drought affects differently forested and shrubland systems (Taylor and Beaty 2005; Westerling and Bryant 2008; Littell et al. 2009; Westerling 2009). Moreover, MC1 only generated predictions for fire (Creutzburg et al. 2014), whereas our bottom-up method also introduced climate variability for drought, invasive upland and riparian species expansion, tree expansion, insect outbreak, flooding, subalpine spring freezing, snow deposition, and tree encroachment (different from tree expansion).

13.4 Management Implications

The examples and case study in this chapter have the following management implications.

1. STSM simulation outcomes presented in this chapter indicate that a program evaluating unlimited budgets for detection, monitoring, and treatments with a long-term scope is worth considering through STSM-based assessments. In landscapes where invasion by species such as *Bromus* is in its early stages allocating resources toward monitoring may increase treatment success. STSMs such as the one discussed for *P. ciliare* can help better understand the trade-off between treatment and monitoring, thereby supporting management investment decisions.
2. Uncertainty among experts over nonnative brome invasion rates and control/restoration success could waste already limited funding for natural resources management. How can diverse opinions be integrated into a robust management strategy for *Bromus* species? At a minimum, the research presented herein indicates that a simple sensitivity analysis of expert opinion on nonnative brome invasion rates and control success should be explored with STSM where experts are able to offer different opinions. In this way, those environmental or management parameters that are most uncertain (i.e., vary most between experts) can be identified for sensitivity analyses. Those uncertain parameters that matter most to management outcomes can then be used to focus efficient monitoring and data collection.
3. Several scenarios can be explored using STSMs. Local managers planning for the conservation of natural resources far into the future (e.g., 50–100 years) can implement programmatic changes in the next decade that might determine whether range shifts will happen sooner or later. For example, managers might have resources to only restore degraded annual grassland to a more resilient state of vegetation that would result in resistance to range shifts or to only control wildfires while maintaining the age diversity of phases in the reference state, thus resulting in plant communities more similar to the reference condition. These two scenarios could readily be explored with STSM.

As STSMs inform agency land management decisions and become more visible as a planning tool to other users, future applications will become more complicated and address increasingly larger landscapes. Future applications will require investments in software enhancements to accelerate processing and to accomplish new processes. Given that the most commonly used STSM platforms are freeware, these investments must come from users, which implies that agency and private users should budget for software enhancements. In addition, while there are various examples of STSMs being used by public land management agencies (Forbis et al. 2006; Provencher et al. 2007, 2013; Low et al. 2010), in most of these applications funding for training and technical support in applying the models has been a key requirement for the success of the initiative. Land management agencies often lack the technical expertise required to be able to apply STSMs without such support (Blankenship et al. 2013).

13.5 Ecological Research Needs

The following are some concepts that need additional consideration in order to move forward with applying STSMs to *Bromus* management.

1. Evaluate how effective alternative inventory approaches are at detecting *Bromus* invasion at different stages. How costly are these different approaches? What is the most effective allocation of resources between management activities including preventive restoration, inventory, treatment, and posttreatment maintenance? A combination of field experiments and STSM development is required to answer these questions at the landscape scale.
2. Explore the reasons for divergence among experts, to refine expert estimates via discussion and comparison to new monitoring data, and to weight expert opinion using established datasets.
3. Determine the rate and fate of range shifts of ecological systems potentially invaded by *Bromus* using field studies coupled with modeling approaches that do not assume infinite species dispersal rates and no resistance to drought.
4. Empirically demonstrate that wildfires primarily fueled by *Bromus* will accelerate range shifts (e.g., from *A. tridentata* spp. *vaseyana* to *A. tridentata* spp. *wyomingensis* and *A. nova*) compared to unburned vegetation
5. Standardize the methodology and science for creating multiple and potentially correlated temporal multipliers in STSMs that reflect different hypotheses between environmental variability and model disturbance rates.
6. Determine to what extent we are uncertain about the rate of spread of invasive *Bromus* species across different ecological systems and about invasive *Bromus* species control success. What field data are already available to reduce uncertainty in estimates for *Bromus* invasion rates (probability of spread and successful establishment of new areas per year)? How might these sources of uncertainty affect vegetation management decisions?

With the existing data available to land managers and the STSM framework as a tool to investigate alternative scenarios and management actions, the above research needs can be addressed. With time, it is hoped that these modeling strategies will assist management decisions and result in desired outcomes with higher efficiency and reduced resources.

Acknowledgments Any use of trade, product, or firm names is for descriptive purposes only and does not imply endorsement by the US Government.

Appendix

Calculating the Palmer Drought Severity Index

The Palmer Drought Severity Index (PDSI) time series was used to calculate the temporal multipliers for replacement fire, drought, annual grass invasion, and tree invasion. Drought is a major influence for these disturbances. PDSI measures long-term soil drought and is updated monthly (Palmer 1965; Heddinghaus and Sabol 1991). Positive values indicate above average soil moisture (>3 is very wet), whereas negative values represent droughty soil (<–3 is very dry). A PDSI of zero is average soil moisture. The formula for PDSI at time *t* (month) is as follows:

$$\mathrm{PDSI}_t = 0.897 \times \mathrm{PDSI}_{t-1} + (k_t / 3) \times (P_t - \underline{P}_t) \tag{13.1}$$

where P_t is precipitation during month *t*, $\underline{P}_t$ is average (historic) precipitation for month *t*, and *k* is a monthly climatic coefficient that weighs the local importance of $(P_t - \underline{P}_t)$ (Palmer 1965). For example, *k* might imply that $(P_t - \underline{P}_t)$ in January does not contribute as much to PDSI as the same deviation in precipitation observed in August (Palmer 1965). Although we downloaded monthly precipitation values and obtained monthly $\underline{P}_t$ from historic precipitation data (respectively, month, precipitation [mm/day]: January, 0.8004; February, 0.8368; March, 1.0234; April, 0.9310; May, 0.9612; June, 0.6130; July, 0.6356; August, 0.7394; September, 0.6876; October, 0.7502; November, 0.7476; December, 0.6858), the value of k_t is unknown and requires complicated field estimation based, among others, on evapotranspiration (Palmer 1965). (To remove this complication and need for a heuristic equation, future projects will use the Standard Precipitation Index [Hayes et al. 1999]). Therefore, we made several arbitrary assumptions to imitate *k* using the month's temperature differential. Specifically,

$$k_t / 3 = 1.5 \times (1 - e^{-0.15 \times (MaxT - Tt)}) \tag{13.2}$$

where $MaxT = 31$ (°C) is the maximum temperature observed, and T_t is the average temperature during month *t*. In this heuristic equation, higher temperatures cause smaller values to multiply $(P_t - \underline{P}_t)$ when monthly precipitation is higher and thus

PDSI becomes smaller (more evapotranspiration). The coefficients 1.5 and −0.15 are fitting constants we iteratively selected that allow the PDSI to vary within the observed range and be responsive to changes in precipitation, primarily, and secondarily to temperature. Using the latest observed monthly PDSI from March 2012 as the first $PDSI_{t-1}$, we estimated future monthly PSDIs per replicate for 75 years using Eqs. (13.1) and (13.2) for both without and with climate change. Compared to the PDSI replicates without climate change, it is noticeable that three temporal replicates of PDSI estimated for climate change effects were drier during certain decades only (replicates #1, 4, and 5), whereas the third replicate was wetter and the second replicate neutral (Fig. 13.6).

Because PDSI can be negative and the STSM software requires positive values, heuristic functions (arbitrary coefficients) were developed for drought, replacement fire, invasive annual grass expansion, and tree expansion that transformed negative values into positive values while maintaining the role of PDSI on the intensity of the disturbance. Not many flexible functions allow the conversion of negative values into positive ones while also accepting positive values; therefore, these curve fitting requirements led us to adopt functions with exponential components that could be easily calibrated. These functions do not calculate the rate of the disturbance, which is found in the STSM, but the temporal variability of the disturbance. All equations generated non-dimensional values and the final temporal multipliers were also non-dimensional.

Drought Disturbance

Because PDSI can be negative, therefore incompatible with PATH's format for temporal multipliers, we chose a negative exponential function for drought to create positive values that increased exponentially with more negative (drier) PDSI values:

$$\text{Yearly drought variability factor} = 0.6 \times e^{-0.6 \times \text{PDSI}} \tag{13.3}$$

The parameters of this function (0.6 and −0.6) were chosen such that PDSI values close to −3 (very dry) were slightly greater than 3 (actually, 3.63) and that very severe droughts with PDSI of −5 (extreme drought) translated into slightly more than doubling of the function (12). Another consideration for curve fitting was that a mild drought characterized by a PDSI of −1 would be about equal to a neutral value of 1. Equation 13.3 is not the final temporal multiplier, however, because it is not divided by its average. In the absence of climate change effects, yearly values of Eq. (13.1) were divided by their temporal average over 75 years, whereas each yearly value of Eq. (13.3) with climate change was divided by the no-climate change average to reflect the hypothesis of altered levels.

Annual Grass Invasion and Tree Invasion Disturbances

The temporal multipliers for invasive annual grass expansion and tree expansion were calculated from two heuristic Gompertz equations (not including the CO_2 fertilization). The Gompertz equation is highly flexible for curve fitting and a special case of it is the negative exponential:

$$\text{Yearly annual grass expansion variability factor} = 4.5 \times e^{-2\times\exp(-0.75\times(\text{PDSI}+1))} \times \text{TMCO}_2 \tag{13.4}$$

$$\text{Yearly tree expansion variability factor} = 2.5 \times e^{-2\times\exp(-0.75\times(\text{PDSI}+1))} \times (\text{TMCO}_2)^{0.5} \tag{13.5}$$

where $TMCO_2$ is the temporal multiplier for CO_2 levels, which is <2 for any yearly value with climate change and equal to one without climate change. In accordance with our hypothesized relationship between species expansion and soil moisture and CO_2 levels, the effect of CO_2 levels as expressed by its temporal multiplier (between 0 and 1) on variability is proportional, whereas the effect of PDSI is exponential (i.e., greater). We arbitrarily dampened the effect of CO_2 fertilization on trees by taking the square root of the CO_2 temporal multiplier. The Gompertz equations allow for some expansion during even dry years ($\text{PDSI} < 0$), average expansion (temporal multiplier close to 1) during average moisture years, and a rapid rise of expansion (multiplier increasing to 4.5 and 2.5), respectively, for invasive annual grass expansion and tree expansion during very wet years. The parameters 4.5 and 2.5 were chosen to match values from the initial Park's study by Provencher et al. (2013). Equations 13.4 and 13.5 are not temporal multipliers, however, because they are not divided by their averages. In the absence of climate change effects, yearly values of Eqs. (13.4) and (13.5), respectively, were each divided by their temporal average over 75 years, whereas each yearly value of Eqs. (13.4) and (13.5) with climate change, respectively, was divided by the no-climate change average to reflect the hypothesis of altered levels.

Fire

The shrubland–woodland fire temporal multipliers considered the roles of 3 years of PDSI, more specifically that fine fuels will more likely burn in the current dry year immediately following two previous and consecutive wetter-than-average years where fine fuels accumulated. The equation to calculate the temporal multipliers from shrubland fire contained two Gompertz functions to account for 3 years of PDSI:

$$\text{Yearly shrubland–woodland area burned variability factor} = \text{MaxFire} \times e^{-3\times\exp-(0.7\times\text{PDSI}t-1+(1-0.7)\times\text{PDSI}t-2)} \times (1 - e^{-3\times\exp(-2\times\text{PDSI}t)}) \tag{13.6}$$

where MaxFire = 1547 hectares and is 10 % of the area sum of all shrubland–woodland ecological systems. Equation 13.6 combines two Gompertz functions to accommodate negative and positive values of PDSI. The first part of Eq. (13.6) after MaxFire, representing fine fuels production, is a classic Gompertz function where a weighted sum is applied to soil moisture during 2 previous years (70 % of PDSI in year $t-1$ and 30 % of PDSI in year $t-2$). Wetter years (PDSI > 0) increase the value of this function (fine fuels accumulation) to a maximum of one. The first part is multiplied by the second function representing the current year, which is one minus another Gompertz function bound between zero and one. Increasingly drier soil moisture (PDSI < 0) causes the second part of Eq. (13.6) to increase to a maximum of one (maximum ignition probability). The PDSI values from the scenarios without and with climate change were used to calculate future area burned. Equation 13.6 is not the final temporal multiplier, however, because it is not divided by its average. In the absence of climate change effects, yearly values of Eq. (13.6) were divided by their temporal average over 75 years, whereas each yearly value of Eq. (13.6) with climate change was divided by the no-climate change average to reflect the hypothesis of altered levels.

References

Abatzoglou JT, Kolden CA (2011) Climate change in western US deserts: potential for increased wildfire and invasive annual grasses. Rangel Ecol Manag 64(5):471–478

Bagchi S, Briske DD, Bestelmeyer BT et al (2013) Assessing resilience and state-transition models with historical records of cheatgrass *Bromus tectorum* invasion in North American sagebrush-steppe. J Appl Ecol 50(5):1131–1141

Barrett T (2001) Models of vegetation change for landscape planning: a comparison of FETM, LANDSUM, SIMPPLLE, and VDDT. Gen Tech Rep RMRS-GTR-76-WWW. USDA Forest Service, Rocky Mountain Research Station, Ogden, UT, p 14

Bestelmeyer BT, Herrick JE, Brown JR et al (2004) Land management in the American Southwest: a state-and-transition approach to ecosystem complexity. Environ Manag 34(1):38–51

Blankenship K, Smith J, Swaty R et al (2012) Modeling on the grand scale: LANDFIRE lessons learned. In: Kerns B, Shlisky AJ, Daniel CJ (eds) First landscape state-and-transition simulation modeling conference, 14–16 June 2011. Gen Tech Rep PNW-GTR-869. USDA, Forest Service, Pacific Northwest Research Station, Portland, OR, pp 43–56

Blankenship K, Provencher L, Frid L et al (2013) Human dimensions of state-and-transition simulation model applications to support decisions in wildland fire management. In: Proceedings of 3rd Human Dimensions of Wildland Fire Conference, Seattle, WA, 17–19 April 2012. International Association of Wildland Fire, Missoula, MT, pp 28–33

Bradley BA (2009a) Regional analysis of the impacts of climate change on cheatgrass invasion shows potential risk and opportunity. Glob Change Biol 15(1):196–208

Bradley BA (2009b) Assessing ecosystem threats from global and regional change: hierarchical modeling of risk to sagebrush ecosystems from climate change, land use and invasive species in Nevada, USA. Ecography 33(1):198–208

Bradley BA, Curtis CA, Chambers JC (2015) *Bromus* response to climate and projected changes with climate change. In: Germino MJ, Chambers JC, Brown CS (eds) Exotic brome-grasses in arid and semi-arid ecosystems of the Western US: causes, consequences and management implications. Springer, New York, NY (Chapter 9)

Briske DD, Bestelmeyer BT, Stringham TK et al (2008) Recommendations for development of resilience-based state-and-transition models. Rangel Ecol Manag 61(4):359–367

Brown TJ, Hall BL, Westerling AL (2004) The impact of twenty-first century climate change on wildland fire danger in the western United States: an applications perspective. Clim Change 62(1–3):365–388

Caudle D, DiBenedetto J, Karl M, Sanchez H, Talbot C (2013) Interagency ecological site handbook for rangelands. http://jornada.nmsu.edu/files/InteragencyEcolSiteHandbook.pdf. Accessed 2 Jun 2015

Chambers J (2008) Invasive plant species and the Great Basin. In: Chambers J, Devoe N, Evenden A (eds) Collaborative management and research in the Great Basin – examining the issues and developing a framework for action. Gen Tech Rep RMRS-GTR-204. USDA, Forest Service, Rocky Mountain Research Station, Fort Collins, CO, pp 38–41

Chambers JC, Miller RF, Board DI et al (2014) Resilience and resistance of sagebrush ecosystems: implications for state and transition models and management treatments. Rangel Ecol Manag 67(5):440–454

Creutzburg MK, Halofsky JE, Halofsky JS et al (2014) Climate change and land management in the rangelands of central Oregon. Environ Manag 55:1–13

Creutzburg MK, Halofsky JS, Hemstrom MA (2012) Using state-and-transition models to project cheatgrass and juniper invasion in southeastern Oregon sagebrush steppe. In: Kerns BK, Shlisky AJ, Daniel CJ (eds) First landscape state-and-transition simulation modeling conference, 14–16 June 2011. USDA, Forest Service, Pacific Northwest Research Station, Portland, OR, pp 73–84

Czembor CA, Morris WK, Wintle BA et al (2011) Quantifying variance components in ecological models based on expert opinion. J Appl Ecol 48(3):736–745

Czembor CA, Vesk PA (2009) Incorporating between-expert uncertainty into state-and-transition simulation models for forest restoration. Forest Ecol Manag 259(2):165–175

Daniel CJ, Frid L (2012) Predicting landscape vegetation dynamics using state-and-transition simulation models. In: Kerns B, Shlisky AJ, Daniel CJ (eds) First landscape state-and-transition simulation modeling conference, 14–16 June 2011. Gen Tech Rep PNW-GTR-869. USDA, Forest Service, Pacific Northwest Forest and Range Experiment Station, Portland, OR, pp 5–22

ESSA Technologies Ltd. (2007) Vegetation dynamics development tool user guide, Version 6.0. ESSA Technologies Ltd, Vancouver

Evers L, Miller RF, Hemstrom M et al (2011) Estimating historical sage-grouse habitat abundance using a state-and-transition model. Nat Resour Environ Iss 17(1):115–129

Evers LB, Miller RF, Doescher PS et al (2013) Simulating current successional trajectories in sagebrush ecosystems with multiple disturbances using a state-and-transition modeling framework. Rangel Ecol Manag 66(3):313–329

Faraway JJ (2006) Extending the linear model with R: generalized linear, mixed effects and non-parametric regression models. Chapman Hall/CRC, Boca Raton, FL

Forbis TA, Provencher L, Frid L et al (2006) Great Basin land management planning using ecological modeling. Environ Manag 38(1):62–83

Frid L, Hanna D, Korb N et al (2013a) Evaluating alternative weed management strategies for three Montana landscapes. Invasive Plant Sci Manag 6(1):48–59

Frid L, Holcombe T, Morisette JT et al (2013b) Using state-and-transition modeling to account for imperfect detection in invasive species management. Invasive Plant Sci Manag 6(1):36–47

Frid L, Wilmshurst JF (2009) Decision analysis to evaluate control strategies for crested wheatgrass (*Agropyron cristatum*) in Grasslands National Park of Canada. Invasive Plant Sci Manag 2(4):324–336

Gelman A, Hill J (2007) Data analysis using regression and multilevel/hierarchical models. Cambridge University Press, New York, NY

Grant CD (2006) State-and-transition successional model for bauxite mining rehabilitation in the Jarrah forest of western Australia. Restor Ecol 14(1):28–37

Halofsky JE, Hemstrom MA, Conklin DR et al (2013) Assessing potential climate change effects on vegetation using a linked model approach. Ecol Model 266:131–143

Hann WJ, Bunnell DL (2001) Fire and land management planning and implementation across multiple scales. Int J Wildl Fire 10(3–4):389–403

Hardesty J, Adams J, Gordon D et al (2000) Simulating management with models: lessons from ten years of ecosystem management at Eglin Air Force Base. Conserv Pract 1(1):26–32

Hayes MJ, Svoboda MD, Wilhite DA et al (1999) Monitoring the 1996 drought using the standardized precipitation index. Bull Am Meteorol Soc 80(3):429–438

Heddinghaus T, Sabol P (1991) A review of the Palmer drought severity index and where do we go from here? 7th Conference on Applied Climatology, Salt Lake City, UT, 10–13 September 1991. American Meteorological Society, Boston, MA, pp 242–246

Hemstrom M, Ager AA, Vavra M et al (2004) A state and transition approach for integrating models. In: Hayes J, Ager AA, Barbour RJ (eds) Methods for integrated modeling of landscape change: interior northwest landscape analysis system. Gen Tech Rep PNW-GTR-610. USDA, Forest Service, Pacific Northwest Research Station, Portland, OR, p 218

Horn HS (1975) Markovian properties of forest successions. In: Cody M, Diamond JM (eds) Ecology and the evolution of communities. Harvard University Press, Cambridge, pp 196–211

Intergovernmental Panel on Climate Change (IPCC) (2013) Climate change 2013: the physical science basis: working group I contribution to the fifth assessment report of the intergovernmental panel on climate change. Cambridge University Press, Cambridge, UK and New York

Johnson CJ, Gillingham MP (2004) Mapping uncertainty: sensitivity of wildlife habitat ratings to expert opinion. J Appl Ecol 41(6):1032–1041

Keane RE, Cary GJ, Davies ID et al (2004) A classification of landscape fire succession models: spatial simulations of fire and vegetation dynamics. Ecol Model 179(1):3–27

Kelly AE, Goulden ML (2008) Rapid shifts in plant distribution with recent climate change. Proc Natl Acad Sci U S A 105(33):11823–11826

Knapp CN, Fernandez-Gimenez M, Kachergis E et al (2011a) Using participatory workshops to integrate state-and-transition models created with local knowledge and ecological data. Rangel Ecol Manag 64(2):158–170

Knapp CN, Fernandez-Gimenez ME, Briske DD et al (2011b) An assessment of state-and-transition models: perceptions following two decades of development and implementation. Rangel Ecol Manag 64(6):598–606

Knapp PA (1996) Cheatgrass (*Bromus tectorum* L.) dominance in the Great Basin Desert – History, persistence, and influences to human activities. Glob Environ Change 6(1):37–52

Kurz WA, Beukema SJ, Klenner W et al (2000) TELSA: the tool for exploratory landscape scenario analyses. Comput Electron Agric 27(1–3):227–242

Littell JS, McKenzie D, Peterson DL et al (2009) Climate and wildfire area burned in western U. S. ecoprovinces, 1916–2003. Ecol Appl 19(4):1003–1021

Low G, Provencher L, Abele S (2010) Enhanced conservation action planning: assessing landscape condition and predicting benefits of conservation strategies. J Conserv Plan 6:36–60

Maxwell BD, Lehnhoff E, Rew LJ (2009) The rationale for monitoring invasive plant populations as a crucial step for management. Invasive Plant Sci Manag 2(1):1–9

McCarthy M (2007) Bayesian methods for ecology. Cambridge University Press, Cambridge

Merzenich J, Kurz WA, Beukema SJ et al (1999) Long-range modeling of stochastic disturbances and management treatments using VDDT and TELSA. Society of American Foresters National Convention: Landscape Analysis session, Portland, OR, 14 September 1999

Miller C (2007) Simulation of the consequences of different fire regimes to support wildland fire use decisions. Fire Ecol 3:83–102

Monsen S, Stevens R, Shaw NL (2004) Restoring western ranges and wildlands. Gen Tech Rep RMRS-GTR-136-vol. 1. USDA, Forest Service, Rocky Mountain Research Station, Fort Collins, CO, pp 89–100

Moody ME, Mack RN (1988) Controlling the spread of plant invasions – the importance of nascent foci. J Appl Ecol 25(3):1009–1021

National Environmental Policy Act (NEPA) (1969) Enacted by the senate and house of representatives of the United States of America in Congress assembled: as amended, Pub. L. 91–190, 42 U.S.C. 4321–4347, 1 Jan 1970, as amended by Pub. L. 94–52, 3 Jul 1975, Pub. L. 94–83, 9 Aug 1975, and Pub. L. 97–258, § 4(b), 13 Sept 1982

Natural Resources Conservation Service (NRCS) (1998) National Forestry Manual. USDA. http://www.nrcs.usda.gov/Internet/FSE_DOCUMENTS/nrcs142p2_050983.pdf. Accessed 2 Jun 2015

Natural Resources Conservation Service (NRCS) (2003) National Range and Pasture Handbook, revision 1. USDA, Grazing Lands Technology Institute. http://www.nrcs.usda.gov/wps/portal/nrcs/main/national/landuse/rangepasture/. Accessed 2 Jun 2015

Nixon K, Silbernagel J, Price J et al (2014) Habitat availability for multiple avian species under modeled alternative conservation scenarios in the Two Hearted River watershed in Michigan, USA. J Nat Conserv 22(4):302–317

Nowak RS, Ellsworth DS, Smith SD (2004) Functional responses of plants to elevated atmospheric CO_2 – do photosynthetic and productivity data from FACE experiments support early predictions? New Phytol 162(2):253–280

Olsson AD, Betancourt JL, Crimmins MA et al (2012) Constancy of local spread rates for buffelgrass (Pennisetum ciliare L.) in the Arizona Upland of the Sonoran Desert. J Arid Environ 87:136–143

Palmer WC (1965) Meteorological drought. Research Paper No 45. US Department of Commerce, Office of Climatology, Weather Bureau Washington, DC, USA

Pennisi E (2010) Western US forests suffer death by degrees. Science 323(5913):447

Pimentel D, Zuniga R, Morrison D (2005) Update on the environmental and economic costs associated with alien-invasive species in the United States. Ecol Econ 52(3):273–288

Price J, Silbernagel J, Miller N et al (2012) Eliciting expert knowledge to inform landscape modeling of conservation scenarios. Ecol Model 229:76–87

Provencher L, Anderson T, Low G et al (2013) Landscape conservation forecasting™ for Great Basin National Park. Park Sci 30:56–67

Provencher L, Forbis TA, Frid L et al (2007) Comparing alternative management strategies of fire, grazing, and weed control using spatial modeling. Ecol Model 209(2–4):249–263

Quinn GP, Keough MJ (2002) Experimental design and data analysis for biologists. Cambridge University Press, Cambridge

Regan HM, Colyvan M, Burgman MA (2002) A taxonomy and treatment of uncertainty for ecology and conservation biology. Ecol Appl 12(2):618–628

Rehfeldt GE, Crookston NL, Warwell MV et al (2006) Empirical analyses of plant-climate relationships for the western United States. Int J Plant Sci 167(6):1123–1150

Rollins MG (2009) LANDFIRE: A nationally consistent vegetation, wildland fire, and fuel assessment. Int J Wildl Fire 18(3):235–249

Rumpff L, Duncan DH, Vesk PA et al (2011) State-and-transition modelling for adaptive management of native woodlands. Biol Conserv 144(4):1224–1236

Scheller RM, Domingo JB, Sturtevant BR et al (2007) Design, development, and application of LANDIS-II, a spatial landscape simulation model with flexible temporal and spatial resolution. Ecol Model 201(3–4):409–419

Smith SD, Huxman TE, Zitzer SF et al (2000) Elevated CO_2 increases productivity and invasive species success in an arid ecosystem. Nature 408(6808):79–82

Speirs-Bridge A, Fidler F, McBride M et al (2010) Reducing overconfidence in the interval judgments of experts. Risk Anal 30(3):512–523

Stringham TK, Krueger WC, Shaver PL (2003) State and transition modeling: an ecological process approach. J Range Manag 56(2):106–113

Tausch R, Nowak RS (1999) Fifty years of ecotone change between shrub and tree dominance in the Jack Springs pinyon research natural area. In: McArthur E, Ostler WK, Wambodt CL (eds) Shrubland Ecotones, Ephraim, UT. USDA, Forest Service, Rocky Mountain Research Station, Ogden, UT, pp 71–77

Taylor AH, Beaty RM (2005) Climatic influences on fire regimes in the northern Sierra Nevada Mountains, Lake Tahoe Basin, Nevada, USA. J Biogeogr 32(3):425–438

Thacker ET, Ralphs MH, Call CA et al (2008) Invasion of broom snakeweed (*Gutierrezia sarothrae*) following disturbance: evaluating change in a state-and-transition model. Rangel Ecol Manag 61(3):263–268

Twidwell D, Allred BW, Fuhlendorf SD (2013) National-scale assessment of ecological content in the world's largest land management framework. Ecosphere 4(8):94

Vavra M, Hemstrom MA, Wisdom M (2007) Modeling the effects of herbivores on the abundance of forest overstory states using a state-transition approach in the upper Grande Ronde River Basin, Oregon, USA. Landscape Urban Plan 80(3):212–222

Westerling A (2009) Climate change impacts on wildfire. In: Schneider M, Rosencranz A, Mastrandrea MD et al (eds) Climate change science and policy. Island Press, Washington, DC

Westerling AL, Bryant BP (2008) Climate change and wildfire in California. Clim Change 87:S231–S249

Westoby M, Walker B, Noy-Meir I (1989) Opportunistic management for rangelands not at equilibrium. J Range Manag 42:266–274

Chapter 14
Human Dimensions of Invasive Grasses

Mark W. Brunson and Halley Kartchner

Abstract Invasive species are problems because of people. Whether these species are introduced accidentally or purposefully, human activities inevitably influence their spread. Disturbance processes and control options are affected by economic, political, and social factors as well as by biological ones. To understand the dynamics of invasion and potential for resilience and resistance, one must also understand the role of human processes. In this chapter, we focus first on individual perceptions and behaviors, exploring how perceptions of exotic invasive species can vary such that one group of stakeholders may actively seek to eliminate an invader while others find it useful. We also describe how people perceive practices used to control exotic annual grasses or reduce their spread. We then shift from individual-level to institutional concerns, reviewing how exotic annual grasses have been treated in US laws and regulations and how environmental policies and politics may complicate restoration efforts. We also explore how voluntary control efforts operate alongside regulatory efforts. We discuss how education activities have affected perceptions, review strengths and weaknesses of different outreach approaches, and describe an educational approach that may prove useful for shifting attention toward annual invasive grasses from curiosity to concern to action. Finally, we discuss the critical role of trust and trust-building efforts in addressing invasive species issues across landscapes.

Keywords Attitudes • Collaborative management • Education • Regulatory policy • Trust

14.1 Introduction

Invasive species management is ultimately a social problem. It is humans who classify impacts of exotic annual grasses as "negative," and humans who have concluded that exotic species should be controlled because they are not native to a region. Society invests in invasive species management and control only when

M.W. Brunson (✉) • H. Kartchner
Department of Environment and Society, Utah State University,
Logan, UT 84322-5215, USA
e-mail: Mark.Brunson@usu.edu

M.J. Germino et al. (eds.), *Exotic Brome-Grasses in Arid and Semiarid Ecosystems of the Western US*, Springer Series on Environmental Management,
DOI 10.1007/978-3-319-24930-8_14

the costs of a species' continued presence and spread—including not only economic impacts but also less easily quantified impacts such as esthetics, wildfire risk, and loss of biodiversity—are great enough to warrant policy and management intervention. Humans also created those impacts in the first place. Humans have facilitated, if not intentionally initiated, the spread of virtually all exotic invasive grass species. They also directly or indirectly affect virtually all of the disturbance processes that influence the abiotic and biotic processes governing resilience and resistance to invasion (Fig. 1.1 in Germino et al. 2015b). Therefore to understand the dynamics of invasion and the options for management of exotic annual grasses, it is necessary to understand the human processes that influence invasion processes.

Science provides many insights regarding the biology and management of exotic invasive species. These insights have been helpful for developing strategies not only to treat infestations but also to reduce the likelihood of invasion in the first place. These strategies, however, require human action that is influenced by societal factors. Humans can choose to control weeds, but they also erect many of the greatest barriers to control—even when they may actually hope to see the elimination of exotic species. These barriers may occur via activities that unintentionally or unavoidably foster continued spread of an invader or through efforts to block implementation of control options because of real or perceived side effects. It is therefore important to understand the psychological and institutional factors that may lead to such paradoxical outcomes.

Exotic species invasions and management are not commonly studied by social scientists. Nonetheless, some published studies do exist, mainly by researchers working within agricultural and natural resource academic programs. These researchers have viewed the problem through various disciplinary lenses. Social-psychological research explores individual-level processes such as the relationships between weed knowledge, attitudes toward control, and personal behaviors toward exotic species. Sociology, the study of groups and institutions, offers insights into the functioning of voluntary organizations formed to address weed issues. Policy sciences help us understand how invasive species laws and regulations are formed and implemented by governmental bodies and why environmental and industry groups respond as they do to invasive species management proposals. In this chapter, we draw upon social science literature to improve understanding about the connections between social and environmental processes that influence invasive plant spread or control. We discuss how that understanding can inform efforts to improve control of exotic annual grasses through education programs and collaborative adaptive management. Our emphasis is on two exotic annual *Bromus* (*Bromus* hereafter) species, *Bromus tectorum* L. (cheatgrass or downy brome) and *Bromus rubens* L. (red brome), as we found no social science information specifically relevant to *Bromus diandrus* Roth. (ripgut brome) or *Bromus hordeaceus* L. (soft brome).

14.2 Beliefs, Attitudes, and Behaviors

A common topic for researchers who study the human dimensions of natural resource management centers on the beliefs, attitudes, and behaviors of people whose opinions or actions affect how land is managed. In the United States (USA), where citizens legally have a stake in the management of public lands, belief-attitude-behavior research is often intended to help land managers understand stakeholders' support or opposition to proposed actions, as well as to guide development of outreach efforts regarding the use of public lands. Because US Department of Interior agencies or the USDA Forest Service manage so much of the land affected by invasion of *B. tectorum* in the Intermountain West and *B. rubens* in the Southwest, researchers have studied how citizens or specific stakeholder groups view the types of management practices used to reduce impacts of exotic grass invasion. Generally, this research has been driven by the need to address concerns about the effect of *Bromus* invasion on wildfire cycles and conversion of native ecosystems to *Bromus* dominance. However, most research on exotic invasive plants has been broad in scope or focused on forb species. In this chapter, we discuss studies in the latter category that are potentially applicable to exotic annual grass management.

Before describing this research, it may be useful to review a few key concepts. Often a goal of applied social-psychological research is to promote a change in behavior toward a particular object (e.g., a person, place, concept, or activity). Because humans generally try to behave in ways that reflect their thoughts (Petty and Briñol 2010; Fishbein and Ajzen 2011), behavior change strategies often focus on influencing people's attitudes—i.e., favorable or unfavorable thought orientations—toward the object and/or behavior of interest. Attitudes, in turn, are expressions of people's values and beliefs. Values broadly express the importance that individuals place on appropriate outcomes or courses of action (what "should" be true), while beliefs are more specific expressions describing what an individual thinks "is" true. Beliefs can be *cognitive*, what we typically think of as factual knowledge; *normative*, related to one's perceptions of what other people believe should occur; or *behavioral*, pertaining to the perceived consequences of behavior (Trafimow and Sheeran 1998; Petty and Briñol 2010; Fishbein and Ajzen 2011). Because values are broadly focused and in adults tend to be stable and difficult to change, attitude change strategies typically work in either of two ways: (1) to convince people that a change in attitude would be more consistent with their values or (2) to change or add to their stockpile of beliefs about the object of interest. A necessary first step in the latter approach is to determine what people already believe about the object and why. Such research often, but not always, is combined with studies of attitudes and potential determinants of attitude change.

14.2.1 Beliefs about Exotic Invasive Plants and Management Options

Some scholars have argued that exotic invasive species impose huge conservation or economic costs on society (Wilcove et al. 1998; Mack et al. 2000; Pimentel et al. 2005, but see also Davis et al. 2011). However, when citizens consider the full range of risks to natural environments, exotic invasive species often do not rank very high. Slimak and Dietz (2006) surveyed members of the public as well as selected US environmental professionals and asked them to rank 24 ecological risk items from climate change to hazardous wastes to sport hunting and fishing. Among the lay public, invasive species ranked 19th, just behind overgrazing and ahead of damming rivers but well below the greatest perceived risks: hazardous waste sites and persistent organic pesticides. The mean rank was 15 for an "experienced public" sample who had participated in US Environmental Protection regional assessments. Professional risk assessors ranked invasive species as ninth most critical of the 24 risks, suggesting that beliefs about the threat of exotic invasive species are highly influenced by knowledge and experience. In another survey of persons to whom the exotic invasive species issue should be especially salient, just 62 % of Minnesota nursery and landscape industry professionals said invasive plants are an important issue (Peters et al. 2006). Economic concerns may have tempered risk assessments in this group, as just 69 % said they would not sell a plant if they knew it was potentially invasive, and 57 % said they would not sell an invasive plant even if they knew a competitor was selling it (Peters et al. 2006).

Looking more specifically at exotic species affecting wildlands in the Western USA, where *Bromus* invasion is of specific concern, Tidwell (2005) surveyed residents of 11 counties with active weed education programs in Arizona, Colorado, New Mexico, and Utah about their beliefs regarding invasive plants and their management. In this survey, 94 % of respondents indicated that they were concerned about invasive plants. When asked about specific threats posed by exotic species, the impacts of greatest concern were increased wildfire danger, reduced quality of wildlife habitat, and loss of native plants. A series of surveys in the Great Basin states of Idaho, Nevada, Oregon, and Utah (Shindler et al. 2011; Gordon et al. 2014) found that in 2010 more than 83 % of respondents believed invasive species pose a threat to healthy rangelands, a 9 % increase over 2006. Another survey in Utah and Idaho found that nearly all respondents were at least moderately concerned about negative effects of invasive plants, and most said controlling plants on their own properties was important (Christensen 2010).

In one of the few studies to ask specifically about *Bromus*, Kelley et al. (2013) measured ranchers and natural resource professionals' perspectives on managing *B. tectorum* in Colorado and Wyoming. Both groups thought the species is a problem in that area, although for the most part ranchers were less concerned than natural resource professionals. Overall, 51 % of ranchers in the sample perceived *B. tectorum* to be a moderate to severe problem, while 18 % felt it was not a problem at all; among natural resource professionals, 66 % perceived *B. tectorum* to be a moderate to severe problem, and just 6 % felt it was not at all a problem. The extent to which ranchers perceived *B. tectorum* to be problematic was aligned with the current distribution of the species, with regional differences linked to elevation, climate, and historic data. In all but one region, natural resource professionals expressed greater concern about *B. tectorum* than did ranchers (Kelley et al. 2013).

14.2.2 Attitudes concerning Exotic Invasive Plants and Their Management

Attitudes can derive from beliefs formed as a result of direct exposure to environmental changes or events or of indirect exposure (via personal contacts or media) that causes people to reevaluate their ideas or actions. The intensity and direction of attitudes are likely to vary depending on a number of personal and contextual factors, including the individual's assessment of catastrophic risk, attributions of blame for undesirable events, regularity of interaction with wildlands, and value assessments about society's right to manipulate natural environments and the nonhuman organisms found there.

The most common way to assess attitudes is to ask people what they think in surveys or interviews. Interviews are useful for obtaining nuanced information, since respondents can be asked on the spot to elaborate on answers they are given. However, they can be time-consuming so that it is difficult to gain information from larger numbers of people. Surveys are especially valuable for decision makers whose jobs require that they be responsive to a range of citizens or interest groups, such as elected officials or public land managers, because they make it possible to distinguish between the views of a typical member of a constituency vs. those of the most persistent or vocal constituents. Surveys are economical, straightforward, and if done by experienced survey writers using scientific sampling methods can be generalized to describe a population of interest such as a stakeholder group (e.g., farmers and ranchers, environmental group members) or region. Notable drawbacks of surveys are that they capture what people are thinking at a single moment in time and may not reflect what people think or do in the long term when faced with economic or time constraints, peer influences, and other limiting factors. This may be especially true in the case of a topic that is not central to their everyday lives, so that their attitudes are not well formed. In such cases, survey respondents may be prone to "social desirability bias," in which they provide the answers they believe the questioner wants to hear or that they perceive are more aligned with a relevant social group. Also, it is important when assessing survey data not to extrapolate results beyond the limited focus of the survey itself.

Tidwell (2005) asked his Southwestern US respondents whether they agreed or disagreed with a statement that "the government isn't doing enough to control invasive plants" and found that 71 % agreed. Perhaps surprisingly, when asked whether this should occur even if it means higher taxes or increased regulation, more people agreed than disagreed. However, he also found that respondents believed control should only be attempted if the control method poses little or no risk to the environment. Respondents were 3.5 times more likely to believe safety should be the most important factor in choosing a control method than to believe effectiveness should be foremost.

Debates about exotic invasive plant management often revolve around the use of specific control options, especially herbicides. Different publics (i.e., specific communities of people who share common interests or values, whether organized as groups or not) view herbicide application differently. For example, Norgaard (2007) examined a controversy over control of *Centaurea maculosa* (spotted knapweed; syn. *Centaurea stoebe* L. ssp. *micranthos* (Gugler) Hayek) in Northern California and found that USDA Forest Service employees considered herbicide application to be safe while community members did not. Similarly, Shindler et al. (2011) reported that residents of rural parts of the Great Basin were more than twice as likely as urban respondents to believe herbicide application is "a legitimate tool that managers should be able to use [for sagebrush ecosystem protection] wherever they see fit."

Tidwell (2005) asked respondents to his Southwest survey about the acceptability of using chemical, mechanical, biological, and cultural controls on invasive plants. Cultural and mechanical approaches were the most likely to be judged acceptable, with biological controls next. Chemical control was the only approach judged to be moderately or highly acceptable by fewer than half of respondents. Tidwell also asked about whether each of those approaches was acceptable in different contexts: in protected areas such as national parks or wilderness areas, on multiple use public lands, on agricultural lands, and on properties adjacent to residential areas. Chemical control was judged more acceptable for use on multiple use or agricultural lands than on protected or residential lands, while there was no difference in support across contexts for biological, cultural, or mechanical approaches.

The two Great Basin surveys found that there is greater support among both urban and rural publics for using livestock grazing for controlling *B. tectorum* than herbicide application. There are no programs currently in the Great Basin to widely spray herbicides to kill existing patches of *B. tectorum*, but herbicide is sometimes applied during reseeding after a wildfire to suppress annual grasses while allowing other plants to take hold. Between 2006 and 2010, support for widespread use of livestock grazing to keep *B. tectorum* in check and reduce fuel loads rose slightly from 57 % to 61 % of respondents, while support for widespread use of herbicides remained under 20 % (Gordon et al. 2014). Because the same individuals completed both surveys, Gordon et al. (2014) could measure the stability of attitudes over time. Attitudes toward livestock grazing were more stable, with 64 % of individuals giving the same response each time. Attitudes toward herbicide application were less so; while 44 % of respondents gave the same answer in 2010 as in 2006, 32 % had a lower opinion of herbicides in 2010, while 24 % had grown more supportive of their use.

Gordon et al. (2014) also identified predictors of support for livestock grazing and herbicide application as vegetation management tools. In 2006, those who supported grazing were more concerned about the health of Great Basin rangelands and the threat posed by exotic invasive plants and less likely to believe that overgrazing and wildfire pose threats to rangelands. They also were more likely to hold utilitarian versus protectionist values toward environmental stewardship. But the most important influence was whether they trusted federal agencies' ability to use grazing effectively as a tool. Respondents in 2006 were more likely to find herbicide application acceptable if they were less concerned about livestock impacts, held values that emphasize the use of public lands over preservationist management, lived in rural areas, and trusted agencies' ability to use herbicides effectively. In 2010, the influence of trust in agencies became stronger, and environmental values remained an influence on acceptance of both practices, but beliefs about condition of and threats to rangelands were no longer influential.

The finding that people are more likely to support using vegetation management tools if they lean toward a utilitarian perspective on land management points to an important influence on attitudes toward exotic invasive grasses—whether one believes they have economic benefit. For example, Marshall et al. (2011) interviewed pastoralists across Australia about the exotic invader *Pennisetum ciliare* (L.) Link (buffel grass) and found they were highly dependent on the species as forage. Eighty percent of those surveyed believed "there is no need to worry about buffel grass from an environmental point of view." The authors concluded that the potential for changing cattle growers' attitudes toward *P. ciliare* grass was low. Similarly, *B. tectorum* may provide the vast majority of forage for cattle in the Great Basin at certain times of year (Brunson and Tanaka 2011). While a study similar to Marshall and colleagues' has not been conducted in the USA, Idaho ranchers have stated in workshop settings that they believe *B. tectorum* can be useful as an early season forage and thus may not be interested in controlling the species at all locations (L. Salo, pers. comm.).

14.2.3 Attitude Change Research

Typically land managers want to know about attitudes because they know or suspect that they may want to try to change those attitudes. Invasive species program managers often seek ways to heighten awareness and concern about exotic invasive plants in hopes of promoting willingness to support or engage in control activities. Public land managers anticipating citizen opposition to a control action such as herbicide application likewise may want to gauge attitudes and their underlying causes in order to launch efforts to reduce that opposition. Once such efforts have occurred, it is useful to measure whether people's attitudes have, in fact, changed.

One way to assess attitude change is to infer it through behavior change. Marler et al. (2005) noted that active public education efforts preceded passage of a property tax levy in Missoula County, Montana, for exotic species control. They suggested that "[t]he fact that Missoula residents would vote to tax themselves in support of invasive plant management reflects the amount of public support that has developed." Participation in volunteer weed control efforts is another example, although these typically have revolved around broadleaf herbaceous weeds rather than exotic invasive grasses.

Another approach to measuring attitude change is to assess attitudes expressed by the same individuals at two points in time. Gordon et al. (2014) completed the only study we could find that has done so in the context of invasive species management. As noted previously, the authors found that while overall acceptance of the use of vegetation management practices remained relatively steady over time, many individuals gave different responses in 2010 than they had in 2006. To assess influences on those changes, the authors statistically identified factors that could predict whether a person's acceptance of livestock grazing or herbicide application would change over that period. For livestock grazing, acceptance increased among respondents who also gained greater trust in land managers' ability to use grazing as an effective management tool, but no other factor was influential. Conversely, among respondents whose trust in the management ability of public land agencies had declined between 2006 and 2010, acceptance of grazing as a way to reduce unwanted plants also declined. However, changes in acceptance for grazing were not related to changes in beliefs about the threat posed by exotic plant invasion. For herbicide application, attitude change also was positively correlated with change in trust in managers' ability to use herbicides effectively, but respondents were also more likely to increase their support for herbicide application if they had grown more concerned about the threat posed by exotic plant invasion.

14.2.4 Reported Behaviors and Behavioral Intentions with Respect to Invasive Annual Grasses

Tidwell asked his Southwestern US respondents about their willingness to volunteer for weed management activities including control, monitoring, education, and restoration (Tidwell and Brunson 2008). Ten percent of respondents said they had engaged previously in weed-related volunteer activities—for example, as members of the Tucson-based Sonoran Desert Weedwackers, which conducts mechanical control of *P. ciliare* and engages in education and restoration—and 43 % expressed willingness to participate. Respondents who displayed higher knowledge about weed impacts, weed species, and factors influencing weed spread were more likely to express willingness to volunteer, especially for control activities such as hand-pulling weeds. When asked about concern for specific impacts, respondents who were willing to volunteer had slightly higher levels of concern about impacts on wildlife and recreation, but not about impacts on native plants, crop values, wildlife danger, or forage for livestock. Those willing to volunteer were 8 years younger on average, but did not differ from those who were not willing in terms of gender, education, property ownership, length of local residency, or income.

Tidwell and Brunson (2008) also asked which sorts of activities the willing-to-volunteer group would prefer to join. Direct participation in control (57 %) and monitoring of weed spread (55 %) were preferred activities over education (39 %) or restoration (38 %). Different activities attracted different individuals. For example, persons interested in control tended to be younger and quite knowledgeable;

those interested in education also tended to be more knowledgeable, with at least one university degree; those interested in monitoring were more concerned about weeds than disinterested respondents, but did not differ in terms of demographics or knowledge; and those interested in restoration were well educated and more likely to have urban backgrounds.

A recent study of federal land managers assessed influences on their willingness to use volunteers in ecosystem restoration, including activities to control and prevent exotic plant invasion (Bruce et al. 2014). About two-thirds of respondents were either positive or neutral toward the idea of using volunteers in restoration. Those who intended to use volunteers were more interested in increasing community support for their efforts, liked working with motivated individuals, and believed they could get more work done that way. They also were less likely to be concerned about constraints such as time, access to trained volunteers, safety, and funding.

Little is known about the extent to which people attempt to control exotic annual grasses on their own properties. Christensen (2010) asked respondents in four small farm communities in Northern Utah and Southern Idaho about prevention strategies they used on their own properties. The most commonly used were limiting soil disturbance (44–54 % depending on the community), limiting off-road vehicle access to their properties (44–51 %), and cleaning vehicles and equipment (28–56 %). Christensen also asked landowners about barriers to their use of prevention strategies, finding that the most important factors were lack of financial resources (41–50 %), knowledge (30–50 %), and lack of ability to see visible results (30–47 %). This is consistent with theories about innovation adoption in agriculture which suggest that landowners are less likely to adopt new practices if time or income are limited and if benefits of adoption are not easily observed (Peterson and Coppock 2001; Didier and Brunson 2004; Pannell et al. 2006).

In a survey of Utah "ranchette" landowners who own 2–50 acres of rangeland, Brunson and Kalnicky (unpublished) found that the likelihood of reporting actions to control exotic weeds was higher in agricultural counties (61 %) than areas with higher proportions of respondents who commuted daily to urban areas or who owned their property chiefly for recreation use. About half of those who reported controlling exotic weeds had done so by using grazing to reduce "fine fuels" (i.e., plant materials that dry readily, such as grass, twigs, or pine needles, and thus allow a fire to spread). Respondents were more likely to control exotic plants if they had larger properties and had lived on the land for more than 10 years.

14.3 Policies and Institutions

Management and control of exotic annual grasses require collective action. While individuals make decisions about their own lands and their beliefs may influence decisions by government officials, the success of weed management ultimately rests with institutions (governments, professional organizations, advocacy and education groups) whose decisions are influenced by social and political forces. Therefore, policy scientists and sociologists also have provided valuable insights that help us understand the human dimensions of invasive annual grasses.

14.3.1 Policy Influences on Exotic Annual Grass Management

A number of US federal laws direct government action against exotic invasive species. Many of these are directed toward particular species or purposes, for example, the Asian Carp Prevention and Control Act of 2010 or the provision in the Safe, Accountable, Flexible, Efficient (SAFE) Transportation Equity Act of 2005 that authorized use of federal highway grants to states for activities that control invasive weeds. The Noxious Weed Control and Eradication Act (Public Law 108-412, 30 Oct 2004) directed the Secretary of Agriculture to establish a program that provides assistance to eligible weed management entities for control and eradication of noxious weeds on both public and private land. These entities can include Indian tribes, states, local governments, specially created districts, or private organizations that were created to control or eradicate weeds and/or have demonstrated expertise and experience in weed management or education.

Not only is it federal policy to provide funds to other entities that have weed management responsibilities, but federal land management agencies were directed by a presidential order (Executive Order 13148 Section 601(a), April 21, 2000, "Greening the Government Through Leadership in Environmental Management") to use relevant programs on their own lands to prevent invasive species introductions, detect and respond rapidly to new infestations, monitor invasive species populations, restore native species in invaded ecosystems, conduct research on invasive species, and promote public education about them. This is particularly important for the management of *B. tectorum*, which has invaded tens of millions of acres of federal land, especially those managed by the Bureau of Land Management (BLM), a US Department of Interior (USDI) agency. Within USDI, every bureau has a responsibility for managing invasive species. The US Geological Survey conducts and supports research on invasive species, including *Bromus* species. The Bureau of Indian Affairs supports tribal government efforts. The BLM, National Park Service, Bureau of Reclamation, and US Fish and Wildlife Service all maintain invasive species management programs and also partner with states, tribes, and the private sector to support prevention and control efforts. Similarly, the Forest Service, an agency of the US Department of Agriculture, maintains its own detection, monitoring, control, eradication, education, and research programs while entering into partnerships with other entities. Also, the Department of Defense maintains an environmental branch that engages in invasive species control on military installations. All of these agencies take action to reduce the spread and influence of invasive annual grasses, especially *B. tectorum.*

One of the most common policy mechanisms to direct attention toward invasive plants in the USA is to declare a species as a *noxious weed*, defined in the Plant Protection Act of 2000 (Public Law 106–224, 20 Jun 2000) as "any plant or plant product that can directly or indirectly injure or cause damage to crops (including nursery stock or plant products), livestock, poultry, or other interests of agriculture, irrigation, navigation, the natural resources of the United States, the public health, or the environment." Under this law, the Secretary of Agriculture has the authority to inspect, seize, and destroy products and quarantine areas to contain or limit the spread of such plants, including authorization of biological control.

In addition, all but four states have developed their own noxious weed lists, as have hundreds of US counties. The federal government, as of this writing, had designated 104 species as noxious, but no *Bromus* species is among them. Of the 46 states with noxious weed lists, only three have declared a *Bromus* species as noxious: *B. tectorum* is classified as noxious in Colorado and Connecticut, and Arkansas has listed *Bromus racemosus* L. (bald brome) and *Bromus secalinus* L. (rye brome).

The lack of noxious weed designation for *Bromus* species can create a barrier to control efforts. Kelley et al. (2013) surveyed land managers in Wyoming and found that the lack of a state noxious weed designation for *B. tectorum* was cited as a barrier to management by 34 % of ranchers and 39 % of natural resource professionals. Lack of a designation means funds cannot be allocated under the Noxious Weed Control and Eradication Act of 2004 if they are to be used solely for control of *B. tectorum* or other *Bromus* species. As one manager in Wyoming stated in a focus group conducted by Kelley et al., "One interesting thing with regard to cheatgrass for me is that my budget is such that, I get money to treat noxious weeds and as of yet, cheatgrass isn't a noxious weed" (note: *B. tectorum* is listed as a noxious weed by five Wyoming counties). The same study found that 77 % of ranchers felt other weeds were a higher priority than *B. tectorum.*

When it comes to regulatory policy, all exotic invasive grass species are not created equal. *B. tectorum* is almost universally recognized as an undesirable species. Novak and Mack (2001) state that the first report of *B. tectorum* in North America came in 1790 from Pennsylvania. Source populations were largely European, and introduction appears to have been unintentional, with spread closely matching patterns of movement by European immigrant farmers. Likewise, *B. rubens* first appeared in herbarium collections in California in 1879 and appears to have been introduced through seeds brought unintentionally from Europe by immigrants during the Gold Rush era (Salo 2005). It was seeded near Tucson from 1906 to 1908 for evaluation of its use as a forage crop, but soon escaped and became established along the Santa Cruz River (Reid et al. 2008). Like *B. tectorum*, *B. rubens* has been recognized as invasive for many decades.

On the other hand, the perennial C3 grass, *Bromus inermis* Leyss. (smooth brome), was brought to North America in the 1880s from Hungary and Russia for soil retention and to provide forage for livestock (Larson et al. 2001). Many authors recognize its tendency to escape from planted habitats in some ecosystems and now consider it as invasive (e.g., D'Antonio and Vitousek 1992; Blankespoor and Larson 1994; Larson et al. 2001; Dillemuth et al. 2009). It is categorized as an invasive species in some states and provinces as well as in the Global Invasive Species Database (www.issg.org/database/welcome). Yet it continues to be recommended as a forage grass and for erosion control by authorities in some other states;

for example, a publication downloaded in February 2014 from the University of Missouri Extension website notes that *B. inermis* "can be a valuable cool-season grass, especially in conjunction with a legume such as alfalfa" (Roberts and Kallenbach 2000). Current US Department of Agriculture information takes a middle path, offering advice on how to establish and manage *B. inermis* while cautioning "this plant may become invasive" and advising readers to consult local specialists before planting (USDA 2002).

The reasons for these policy discrepancies can likely be ascribed to either of two factors: utility and futility. The utility explanation acknowledges that *Bromus* species, while often invasive in North America, also can have beneficial uses under specific circumstances. While *B. rubens* is generally considered of little value to livestock and big game, *B. inermis* has forage value as noted above. *B. tectorum* is nutritious when young and palatable to a wide range of ungulates, although its short green period and variable productivity make it a poor choice for forage and its flammability offsets its value for many users (Austin et al. 1999; Germino et al. 2015a). The futility explanation acknowledges that when a species becomes too common, it may be more useful to focus precious control resources on species that are more amenable to local eradication. *B. tectorum*, in particular, is so widespread that eradication is often impracticable, even on a local basis, and control strategies instead focus on minimizing its impact and dominance.

14.3.2 Emerging Institutions for Exotic Invasive Species Management

A number of collaborative groups have arisen in recent years around exotic plant management. These volunteer groups (e.g., Nevada's Spring Creek Weed Action Team (SWAT), Arizona's Sonoran Desert Weedwackers, and Montana's Gallatin/Big Sky Noxious Weed Committee) typically involve multiple public agencies as well as citizens groups that collaboratively engage volunteers in mapping, monitoring, and controlling exotic invasive plant infestations. Some are focused on all species in a specific locale, while others are organized around particular taxa, but nearly all are geographically limited and supported by a combination of public and private funds. Such collaboratives are increasingly common to address all sorts of land management issues. Fernandez-Gimenez et al. (2004) identified more than 100 collaboratives working in Arizona's rangelands alone, although only a small subset of those focused on exotic plants. Such groups are increasingly popular because they can accomplish tasks that are not covered by limited tax revenues, can inform management decisions with stakeholder perspectives, and can breach communications barriers between groups of people (e.g., ranchers and environmentalists) to achieve goals of common interest. However, sustaining volunteer effort over time can be challenging, and groups often must confront barriers due to inadequate funding and problems negotiating the complex rules of agency bureaucracies.

Weed programs can take on various organizational structures. Working alongside federal and state government entities and grassroots groups such as those described above, most county governments maintain weed programs, and some locations have cooperative weed management areas that encompass multiple government entities and/or weed districts created by public vote. Hershdorfer et al. (2007) surveyed coordinators of 53 local weed programs in Arizona, Colorado, New Mexico, and Utah to determine how attributes of the programs were linked to performance of control, education, monitoring, and integrated weed management. They found that programs that used volunteers did more monitoring but less direct control than those that relied entirely on paid employees. Contrary to the researchers' expectations, more regulatory action did not translate to better control. In fact, groups that had regulatory authority but generally refrained from punitive enforcement treated more infestations, partly because staffs are typically small and enforcement takes time and partly because a gentler approach seems to yield better results with private landowners.

One potential drawback of mobilizing citizen groups in addition to government entities is that citizen-driven collaborative efforts are more likely to address problems that are most visible to citizens. For example, Stevens and Falk (2009) have noted that efforts to treat *P. ciliare* populations are concentrated in high-visibility locations such as national parks and research areas. Such locations tend to be relatively low-productivity sink habitats, while higher-productivity source locations such as roadways or urban unmanaged sites receive relatively little attention even though they may contribute the most to future invasions. One reason for the focus on national parks is likely to be the continued involvement of the Sonoran Desert Weedwackers organization which began as a partnership with Saguaro National Park and Pima County's Tucson Mountain Park, although other weedwackers' groups have since emerged in the Tucson and Phoenix metropolitan areas.

14.4 Education

Education is invariably part of exotic invasive plant management strategies. Heightened public awareness not only leads to direct participation in management efforts (Marler et al. 2005) but ultimately influences policy, either by increasing pressure on existing authorities to improve performance or by leading to creation of new organizations (Hershdorfer et al. 2007). Public outreach is relatively low cost and can take a variety of forms, from printed materials and electronic resources such as websites or videos available to direct engagement of citizens in activities (DiTomaso 2000). For example, Marler et al. (2005) report using a suite of educational and citizen engagement efforts in Missoula that included stewardship opportunities (Adopt-a-Switchback on a popular trail; a Prairie Keepers program that organized activities that included weed pulls, seed collecting, and K-12 education); an annual "weed fair" education project drew significant attention in the community, and a Grow Native project engaged junior high school students in restoration activities. Strategies that directly engage learners, either through active participation or involvement in discussion, have been shown to work better than unidirectional or rote-learning approaches for increasing knowledge (DiEnno and Hilton 2005).

Jordan et al. (2011) found participation in a 3-day program that included exotic invasive plant education and data gathering led to an increase in knowledge but not to a change in reported behavior toward invasive plants. Crall et al. (2013) found higher levels of content knowledge among persons who had attended an 8-hour training program conducted by the National Institute for Invasive Species Science, as well as significant changes in intended behavior regarding invasive species, but they were not able to measure whether there were actual changes in behavior. This suggests that to achieve the ultimate goal of increasing management actions, careful attention must be paid to audience and curriculum design. Two such curricula have been created as part of the Ecologically-Based Invasive Plant Management (EBIPM) program, instituted by the USDA Agricultural Research Service to provide science-based solutions to annual grass invasions. EBIPM has held field schools at research and demonstration sites throughout the Great Basin since 2009 (Call et al. 2012). These events are held on ranches as well as public lands and are designed to help participants directly observe the effects of different management strategies. Additional educational opportunities include workshops and both onsite and virtual field tours. Participants have created high school and university curricula that may be found online at http://www.ebipm.org/.

The university curriculum (Kartchner 2013) offers an example of how an education program can be structured when direct engagement is not always feasible. EBIPM offers a decision-making framework for landowners and managers; therefore, the curriculum covers each of the decision steps in separate modules (Table 14.1) that include a synoptic reading, case studies, in-class and field activities, review questions, additional resources, and a PowerPoint presentation. The modular curriculum was developed with collaborators who included plant ecologists, weed scientists, social scientists, economists, range managers, and media developers. As of this writing, implementation of the curriculum had only just begun, so there was not yet any opportunity to assess its effectiveness.

14.5 Management Implications

The studies reviewed in this chapter focus on the potential to mobilize members of the public in ongoing efforts to support agencies' monitoring, control, and restoration practices for exotic invasive species and to stimulate active citizen participation in management activities on public lands or their own properties. Most of this research focuses on exotic forb (herbaceous broadleaf) species that are more visible to the public and often more easily controlled. Although exotic annual grass control can be more difficult in some respects, reducing the extent and impact of *Bromus* invasion is an especially important issue for citizens as well as land managers

Table 14.1 Subject matter covered in six modules of the EBIPM university curriculum

Module	Content elements
1: Introduction to ecologically based invasive plant management	□ Traditional vs. ecologically based weed management □ Overview of EBIPM approach and framework
2: Assessment of ecological conditions and processes in need of repair	□ Importance of rangeland health assessment (RHA) in planning process □ RHA protocol and underlying ecological concepts □ Integration of RHA and successional weed management
3: Identifying the underlying causes of plant community change	□ Overview of causes of succession/ecological processes □ How causes/processes influence invasion/restoration
4: Ecological principles for invasive plant management	□ Principles provide a bridge between theory and practice □ Principles guide selection of tools and strategies
5: Tools and strategies: managing site availability, species availability, and species performance	□ Prevention, control, and restoration strategies □ Biological, chemical, mechanical, and cultural tools □ Integration of tools and strategies
6: Adaptive management	□ Managing complex problems in the face of uncertainty □ Management as an experiment: an eight-step process

because of the role these species play in increasing the size, frequency, and intensity of wildfires. The research described in this chapter offers insights into how citizens currently perceive exotic species management and also illuminates potential avenues for enlisting the public more effectively in invasive species programs.

An initial conclusion one can draw from this research is that, while there is continued need for improved education and awareness, Americans do agree that a problem exists. The American public is concerned about invasive species—especially in locations such as the Great Basin where the impacts of *B. tectorum* invasion pose a particularly grave threat—but exotic invasion does not rise to the top of most people's lists of environmental threats. Accordingly, it may prove more difficult to keep public attention focused on invasion risks without specific evidence of immediate threats, as may be seen in the Great Basin, where *Bromus*-fueled wildfires provide vivid reminders of that threat each summer.

Surveys in the Southwestern USA found that people are concerned enough to want to see authorities take more action against invasive species. They do not always agree, however, on what action is appropriate. Many Americans are greatly concerned about the use of herbicides—often the most effective tool in the invasive plant management in terms of both cost and control efficacy—and even those who recognize the value of chemicals may have strong opinions about where those can be used safely. A problem specific to exotic grasses is that some stakeholders, most notably livestock operators, may see a benefit to *Bromus* invasion because livestock readily forage on *B. tectorum* and *B. inermis* under some conditions.

A factor that strongly influences citizen support for invasive plant management strategies is their trust in the organizations responsible for implementing those strategies. Willingness to accept a practice (e.g., herbicide application) in principle does not necessarily imply that people believe agencies can implement it safely or effectively. Trust in government has been declining in the USA for decades (Alford 2001), and that growing distrust can be directed toward public officials at all levels of government. Efforts to increase public support for invasive species management almost certainly must include activities designed to increase trust in the agencies responsible for that management.

Furthermore, cost is an issue. Although citizens may want government to "do more" about exotic invasive plants and in some cases have been willing to increase their tax burden to address weed problems, as was the case in Missoula, Montana, invasive species management must compete with many other demands for tax dollars. When revenues are scarce, managers are likely to focus on species that carry state, federal, or local designations as noxious weeds. This can negatively affect efforts to control exotic annual grasses, as these typically are not found on noxious weed lists, perhaps because in some places they are too ubiquitous and in other places not all stakeholders consider them to be sufficiently problematic.

An obvious part to the solution to this dilemma is education at all levels, from secondary schools to universities to extension programs and field schools. However, the best education strategies are not always the easiest to implement. It may prove easier to find one-time funds for printing a brochure than to secure ongoing resources to support efforts involving direct interaction with the public. However, educational efforts will be most successful if they directly engage the audience through discussion or, even better, active participation. The latter sort of educational activity can be time-intensive but is more likely to pay off. Careful attention to the design of educational strategies is important. Success in education also is likely to be greater when there are multiple pathways to reach citizens, including both asynchronous or unidirectional media (websites, videos) as well as opportunities for active engagement.

How does one surmount these barriers of cost, distrust, lack of knowledge, competition for resources, and competition for public attention? The best answer may lie with the emergence of multiparty collaborative efforts for exotic invasive plant management. These efforts allow for cost-sharing across institutions (different levels of government as well as nongovernmental organizations [NGOs]), can leverage funds that may not otherwise be available to government but are provided to NGOs, and often reduce costs by relying partially on volunteer labor.

One of the more encouraging findings from the studies we reviewed is that citizens want to volunteer to engage in invasive plant management. The voluntary sector is one of the fastest growing in North America (Rekart 2011), and research suggests that volunteering to benefit natural environments is especially popular. Survey respondents expressed willingness to participate in all aspects of a cooperative adaptive weed management strategy, from planning to monitoring to control to restoration to education. It is worth noting, however, that volunteer labor is not free. Organizations must be willing to invest in volunteer coordination and nonmonetary rewards and to find employees who enjoy working with citizens to achieve mutual goals.

One benefit of working within collaborative groups is that they provide the sort of participatory engagement that is said to be most effective for education. While we have noted that active education is more time-consuming and possibly more costly than unidirectional approaches, those costs can be offset to some extent by engagement of volunteers as educators as well as targets of educational efforts. Compared to individuals who receive information through casual encounters, as at a museum, those who are willing to allocate time to participatory engagement are more likely also to be attuned to the educational messages, increasing the likelihood that those messages will lead to changes in behavior.

A particularly valuable aspect of collaborative group processes is that they build trust. There is growing evidence that trust in people and in institutions depends more on how citizens feel about process issues (fairness, equity, or ability to participate) rather than about outcomes (Shindler et al. 2002; Van Ryzin 2011). Negative judgments about a decision—for example, to use herbicides or remove existing vegetation as part of a restoration project—are most likely to change when the decision process is transparent and the affected parties understand the trade-offs considered by the managers. Lack of trust in agencies often results from poor communication between agencies and the local community (McCaffrey 2004). Forming and maintaining collaborative management groups can address these issues by engaging interested and affected parties in decisions at all stages of the weed management process and by educating stakeholders through active engagement both prior to and during management activities.

14.6 Research Needs

Relatively little research has been conducted on the human dimensions of exotic invasive plants in North America. Only a small proportion of that work addressed the specific issue of exotic annual grasses such as *Bromus* species. In fact, only the recently published study by Kelley et al. (2013) focused specifically on attitudes toward management of an invasive annual grass (*B. tectorum*), and that work took place in a region where concern is increasing along with the spread of exotic species. Research in the Great Basin, where the problem is more widespread, addressed invasion more generally, although we can anticipate that most survey respondents answered with *B. tectorum* in mind because that species poses the greatest threat.

The most obvious human dimensions research need, therefore, is for studies that focus on public knowledge and attitudes regarding exotic annual grasses. We know less than we should about how stakeholders perceive options for restoration of exotic grass-invaded areas and virtually nothing about public awareness of other problematic species such as *B. rubens*, *B. hordeaceus*, *B. diandrus*, and *P. ciliare*, each of which poses as grave a threat as *B. tectorum* over their more restricted invasion ranges, nor do we know how stakeholders perceive *B. inermis*, which is considered invasive by some scientists and managers while it continues to be recommended for planting by others. Further research is needed as well about factors that underlie the knowledge and attitudes of natural resource professionals, who may differ among themselves on issues such as the role of livestock grazing in invasion and restoration or the advisability of using nonnative plants in post-wildfire rehabilitation.

We suggested that education and collaboration are important components of an integrated exotic invasive grass management strategy. However, our recommendations are based more on theory and extrapolation than on direct observation of the effectiveness of such efforts. Evaluations of educational strategies such as the EBIPM curricula are needed. Also, while active engagement in invasive plant management has been shown to improve citizen's knowledge, results of specific programs are mixed—Jordan et al. (2011) did not find improvement in the use of practices that could reduce invasion or control infestations, whereas Crall et al. (2013) found after a different training program that participants intended to change practices and hopefully did so. It is important to understand which sorts of educational efforts are more likely to lead to behavior change, as well as the degree to which external barriers prevent behavior change even when knowledge increases.

Similarly, while it is often argued that participation in collaborative natural resource management efforts build trust, this has not been tested in the specific situation of exotic invasive plant management. Research is needed that can identify the elements or procedures of collaborative programs that are most closely associated with improved trust, as well as improved success in detection, control, and restoration.

References

Alford JR (2001) We're all in this together: the decline in trust in government, 1958–1996. In: Hibbing JR, Theiss-Morse E (eds) What is it about government that Americans dislike? Cambridge University Press, Cambridge, pp 28–46

Austin DD, Stevens R, Jorgensen KR et al (1999) Preferences of mule deer for 16 grasses found on Intermountain winter ranges. J Range Manag 47:308–311

Blankespoor GW, Larson EA (1994) Response of smooth brome (*Bromus inermis* Leyss.) to burning under varying soil moisture conditions. Am Midl Nat 131:266–272

Bruce MC, Newingham BA, Harris CC et al (2014) Opinions toward using volunteers in ecological restoration: a survey of federal land managers. Restor Ecol 22:5–12

Brunson MW, Tanaka J (2011) Economic and social impacts of wildfires and invasive plants in American deserts: lessons from the Great Basin. Rangel Ecol Manag 64:463–470

Call C, Smith B, Kartchner H et al (2012) Advancing EBIPM through education. Rangelands 34:57–63

Christensen SD (2010) Establishing weed prevention areas and evaluating their impact. MS thesis, Utah State University

Crall AW, Jordan R, Holfelder K et al (2013) The impacts of an invasive species citizen science training program on participant attitudes, behavior, and science literacy. Public Underst Sci 22:745–764

Davis MA, Chew MK, Hobbs RJ et al (2011) Don't judge species on their origins. Nature 474:153–154

D'Antonio CM, Vitousek PM (1992) Biological invasions by exotic grasses, the grass/fire cycle, and global change. Ann Rev Ecol Syst 23:63–87

Didier EA, Brunson MW (2004) Adoption of range management innovations by Utah ranchers. J Range Manag 57:330–336

DiEnno CM, Hilton SC (2005) High school students' knowledge, attitudes, and levels of enjoyment of an environmental education unit on nonnative plants. J Environ Educ 37:13–25

Dillemuth FP, Rietschier EA, Cronin JT (2009) Patch dynamics of a native grass in relation to the spread of invasive smooth brome (*Bromus inermis*). Biol Invasions 11:1381–1391

DiTomaso JM (2000) Invasive weeds in rangelands: species, impacts, and management. Weed Sci 48:255–265

Fernandez-Gimenez ME, LeFebre S, Conley A et al (2004) Collaborative stewardship of Arizona's rangelands: making a difference in resource management. Rangelands 26:24–30

Fishbein M, Ajzen I (2011) Predicting and changing behavior: the reasoned action approach. Taylor & Francis, London

Germino MJ, Belnap J, Stark JM et al (2015a) Ecosystem impacts of exotic annual invaders in the genus *Bromus*. In: Germino MJ, Chambers JC, Brown CS (eds) Exotic brome-grasses in arid and semiarid ecosystems of the Western USA: causes, consequences, and management implications. Springer, New York, NY (Chapter 3)

Germino MJ, Chambers JC, Brown CS (2015b) Introduction: exotic annual *Bromus* in the Western US. In: Germino MJ, Chambers JC, Brown CS (eds) Exotic brome-grasses in arid and semiarid ecosystems of the Western USA: causes, consequences, and management implications. Springer, New York, NY (Chapter 1)

Gordon R, Brunson MW, Shindler B (2014) Acceptance, acceptability and trust for sagebrush restoration options in the Great Basin: a longitudinal response. Rangel Ecol Manag 67:573–583

Hershdorfer ME, Fernandez-Gimenez ME, Howery LD (2007) Key attributes influence the performance of local weed management programs in the Southwest United States. Rangel Ecol Manag 60:225–234

Jordan RC, Gray SA, Howe DV et al (2011) Knowledge gain and behavioral change in citizen-science programs. Conserv Biol 25:1148–1154

Kartchner H (2013) Development of ecologically-based invasive plant management curriculum for university audiences. Thesis, Utah State University

Kelley WK, Fernandez-Gimenez ME, Brown CS (2013) Managing downy brome (*Bromus tectorum*) in the Central Rockies: land manager perspectives. Invasive Plant Sci Manag 6:521–535

Larson DL, Anderson PJ, Newton W (2001) Alien plant invasion in mixed-grass prairie: effects of vegetation type and anthropogenic disturbance. Ecol Appl 11:128–141

Mack RN, Simberloff D, Lonsdale WM et al (2000) Biotic invasions: causes, epidemiology, global consequences, and control. Ecol Appl 10:689–710

Marler MJ, Supplee K, Wessner M et al (2005) Changing attitudes about grassland conservation in Missoula, Montana – "Weed Capital of the West". Ecol Restor 23:29–34

Marshall NA, Friedel M, van Klinken RD et al (2011) Considering the social dimensions of invasive species: the case of buffel grass. Environ Sci Pol 14:327–338

McCaffrey S (2004) Fighting fire with education: what is the best way to reach out to homeowners? J For 102(5):12–19

Norgaard KM (2007) The politics of invasive weed management: gender, race, and risk perception in northern California. Rural Sociol 72:450–477

Novak SJ, Mack RN (2001) Tracing plant introduction and spread: genetic evidence from *Bromus tectorum* (cheatgrass). BioScience 51:114–122

Pannell DJ, Marshall GR, Barr N et al (2006) Understanding and promoting adoption of conservation practices by rural landholders. Aust J Exp Agric 46:1407–1424

Peters WL, Meyer MH, Anderson NO (2006) Minnesota horticultural industry survey on invasive plants. Euphytica 148:75–86

Peterson R, Coppock DL (2001) Economics and demographics constrain investment in Utah private lands. J Range Manag 45:106–114

Petty RE, Briñol P (2010) Attitude change. In: Baumeister RF, Finkel EJ (eds) Advanced social psychology: the state of the science. Oxford University Press, London, pp 217–259

Pimentel D, Zuniga R, Morrison D (2005) Update on the environmental and economic costs associated with alien-invasive species in the United States. Ecol Econ 52:273–288

Reid CR, Goodrich S, Bowns JE (2008) Cheatgrass and red brome: history and biology of two invaders. In: Kitchen SG, Pendleton RL, Monaco TA et al (eds) Shrublands under fire: disturbance and recovery in a changing world, 6–8 Jun 2006, Cedar City, UT. Gen Tech Rep RMRS-P-52. USDA Forest Service, Rocky Mountain Research Station, pp 27–32

Rekart J (2011) Public funds, private provision: the role of the voluntary sector. UBC Press, Vancouver, Canada

Roberts C, Kallenbach RL (2000) Smooth bromegrass. University of Missouri Extension Agricultural Guide G4672. http://extension.missouri.edu/p/g4672. Accessed 13 Feb 2014

Salo LC (2005) Red brome (*Bromus rubens* subsp. *Madritensis*) in North America: possible modes for early introductions, subsequent spread. Biol Invasions 7:165–180

Shindler BA, Brunson MW, Stankey GH (2002) Social acceptability of forest conditions and management practices: a problem analysis. Gen Tech Rep PNW-GTR-537. USDA, Forest Service, Pacific Northwest Research Station

Shindler B, Gordon R, Brunson MW (2011) Public perceptions of sagebrush ecosystem management in the Great Basin. Rangel Ecol Manag 64:335–343

Slimak MW, Dietz T (2006) Personal values, beliefs, and ecological risk perception. Risk Anal 26:1689–1705

Stevens J, Falk DA (2009) Can buffelgrass invasions be controlled in the American Southwest? Using invasion ecology theory to understand buffelgrass success and develop comprehensive restoration and management. Ecol Restor 27:417–427

Tidwell LS (2005) Information sources, willingness to volunteer, and attitudes toward invasive plants in the Southwestern United States. MS thesis, Utah State University

Tidwell LS, Brunson MW (2008) Volunteering to manage rangeland weeds: results of a citizen survey in the southwestern United States. Rangelands 30(4):19–24

Trafimow D, Sheeran P (1998) Some tests of the distinction between cognitive and affective beliefs. J Exp Soc Psychol 34:378–397

USDA-Natural Resources Conservation Service (2002) Plant fact sheet: smooth brome. https://plants.usda.gov/factsheet/pdf/fs_brin2.pdf. Accessed 13 Feb 2014

Van Ryzin GG (2011) Outcomes, process, and trust of civil servants. J Public Adm Res Theory 21:745–760

Wilcove DS, Rothstein D, Dubow J et al (1998) Quantifying threats to imperiled species in the United States. BioScience 48:607–615

Chapter 15
Economic Modeling and the Management of Exotic Annual *Bromus* Species: Accounting for Ecosystem Dynamics, Ecological Thresholds, and Spatial Interdependencies

Mark Eiswerth, Rebecca Epanchin-Niell, Kimberly Rollins, and Michael H. Taylor

Abstract This chapter describes how economic models can inform management of exotic annual *Bromus* species on rangelands in the Western United States. It surveys published studies that develop bio-economic models of the management of *Bromus* species and other exotic annual invasive grasses, focusing on the challenges of representing the complex dynamics of rangeland ecosystems within tractable models of economic decision-making. The discussion starts with elements that are common to most economic models of *Bromus* management, then turns to contributions from the literature that have developed bio-economic models that capture three salient features of *Bromus* invasion: the dynamics of *Bromus* invasion, ecological thresholds related to *Bromus*, and spatial interdependencies in biophysical and human systems. The chapter synthesizes insights gained from this literature for managing *Bromus* in the Western United States, including insights on where to direct *Bromus* management resources on the landscape to achieve the greatest benefit given limited funds for management and on how to improve the design of policies that encourage socially efficient *Bromus* management by private land managers. The chapter concludes by identifying key areas where further research into the economics of *Bromus* management is needed.

M. Eiswerth (✉)
Department of Economics, University of Northern Colorado, Greeley, CO 80639, USA
e-mail: Mark.Eiswerth@unco.edu

R. Epanchin-Niell
Resources for the Future, Washington, DC 20036, USA
e-mail: epanchin-niell@rff.org

K. Rollins • M.H. Taylor
Department of Economics, University of Nevada, Reno, Reno, NV 89557, USA
e-mail: krollins@unr.edu; mhtaylor@unr.edu

M.J. Germino et al. (eds.), *Exotic Brome-Grasses in Arid and Semiarid Ecosystems of the Western US*, Springer Series on Environmental Management,
DOI 10.1007/978-3-319-24930-8_15

Keywords Economics • Policy • *Bromus tectorum* • Invasive species • Invasive plant management

15.1 Introduction

Public and private land managers regularly make decisions that influence the prevalence and spread of exotic annual *Bromus* species (*Bromus* hereafter) and other exotic annual invasive grasses on semiarid rangelands in the Western United States. Ranchers affect *Bromus* through weed control efforts and livestock management (inappropriate livestock grazing can reduce the ability of native plants to compete with *Bromus*). Public land management agencies affect *Bromus* through weed management; post-wildfire rehabilitation; education, outreach, and incentive programs that target land managers; and regulation of the activities of ranchers and other private entities that operate on public land. Understanding how public and private land managers make decisions, and how their decisions are influenced by the financial and ecological constraints they face, is essential for designing policies and regulations that encourage private managers to effectively manage *Bromus* invasions, as well as for determining the efficient use of limited resources available to public agencies for *Bromus* management. This chapter surveys published economic studies that have developed and used integrated ecological and economic models (henceforth, bio-economic models) to analyze public and private decision-making on semiarid rangelands affected by *Bromus* and other exotic annual invasive grasses and discusses the implications of these studies for the management of *Bromus* on rangelands in the Western United States.

The chapter focuses on simulation and optimization bio-economic models that take into account one of three attributes of *Bromus* invasions that pose particular challenges to economic analysis: the temporal dynamics of invasion, ecological thresholds, and spatial interdependencies in biophysical and human systems. All these bio-economic models of *Bromus* grass management include an ecological component to represent how *Bromus* species behave and are likely to respond to management. A primary challenge when developing bio-economic models is to balance the competing imperatives of (1) accurately representing the ecology of a *Bromus* invasion while (2) precisely and parsimoniously describing the elements of the economic decision problem. Ultimately, a modeler chooses which aspects of the complex ecology of *Bromus* invasion are necessary to include in the model—and what aspects can be safely ignored—to address a specific management or policy question. (For the studies reviewed in this chapter, the ecological complexity relates to the dynamics of *Bromus* invasion, including movements between stable ecological states separated by thresholds and to the spatial spread of *Bromus*.) Similarly, choices must be made about which aspects of the economic decision problem to include. As discussed in the next section, economic decision problems may include the decision-maker's objective function, constraints on the ability to meet the objective, and uncertainty about future economic and ecological conditions. The discussion examines how previous studies have balanced these competing imperatives in developing bio-economic models that are both ecologically sound and capable of addressing practical management issues related to *Bromus*.

The chapter starts by describing several elements that are common to most of the bio-economic models of *Bromus* management surveyed in this chapter. It then turns to the various approaches these existing models have used to address the dynamics of *Bromus* invasions, ecological thresholds related to *Bromus*, and spatial interdependencies in biophysical and human systems where *Bromus* are present. Next is a summary of insights from the studies reviewed for management and regulation of *Bromus* on semiarid rangelands in the Western United States. The chapter concludes by identifying areas for further research.

15.2 Common Bio-Economic Modeling Elements

Most of the bio-economic studies reviewed in this chapter share common elements. At the most basic level is the recognition that economic problems are posed from the perspective of a decision-maker with specific objectives and constraints. The studies reviewed in this chapter consider decisions from two perspectives. First is the perspective of an individual (or firm) whose primary concern is to meet private objectives, which may be to maximize profits from a ranch operation, or to ensure the ecological sustainability of land under his or her stewardship, or a combination of both. Second is the perspective of a "public decision-maker" concerned with a broader range of benefits and costs that accrue to society (where "public" refers not to a particular public agency, such as the Bureau of Land Management, but to the context in which decisions are made, taking into account the full range of social benefits and costs). In studies of *Bromus* management, public decision-makers are typically assumed to balance the costs of *Bromus* control with the full range of costs associated with *Bromus* invasion, which include reduced livestock forage, increased wildfire frequency and costs, and degradation of ecosystem services such as wildlife habitat, biodiversity, carbon sequestration, and hydrologic functions that reduce soil erosion and flooding costs. (See Havstad et al. 2007 for a comprehensive list of ecosystem services from arid rangeland ecosystems.)

The public-good nature of many of the costs of *Bromus* invasion justifies a role for rangeland policy to align incentives of private decision-makers with social management objectives. For example, although many private ranchers may be motivated to earn profits through their ranch operation and also be "good stewards" who protect the ecological health of the land under their management, they are unlikely to take into account all the ways in which their management decisions generate benefits and costs for other segments of society through their effects on *Bromus*. For this reason, many private decision-making models are designed to address how public programs or policies (e.g., cost sharing of rangeland management treatments) can align private incentives with social goals.

The studies reviewed here, as well as in economics in general, use mathematics to define a decision-maker's objective function and the constraints on the decision-maker's ability to maximize this objective. In the case of *Bromus* management, constraints include those imposed by the ecology of semiarid rangeland ecosystems that have been affected by *Bromus*. For example, a land manager's ability to rehabilitate a site dominated by *Bromus* is constrained by the biology of *Bromus* and the biophysical features of the site (e.g., precipitation, elevation, soil characteristics), among other factors. Similarly, a rancher's profits are constrained by cattle herd growth dynamics and forage availability. Constraints such as those imposed by regulations, limits to public land access, effectiveness of rehabilitation treatments and *Bromus* management technologies, and limited budgets are all incorporated into bio-economic models as mathematical relationships.

The objective function in a bio-economic model should represent the decision-making criteria used by the decision-maker being modeled so that the model's predictions comport with reality. Several decision-making criteria are represented in the studies reviewed in this chapter. Objectives of public decision-makers include minimizing the sum of treatment costs and damages from exotic invasive plants (e.g., Olson and Roy 2002; Eiswerth and Johnson 2002; Finnoff et al. 2010; Epanchin-Niell and Wilen 2012) and maximizing the flow of future benefits from controlling an invader, minus management costs (e.g., Polasky 2010). Models of private decision-makers largely focus on ranchers and assume either that the rancher's objective is to maximize the present value of profits from the ranch (e.g., Huffaker and Cooper 1995; Kobayashi et al. 2014) or that the rancher follows a decision heuristic, or rule of thumb, that determines how to adjust stocking rates and management in response to exotic invasive plant encroachment (e.g., Janssen et al. 2004).

Uncertainty and risk are inherent to *Bromus* invasion and management. Many studies reviewed in this chapter incorporate risk by including stochastic parameters to characterize sources of uncertainty. Studies have used stochastic parameters for rainfall variability and drought (e.g., Janssen et al. 2004; Ritten et al. 2010), wildfire (e.g., Huffaker and Cooper 1995; Epanchin-Niell et al. 2009), the success or failure of management treatments (e.g., Eiswerth and van Kooten 2002; Epanchin-Niell et al. 2009; Taylor et al. 2013a), and market volatility (Karp and Pope 1984; Carande et al. 1995). Some sources of uncertainty are exogenous to the efforts of the decision-maker (e.g., lightning strikes, drought, market variability), while others are at least partially endogenous, in that the decision-maker's actions influence the likelihoods of particular outcomes. The probability of catastrophic wildfire, for example, is a function of fuel loading, which can be managed with fuel removal treatments (Taylor et al. 2013a). Models that include uncertainty and risk produce ranges of outcomes that depend on the realization of stochastic parameters. The determination of which sources of risk to include in a model depends on the management question(s) being considered.

Bio-economic models of private decision-making under risk require two additional assumptions relative to models that do not consider risk. First, models that consider decision-making over time require an assumption to describe how the decision-maker forms expectations about the outcomes of his or her actions given uncertain future ecological and market conditions. Most studies in this review assume that decision-makers have "perfect foresight." Although decision-makers do not know precise outcomes of their actions given uncertainty about future conditions, they do know the range of possible outcomes and the probability that each outcome will occur from any course of action, and they use this knowledge to pursue actions that have the greatest expected value given their objective function and set of constraints. Even though the assumption of perfect foresight is unrealistic, several ranch models that assume perfect foresight generate realistic predictions for cattle stocking rates, average annual profits, and other relevant variables (e.g., Karp and Pope 1984; Passmore and Brown 1991; Carande et al. 1995; Wang and Hacker 1997; Ritten et al. 2010).

The second assumption required by models that analyze private decision-making under risk concerns the decision-maker's attitude toward risk. In particular, the modeler must decide whether to portray the decision-maker as risk averse, risk neutral, or risk loving. A risk-averse decision-maker faced with the option of participating in a lottery versus receiving the expected value of the lottery with certainty would choose the latter, a risk-neutral decision-maker would be indifferent between the two, and a risk-loving decision-maker would prefer to participate in the lottery. The majority of studies of private decision-makers reviewed in this chapter assume a risk-neutral decision-maker (e.g., Huffaker and Cooper 1995; Finnoff et al. 2008; Kobayashi et al. 2014). However, Passmore and Brown (1991) and Carande et al. (1995) find that, faced with uncertain prices for cattle, risk-averse ranchers would maintain lower average annual stocking rates than risk-neutral ranchers to achieve lower but less volatile streams of profits. A finding that greater levels of risk aversion by ranchers can lead to substantial reductions in stocking rates suggests that risk attitudes are also likely to matter for decisions concerning *Bromus* management.

Dynamic economic models generally assume some form of discounting such that future benefits and costs are valued less, and have less weight in decision-making, than benefits and costs that occur today. The appropriate discount rate for economic studies of *Bromus* depends crucially on the decision-maker being modeled. Previous studies that have considered the decision-making of a private rancher or land manager have elected to use the market interest rate to discount future benefits and costs (e.g., Kobayashi et al. 2014). The market rate of interest is used because it reflects the opportunity cost of the funds used for *Bromus* management (i.e., the lost opportunity from not being able to invest the funds used for *Bromus* management elsewhere in the economy). Conversely, when the decision-maker is a public sector agency, previous studies have used the social rate of time preference to discount future benefits and costs (e.g., Taylor et al. 2013a). The social rate of time preference is the rate that society is willing to trade off future consumption for current consumption and is generally lower than the private discount rate. In addition, although all the bio-economic models reviewed in this chapter assume linear discounting (i.e., all future years are discounted using the same discount rate), several previous studies have suggested that nonlinear or hyperbolic discounting (i.e., the near future is discounted at a higher rate than the more distant future) provides a more accurate description of how people evaluate trade-offs between the present and future benefits and costs (Karp 2005). It is likely that the bio-economic models reviewed in this chapter all assume linear discounting in part because it is a straightforward assumption to implement in dynamic economic models.

Finally, the specific management questions addressed in an economic study of *Bromus* species depend in part on the "stage" of invasion being analyzed. Biological invasions generally have four stages: introduction, establishment, spread, and saturation. Because *Bromus*, in particular *Bromus tectorum* L. (cheatgrass or downy brome), is present to some extent in much of its potential nonnative range in the Western United States, the *Bromus* studies reviewed in this chapter focus primarily on the spread and saturation stages of invasion. These studies analyze management issues that include ecological rehabilitation, preventing partially invaded land from transitioning across an ecological threshold to an exotic invasive-dominated ecological state and minimizing the damages from *Bromus* invasion at a site. Nonetheless, in regions where there remain areas of relatively uninvaded rangeland, management strategies such as detection and quarantine, which aim to slow the spread of the invasion across the landscape, are paramount. Hence, several bio-economic models designed to analyze such management strategies also are reviewed.

15.3 Economic and Ecological Dynamics

This section reviews how dynamic optimization has been used to analyze the dynamic aspects of *Bromus* management and discusses insights yielded by such studies. Most of the studies reviewed in this chapter incorporate dynamics of natural processes (e.g., species dynamics, climate, and fire) to generate benefits and costs of invasive species management strategies in units that are comparable over time. Management actions may intentionally or unintentionally cause ecological and economic processes to speed up, slow, reverse, or be shifted to move along completely different pathways, depending on the timing of an action and the nature of the dynamic processes that occur within and between ecological and economic systems. Only a few economic studies have used dynamic optimization methods to analyze management of *Bromus* in particular (Kobayashi et al. 2014) or exotic annual invasive grasses more generally (Huffaker and Cooper 1995; Finnoff et al. 2008), but several dynamic bio-economic studies of other exotic invasive species can provide important insights.

It is important to distinguish between dynamic optimization and dynamic simulation models. This section focuses on optimization models, which analyze the behavior of decision-makers whose goal is maximizing (or minimizing) an objective function given one or more constraints. Because optimization models describe the incentives and constraints faced by real-world decision-makers, they are generally preferred over simulation models for most economic analysis. Simulation models are often used for economic analyses of *Bromus* and other exotic annual invasive grasses because ecological thresholds, spatial dynamics, and other factors complicate the mathematical description of the economic decision problem. Although simulation methods are not designed to identify optimal outcomes, they can be used to compare outcomes from different management options and identify those that are more efficient.

15.3.1 General Framework for Dynamic Optimization

Dynamic optimization is an approach used to determine which actions from a defined set of possible actions a decision-maker should choose to maximize (or minimize) an objective function over a defined time horizon. Because the decision-maker's chosen actions affect the future trajectory of the entire dynamic system, these actions are referred to as "control variables." Examples of control variables are (1) a private rancher's choice of which series of weed treatments and herd management actions will maximize the present-valued net worth of a ranch operation affected by *Bromus* and (2) a land management agency's choice of the timing of landscape-level treatments to result in a cost-minimizing strategy to maintain an area's resistance to *Bromus* invasion over a 20-year time horizon.

Dynamic optimization models integrate time and system dynamics through state equations, which define how "state variables" change over time as a function of both natural processes and the application of control variables. A state variable describes the level (or quantity) at a given time of a variable that can change over time. Examples of state variables are (1) the size (measured as density or spatial coverage) of a *Bromus* infestation, (2) the stock of forage for livestock, and (3) the size of a herd. A state equation that describes how the size of an invasion evolves over time depends on the natural (or "intrinsic") rate of growth of the species, as well as the influence of human efforts to manage it through control variables. The goal is to choose values for control variables strategically to influence how the state variables evolve over time to either maximize or minimize the decision-maker's objective function. The solution to a dynamic optimization describes the values of all the state variables and control variables at each point along the optimum path and is a function of the starting points of the state variables. For example, the level of a *Bromus* infestation at the start of the management problem would affect the solution to the optimization problem.

Two methods used for solving dynamic constrained optimization problems are optimal control (OC) and stochastic dynamic programming (SDP). OC methods are typically characterized by models that yield mathematical equations broadly characterizing analytical properties of solutions to the optimization problem, thereby allowing one to reach general conclusions. These models often are of the "continuous-time" variety, meaning that the state variables are modeled as changing continuously over time (using differential equations), rather than once every specified period (e.g., once each year), as in a discrete-time model.

In contrast, SDP is a means of numerically solving a constrained optimization problem to provide approximate solutions to systems that do not result in analytical solutions because of their mathematical complexity. In addition, SDP offers a more tractable approach than OC alone for accommodating uncertainty, including uncertainty regarding the performance of control measures, how an infestation evolves in response to natural conditions and management, and the occurrence of fire and other stochastic disturbances. Finally, because SDP models are based on discrete rather than continuous time, parameterizing economic models of *Bromus* for data such as livestock herd size, end-of-year cattle prices, acreage treated, and other parameters that tend to be considered in discrete or annual increments is more straightforward than with the continuous formulations used in OC methods. OC and SDP models applied in the exotic invasive context also typically maximize an objective function over an "infinite time horizon." Alternatively, OC and SDP models may maximize an objective function over a finite time horizon, taking into account what the predicted "terminal value" of a specified state variable will be at the end of that finite time horizon.

15.3.2 Cost Minimization Models

Several studies formulate the invasive species management problem as minimizing the sum of the costs of invasion (damages) and management costs. Cost minimization approaches are particularly useful when evaluating alternative management strategies. An example of a cost minimization model is Eiswerth and Johnson (2002), who use an OC model to minimize the sum of damages and management costs following the introduction and establishment of exotic invasive weed and grass species on arid lands. Olson and Roy (2002) similarly use an OC model to minimize the sum of damages and management costs for an invasive species; however, they also model uncertainty in how an infestation grows and spreads over time.

In a useful extension of previous work, Ranjan et al. (2008) use a cost minimization approach to determine optimal combinations of strategies for allocating a fixed pool of financial resources between efforts to prevent introduction of an exotic invasive species into a new location and efforts to mitigate adverse impacts once establishment takes place. Buhle et al. (2005) combine data on population dynamics with control costs to identify the least-cost approaches for preventing *Ocinebrellus inornatus* (Japanese oyster drill), an established aquatic invasive snail, from spreading. Their methods are applicable to exotic annual invasive grass management where (1) managers wish to minimize the total costs of exotic invader control, and (2) it is relevant to ask at what stage of the life cycle of an exotic invader it is most cost-effective for managers to apply a control measure.

15.3.3 Ranch Management Models Using Dynamic Optimization

Several studies use SDP methods to analyze the effect of ranch management on rangeland ecosystems (Karp and Pope 1984; Passmore and Brown 1991; Carande et al. 1995; Wang and Hacker 1997; Ritten et al. 2010). These studies are calibrated to quantitatively match observed ranch outcomes and have typically used livestock stocking rates and the intensity of rangeland vegetation treatments as control variables (e.g., Karp and Pope 1984; Kobayashi et al. 2014). Kobayashi et al. (2014) is the only study to date that uses SDP to analyze a rancher's decision-making in the context of exotic annual invasive grasses; their model considers stochastic wildfire and effectiveness of rangeland rehabilitation treatments and incorporates ecological thresholds. Because their model is calibrated to quantitatively match observed ranch outcomes, Kobayashi et al. (2014) are able to analyze whether realistic and specific changes in rangeland policy (e.g., expansion of cost sharing for vegetation treatments) will induce changes in ranch management that translate into empirically meaningful improvements in the economic viability of ranching, rangeland ecological health, and the likelihood that privately managed rangeland will cross an ecological threshold to an annual grass-dominated ecological state.

15.3.4 Incorporating Stochastic Processes and Uncertainty into Dynamic Optimization

Studies that use stochastic dynamic optimization methods must address the issue of how to parameterize the stochastic elements of the problem. Eiswerth and van Kooten (2002) use a discrete-time, infinite-horizon SDP model to identify preferred approaches for managing *Centaurea solstitialis* L. (yellow star thistle), an exotic invasive rangeland weed. The researchers include a random variable in the state equation to capture the stochastic nature by which the size of the infestation evolves over time. To parameterize their stochastic state equation, they use data collected via a survey of weed and plant experts to develop infestation transition probability matrices for each control option under consideration.

Kobayashi et al. (2014) use historical data for wildfire ignitions on rangeland systems distinguished by ecological states to parameterize the probability of stochastic fire events and the role of *Bromus* in crossing ecological thresholds. Other studies based on SDP models that are relevant for *Bromus* management include studies performed in the contexts of cropland weeds (e.g., Pandey and Medd 1991) and invasive species in general (Leung et al. 2002; Bogich and Shea 2008; Polasky 2010).

In another study employing SDP methods potentially applicable to the context of *Bromus*, Hyytiäinen et al. (2013) develop an SDP model to simultaneously examine the optimal magnitudes and timing of prevention, eradication, control, and adaptation for an aquatic exotic invader, *Corbicula fluminea* L. (Asian clam). Such an approach may apply especially well to exotic annual invasive grasses in cases where managers are free to choose among (1) a prevention strategy that stops or delays the invasion at a particular site, (2) a mitigation strategy that focuses on early detection and control of an infestation once established, and (3) an adaptation strategy that seeks to minimize economic losses without controlling the infestation.

In a study modeling *Dreissena polymorpha* L. (zebra mussel), Timar and Phaneuf (2009) use methods that also could be translated to *Bromus*. They estimate the probability of a dispersal event occurring at a given point in time, originating from an infested site and spreading to an uninfested site, to parameterize the likelihood of observing an infestation of zebra mussels as a function of the anthropogenic and natural factors that influence spread. An application or adaptation of this approach to exotic annual invasive grasses would be useful for developing a time- and site-dependent invasion hazard index as a function of (1) variables that indicate naturally occurring threats of invasion to each site (i.e., factors that influence the natural dispersal of seeds) and (2) a human threat variable (e.g., incorporating livestock stocking rates and management practices and rangeland fire prevention and restoration practices).

15.3.5 Imperfect Information and Dynamic Optimization in Invasive Species Management

Most dynamic optimization models assume decision-makers have a rather sophisticated understanding of the ecological conditions on the land that they manage, can observe or monitor changes in ecological conditions without incurring costs, and are aware of the impact of their management actions on future ecological conditions. Several recent studies relax these stringent assumptions to develop models of how decision-makers adapt to imperfect information in the context of making multiple decisions over time. In an application to the exotic invasive rangeland weed *C. solstitialis*, Eiswerth and van Kooten (2007) compare the results of an SDP model to those of a "reinforcement-based, experience-weighted attraction learning model" (for background, see Camerer and Ho 1999; Hanaki et al. 2005), which is a formulation from game theory of a model describing adaptive management. This type of model simulates how a decision-maker incorporates additional information over time as more is learned about the net benefits of alternative management strategies, based on observing outcomes from implementing different strategies in each previous time period. The decision-maker adapts by adjusting the value of selecting a particular management strategy in a given time period based on how well different strategies have worked in the past, thereby allowing for efficient use of management resources over time.

Another approach to modeling how decision-makers handle imperfect information is to assume that a land manager with imperfect knowledge characterizes ecological conditions (and, hence, the level of infestation) into broad categories (e.g., good, fair, poor) that are used along with decision heuristics, or rules of thumb, to make management decisions (e.g., Eiswerth and van Kooten 2002). Such approaches use "fuzzy" methods to model decision-making in the context where ecological conditions and other elements of the problem are classified by the decision-maker into discrete categories, rather than treating these as continuous data. Such approaches must address two aspects of decision-maker subjectivity. First, two managers may label a given infestation differently depending on differences in their experience, knowledge, and judgment. For example, one manager may think of a specific infestation as "minimal," while another labels it as "moderate" or even "high." Second, depending on differences in characteristics and human uses of the landscapes at two locations, a manager may classify infestations of similar size and type (e.g., same species, infestation size, and density) as "minimal" at one location but "moderate" at another. As a result, infestations display characteristics associated with fuzzy variables that can be analyzed using fuzzy membership functions (e.g., Zadeh 1965), which are different from conventional probability distributions (Kosko 1992). Other variables in dynamic models of exotic annual invasive grasses or other rangeland exotic invaders (e.g., the intrinsic rate of growth) also may be treated as fuzzy variables. Fuzzy methods offer an approach for dealing with issues related to bounded rationality, which may more realistically represent how private and public actors make decisions regarding *Bromus*. It is important to note that fuzzy methods are not used to represent uncertainty on the part of the economic modeler concerning parameters and other model assumptions; rather, fuzzy methods capture the uncertainty of the decision-maker in the problem that is being modeled.

15.4 Ecological Thresholds

An ecological threshold is a critical point at which small changes in one or more ecosystem variables can lead to sudden, extreme changes in ecosystem condition (Holling 1973). More formally, thresholds are boundaries in conditions that lead to alternative stable states. Thresholds are crossed when an ecosystem does not return to the original state via self-organizing processes after stress or disturbance but instead transitions to a new, alternative state that has altered attributes and primary ecological processes (Beisner et al. 2003; Stringham et al. 2003).

Thresholds are a feature of many natural resource management problems, including rangeland management in the presence of *Bromus* and other exotic annual invasive grasses (Stringham et al. 2003). The prevalence of exotic annual invasive grasses can play a role, along with other biotic and abiotic factors (e.g., temperature, moisture regime, large deep-rooted perennial bunchgrasses), in determining whether rangeland ecosystems will transition to an exotic annual invasive grass-dominated state after a disturbance such as wildfire or drought (McIver et al. 2010). In many cases, transitions across thresholds can only be reversed through costly management interventions or are irreversible with current restoration technology. Ecological thresholds are relevant for rangeland policy because desired ecological states (e.g., states dominated by native perennial grasses and sagebrush with a small presence of *Bromus*) and exotic annual invasive grass-dominated states are very different in terms of livestock forage productivity and effects on ecosystem services such as the frequency and severity of wildfires, wildlife habitat for game animals and sensitive species, and the extent of soil erosion (Havstad et al. 2007).

A large and growing literature in economics analyzes the management of ecosystems in the presence of ecological thresholds in a variety of contexts, including eutrophication of freshwater lakes (Carpenter et al. 1999; Naevdal 2001; Brock and Starrett 2003; Mäler et al. 2003), infectious wildlife disease (Horan and Wolf 2005), environmentally sustainable economic development (Common and Perrings 1992), wildlife management (Rondeau 2001), and rangeland management (Perrings and Walker 1997, 2004). Most economic studies that analyze thresholds in terrestrial ecosystems focus on rangeland ecosystems (Huffaker and Cooper 1995; Perrings and Walker 1997, 2004; Anderies et al. 2002; Janssen et al. 2004; Finnoff et al. 2008), likely because an estimated 10–20 % of rangelands worldwide are degraded (Millennium Ecosystem Assessment 2005), and this degradation is often associated with crossing thresholds to undesirable ecological states, including ecological states dominated by exotic invasive plants.

15.4.1 Optimization Models and Ecological Thresholds

Several studies use optimal control (OC) to analyze interactions between ecological thresholds and livestock management in rangeland ecosystems (Huffaker and Cooper 1995; Perrings and Walker 1997, 2004; Anderies et al. 2002; Janssen et al. 2004; Finnoff et al. 2008). Two of these studies—Huffaker and Cooper (1995) and Finnoff et al. (2008)—consider exotic annual invasive grasses. In these models, ecological thresholds arise endogenously as a result of interspecies plant competition and are characterized by the inherent properties of an ecosystem. These studies use OC methods to generate qualitative analytic results that describe generally how economic factors interact with ecological factors to determine whether management will result in crossing of an ecological threshold. Results demonstrate that a combination of economic factors (cattle prices, land management treatment costs, interest rates) and the initial ecosystem condition determine whether it is in the economic best interest of the decision-maker to maintain an ecosystem in a desired ecological state (e.g., a state dominated by native perennial grasses and sagebrush with a small presence of *Bromus*) or to allow it to cross a threshold to a degraded ecological state (e.g., a state dominated by *Bromus* and other exotic annual invasive grasses).

Kobayashi et al. (2014) incorporate ecological thresholds and exotic annual invasive grasses into a stochastic dynamic programming model of a ranch typical of northern Nevada. They use their model to analyze when and if subsidies that offset the cost of rehabilitation treatments and/or improved treatment success rates will result in changes in ranch management that make crossing ecological thresholds to an exotic annual invasive grass-dominated state less likely. The authors find that on rangeland dominated by native perennial grasses and sagebrush, lower treatment costs and improved success rates lead to larger herd sizes and higher profits but do not reduce the likelihood that the ranch will cross an ecological threshold to an exotic annual invasive grass-dominated state. The explanation is that the rancher has sufficient private incentive to maintain the land in the ecological state dominated by native perennial grasses and sagebrush through herd management and periodic vegetation treatments at current treatment costs and success rates. Conversely, on rangeland that has been invaded by exotic annual invasive grasses, lower treatment costs and/or improved treatment success rates cause the rancher to increase the frequency and intensity of restoration treatments, making it less likely that the land will convert to the exotic annual invasive grass-dominated state. These results suggest that subsidies to lower restoration treatment costs are most effective if directed toward ranchers whose land has been invaded by exotic annual invasive grasses but has not yet crossed the ecological threshold to the exotic annual invasive grass-dominated state.

15.4.2 Ecological Resilience

Several of the studies using OC methods mentioned thus far—in particular, Perrings and Walker (1997, 2004), Anderies et al. (2002), and Janssen et al. (2004)—consider the role of ecosystem resilience on the optimal management of a livestock operation in the presence of exotic invasive grasses and ecological thresholds. Consistent with the other chapters in this book, we define resilience as the capacity of an ecosystem to regain its fundamental structure, processes, and functioning when altered by stressors such as increased CO_2, nitrogen deposition, and drought and by disturbances including land development and fire (Holling 1973; Allen et al. 2005). Understanding how alternative management strategies influence the resilience of rangeland ecosystems is important because, as Scheffer et al. (2001) write, in rangeland ecosystems "a loss in resilience usually paves the way for a switch to an alternative state."

Anderies et al. (2002) develop a mathematical model to characterize the dynamic interactions among grass, shrubs, fire, and livestock in a setting with stochastic rainfall and ecological thresholds. These authors use their model to explore how ecological, economic, and management factors influence the resilience of a rangeland ecosystem. In a follow-up paper using the same analytical framework, Janssen et al. (2004) develop a methodology to characterize "robust" management strategies that determine when a rancher should reduce sheep stocking densities to allow the ecosystem to maintain ecological resilience. The authors of these studies conclude that economic conditions (such as high livestock prices) can cause ranchers to adopt management strategies (such as high livestock stocking rates) that compromise ecological resilience and make transition across ecological thresholds to degraded states more likely, in the event of disturbances such as drought or wildfire.

15.4.3 Computer Simulation Models

An alternative approach for analyzing how ecological thresholds influence the management of exotic annual invasive grasses in rangeland ecosystems is to develop computer simulation models that integrate state-and-transition models (STM) from rangeland ecology with economic models. This approach is used by Epanchin-Niell et al. (2009) and Taylor et al. (2013a). An STM describes an ecosystem as being in one of several alternative states separated by ecological thresholds, where transitions between states are often triggered by disturbances including natural events (e.g., drought, wildfire) and management actions (e.g., grazing, prescribed burns) (Stringham et al. 2003). The STM framework allows for the incorporation of complex ecosystem dynamics into bio-economic simulation models, including the role of ecological disturbances such as wildfire and drought as catalysts for transitions across ecological thresholds to an exotic annual invasive grass-dominated state.

In contrast to studies based on optimization models, simulation models do not analyze a self-interested decision-maker's optimal response to changes in ecological or economic conditions or policy. Rather, simulation models are used to analyze and compare the economic efficiency of alternative management regimes in a setting where the effects of stochastic factors such as wildfire, post-disturbance ecological transition, and treatment success on the distribution of outcomes can be explicitly analyzed.

Bio-economic simulation models that incorporate STMs of exotic annual invasive grasses in rangeland ecosystems have reached several novel management conclusions. Epanchin-Niell et al. (2009) demonstrate that in the sagebrush steppe ecosystem in the Western United States (specifically, the Wyoming big sagebrush community in the Great Basin 8–10-inch precipitation zone), investment in post-wildfire revegetation of degraded sites (i.e., sites with sagebrush and *B. tectorum* but sparse to no perennial bunchgrasses) using either native or nonnative perennial bunchgrasses can reduce long-term fire management costs over a 50-year time horizon by more than enough to offset the costs of treatment. These reductions in fire management costs are accomplished in part by reducing the amount of land that transitions to a state entirely dominated by *B. tectorum*, where wildfires occur more frequently. The study also quantifies the economic and biodiversity trade-offs of revegetating using native versus nonnative perennial bunchgrasses and derives a cost function for the long-term maintenance of native sagebrush steppe vegetation on the landscape via postfire revegetation. Epanchin-Niell et al. (2009) also reach the significant if not surprising conclusion that although greater federal funding levels for post-wildfire restoration in the sagebrush steppe are warranted based on the expected economic returns and biodiversity benefits, this management strategy alone is insufficient to reverse the continued transition of rangelands across thresholds to exotic annual invasive grass-dominated states.

Taylor et al. (2013a) compare the economic efficiency of prevention versus rehabilitation treatments for two rangeland ecosystems (Wyoming sagebrush steppe and mountain big sagebrush ecosystems) affected by invasive grasses. They find that over a 200-year time horizon, prevention treatments applied to contain exotic annual invasive grasses on lands that have not yet become dominated by the invasive grasses yield benefits in terms of expected future wildfire suppression cost savings that are several times greater than the cost of treatment and that the savings in wildfire suppression costs pay for treatment costs within 20 years. Conversely, they find that for systems dominated by exotic annual invasive grasses, the wildfire suppression cost savings associated with rehabilitation are not sufficient to justify the high per acre treatment costs and low restoration success rates. Together, these results suggest that limited budgets for exotic annual invasive grass management are most efficiently directed toward preventing rangeland that has not yet become dominated by *Bromus* from crossing an ecological threshold to *Bromus* dominance.

15.4.4 Uncertain Ecological Thresholds

It is often difficult for experienced rangeland ecologists to determine with certainty whether an ecosystem has crossed a threshold between states (McIver et al. 2010). This uncertainty can be costly because treatment methods that are appropriate before a threshold has been reached may be ineffective, or could even hasten exotic annual invasive grass domination, after the threshold has been crossed. Taylor et al. (2013a) find that the expected economic benefits of restoration-based hazardous fuel reduction treatments on rangelands increase with the land manager's ability to determine whether the land has crossed an ecological threshold related to exotic annual invasive grasses. The increased expected economic benefit of treatment occurs because uncertainty about whether the threshold has been crossed causes land managers to treat land that is not at immediate risk of crossing a threshold to an exotic annual invasive grass-dominated state in the event of wildfire or other disturbance, and hence where treatment could have been delayed at no cost. Further, reduced uncertainty makes it less likely that land managers will treat in areas that have already crossed a threshold to a state where treatment is a disturbance that moves the land to an exotic annual invasive grass-dominated state. In this manner, Taylor et al. (2013a) quantify the economic benefits of rangeland ecology research and outreach that improves accuracy in assessing whether *Bromus*-affected rangeland has crossed a threshold.

15.5 Spatial Considerations

The question of how to prioritize among locations is paramount when a fixed set of management resources is to be allocated across a number of locations that have different economic and ecosystem characteristics or when there are strong interdependencies across sites that affect the costs or benefits from investments at each location. Because incorporating spatial interactions into decision models with dynamic ecological and economic processes introduces another level of mathematical complexity, modelers use simplifying assumptions to keep models tractable for practical application. Ultimately, the modeler must decide whether to ignore spatial interdependencies because the benefits from accounting for them are very small or are not important in the particular context or whether the benefits are large and important enough to justify more involved modeling approaches. The existing literature contains a variety of approaches to spatial bio-economic modeling of landscape-level economic decision-making problems that could be adapted to *Bromus* and invaded rangelands. This literature includes spatial models that can facilitate development of management rules of thumb or management strategies to be tested within adaptive management frameworks, to confirm or reject existing intuition about best management strategies, and to identify specific cost-effective management strategies. This section reviews studies that are most relevant for *Bromus* management scenarios.

Two classes of problems involving decision-making that affects multiple sites are relevant for *Bromus* management. In the first class of problems, the level and importance of spillover effects from one site to other sites is minor or relatively unimportant to the management scenarios being considered. In this case, decision problems are linked spatially through the need to allocate a fixed set of management resources across a number of locations with different characteristics (such as different ecological states). Management benefits and costs may also vary across sites because of travel distances, spatial configurations, proximity to residential areas, and presence of critical wildlife habitat or highly valued cultural features. In cases where activities on one site do not generate important changes in the ecological processes and responses to management activities on other sites—that is, where the interactions between locations can be treated as independent—the spatial modeling problem is relatively straightforward.

The second class of problems involves situations where it is not reasonable to ignore spatial interdependencies between locations. Spatial interdependencies in biophysical systems occur when underlying dynamics are interdependent across sites, such as through the spread of exotic invasive species, spread of wildfire, and habitat linkages. Management actions at one location could affect adjacent sites through these biophysical interdependencies. Similarly, spatial interdependencies in human systems arise from the choices made by a manager in one area influencing the effectiveness of options available to a manager in an adjacent area. For example, treating a *Bromus* monoculture in one location may reduce fire risk to adjoining locations, thereby altering expectations regarding habitat, and thus management decision-making, on these adjacent lands.

15.5.1 Management Efforts across Space with Resource Constraints

If the assumption of spatial independence is reasonable, the returns from a fixed amount of resources to invest in rehabilitation treatment are maximized across a heterogeneous landscape by targeting effort first to the site with the highest benefit–cost ratio from treatment, then moving sequentially among sites in decreasing order of the benefit–cost ratios until the budget is exhausted or the costs outweigh benefits (Broadman et al. 2006; Pearce et al. 2006; Boyd et al. 2012). The practical issue is how to assess benefits and costs of treatment to account for varying levels of effort and alternative suites of management actions for each location. For large landscapes, the process of measuring benefit–cost ratios for relevant ranges of treatments for every individual site is likely to be impractical for most landscape-level management scenarios. An alternative approach to simplify the process is to identify a set of site attributes (using ecology and other criteria) that allow for classification of all sites into a smaller number of site "types." The sets of attributes can be used to estimate a standard unit area benefit–cost ratio for each type. The standardized benefit–cost ratios are then applied to all sites matching each type. Furthermore, the attributes for site types are ready for subsequent policy and programmatic decisions. This approach requires consideration of the relationship between definitions of site type attributes and the delineation of actual sites, and how attributes can be defined to facilitate application to different regions.

The development and application of standardized benefit–cost ratios per unit area by landscape type is practical where minor spatial interdependencies and unique features for specific sites can be taken into account after the fact, during the decision process. This approach may be particularly useful for allocating rehabilitation treatments across sites to prevent rangeland from transitioning to an exotic annual invasive grass-dominated state. Taylor et al. (2013a) use benefit–cost ratios (where benefits of treatment are measured as wildfire suppression cost savings) and the assumption of independence of locations to evaluate which *Bromus*-influenced ecological states should be prioritized for restoration-based fuel treatment. They quantify benefits and costs and quantify the differences in returns on treatment between lands where exotic annual invasive grasses are present but not yet a dominant component of the understory versus lands that have already crossed an ecological threshold and require rehabilitation treatments.

Similar intuition applies when deciding among locations for investing in postfire restoration. Epanchin-Niell et al. (2009) demonstrate that returns from postfire restoration of degraded sagebrush and exotic annual grass-dominated sites are greatest on sites with higher restoration success rates (e.g., due to soil type, climate), lower costs (e.g., due to terrain, accessibility), greater benefits from preventing a transition to an exotic annual invasive grass-dominated state (e.g., due to averting high fire suppression costs near developed areas), and in close proximity to high-valued natural resources at risk (e.g., sage grouse habitat). While each of these attributes contributes to greater "bang for the buck" from rehabilitation, all else equal, in practical application these attributes are present in various combinations at different sites across a landscape. Studies such as this suggest which attributes (restoration success rate, fire suppression cost, at-risk resources) are most useful for developing a classification system of "types" for standardized benefit–cost ratios.

In addition to rehabilitation treatments, spatial prioritization is important for decisions about optimal locations for monitoring effort using a fixed set of monitoring resources. For example, studies of exotic invasive plants (Hauser and McCarthy 2009) and exotic invasive forest pests (Epanchin-Niell et al. 2012, 2014) have developed approaches to cost-effectively allocate monitoring resources across sites that vary in the likelihood of exotic invasive introduction, ecosystem values at risk, and monitoring costs. These could be adapted to design cost-effective survey and control strategies for new incursions of *Bromus*.

15.5.2 Spatial Interdependency of Biophysical Processes

Spatial interdependencies in exotic annual invasive grass management exist where interconnected biophysical processes cause management effort at one location to influence benefits and costs or management outcomes at other locations. Invasions generally begin with introductions into limited locations, which then spread over time through reproduction and dispersal. Damages increase over time as more area is affected. Interdependencies can result when controls applied at one location affect the rate of spread at other locations. In the extreme, eradication of an invasive species from a single key area could prevent spread and damage across a much larger area. A review of studies that integrate ecology and economics to analyze optimal management of the geographic spread of established invasions (Epanchin-Niell and Hastings 2010) finds that models used to incorporate spatial interdependencies to represent the spread of invasive species are generally of two main types: spatially implicit models predict the total invaded area over time without considering specific invaded area locations, and spatially explicit models take into account the details of specific locations. These models use optimization as well as simulation approaches to compare management strategies.

Spatially implicit models can be used to determine cost-effective levels of effort for controlling invasions in cases where it is not necessary to model precise locations for where controls should be targeted. The general findings of the studies reviewed in Epanchin-Niell and Hastings (2010) are fairly intuitive: features that have the greatest influence on whether control is cost-effective include the speed of an invasion's spread, the potential size of area that could be affected, the degree of potential damages, and control cost. Features that increase the likelihood that eradication is an optimal strategy include invasions being small when detected and reintroduction being unlikely and/or infrequent. Although these generalized findings confirm intuition, many of the spatially implicit studies reviewed by Epanchin-Niell and Hastings (2010) provide modeling approaches that can be parameterized (calibrated) for application to specific management contexts and locations, thereby yielding quantitative results for critical management questions, such as how much to invest in control effort, whether the economics support pursuing an eradication strategy, and whether it is more cost-effective to strive for containment or slow the spread of an infestation. Some of the spatially implicit models most applicable to informing *Bromus* management are described in Sect. 15.3.

A spatially explicit modeling approach is necessary when a problem requires determining precisely which combination of specific locations is optimal for applications of controls to minimize the costs of exotic annual invasive grass spread. Models that use spatially explicit methods to account for spatial interdependencies generally reduce computational challenges by making simplifying assumptions about the underlying ecological and economic processes (Epanchin-Niell and Hastings 2010). They nonetheless provide important guidance about where, when, and how much to apply controls across heterogeneous landscapes—guidance that often cannot come from intuition or experience alone. For example, Cacho et al. (2010)

and Cacho and Hester (2011) use simulation models to compare decision heuristics about where to search for and control a reproducing and dispersing weed to minimize the area of the weed's spread, subject to an effort constraint. Their spatial-dynamic model showed, for a hypothetical weed invasion, how cost-effective search strategies change in response to increases in annual budgets: as budgets increase, strategies shift first to sampling sites more intensively, then to increasing the duration of the control program, and finally to applying repeat control treatments to found infestations.

Epanchin-Niell and Wilen (2012) develop a model of invasion spread that accounts for how positioning controls at alternative locations affects invasion spread across a landscape and identify optimal spatial-dynamic strategies for controlling invasions. They applied their model to various hypothetical invasions and showed how long-term invasion costs and damages can be reduced by limiting the length of the spreading invasion front through control or strategic use of landscape features, such as by directing the invasion toward mountain ranges or rivers that act as natural barriers. The study also showed that for certain combinations of control costs, damages, and initial invasion conditions, it is optimal to spatially target controls to slow or prevent the spread of an invasion toward high-value resources. In addition to deriving general spatial control strategies that could be applied to managing *Bromus* spread, the spatially explicit modeling approach developed by Epanchin-Niell and Wilen (2012) could be adapted and parameterized to identify the locations, timing, and amount of resources to cost-effectively manage the spread of exotic invasive annual grasses in specific contexts.

15.5.3 Coordination of Exotic Annual Invasive Grass Management

Management outcomes may depend on the actions and incentives of multiple decision-makers in cases where invasions can spread spatially across property, political, and jurisdictional boundaries (Epanchin-Niell et al. 2010; Epanchin-Niell and Wilen 2015). Decisions regarding rangeland restoration, stocking rates, and fuels management at one location can affect the spread of fire or exotic annual invasive grasses to neighboring locations. However, if managers consider the benefits of their management decisions for reducing fire risk or enhancing forage values only on their own property, they generally underinvest in management relative to what is best for society. Several studies have quantified the often large spatial externalities that arise when decision-makers do not fully include all effects of their exotic invasive species management decisions on others (Bhat et al. 1996; Jones et al. 2000; Wilen 2007; Epanchin-Niell and Wilen 2015; Fenichel et al. 2014).

A few studies model how spatial coordination among private managers and public land managers improves exotic invasive species management outcomes through incorporating more comprehensive considerations of costs and damages, increasing efficacy by reducing the rates of local reinvasion, and inducing control by a wider set of managers (Wilen 2007; Fenichel et al. 2014; Epanchin-Niell and Wilen 2015). Some studies consider spatial coordination simply in terms of the timing of management actions across space to improve outcome efficacy by reducing reinvasion; others consider ways in which targeted transfers of management resources across jurisdictional boundaries increase total landscape-level benefits. For example, Epanchin-Niell and Wilen (2015) model the spread of a hypothetical invasive species across a landscape with many managers and compare outcomes in cases where each manager chooses how much to control the invasion on his or her own property, based on individual benefits and costs, versus a coordinated strategy in which landowners farther from the invasion may contribute resources to invasion control to prevent spread onto their properties. The study finds that even highly localized coordination among small groups of landowners can provide large economic benefits relative to independent management, such that strategies that enhance coordination may have large social payoffs. Strategies that could encourage such coordination and improve landscape-wide management of exotic invasive plants include creating weed management areas or similar institutions that facilitate communication among landowners and reduce the transaction costs of coordination, as well as making the distribution of control incentives (e.g., cost-sharing programs) contingent on local coordination of management efforts (Epanchin-Niell et al. 2010; Epanchin-Niell and Wilen 2015).

The magnitude of benefits of coordinated management can be affected by the strength of spatial interdependencies across properties. Taylor et al. (2013b) find that homeowners' wildfire risk is determined in part by their neighbors' decisions to create defensible space on their properties in pinyon-juniper woodland, sagebrush shrublands, and alpine forest communities but not in *Bromus*-dominated grassland communities in Nevada. This result suggests that spatial interdependencies and coordination among neighboring decision-makers related to wildfire are likely to be more important on rangelands where pinyon-juniper or sagebrush is the dominant vegetation than on rangeland dominated by exotic annual invasive grasses. However, this may depend on the specific management actions considered (e.g., restoration, firebreaks, weed control, defensible space creation).

15.6 Management Implications

The bio-economic models reviewed in this chapter have numerous implications for *Bromus* management, in two respects. First, the results from studies that quantify the economic benefits from *Bromus* management can inform where and how to direct management resources to achieve the greatest economic benefit given limited funds for management. Second, the bio-economic models developed in some studies can be used to improve the design of programs and policies that encourage socially efficient management of *Bromus* by ranchers and other private land managers. As we discuss above, bio-economic models are unique in their ability to analyze how private decision-makers are likely to adjust their management in response to counterfactual changes in policy and to evaluate how these changes in management will affect the prevalence of *Bromus* and other management outcomes.

Before proceeding, it is worth emphasizing that the bio-economic models reviewed in this chapter have implications for *Bromus* management, even though none were expressly constructed to be used as management tools. An important takeaway from this chapter is that economic studies do not have to be expressly constructed as management tools to produce insights and information that are useful for *Bromus* management.

15.6.1 Economically Efficient **Bromus** *Management*

Existing studies suggest strategies for managing *Bromus* on the landscape to achieve the greatest economic benefit given limited funds for management:

- Research into the long-term benefits and costs of treating *Bromus* on sagebrush rangelands has found benefit–cost ratios of 13 to 1 for preventing rangelands with intact native perennial grass cover from becoming *Bromus* dominated. However, low success rates for rehabilitation treatments cause the expected benefits of rehabilitating lands dominated by *Bromus* to be less than the costs of treatment (Taylor et al. 2013a).
- Taylor et al. (2013a) find that on sagebrush rangelands, the benefits of rehabilitation treatments on land dominated by *Bromus* outweigh the costs of treatment for success rates of 52 % or higher when treatment costs of $165 per acre (2010 dollars) are assumed. This result implies that the rehabilitation treatments will become cost-effective if success rates improve and/or costs decline relative to current levels.
- Epanchin-Niell et al. (2009) find that postfire revegetation treatments on sagebrush shrubland sites that lack the necessary perennial grasses and forbs to recover but have not transitioned to *Bromus*-dominated states can reduce long-term management costs while providing biodiversity benefits. For example, postfire revegetation treatments can reduce fire suppression costs by greater than the cost of treatment.
- Current funding levels from federal land management agencies for post-wildfire restoration are insufficient to reverse the continued transition of rangelands in the Western United States across thresholds to exotic annual invasive grass-dominated states (Epanchin-Niell et al. 2009).
- Kobayashi et al. (2010) find that the general public in Nevada has a higher willingness to pay for preventing conversion of rangelands that are currently dominated by native perennial grass and sagebrush to *Bromus*-dominated states than for rehabilitating lands that are currently in *Bromus*-dominated states. This result suggests that over time, as more rangeland transitions to *Bromus*-dominated states, public support for *Bromus* management could decline.

15.6.2 Roles for Incentives and Coordination to Enhance Management

Existing studies suggest how to design public programs and policies to better align private land managers' incentives for *Bromus* management with social goals:

- Kobayashi et al. (2014) find that ranchers operating on rangeland dominated by native perennial grasses and sagebrush have a private incentive to maintain rangeland health through herd management and rehabilitation treatments and that policies to improve the success rates of rehabilitation treatments and lower treatment costs lead to larger herd sizes, more acres receiving treatment, and higher ranch profits but do not affect the long-run ecological condition of the ranch. Conversely, for ranchers operating on rangeland dominated by exotic annual invasive grasses, Kobayashi et al. (2014) find that while it is not optimal for private ranchers to perform rehabilitation treatments, improved success rates or reduced costs could lead them to undertake rehabilitation treatments and that such treatments will improve the long-run ecological health and economic viability of the ranch.
- Kobayashi et al. (2014) demonstrate that market forces, such as high cattle prices, may cause ranchers to place short-term economic gain ahead of the long-run ecological health of their ranches. In particular, high cattle prices may cause ranchers to increase stocking rates, which raises the likelihood that land on the ranch will cross an irreversible ecological threshold to an exotic annual invasive grass-dominated state in the event of a disturbance such as wildfire. This result suggests that in periods of high cattle prices, grazing policies and allotment management plans need to be strictly enforced on public lands to prevent the potential for ecological damage through inappropriate grazing.
- Taylor et al. (2013a) show that uncertainty about whether a sagebrush rangeland ecosystem has crossed an ecological threshold between an ecological state dominated by native perennial grasses and sagebrush and a decadent sagebrush state that will transition to a *Bromus*-dominated state after a disturbance (e.g., wildfire, drought) lowers the expected economic benefits from treatment. This suggests that there may be significant economic benefits to extension and outreach programs that improves the accuracy of land managers' assessments of whether their *Bromus*-affected rangeland has crossed a threshold between ecological states before undertaking *Bromus* treatments.
- Epanchin-Niell and Wilen (2015) demonstrate that coordination of exotic invasive management activities across locations can improve expected outcomes. Public policy can improve the coordination of *Bromus* management across locations by (1) reducing transaction costs of coordinating work across districts, jurisdictions, and agencies; (2) creating and supporting institutions, such as weed management areas, that lower barriers to coordination; and (3) making funds for *Bromus* management contingent on coordination efforts.

15.7 Research Needs

This final section of the chapter discusses areas where further research into the economics of *Bromus* is needed. The research needs are divided into two categories: research that relaxes the economic modeling assumptions made in previous studies to better capture land managers' decision-making regarding *Bromus* and research that considers economic features of the *Bromus* management problem that have not previously been analyzed. Despite the growing literature on the economics of exotic invasive plants, relatively few studies have focused specifically on *Bromus*. Targeted studies are needed because, as discussed above, aspects of *Bromus* invasion—such as the fact that eradication is not a realistic management outcome on most invaded sites—are not shared by many of the other invasive plants analyzed in the previous economics literature.

To ensure that future research into the economics of *Bromus* management is of practical value to land managers, policy-makers, and other stakeholders, economists must continue to coordinate their work with rangeland ecologists and other scientists engaged in *Bromus* research, as well as with ranchers, public land managers, and others involved in on-the-ground *Bromus* management. Input from these various sources is vital to ensuring that the complex ecology of *Bromus* invasion is accurately captured in future economic models and that future economic analysis focuses on timely and relevant *Bromus* management issues.

15.7.1 Research Needs: Modeling Assumptions

Further research is needed that relaxes the modeling assumptions made in previous economic studies of *Bromus* management to better capture land managers' decision-making:

- Further research is needed into how land managers' attitudes toward risk influence their decision-making regarding *Bromus* management. Attitudes toward risk are likely to be relevant because managers must balance the upfront costs of management against uncertain future benefits. Further research could explore whether conventional analysis of decision-making under risk based on expected utility theory can explain observed *Bromus* management, as well as the relevance of concepts such as probability weighting and loss aversion emphasized in prospect theory (Kahneman and Tversky 1979).
- Further research is needed that analyzes the decision-making of land managers who have imperfect knowledge of the ecology of *Bromus* and the consequences of management actions to control *Bromus*, to shed light on the extent to which limited knowledge explains observed management (or lack thereof) of *Bromus* on western rangelands.

- Further work is needed that compares the performance of ranch-level models that assume profit maximization and perfect foresight (all the studies reviewed in this chapter, apart from Janssen et al. 2004, make these assumptions) against models that assume alternative decision-making criteria. As Janssen et al. (2004) have argued, given the complex ecology of *Bromus* invasion and uncertainty inherent in *Bromus* management, assuming that ranchers follow a decision heuristic, or rule of thumb, may provide a more realistic description of rancher decision-making than assuming perfect foresight and profit maximization. In addition, previous research has found that ranchers receive compensation from ranching in the form of "consumptive amenities" related to the "ranching lifestyle" (Torell et al. 2005). Evidence of these consumptive amenities suggests that ranchers are motivated by more than solely maximizing profits.

15.7.2 Research Needs: Management Issues

Further research is needed that considers economic features of the *Bromus* management problem that have not yet been analyzed:

- Economic analyses of the benefits and costs of *Bromus* management have focused primarily on the benefits of management in terms of wildfire suppression cost savings (Epanchin-Niell et al. 2009; Taylor et al. 2013a). A full accounting of the benefits and costs of *Bromus* management requires further research to quantify how the economic value of wildlife habitat, forage for livestock, recreation opportunities, erosion control, and other ecosystem goods and services are influenced by *Bromus* invasion.
- Epanchin-Niell et al. (2009) consider how the level of funding for post-wildfire restoration affects the expected amount of land that will cross ecological thresholds to exotic annual invasive plant-dominated states. Further research is needed that explores the relationship between funding for *Bromus* management and long-run ecological conditions on sagebrush rangelands affected by *Bromus*.
- Although existing studies have analyzed the benefits and costs of pre-fire rehabilitation treatments (e.g., Taylor et al. 2013a) and postfire restoration treatments (e.g., Epanchin-Niell et al. 2009) for *Bromus* management, no previous study has jointly analyzed both management options. Such analysis is needed to enhance understanding of the economic trade-offs and complementarities between the two options, given limited public funds for *Bromus* management.
- To date, economic models of *Bromus* management have not accounted for spatial interdependencies related to the *Bromus* propagation and the spatial spread of wildfire. Further research is needed that accounts for these interdependencies to inform how management options, such as rehabilitation, fuel treatments, and firebreaks, can be most cost-effectively located on the landscape to protect natural resources. Further research also is needed to explore how land managers' decision-making is influenced by *Bromus* management on adjacent land, and

whether strategic interactions between neighboring decision-makers can result in inefficiently low levels of *Bromus* management from a societal perspective. Taylor et al. (2013b) find that homeowners' decisions to invest in mitigating wildfire risk on their property are determined in part by their neighbors' wildfire risk mitigation investment decisions in pinyon-juniper woodland, sagebrush, and alpine forest communities but not in grassland communities. This result suggests that spatial interdependencies between neighboring decision-makers related to wildfire are likely to be important on rangelands where pinyon-juniper or sagebrush is the dominant vegetation and *Bromus* is a component of the understory, rather than on *Bromus*-dominated rangeland.

- It has been suggested that land managers adopt "adaptive management" to effectively manage *Bromus*, given the uncertainty inherent in *Bromus* invasions and management (Morghan et al. 2006). Adaptive management involves deliberate learning-by-doing by land managers to compare the effectiveness of alternative *Bromus* management strategies. Although learning models (Camerer and Ho 1999; Hanaki et al. 2005) have been applied to the problem of managing exotic invasive plants (e.g., Eiswerth and van Kooten 2007), further economic research is needed to analyze the economic benefits of adaptive management for *Bromus* and to design programs and policies to encourage adaptive management.
- As discussed in the previous section, the bio-economic models of *Bromus* management reviewed in this chapter were not constructed expressly to inform on-the-ground management. Rather, they were constructed to analyze and better understand a complex problem that involves capturing the incentives and constraints faced by decision-makers managing *Bromus*, while taking into account the complex ecological features of *Bromus* (e.g., dynamics, ecological thresholds, and spatial considerations). An important goal of future research is to tailor and refine these bio-economic models so that they are better suited to provide decision support for land managers deciding how to deal with *Bromus* on their land.

References

Allen CR, Gunderson L, Johnson AR (2005) The use of discontinuities and functional groups to assess relative resilience in complex systems. Ecosystems 8:958–966

Anderies JM, Janssen MA, Walker BH (2002) Grazing management, resilience, and the dynamics of a fire-driven rangeland system. Ecosystems 5:23–44

Beisner BE, Haydon DT, Cuddington K (2003) Alternative stable states in ecology. Front Ecol Environ 1:376–382

Bhat MG, Huffaker RG, Lenhart SM (1996) Controlling transboundary wildlife damage: modeling under alternative management scenarios. Ecol Model 92:215–224

Bogich T, Shea K (2008) A state-dependent model for the optimal management of an invasive metapopulation. Ecol Appl 18:748–761

Boyd J, Epanchin-Niell R, Siikamaki J (2012) Conservation return on investment analysis: a review of results, methods, and new directions. Resources for the Future, Washington, DC

Broadman A, Greenberg DH, Vining AR et al (2006) Cost-benefit analysis: concepts and practice, 3rd edn. Prentice Hall, Upper Saddle River, NJ

Brock W, Starrett DA (2003) Non-convexities in ecological management problems. Environ Resour Econ 26:575–602
Buhle ER, Margolis M, Ruesink JL (2005) Bang for buck: cost-effective control of invasive species with different life histories. Ecol Econ 52:355–366
Cacho OJ, Hester SM (2011) Deriving efficient frontiers for effort allocation in the management of invasive species. Aust J Agric Resour Econ 55:72–89
Cacho OJ, Spring D, Hester S et al (2010) Allocating surveillance effort in the management of invasive species: a spatially-explicit model. Environ Model Softw 25:444–454
Camerer C, Ho TH (1999) Experience-weighted attraction learning in normal form games. Econometrica 67:827–874
Carande VG, Bartlett ET, Gutierrez PH (1995) Optimization of rangeland management strategies under rainfall and price risks. J Range Manag 48:68–72
Carpenter SR, Ludwig D, Brock WA (1999) Management of eutrophication for lakes subject to potentially irreversible change. Ecol Appl 9:751–771
Common M, Perrings C (1992) Towards an ecological economics of sustainability. Ecol Econ 6:7–34
Eiswerth ME, van Kooten GC (2007) Dynamic programming and learning models for management of a nonnative species. Can J Agric Econ 55:487–500
Eiswerth ME, Johnson WS (2002) Managing nonindigenous invasive species: insights from dynamic analysis. Environ Resour Econ 23:319–342
Eiswerth ME, van Kooten GC (2002) Uncertainty, economics, and the spread of an invasive plant species. Am J Agric Econ 84:1317–1322
Epanchin-Niell R, Englin J, Nalle D (2009) Investing in rangeland restoration in the Arid West, USA: countering the effects of an invasive weed on the long-term fire cycle. J Environ Manag 91:370–379
Epanchin-Niell RS, Brockerhoff EG, Kean JM et al (2014) Designing cost-efficient surveillance for early detection and control of multiple biological invaders. Ecol Appl 24:1258–1274
Epanchin-Niell RS, Haight RG, Berec L et al (2012) Optimal surveillance and eradication of invasive species in heterogeneous landscapes. Ecol Lett 15:803–812
Epanchin-Niell RS, Hastings A (2010) Controlling established invaders: integrating economics and spread dynamics to determine optimal management. Ecol Lett 13:528–541
Epanchin-Niell RS, Hufford MB, Aslan CE et al (2010) Controlling invasive species in complex social landscapes. Front Ecol Environ 8:210–216
Epanchin-Niell RS, Wilen JE (2012) Optimal spatial control of biological invasions. J Environ Econ Manag 63:260–270
Epanchin-Niell RS, Wilen JE (2015) Individual and cooperative management of invasive species in human-mediated landscapes. Am J Agric Econ 97:180–198
Fenichel EP, Richards TJ, Shanafelt DW (2014) The control of invasive species on private property with neighbor-to-neighbor spillovers. Environ Resour Econ 59:231–255
Finnoff D, Potapov A, Lewis MA (2010) Control and the management of a spreading invader. Resour Energy Econ 32:534–550
Finnoff D, Strong A, Tschirhart J (2008) A bioeconomic model of cattle stocking on rangeland threatened by invasive plants and nitrogen deposition. Am J Agric Econ 90:1074–1090
Hanaki N, Sethi R, Erev I et al (2005) Learning strategies. J Econ Behav Organ 56:523–542
Hauser CE, McCarthy MA (2009) Streamlining 'search and destroy': cost-effective surveillance for invasive species management. Ecol Lett 12:683–692
Havstad KM, Peters DPC, Skaggs R et al (2007) Ecological services to and from rangelands of the United States. Ecol Econ 64:261–268
Holling CS (1973) Resilience and stability of ecological systems. Ann Rev Ecol Syst 4:1–23
Horan RD, Wolf CA (2005) The economics of managing infectious wildlife disease. Am J Agric Econ 87:537–551
Huffaker R, Cooper K (1995) Plant succession as a natural range restoration factor in private livestock enterprises. Am J Agric Econ 77:901–913

Hyytiainen K, Lehtiniemi M, Niemi JK et al (2013) An optimization framework for addressing aquatic invasive species. Ecol Econ 91:69–79
Janssen MA, Anderies JM, Walker BH (2004) Robust strategies for managing rangelands with multiple stable attractors. J Environ Econ Manag 47:140–162
Jones RE, Vere DT, Campbell MH (2000) The external costs of pasture weed spread: an economic assessment of serrated tussock control. Agric Econ 22:91–103
Kahneman D, Tversky A (1979) Prospect theory: an analysis of decision under risk. Econometrica 47:263–291
Karp L (2005) Global warming and hyperbolic discounting. J Public Econ 89:261–282
Karp L, Pope A (1984) Range management under uncertainty. Am J Agric Econ 66:437–446
Kobayashi M, Rollins K, Evans M (2010) Sensitivity of WTP estimates to definition of 'yes': reinterpreting expressed response intensity. Agric Resour Econ Rev 39:37–55
Kobayashi M, Rollins K, Taylor MH (2014) Optimal livestock management on sagebrush rangeland with ecological thresholds, wildfire, and invasive plants. Land Econ 90:623–648
Kosko B (1992) Neural networks and fuzzy systems: a dynamical systems approach to machine intelligence/book and disk. Prentice Hall, Englewood Cliffs, NJ
Leung B, Lodge DM, Finnoff D et al (2002) An ounce of prevention or a pound of cure: bioeconomic risk analysis of invasive species. Proc R Soc Lond B Biol 269:2407–2413
Mäler K, Xepapadeas A, deZeeuw A (2003) The economics of shallow lakes. Environ Resour Econ 26:603–624
McIver J, Brunson M, Bunting SC et al (2010) The sagebrush steppe treatment evaluation project (SageSTEP): a test of state-and-transition theory. Gen Tech Rep RMRS-GTR-237:16. USDA, Forest Service, Rocky Mountain Research Station, Fort Collins, CO
Millennium Ecosystem Assessment (2005) Ecosystems and human well-being: desertification synthesis. World Resources Institute, Washington, DC
Morghan KJR, Sheley RL, Svejcar TJ (2006) Successful adaptive management: the integration of research and management. Rangel Ecol Manag 59:216–219
Naevdal E (2001) Optimal regulation of eutrophying lakes, fjords, and rivers in the presence of threshold effects. Am J Agric Econ 83:972–984
Olson LJ, Roy S (2002) The economics of controlling a stochastic biological invasion. Am J Agric Econ 84:1311–1316
Pandey S, Medd RW (1991) A stochastic dynamic-programming framework for weed-control decision-making: an application to *Avena fatua* L. Agric Econ 6:115–128
Passmore G, Brown C (1991) Analysis of rangeland degradation using stochastic dynamic-programming. Aust J Agric Econ 35:131–157
Pearce D, Atkinson G, Mourato S (2006) Cost-benefit analysis and the environment. Recent developments, organisation for economic co-operation and development publishing, Paris, France
Perrings C, Walker B (1997) Biodiversity, resilience and the control of ecological-economic systems: the case of fire-driven rangelands. Ecol Econ 22:73–83
Perrings C, Walker B (2004) Conservation in the optimal use of rangelands. Ecol Econ 49:119–128
Polasky S (2010) A model of prevention, detection, and control for invasive species. In: Perrings C, Mooney H, Williamson M (eds) Bioinvasions & globalization: ecology, economics, management, and policy. Oxford University Press, Oxford, pp 100–109
Ranjan R, Marshall E, Shortle J (2008) Optimal renewable resource management in the presence of endogenous risk of invasion. J Environ Manag 89:273–283
Ritten JP, Frasier WM, Bastian CT et al (2010) Optimal rangeland stocking decisions under stochastic and climate-impacted weather. Am J Agric Econ 92:1242–1255
Rondeau D (2001) Along the way back from the brink. J Environ Econ Manag 42:156–182
Scheffer M, Carpenter S, Foley JA et al (2001) Catastrophic shifts in ecosystems. Nature 413:591–596
Stringham TK, Krueger WC, Shaver PL (2003) State and transition modeling: an ecological process approach. J Range Manag 56:106–113

Taylor M, Rollins K, Kobayashi M et al (2013a) The economics of fuel management: wildfire, invasive plants, and the evolution of sagebrush rangelands in the Western United States. J Environ Manag 126:157–173
Taylor MH, Christman L, Rollins K (2013b) Risk externalities, wildfire hazard, and private investment to mitigate wildfire risk in the wildland-urban interface. UNR Economics Working Paper Series, Working Paper No. 13-003
Timar L, Phaneuf DJ (2009) Modeling the human-induced spread of an aquatic invasive: the case of the zebra mussel. Ecol Econ 68:3060–3071
Torell LA, Rimbey NR, Ramirez OA et al (2005) Income earning potential versus consumptive amenities in determining ranchland values. J Agric Resourc Econ 30:537–560
Wang KM, Hacker RB (1997) Sustainability of rangeland pastoralism – a case study from the west Australian arid zone using stochastic optimal control theory. J Environ Manag 50:147–170
Wilen JE (2007) Economics of spatial-dynamic processes. Am J Agric Econ 89:1134–1144
Zadeh LA (1965) Fuzzy sets. Inf Control 8:338–353

Index

M.J. Germino et al. (eds.), *Exotic Brome-Grasses in Arid and Semiarid Ecosystems of the Western US*, Springer Series on Environmental Management,
DOI 10.1007/978-3-319-24930-8

Q

R

T

MIX
Papier aus verantwortungsvollen Quellen
Paper from responsible sources
FSC® C105338

If you have any concerns about our products,
you can contact us on
ProductSafety@springernature.com

In case Publisher is established outside the EU,
the EU authorized representative is:
Springer Nature Customer Service Center GmbH
Europaplatz 3, 69115 Heidelberg, Germany

Printed by Libri Plureos GmbH
in Hamburg, Germany